Springer Series in

SOLID-STATE SCIENCES 170

AF786757

Springer Series in
SOLID-STATE SCIENCES

The Springer Series in Solid-State Sciences consists of fundamental scientific books prepared by leading researchers in the field. They strive to communicate, in a systematic and comprehensive way, the basic principles as well as new developments in theoretical and experimental solid-state physics.

Please view available titles in *Springer Series in Solid-State Sciences* on series homepage http://www.springer.com/series/682

Sitangshu Bhattacharya
Kamakhya Prasad Ghatak

Fowler-Nordheim Field Emission

Effects in Semiconductor Nanostructures

With 79 Figures

Springer

Dr. Sitangshu Bhattacharya
Indian Institute of Science, Ctr. Electronics Design and Technology
Nano Scale Device Research Laboratory
Bangalore, India
isbsin@yahoo.co.in

Professor Dr. Kamakhya Prasad Ghatak
University of Calcutta, Department of Electronic Science
Acharya Prafulla Chandra Rd. 92, 700009 Kolkata, India
kamakhyaghatak@yahoo.co.in

Series Editors:

Professor Dr., Dres. h. c. Manuel Cardona
Professor Dr., Dres. h. c. Peter Fulde*
Professor Dr., Dres. h. c. Klaus von Klitzing
Professor Dr., Dres. h. c. Hans-Joachim Queisser
Max-Planck-Institut für Festkörperforschung, Heisenbergstrasse 1, 70569 Stuttgart, Germany
* Max-Planck-Institut für Physik komplexer Systeme, Nöthnitzer Strasse 38
01187 Dresden, Germany

Professor Dr. Roberto Merlin
Department of Physics, University of Michigan
450 Church Street, Ann Arbor, MI 48109-1040, USA

Professor Dr. Horst Störmer
Dept. Phys. and Dept. Appl. Physics, Columbia University, New York, NY 10027 and
Bell Labs., Lucent Technologies, Murray Hill, NJ 07974, USA

Springer Series in Solid-State Sciences ISSN 0171-1873
ISBN 978-3-642-44515-6 ISBN 978-3-642-20493-7 (eBook)
DOI 10.1007/978-3-642-20493-7
Springer Heidelberg Dordrecht London New York

Printed on acid-free paper

Springer is part of Springer Science+Business Media (www.springer.com)

This book is dedicated to Mr. Ishwar Prasad Bhattacharya and Mrs. Bela Bhattacharya, parents of the first author, and
Late Dr. Abhoyapada Ghatak and Mrs. Mira Ghatak, parents of the second author

Preface

With the advent of modern quantized structures in one, two, and three dimensions (such as quantum wells, nipi structures, inversion and accumulation layers, quantum well superlattices, carbon nanotubes, quantum wires, quantum wire superlattices, quantum dots, magneto inversion and accumulation layers, quantum dot superlattices, etc.), there has been a considerable interest to investigate the different physical properties of not only such low-dimensional systems but also the different nanodevices made from them and they unfold new physics and related mathematics in the whole realm of solid state sciences in general. Such quantum-confined systems find applications in resonant tunneling diodes, quantum registers, quantum switches, quantum sensors, quantum logic gates, quantum well and quantum wire transistors, quantum cascade lasers, high-resolution terahertz spectroscopy, single electron/molecule electronics, nanotube-based diodes, and other nanoscale devices.

At field strengths of the order of 10^8 V/m (below the electrical breakdown), the potential barriers at the surfaces of different materials usually become very thin resulting in field emission of the electrons due to the tunnel effect. With the advent of Fowler–Nordheim field emission (FNFE) in 1928 [1, 2], the same has been extensively studied under various physical conditions with the availability of a wide range of materials and with the facility for controlling the different energy band constants under different physical conditions and also finds wide applications in solid state and related sciences [3–39]. It appears from the detailed survey of almost the whole spectrum of the literature in this particular aspect that the available monographs, hand books, and review articles on field emission from different important semiconductors and their quantum-confined counterparts have not included any detailed investigations on the FNFE from such systems having various band structures under different physical conditions.

The research group of A.N. Chakravarti [38, 39] has shown that the FNFE from different semiconductors depends on the density of states function (DOS), velocity of the electrons in the quantized levels, and the transmission coefficient of the electron. Therefore, it assumes different values for different systems and varies with the electric field, the magnitude of the reciprocal quantizing magnetic field under magnetic quantization, the nanothickness in quantum wells, wires, and dots, the quantizing electric field as in inversion layers, the carrier statistics in various types of quantum-confined superlattices having different carrier energy spectra and other types of low-dimensional field-assisted systems.

The present monograph is divided into three parts. The first part consists of four chapters. In Chap. 1, the FNFE has been investigated for quantum wires of nonlinear optical, III–V, II–VI, bismuth, IV–VI, stressed materials, Te, n-GaP, $PtSb_2$, Bi_2Te_3, n-Ge, GaSb, and II–V semiconductors on the basis of respective carrier energy spectra. Chapter 2 deals with the field emission from III–V, II–VI, IV–VI, and HgTe/CdTe quantum wires superlattices with graded interfaces have been studied. The same chapter also explores the FNFE from quantum wire effective mass superlattices of aforementioned constituent materials. In Chap. 3, the FNFE from nonlinear optical, III–V, II–VI, bismuth, IV–VI, stressed semiconductors, Te, n-GaP, $PtSb_2$, Bi_2Te_3, n-Ge, GaSb, and II–V compounds under strong magnetic quantization has been studied. In Chap. 4, the FNFE from III–V, II–VI, IV–VI, and HgTe/CdTe superlattices with graded interfaces and effective mass superlattices of the aforementioned constituent materials under magnetic quantization have also been investigated.

The Part II contains the solo Chap. 5 and investigates the influence of light waves on the FNFE from III–V compounds covering the cases of magnetic quantization, quantum wires, effective mass superlattices under magnetic quantization, superlattices with graded interfaces in the presence of quantizing magnetic field, quantum wire effective mass superlattices, and also quantum wire superlattices of the said materials with graded interfaces on the basis of newly formulated carrier energy spectra. Chapter 6 of the last part deals with the FNFE from quantum confined optoelectronic semiconductors in the presence of external intense electric fields. It appears from the literature that the investigations have been carried out on the FNFE under the assumption that the band structures of the semiconductors are invariant quantities in the presence of intense electric fields, which is not fundamentally true. The physical properties of nonparabolic semiconductors in the presence of strong electric field which changes the basic dispersion relation have relatively been less investigated [40]. Chapter 6 explores the FNFE from ternary and quaternary compounds in the presence of intense electric fields on the basis of electron dispersion laws under strong electric field covering the cases of magnetic quantization, quantum wires, effective mass superlattices under magnetic quantization, quantum wire effective mass superlattices, superlattices with graded interfaces in the presence of quantizing magnetic field, and also quantum wire superlattices of the said materials with graded interfaces.

Chapter 7 contains different applications and brief review of the experimental results. In the same chapter, the FNFE from carbon nanotubes in the presence of intense electric field and the importance of the measurement of band-gap of optoelectronic materials in the presence of light waves have also been discussed. Chapter 8 contains conclusion and future research. Besides, 200 open research problems have been presented which will be useful for the researchers in the fields of solid state and allied sciences, in general, in addition to the graduate courses on electron emission from solids in various academic departments of many

Institutes and Universities. We expect that the readers of this monograph will not only enjoy the investigations of the FNFE for a wide range of semiconductors and their nanostructures having different energy-wave vector dispersion relation of the carriers under various physical conditions as presented in this book but also solve the said problems by removing all the mathematical approximations and establishing the appropriate uniqueness conditions, together with the generation of all together new research problems, both theoretical and experimental. Each chapter except the last two contains a table highlighting the basic results pertaining to it in a summarized form.

It is needless to say that this monograph is based on the iceberg principle [41] and the rest of which will be explored by the researchers of different appropriate fields. It has been observed that still new experimental investigations of the FNFE from different semiconductors and their nanostructures are needed since such studies will throw light on the understanding of the band structures of quantized structures which, in turn, control the transport phenomena in such **k** space asymmetric systems. We further hope that the readers will transform this book into a standard reference source in connection with the field emission from solids to probe into the investigation of this particular research topic.

Acknowledgments

Acknowledgment by Sitangshu Bhattacharya: I express my gratitude to my teacher S. Mahapatra at the Centre for Electronics Design and Technology at Indian Institute of Science, Bangalore, for his academic advices. I offer special thanks for having patient to my sister Ms. S. Bhattacharya and my beloved friend Ms. R. Verma and for standing by my side at difficult times of my research life. I am indebted to the Department of Science and Technology, India, for sanctioning the project and the fellowship under "SERC Fast Track Proposal of Young Scientist" scheme 2008-2009 (SR/FTP/ETA-37/08) under which this monograph has been completed. As always, I am immensely grateful to the second author, my friend, philosopher, and PhD thesis advisor.

Acknowledgment by Kamakhya Prasad Ghatak: I am grateful to A.N. Chakravarti, my PhD thesis advisor and mentor who convinced an engineering graduate that theoretical semiconductor physics is the confluence of quantum mechanics and statistical mechanics, and even to appreciate the incredible beauty, he placed a stiff note for me to understand deeply the Course of Theoretical Physics, the Classics of Landau–Lifshitz together with the two volume Classics of Morse–Feshbach 35 years ago. I am also indebted to P.K. Choudhury, M. Mitra, T. Moulick, and S. Sarkar for creating the interest in various topics of Applied Mathematics in general. I consider myself to be rather fortunate to learn quantum mechanics directly from the Late C.K. Majumdar of the Department of Physics of the University of Calcutta. I express my gratitude to M. Green, D.J. Lockwood, A.K. Roy, H.L. Hartnagel, Late P.N. Robson, D. Bimberg, W.L. Freeman, and W. Schommers for various academic interactions spanning over the last two decades. I am grateful to S.N. Sarkar, EX-Head of my Department, who has been playing a supportive role in my academic career. The well-known scientist Late P.N. Butcher has been the main driving force since 1985 before his demise with respect to our scripting the new series in band structure-dependent properties of nanoscale materials. He strongly influenced me regarding it and to satisfy his desire, myself with the prominent members of my research team wrote the *Einstein Relation in Compound Semiconductors and Their Nanostructures*, Springer Series in Materials Science, vol. 116, 2009, as the first one, *Photoemission from Optoelectronic Materials and Their Nanostructures*, Springer Series in Nanostructure Science and Technology, 2009, as the second one, *Thermoelectric Power in Nanostructured Materials: Strong Magnetic Fields*, Springer Series in Materials Science, vol. 137, 2010, as the third one, and the present monograph as the fourth one.

I am grateful to R.K. Poddar, Ex-Vice Chancellor of the University of Calcutta, S.K. Sen, Ex-Vice Chancellor of Jadavpur University, and D.K. Basu, Ex-Vice Chancellor of Burdwan and Tripura Universities, the three pivotal persons in my academic career, Myself and D.De are grateful to UGC, India for sanctioning the research project (No-F 40-469/2011(SR)) in this context. Besides, D. De and myself are grateful to DST, India for sanctioning the fellowship SERC/ET-0213/2011. My wife and daughter deserve a very special mention for forming the backbone of my long unperturbed research career. I offer special thanks to P.K. Sarkar of Semiconductor Device Laboratory, S. Bania of Digital Electronics Laboratory, and my life-long time tested friend B. Nag of Applied Physics Department for always standing by me and consoling me at turbulent times.

Joint acknowledgments: Dr. C. Ascheron, Executive Editor Physics, Springer-Verlag, is our constant source of inspiration since the inception of our first book from Springer-Verlag and we are grateful to him for his priceless technical assistance and right motivation. We owe truly a lot to Ms. A. Duhm, Associate Editor Physics, Springer, and Mrs. E. Suer, assistant to Dr. Ascheron. We are indebted to Dr D. De, the prominent member of our research group for his academic help. We offer special thanks to S.M. Adhikari, S. Ghosh, and N. Paitya for critically reading the manuscript. The production of error-free first edition of any book from every point of view permanently enjoys the domain of impossibility theorems although we are open to accept constructive criticisms for the purpose of their inclusion in the future edition, if any.

Bangalore
Kolkata
July 2011

S. Bhattacharya
K.P. Ghatak

References

1. R.H. Fowler, L. Nordheim, Proc. Roy. Soc. London A **119**, 173 (1928)
2. A. Van Der Ziel, *Solid State Physical Electronics* (Prentice-Hall, Englewood Cliffs, 1957), p. 176
3. B. Mitra, K.P. Ghatak, Phys. Lett. A, **357**, 146 (1990)
4. B. Mitra, K.P. Ghatak, Phys. Lett. A **142A**, 401 (1989)
5. K.P. Ghatak, M. Mondal, J. Mag. Mag. Mat. **74**, 203 (1988)
6. K.P. Ghatak, B. Mitra, Phys. Lett. **156A**, 233 (1991)
7. K.P. Ghatak, A. Ghosal, S.N. Biswas, M. Mondal, Proc. SPIE **1308**, 356 (1990)
8. V.T. Binh, Ch. Adessi, Phys. Rev. Lett. **85**, 864 (2000)
9. R.G. Forbes, Ultramicroscopy **79**, 11 (1999)
10. J.W. Gadzuk, E.W. Plummer, Rev. Mod. Phys. **45**, 487 (1973)
11. J.M. Beebe, B. Kim, J.W. Gadzuk, C.D. Frisbie, J.G. Kushmerick, Phys. Rev. Lett. **97**, 026801 (1999)
12. Y. Feng, J.P. Verboncoeur, Phys. Plasmas **12**, 103301 (2005)
13. W.S. Koh, L.K. Ang, Nanotechnology **19**, 235402 (2008)
14. M. Razavy, *Quantum Theory of Tunneling* (World Scientific, Singapore, 2003)
15. S.I. Baranchuk, N.V. Mileshkina, Sov. Phys. Solid State **23**, 1715 (1981)
16. P.G. Borzyak, A.A. Dadykin, Sov. Phys. Docklady **27**, 335 (1982)
17. S. Bono, R.H. Good, Jr., Surf. Sci. **134**, 272 (1983)
18. S.M. Lyth, S.R.P. Silva, Appl. Phys. Lett. **90**, 173124 (2007)
19. C. Xu, X. Sun, Int. J. Nanotechnol. **1**, 452 (2004)
20. S.D. Liang, L. Chen, Phys. Rev. Lett. **101**, 027602 (2008)
21. E.C. Heeres, E.P.A.M. Bakkers, A.L. Roest, M. Kaiser, T.H. Oosterkamp, N. de Jonge, Nano Lett. **7**, 536 (2007)
22. L. Dong, J. Jiao, D.W. Tuggle, J.M. Petty, S.A. Elliff, M. Coulter, Appl. Phys. Lett. **82**, 1096 (2003)
23. S.Y. Li, P. Lin, C.Y. Lee, T.Y. Tseng, J. Appl. Phys. **95**, 3711 (2004)
24. N.N. Kulkarni, J. Bae, C.K. Shih, S.K. Stanley, S.S. Coffee, J.G. Ekerdt, Appl. Phys. Lett **87**, 213115 (2005)
25. K. Senthil K. Yong, Mat. Chem. Phys. **112**, 88 (2008)
26. R. Zhou, H.C. Chang, V. Protasenko, M. Kuno, A.K. Singh, D. Jena, H. Xing, J. Appl. Phys. **101**, 073704 (2007)
27. K.S. Yeong, J.T.L. Thong, J. Appl. Phys. **100**, 114325 (2006)
28. C.H. Oon, S.H. Khong, C.B. Boothroyd, J.T.L. Thong, J. Appl. Phys. **99**, 064309 (2006)
29. B.H. Kim, M.S. Kim, K.T. Park, J.K. Lee, D.H. Park, J. Joo, S.G. Yu, S.H. Lee, Appl. Phys. Lett. **83**, 539 (2003)
30. Z.S. Wu, S.Z. Deng, N.S. Xu, J. Chen, J. Zhou, J. Chen, Appl. Phys. Lett. **80**, 3829 (2002)
31. Y.W. Zhu, T. Yu, F.C. Cheong, X.J. Xu, C.T. Lim, V.B.C. Tan, J.T.L. Thong, C.H. Sow, Nanotechnology **16**, 88 (2005)
32. Y.W. Zhu, H.Z. Zhang, X.C. Sun, S.Q. Feng, J. Xu, Q. Zhao, B. Xiang, R.M. Wang, D.P. Yu, Appl. Phys. Lett. **83**, 144 (2003)
33. S. Bhattacharjee, T. Chowdhury, Appl. Phys. Lett. **95**, 061501 (2009)
34. S. Kher, A. Dixit, D.N. Rawat, M.S. Sodha, Appl. Phys. Lett. **96**, 044101 (2010)
35. I. Shigeo, W. Teruo, O. Kazuyoshi, T. Masateru, U. Satoshi, N. Norio, J. Vac. Sci. Technol. B: **13**, 487 (2009)
36. C.A. Spindt, I. Brodie, L. Humphrey, E.R. Westerberg, J. Appl. Phys. **47**, 5248 (2009)
37. Q. Fu, A.V. Nurmikko, L.A. Kolodziejski, R.L. Gunshor, J.W. Wu, Appl. Phys. Lett. **51**, 578 (2009)
38. C. Majumdar, M.K. Bose, A.B. Maity, A.N. Chakravarti, Phys. Stat. Sol. (b) **141**, 435 (1987)
39. M.K. Bose, C. Majumdar, A.B. Maity, A.N. Chakravarti, Phys. Stat. Sol. (b) **143**, 113 (1987)
40. P.K. Chakraborty, S. Choudhury, K.P. Ghatak, Phys. B **387**, 333 (2007)
41. A. Pais, J. Robert, *Oppenheimer* (Oxford University Press, Oxford, 2006), p. xviii

Contents

Part I Fowler–Nordheim Field Emission from Quantum Wires and Superlattices of Nonparabolic Semiconductors

1 Field Emission from Quantum Wires of Nonparabolic Semiconductors 3
1.1 Introduction 3
1.2 Theoretical Background 7
1.2.1 The Field Emission from Quantum Wires of Nonlinear Optical Semiconductors 7
1.2.2 The Field Emission from Quantum Wires of III–V Semiconductors 11
1.2.3 The Field Emission from Quantum Wires of II–VI Semiconductors 18
1.2.4 The Field Emission from Quantum Wires of Bismuth 19
1.2.5 The Field Emission from Quantum Wires of IV–VI Semiconductors 24
1.2.6 The Field Emission from Quantum Wires of Stressed Semiconductors 26
1.2.7 The Field Emission from Quantum Wires of Tellurium 28
1.2.8 The Field Emission from Quantum Wires of Gallium Phosphide 29
1.2.9 The Field Emission from Quantum Wires of Platinum Antimonide 31
1.2.10 The Field Emission from Quantum Wires of Bismuth Telluride 32
1.2.11 The Field Emission from Quantum Wires of Germanium... 34
1.2.12 The Field Emission from Quantum Wires of Gallium Antimonide 37
1.2.13 The Field Emission from Quantum Wires of II–V Materials 38

1.3 Result and Discussions ... 39
1.4 Open Research Problems ... 53
References ... 63

2 Field Emission from Quantum Wire Superlattices of Non-parabolic Semiconductors ... 71
2.1 Introduction ... 71
2.2 Theoretical Background ... 72
2.2.1 The Field Emission from III–V Quantum Wire Superlattices with Graded Interfaces ... 72
2.2.2 The Field Emission from II–VI Quantum Wire Superlattices with Graded Interfaces ... 77
2.2.3 The Field Emission from IV–VI Quantum Wire Superlattices with Graded Interfaces ... 82
2.2.4 The Field Emission from HgTe/CdTe Quantum Wire Superlattices with Graded Interfaces ... 87
2.2.5 The Field Emission from Quantum Wire III–V Effective Mass Superlattices ... 92
2.2.6 The Field Emission from Quantum Wire II–VI Effective Mass Superlattices ... 94
2.2.7 The Field Emission from Quantum Wire IV–VI Effective Mass Superlattices ... 96
2.2.8 The Field Emission from Quantum Wire HgTe/CdTe Effective Mass Superlattices ... 99
2.3 Result and Discussions ... 101
2.4 Open Research Problems ... 104
References ... 106

3 Field Emission from Quantum Confined Semiconductors Under Magnetic Quantization ... 109
3.1 Introduction ... 109
3.2 Theoritical Background ... 110
3.2.1 The Field Emission from Nonlinear Optical Semiconductors Under Magnetic Quantization ... 110
3.2.2 The Field Emission from III–V Semiconductors Under Magnetic Quantization ... 112
3.2.3 The Field Emission from II–VI Semiconductors Under Magnetic Quantization ... 119
3.2.4 The Field Emission from Under Bismuth Magnetic Quantization ... 119
3.2.5 The Field Emission from IV–VI Semiconductors Under Magnetic Quantization ... 124
3.2.6 The Field Emission from Stressed Semiconductors Under Magnetic Quantization ... 129

3.2.7 The Field Emission from Tellurium Under Magnetic Quantization 130
3.2.8 The Field Emission from n-Gallium Phosphide Under Magnetic Quantization 131
3.2.9 The Field Emission from Platinum Antimonide Under Magnetic Quantization 133
3.2.10 The Field Emission from Bismuth Telluride Under Magnetic Quantization 135
3.2.11 The Field Emission from Germanium Under Magnetic Quantization 136
3.2.12 The Field Emission from Gallium Antimonide Under Magnetic Quantization 138
3.2.13 The Field Emission from II–V Semiconductors Under Magnetic Quantization 139
3.3 Result and Discussions 140
3.4 Open Research Problems 152
References 153

4 Field Emission of Nonparabolic Semiconductors Under Magnetic Quantization 157
4.1 Introduction 157
4.2 Theoretical Background 157
4.2.1 The Field Emission from III–V Superlattices with Graded Interfaces Under Magnetic Quantization 157
4.2.2 The Field Emission from II–VI Superlattices with Graded Interfaces Under Magnetic Quantization 161
4.2.3 The Field Emission from IV–VI Superlattices with Graded Interfaces Under Magnetic Quantization 165
4.2.4 The Field Emission from HgTe/CdTe Superlattices with Graded Interfaces Under Magnetic Quantization 168
4.2.5 The Field Emission from III–V Effective Mass Superlattices Under Magnetic Quantization 172
4.2.6 The Field Emission from II–VI Effective Mass Superlattices Under Magnetic Quantization 173
4.2.7 The Field Emission from IV–VI Effective Mass Superlattices Under Magnetic Quantization 175
4.2.8 The Field Emission from HgTe/CdTe effective mass superlattices under magnetic quantization 177
4.3 Result and Discussions 178
4.4 Open Research Problems 181
References 184

Part II Fowler–Nordheim Field Emission from Quantum-Confined III–V Semiconductors in the Presence of Light Waves

5 Field Emission from Quantum-Confined III–V Semiconductors in the Presence of Light Waves 187
5.1 Introduction 187
5.2 Theoretical Background 187
5.2.1 Field Emission from III–V Semiconductors Under Magnetic Quantization in the Presence of Light Waves 187
5.2.2 Field Emission from Quantum Wires of III–V Semiconductors 192
5.2.3 Field Emission from Effective Mass Superlattices of III–V Semiconductors in the Presence of Light Waves Under Magnetic Quantization 194
5.2.4 Field Emission from Quantum Wire Effective Mass Superlattices of III–V Semiconductors 200
5.2.5 Field Emission from Superlattices of III–V Semiconductors with Graded Interfaces Under Magnetic Quantization 204
5.2.6 Field Emission from Quantum Wire Superlattices of III–V Semiconductors with Graded Interfaces 210
5.3 Result and Discussions 215
5.4 Open Research Problems 221
References 230

Part III Fowler–Nordheim Field Emission from Quantum-Confined Optoelectronic Semiconductors in the Presence of Intense Electric Field

6 Field Emission from Quantum-Confined Optoelectronic Semiconductors 233
6.1 Introduction 233
6.2 Theoretical Background 234
6.2.1 Field Emission from Optoelectronic Semiconductors Under Magnetic Quantization 234
6.2.2 Field Emission from Quantum Wires of Optoelectronic Semiconductors 248
6.2.3 Field Emission from Effective Mass Superlattices of Optoelectronic Semiconductors Under Magnetic Quantization 249

6.2.4 Field Emission from Quantum Wire Effective Mass Superlattices of Optoelectronic Semiconductors 253
6.2.5 Field Emission from Superlattices of Optoelectronic Semiconductors with Graded Interfaces Under Magnetic Quantization 256
6.2.6 Field Emission from Quantum Wire Superlattices of Optoelectronic Semiconductors with Graded Interfaces 262
6.3 Results and Discussion 267
6.4 Open Research Problem 280
References 280

7 Applications and Brief Review of Experimental Results 281
7.1 Introduction 281
7.2 Applications 281
7.2.1 Debye Screening Length 281
7.2.2 Carrier Contribution to the Elastic Constants 284
7.2.3 Effective Electron Mass 291
7.2.4 Diffusivity–Mobility Ratio 295
7.2.5 Measurement of Bandgap in the Presence of Light Waves 299
7.2.6 Diffusion Coefficient of the Minority Carriers 302
7.2.7 Nonlinear Optical Response 303
7.2.8 Third-Order Nonlinear Optical Susceptibility 303
7.2.9 Generalized Raman Gain 303
7.3 Brief Review of Experimental Works 304
7.3.1 Field Emission from Carbon Nanotubes in the Presence of Strong Electric Field 304
7.3.2 Optimization of Fowler–Nordheim (FN) Field Emission Current from Nanostructured Materials 309
7.3.3 Very Brief Description of Experimental Results of FNFE from Nanostructured Materials 312
7.4 Open Research Problem 323
References 323

8 Conclusion and Future Research 329
References 333

Material Index 335

Subject Index 337

List of Symbols

α	Band nonparabolicity parameter
$\alpha_{11}, \alpha_{22}, \alpha_{33}, \alpha_{23}$	Energy band constants
$\bar{\alpha}_{11}, \bar{\alpha}_{22}, \bar{\alpha}_{33}, \bar{\alpha}_{23}$	System constants
$\beta_1, \beta_2, \beta_4, \beta_5$	System constants
$\psi_1, \psi_2, \psi_3, \psi_4$	Energy band constants
δ	Crystal field splitting constant
$\bar{\delta}_0$	Band constant
$\Delta_{\|}$	Spin–orbit splitting constant parallel to the C-axis
$\Delta_{\perp}$	Spin–orbit splitting constant perpendicular to the C-axis
Δ	Isotropic spin–orbit splitting constant
Δ_0	Interface width in superlattices
Δ_1	Energy band constant
Δ'_c, Δ''_c	Spectrum constants
λ	Wavelength
$\hat{\varepsilon}$	Strain tensor
ε	Trace of the strain tensor
$\overline{\varepsilon}$	Energy as measured from the center of the band gap E_{g_0}
ε_{sc}	Semiconductor permittivity
ε_0	Permittivity of vacuum
$\zeta(2r)$	Zeta function of order $2r$
ζ_0	Constant of the spectrum
$\Gamma(j+1)$	Complete gamma function
ω_0	Cyclotron resonance frequency
υ	Frequency
$\bar{\upsilon}$	System constants
$f_1(k)$	Warping of the Fermi surface
$f_2(k)$	Inversion asymmetry splitting of the conduction band
$\bar{a}$	Constant of the spectrum
a_c	Nearest neighbor C–C bonding distance
a_{13}	Nonparabolicity constant
a_{15}	Warping constant

a_0	The width of the barrier for superlattice structures
b_0	The width of the well for superlattice structures
A_{10}, B_{10}	Energy band constants
B	Quantizing magnetic field
B_2	Momentum matrix element
c	Velocity of light
$\bar{c}$	Constant of the spectrum
C_0	Splitting of the two-spin states by the spin–orbit coupling and the crystalline field
C_1	Conduction band deformation potential
C_2	Strain interaction between the conduction and valance bands
d_x, d_y, d_z	Nanothickness along the x-, y-, and z-directions
e	Magnitude of electron charge
E	Total energy of the carrier as measured from the band edge in the absence of any quantization in the vertically upward direction for electrons or vertically downward directions for holes
E_{ij}	Subband energy
$\overline{E}$	Energy of the hole as measured from the top of the valance band in the vertically downward direction
E_{F_1}	Fermi energy as measured from the mid of the band gap in the vertically upward direction in connection with nanotubes
E_{FB}	Fermi energy in the presence of magnetic quantization as measured from the edge of the conduction band in the absence of any quantization in the vertically upward direction
E_{Fs}	Fermi energy as measured in the presence of intense electric field as measured from the edge of the conduction band in the vertically upward direction in the absence of any field
$\overline{E}_i$	Energy band constant
E_{n_z}	Energy of the nth subband
E_{F1D}	Fermi energy in the presence of two-dimensional quantization as measured from the edge of the conduction band in the vertically upward direction in the absence of any quantization
E_{g0}	Band gap in the absence of any field
F_{sz}	Surface electric field along z-axis
$F_j(\eta)$	One parameter Fermi–Dirac integral of order j
$f(E)$	Fermi–Dirac occupation probability factor
G	Thermoelectric power under strong magnetic field
$\overline{G_0}$	Deformation potential constant
g_v	Valley degeneracy
h	Planck's constant
$\hbar$	Dirac's constant($\equiv h/(2\pi)$)
H	Heaviside step function
I_0	Light intensity
I	System constants

k_0	Constant of the energy spectrum
k_B	Boltzmann's constant
k	Electron wave vector
l_x	Sample length along x-direction
$\bar{l}, \bar{m}, \bar{n}$	Matrix elements of the strain perturbation operator
L_0	Period of the superlattices
L_D	Debye screening length
l	Band constant
m_0	Free electron mass
m_c	Isotropic effective electron mass at the edge of the conduction band
$m_{\|}^*$	Longitudinal effective electron mass at the edge of the conduction band
$m_{\perp}^*$	Transverse effective electron mass at the edge of the conduction band
m_1	Effective carrier mass at the band-edge along x-direction
m_2	Effective carrier mass at the band-edge along y-direction
m_3	The effective carrier mass at the band-edge along z-direction
m_2'	Effective mass tensor component at the top of the valence band (for electrons) or at the bottom of the conduction band (for holes)
m_t^*	The transverse effective mass at $k = 0$
m_l^*	The longitudinal effective mass at $k = 0$
$m_{\perp,1}^*, m_{\|,1}^*$	Transverse and longitudinal effective electron mass at the edge of the conduction band for the first material in superlattice
m_r	Reduced mass
m_v	Effective mass of the heavy hole at the top of the valance band in the absence of any field
m_v^*	Effective mass of the holes at the top of the valence band
$m_t^{\pm}$	Contributions to the transverse effective mass of the external L_6^+ and L_6^- bands arising from the $\vec{k} \cdot \vec{p}$ perturbations with the other bands taken to the second order
$m_l^{\pm}$	Contributions to the longitudinal effective mass of the external L_6^+ and L_6^- bands arising from the $\vec{k} \cdot \vec{p}$ perturbations with the other bands taken to the second order
m_{tc}	Transverse effective electron mass of the conduction electrons at the edge of the conduction band
m_{lc}	Longitudinal effective electron mass of the conduction electrons at the edge of the conduction band
m_{tv}	Transverse effective hole mass of the holes at the edge of the valence band
m_{lv}	Longitudinal effective hole mass of the holes at the edge of the valence band
$N_{1D}(E)$	Density of states function per subbands in 2D quantization
N_c	Effective number of states in the conduction band
n_x, n_y, n_z	Size quantum numbers along the x-, y-, and z-directions

$E_{n_x}, E_{n_y}, E_{n_z}$	The quantized energy levels due to infinity deep potential well along the x-, y-, and z-directions
n_0	Carrier degeneracy
$N_{2D}(E)$	2D density of states function per subband
$N_{2DT}(E)$	Total 2D density of states function
n	Landau quantum number/chiral indices
$\bar{n}$	Band constant
P_0	Momentum matrix element
$(\bar{P})$	Energy band constant
$P_{\parallel}, P_{\perp}$	Momentum matrix elements parallel and perpendicular to the direction of C-axis
$\bar{Q}, \bar{R}$	Spectrum constants
R_G	Generalized Raman gain
r	Set of real positive integers whose upper limit is r_0 or s_0
$\bar{r}_0$	Radius of the nanotube
$\bar{S}_0$	Entropy per unit volume
$\bar{s}$	Spectrum constant
s_0	Upper limit of the summation
T	Temperature
t_c	Tight binding parameter
t_i	Energy band constants
t_{ij}	Transmission coefficient
i	Integer
v	Velocity of the electron
$\bar{v}_0, \bar{w}_0$	Constants of the spectrum
V_0	Equal to the addition of the Fermi energy in the corresponding case and the work function ϕ_w of the material
$\lvert V_G \rvert$	Constant of the energy spectrum
x, y	Alloy compositions
Z_0	The nonlinear response from the optical excitation of the free carriers
$\chi_{NP}(\omega_1, \omega_2, \omega_3)$	Third-order nonlinear optical susceptibility

Part I
Fowler–Nordheim Field Emission from Quantum Wires and Superlattices of Nonparabolic Semiconductors

Chapter 1
Field Emission from Quantum Wires of Nonparabolic Semiconductors

1.1 Introduction

The Fowler–Nordheim field emission (FNFE) is a well-known quantum-mechanical phenomenon that involves tunneling of electrons through a surface barrier due to the application of an intense external electric field. Normally, at field strengths of the order of 10^8 V/m (below the electrical breakdown), the potential barriers at the surfaces of metals and semiconductors usually become very thin and result in field emission of electrons due to the tunnel effect [1, 2]. This has been well investigated with reference to three-dimensional electron gases in metals and semiconductors and the FNFE from quantum-confined structures has also been studied in this context [3–37]. Some of significant features of the FNFE which have emerged from these investigations are as follows:

1. The FNFE increases with increasing electron concentration in bulk materials and are significantly influenced by the carrier energy spectra of different electronic materials.
2. The FNFE increases with increasing electric field.
3. The FNFE oscillates with film thickness for quantum-confined systems.
4. The FNFE oscillates with inverse quantizing magnetic field in the presence of magnetic quantization due to the Shubnikov–de Haas effect.
5. For various types of superlattices of different materials, the FNFE shows composite oscillations with different system variables.

In recent years, with the advent of fine lithographical methods [38, 39], molecular beam epitaxy [40], organometallic vapor-phase epitaxy [41], and other experimental techniques, the restriction of the motion of the carriers of bulk materials in one (quantum wells in ultrathin films, NIPI structures, inversion, and accumulation layers), two (quantum wires), and three (quantum dots, magnetosize quantized systems, magneto-accumulation layers, magneto-inversion layers quantum dot superlattices, magneto-quantum well superlattices, and magneto-NIPI structures) dimensions have in the last few years, attracted much attention not only for their

S. Bhattacharya and K.P. Ghatak, *Fowler–Nordheim Field Emission*, Springer Series in Solid-State Sciences 170, DOI 10.1007/978-3-642-20493-7_1,

potential in uncovering new phenomena in nanoscience, but also for their interesting quantum device applications [42–45]. In ultrathin films, the restriction of the motion of the carriers in the direction normal to the film (say, the z direction) may be viewed as carrier confinement in an infinitely deep 1D rectangular potential well, leading to quantization [known as quantum size effect (QSE)] of the wave vector of the carrier along the direction of the potential well, allowing 2D carrier transport parallel to the surface of the film representing new physical features not exhibited in bulk semiconductors [46–50]. The low-dimensional heterostructures based on various materials are widely investigated because of the enhancement of carrier mobility [51]. These properties make such structures suitable for applications in quantum well lasers [52], heterojunction FETs [53, 54], high-speed digital networks [55–58], high-frequency microwave circuits [59], optical modulators [60], optical switching systems [61], and other devices. The constant energy 3D wavevector space of bulk semiconductors becomes 2D wavevector surface in ultrathin films or quantum wells due to dimensional quantization. Thus, the concept of reduction of symmetry of the wavevector space and its consequence can unlock the physics of low-dimensional structures.

It is well known that in quantum wires (QWs), the restriction of the motion of the carriers along two directions may be viewed as carrier confinement by two infinitely deep 1D rectangular potential wells, along any two orthogonal directions leading to quantization of the wave vectors along the said directions, allowing 1D carrier transport [62–64]. With the help of modern fabrication techniques, such one-dimensional quantized structures have been experimentally realized and enjoy an enormous range of important applications in the realm of nanoscience in quantum regime. They have generated much interest in the analysis of nanostructured devices for investigating their electronic, optical, and allied properties [65–72]. Examples of such new applications are based on the different transport properties of ballistic charge carriers which include quantum resistors [73–75], resonant tunneling diodes and band filters [76, 77], quantum switches [78], quantum sensors [79, 80], quantum logic gates [81, 82], quantum transistors and subtuners [83, 84], heterojunction FETs [85], high-speed digital networks [86, 87], high-frequency microwave circuits [88], optical modulators [89], optical switching systems [90], and other nanoscale devices.

In this chapter, we shall study the FNFE from QWs of nonparabolic semiconductors having different band structures. At first we shall investigate the FNFE from QWs of nonlinear optical compounds which are being used in nonlinear optics and light-emitting diodes [91, 92]. The quasi-cubic model can be used to investigate the symmetric properties of both the bands at the zone center of wavevector space of the same compound.Including the anisotropic crystal potential in the Hamiltonian, and special features of the nonlinear optical compounds, Kildal [93] formulated the electron dispersion law under the assumptions of isotropic momentum matrix element and the isotropic spin–orbit splitting constant, respectively, although the anisotropies in the two aforementioned band constants are the significant physical features of the said materials [94–96]. In Sect. 1.2.1, the FNFE from QWs of nonlinear optical semiconductors has been investigated by considering the combined influence of the anisotropies of the said energy band constants together with the inclusion of the crystal field splitting respectively within the framework of $\mathbf{k} \cdot \mathbf{p}$ formalism.

The III–V compounds find applications in infrared detectors [97], quantum dot light-emitting diodes [98], quantum cascade lasers [99], quantum well wires [100], optoelectronic sensors [101], high electron mobility transistors [102], etc. The electron energy spectrum of III–V semiconductors can be described by the three- and two-band models of Kane [103, 104], together with the models of Stillman et al. [105], Newson and Kurobe [106], and Palik et al. [107], respectively. In this context, it may be noted that the ternary and quaternary compounds enjoy the singular position in the entire spectrum of optoelectronic materials. The ternary alloy $Hg_{1-x}Cd_xTe$ is a classic narrow gap compound. The band gap of this ternary alloy can be varied to cover the spectral range from 0.8 to over 30 μm [108] by adjusting the alloy composition. $Hg_{1-x}Cd_xTe$ finds extensive applications in infrared detector materials and photovoltaic detector arrays in the 8–12 μm wave bands [109]. The above uses have generated the $Hg_{1-x}Cd_xTe$ technology for the experimental realization of high mobility single crystal with specially prepared surfaces. The same compound has emerged to be the optimum choice for illuminating the narrow subband physics because the relevant material constants can easily be experimentally measured [110]. Besides, the quaternary alloy $In_{1-x}Ga_xAs_yP_{1-y}$ lattice matched to InP, also finds wide use in the fabrication of avalanche photodetectors [111], heterojunction lasers [112], light-emitting diodes [113] and avalanche photodiodes [114], field effect transistors, detectors, switches, modulators, solar cells, filters, and new types of integrated optical devices are made from the quaternary systems [115]. It may be noted that all types of band models as discussed for III–V semiconductors are also applicable for ternary and quaternary compounds. In Sect. 1.2.2, the FNFE from QWs of III–V, ternary, and quaternary semiconductors has been studied in accordance with the said band models and the simplified results for wide gap materials having parabolic energy bands under certain limiting conditions have further been demonstrated as a special case and thus confirming the compatibility test.

The II–VI semiconductors are being used in nanoribbons, blue green diode lasers, photosensitive thin films, infrared detectors, ultrahigh-speed bipolar transistors, fiber optic communications, microwave devices, solar cells, semiconductor gamma-ray detector arrays, and semiconductor detector gamma camera and allow for a greater density of data storage on optically addressed compact discs [116–123]. The carrier energy spectra in II–VI compounds are defined by the Hopfield model [124] where the splitting of the two-spin states by the spin–orbit coupling and the crystalline field has been taken into account. Section 1.2.3 contains the investigation of the FNFE from QWs of II–VI compounds.

In recent years, Bismuth (Bi) nanolines have been fabricated and Bi also finds use in array of antennas, which leads to the interaction of electromagnetic waves with such Bi-nanowires [125, 126]. Several dispersion relations of the carriers have been proposed for Bi. Shoenberg [127] experimentally verified that the de Haas–Van Alphen and cyclotron resonance experiments supported the ellipsoidal parabolic model of Bi, although the magnetic field dependence of many physical properties of Bi supports the two-band model [128]. The experimental investigations on the magneto-optical and the ultrasonic quantum oscillations support the Lax ellipsoidal nonparabolic model [129]. Kao [130], Dinger and Lawson [131], and Koch and Jensen [132] demonstrated that the Cohen model [133] is in conformity with the experimental results in a better way. Besides, the hybrid model of bismuth, as developed by Takaoka et al. also finds use in the literature [134]. McClure and Choi [135] derived a new model of Bi and they showed that it can explain the data for a large number of magneto-oscillatory and resonance experiments. In Sect. 1.2.4, the FNFE from QWs of Bi has been formulated in accordance with the aforementioned energy band models for the purpose of relative assessment. Besides, under certain limiting conditions all the results for all the models of 1D systems are reduced to the well-known result of the FNFE from QWs of wide gap materials. This above statement exhibits the compatibility test of our theoretical analysis.

Lead chalcogenides (PbTe, PbSe, and PbS) are IV–VI nonparabolic semiconductors whose studies over several decades have been motivated by their importance in infrared IR detectors, lasers, light-emitting devices, photovoltaics, and high-temperature thermoelectrics [136–140]. PbTe, in particular, is the end compound of several ternary and quaternary high-performance high-temperature thermoelectric materials [141–145]. It has been used not only as bulk but also as films [146–149], quantum wells [150], superlattices [151, 152], nanowires [153], colloidal and embedded nanocrystals [154–157], and PbTe films doped with various impurities have also been investigated [158–165]. These studies revealed some of the interesting features that had been seen in bulk PbTe, such as Fermi level pinning in the case of superconductivity [166]. In Sect. 1.2.5, the FNFE from QWs of IV–VI semiconductors has been studied taking PbTe as an example.

The stressed semiconductors are being investigated for strained silicon transistors, quantum cascade lasers, semiconductor strain gages, thermal detectors and strained-layer structures [167–170]. The FNFE from QWs of stressed compounds (taking stressed n-InSb as an example) has been investigated in Sect. 1.2.6. The vacuum deposited Tellurium (Te) has been used as the semiconductor layer in thin-film transistors (TFT) [171], which is being used in CO_2 laser detectors [172], electronic imaging, strain-sensitive devices [173, 174], and multichannel Bragg cell [175]. Section 1.2.7 contains the investigation of FNFE from QWs of Tellurium.

The n-gallium phosphide (n-GaP) finds applications in quantum dot light-emitting diode [176], high efficiency yellow solid-state lamps, light sources, and high peak current pulse for high gain tubes. The green and yellow light-emitting diodes made of nitrogen-doped n-GaP possess a longer device life at high drive currents [177–179]. In Sect. 1.2.8, the FNFE from QWs of n-GaP has been studied. The Platinum Antimonide ($PtSb_2$) is used in device miniaturization, colloidal nanoparticle synthesis, sensors, detector materials, and thermo-photovoltaic devices [180–182]. Section 1.2.9 explores the FNFE from QWs of $PtSb_2$. Bismuth telluride (Bi_2Te_3) was first identified as a material for thermoelectric refrigeration in 1954 [183] and its physical properties were later improved by the addition of bismuth selenide and antimony telluride to form solid solutions [184–188]. The alloys of Bi_2Te_3 are useful compounds for the thermoelectric industry and have been investigated in the literature [184–188]. In Sect. 1.2.10, the FNFE from QWs of Bi_2Te_3 has been considered.

The usefulness of elemental semiconductor Germanium is already well known since the inception of transistor technology, and it is also being used in memory circuits, single photon detectors, single photon avalanche diode, ultrafast optical switch, THz lasers, and THz spectrometers [189–192]. In Sect. 1.2.11, the FNFE has been studied from QWs of Ge. Gallium Antimonide (GaSb) finds applications in the fiber optic transmission window, heterojunctions, and quantum wells. A complementary heterojunction field effect transistor in which the channels for the p-FET device and the n-FET device forming the complementary FET are formed from GaSb. The band gap energy of GaSb makes it suitable for low power operation [193–198]. In Sect. 1.2.12, the FNFE from QWs of GaSb has been studied. The II–V semiconductors are being used in photovoltaic cells constructed of single crystal semiconductor materials in contact with electrolyte solutions. Cadmium selenide shows an open-circuit voltage of 0.8 V and power conservation coefficient is nearly 6% for 720-nm light [199]. They are also used in ultrasonic amplification [200]. The development of an evaporated TFT using cadmium selenide as the semiconductor has also been reported [201, 202]. In Sect. 1.2.13, we shall study the FNFE from QWs of II–V semiconductors. Section 1.3 contains the result and discussions pertaining to this chapter. Section 1.4 contains open research problems.

1.2 Theoretical Background

1.2.1 The Field Emission from Quantum Wires of Nonlinear Optical Semiconductors

The form of $\mathbf{k}\cdot\mathbf{p}$ matrix for nonlinear optical compounds can be expressed extending Bodnar [94] as

$$H = \begin{bmatrix} H_1 & H_2 \\ H_2^+ & H_1 \end{bmatrix}, \tag{1.1}$$

where

$$H_1 \equiv \begin{bmatrix} E_{g_0} & 0 & P_{\|}k_z & 0 \\ 0 & (-2\Delta_{\|}/3) & (\sqrt{2}\Delta_{\perp}/3) & 0 \\ P_{\|}k_z & (\sqrt{2}\Delta_{\perp}/3) & -(\delta+\frac{1}{3}\Delta_{\|}) & 0 \\ 0 & 0 & 0 & 0 \end{bmatrix}, \quad H_2 \equiv \begin{bmatrix} 0 & -f_{,+} & 0 & f_{,-} \\ f_{,+} & 0 & 0 & 0 \\ 0 & 0 & 0 & 0 \\ f_{,+} & 0 & 0 & 0 \end{bmatrix},$$

in which E_{g_0} is the band gap in the absence of any field, $P_{\|}$ and $P_{\perp}$ are the momentum matrix elements parallel and perpendicular to the direction of crystal axis, respectively, δ is the crystal field splitting constant, and $\Delta_{\|}$ and $\Delta_{\perp}$ are the spin–orbit splitting constants parallel and perpendicular to the C-axis, respectively, $f_{,\pm} \equiv (P_{\perp}/\sqrt{2})(k_x \pm ik_y)$ and $i = \sqrt{-1}$. Thus, neglecting the contribution of the higher bands and the free electron term, the diagonalization of the above matrix leads to the dispersion relation of the conduction electrons in bulk specimens of nonlinear optical semiconductors as

$$\gamma(E) = f_1(E)k_s^2 + f_2(E)k_z^2, \tag{1.2}$$

where

$$\gamma(E) \equiv E\left(E + E_{g_0}\right)\left[(E + E_{g_0})(E + E_{g_0} + \Delta_{\|}) + \delta\left(E + E_{g_0} + \frac{2}{3}\Delta_{\|}\right) + \frac{2}{9}\left(\Delta_{\|}^2 - \Delta_{\perp}^2\right)\right],$$

$$f_1(E) \equiv \frac{\hbar^2 E_{g_0}(E_{g_0} + \Delta_{\perp})}{\left[2m_{\perp}^*\left(E_{g_0} + \frac{2}{3}\Delta_{\perp}\right)\right]}\left[\delta\left(E + E_{g_0} + \frac{1}{3}\Delta_{\|}\right) + \left(E + E_{g_0}\right)\left(E + E_{g_0} + \frac{2}{3}\Delta_{\|}\right) + \frac{1}{9}\left(\Delta_{\|}^2 - \Delta_{\|}^2\right)\right],$$

where E is the total energy of the electron as measured from the edge of the conduction band in the vertically upward direction in the absence of any quantization, $k_s^2 = k_x^2 + k_y^2$, $f_2(E) \equiv \dfrac{\hbar^2 E_{g_0}(E_{g_0} + \Delta_{\|})}{\left[2m_{\|}^*\left(E_{g_0} + \frac{2}{3}\Delta_{\|}\right)\right]}\left[\left(E + E_{g_0}\right)\left(E + E_{g_0} + \frac{2}{3}\Delta_{\|}\right)\right]$, $h = h/2\pi, h$ is the Planck's constant, and $m_{\|}^*$ and $m_{\perp}^*$ are the longitudinal and transverse effective electron masses at the edge of the conduction band, respectively.

For two-dimensional quantization along the x and y directions, (1.2) assumes the form

$$k_z^2 = A_{11}(E, n_x, n_y), \tag{1.3}$$

where $A_{11}(E, n_x, n_y) = [f_2(E)]^{-1}[\gamma(E) - \phi_1(n_x, n_y) f_1(E)]$, $\phi_1(n_x, n_y) = \left(\dfrac{n_x\pi}{d_x}\right)^2 + \left(\dfrac{n_y\pi}{d_y}\right)^2$, $n_x = (1, 2, 3, \ldots)$, $n_y = (1, 2, 3, \ldots)$ are the size quantum numbers along the x and y directions, respectively, and d_x and d_y are the nanothickness along the x and y directions, respectively.

The quantized subband energy (E_{11}) is given by

$$\gamma(E_{11}) = f_1(E_{11})\phi_1(n_x, n_y). \tag{1.4}$$

The electron concentration per unit length can be expressed as

$$n_0 = \frac{2g_v}{\pi}\sum_{n_x=1}^{n_{x\max}}\sum_{n_y=1}^{n_{y\max}}\left[B_{11}\left(E_{\mathrm{F1D}}, n_x, n_y\right) + B_{12}\left(E_{\mathrm{F1D}}, n_x, n_y\right)\right], \tag{1.5}$$

where g_v is the valley degeneracy, $B_{11}(E_{F1D}, n_x, n_y) = [A_{11}(E_{F1D}, n_x, n_y)]^{1/2}$, $B_{12}(E_{F1D}, n_x, n_y) = \sum_{r=1}^{r_0} Z_{1D}(r)[B_{11}(E_{F1D}, n_x, n_y)]$, $Z_{1D}(r) = 2(k_B T)^{2r}(1-2^{1-2r})$ $\xi(2r)\frac{\partial^{2r}}{\partial E_{F1D}^{2r}}$, k_B is the Boltzmann constant, T is the temperature, r is the set of real positive integers whose upper limit is r_0, $\xi(2r)$ is the Zeta function of order $2r$ [203], and E_{F1D} is the Fermi energy in the presence of 2D quantization as measured from the edge of the conduction band in the vertically upward direction in the absence of any quantization.

The current (I) due to Fowler–Nordheim {FN}field emission can be written as

$$I = \frac{1}{2}\sum_{n_x=1}^{n_{x\max}}\sum_{n_y=1}^{n_{y\max}}\left[\int_{E_{11}}^{\infty}[e \cdot v n_1 t_{11}]\right], \tag{1.6a}$$

where e is the magnitude of the electron charge, v is the velocity of the electron and given by $v = (1/\hbar)(\partial E/\partial k_z)$, $n_1 = N_{1D}(E).f(E)dE$, $N_{1D}(E)$ is the 1D density-of-states function per subbands and can be written as $N_{1D}(E) = (2g_v/\pi)(\partial k_z/\partial E)$, $f(E)$ is the Fermi–Dirac occupation probability factor and is given by $f(E) = [1 + \exp(E - E_{F1D}/k_B T)]^{-1}$, and t_{11} is the transmission coefficient. From (1.6a), it appears that I is a function of the product of the carrier velocity, concentration, and transmission coefficient. These three quantities in turn depend totally on the dispersion relation of the material. As the basic **E–k** relation changes, all the aforementioned quantities will change and the current due to FN field emission will be different consequently. Thus, the field emission will change all together in 1D, 2D, and 3D quantization of the wave vector space encompassing the whole arena of quantized structures.

Thus, from (1.6a) one can write

$$I = \frac{1}{2}\sum_{n_x=1}^{n_{x\max}}\sum_{n_y=1}^{n_{y\max}}\left[\int_{E_{11}}^{\infty}\left[\frac{e}{\hbar}\frac{\partial E}{\partial k_z}.\frac{2g_v}{\pi}\frac{\partial k_z}{\partial E}f(E)\mathrm{d}E\right]t_{11}\right]. \tag{1.6b}$$

The term $\partial E/\partial k_z$ from velocity and the term $\partial k_z/\partial E$ from the density-of-states function per subbands cancel each other leaving the constant prefactor.

Therefore, (1.6b) assumes the form

$$I = \frac{e \cdot g_v}{\hbar\pi}\sum_{n_x=1}^{n_{x\max}}\sum_{n_y=1}^{n_{y\max}}\left[\int_{E_{11}}^{\infty}[f(E)\mathrm{d}E]\cdot t_{11}\right]. \tag{1.6c}$$

The term t_{11} in this case can be expressed by using the method as given in [204] as

$$t_{11} = \exp(-\beta_{11}) \tag{1.7}$$

in which

$$\beta_{11} = \frac{4\left[A_{11}\left(V_0, n_x, n_y\right)\right]^{3/2}}{3eF_{sz}\left[A'_{11}\left(V_0, n_x, n_y\right)\right]}. \tag{1.8}$$

F_{sz} is the electric field along the z-axis,

$A_{11}(V_0, n_x, n_y) = [f_2(V_0)]^{-1}[\gamma(V_0) - \phi_1(n_x, n_y) f_1(V_0)]$, V_0 is equal to the addition of the Fermi energy in the corresponding case and the work function ϕ_w of the material,

$$f_2(V_0) \equiv \frac{\hbar^2 E_{g0}(E_{g0} + \Delta_{\|})}{\left[2m_{\|}^*\left(E_{g0} + \frac{2}{3}\Delta_{\|}\right)\right]}\left[(V_0 + E_{g0})\left(V_0 + E_{g0} + \frac{2}{3}\Delta_{\|}\right)\right],$$

$$\gamma(V_0) \equiv V_0\left(V_0 + E_{g0}\right)\Big[\left(V_0 + E_{g0}\right)\left(V_0 + E_{g0} + \Delta_{\|}\right) + \delta\left(V_0 + E_{g0} + \frac{2}{3}\Delta_{\|}\right) + \frac{2}{9}\left(\Delta_{\|}^2 - \Delta_{\perp}^2\right)\Big],$$

$$f_1(V_0) \equiv \frac{\hbar^2 E_{g0}(E_{g0} + \Delta_{\perp})}{\left[2m_{\perp}^*\left(E_{g0} + \frac{2}{3}\Delta_{\perp}\right)\right]}\Big[\delta\left(V_0 + E_{g0} + \frac{1}{3}\Delta_{\|}\right) + (V_0 + E_{g0})\left(V_0 + E_{g0} + \frac{2}{3}\Delta_{\|}\right) + \frac{1}{9}\left(\Delta_{\|}^2 - \Delta_{\|}^2\right)\Big],$$

$$A'_{11}(V_0, n_x, n_y) = \Big[\frac{-A_{11}(V_0, n_x, n_y) f'_2(V_0)}{f_2(V_0)} + [f_2(V_0)]^{-1}[\gamma'(V_0) - f'_1(V_0)\phi_1(n_x, n_y)]\Big],$$

$$f'_2(V_0) = \left[[\hbar^2 E_{g0}(E_{g0} + \Delta_{\|})]\left[2m_{\|}^*\left(E_{g0} + \frac{2}{3}\Delta_{\|}\right)\right]^{-1} \times \left[2V_0 + 2E_{g0} + \frac{2}{3}\Delta_{\|}\right]\right],$$

$$f'_1(V_0) = \left[[\hbar^2 E_{g0}(E_{g0} + \Delta_{\perp})]\left[2m_{\perp}^*\left(E_{g0} + \frac{2}{3}\Delta_{\perp}\right)\right]^{-1} \times \left[2V_0 + 2E_{g0} + \frac{2}{3}\Delta_{\|} + \delta\right]\right],$$

and

$$\gamma'(V_0) = \left[\frac{\gamma(V_0)(2V_0 + E_{g0})}{V_0(V_0 + E_{g0})} + (V_0(V_0 + E_{g0}))\left[2V_0 + 2E_{g0} + \Delta_{\|} + \delta\right]\right].$$

Thus, we can write

$$I = \frac{2g_v e k_B T}{h} \sum_{n_x=1}^{n_{x\max}} \sum_{n_y=1}^{n_{y\max}} F_0(\eta_{11})\, t_{11}, \tag{1.9}$$

where $\eta_{11} = (E_{F1D} - E_{11})/k_B T, F_0(\eta_{11})$ is the special case of the one-parameter Fermi–Dirac integral of order j which can be written as

$$F_j(\eta) = \left(\frac{1}{\Gamma(j+1)}\right) \int_0^\infty \frac{x^j \, dx}{1 + \exp(x - \eta)}, \qquad j > -1 \tag{1.10}$$

or for all j, analytically continued as a complex contour integral around the negative x-axis

$$F_j(\eta) = \left(\frac{\Gamma(-j)}{2\pi\sqrt{-1}}\right) \int_{-\infty}^{+0} \frac{x^j \, dx}{1 + \exp(-x - \eta)}, \tag{1.11}$$

where η is the dimensionless$x$ independent variable,

$$\Gamma(j+1) = j\Gamma(j),\ \Gamma\left(\frac{1}{2}\right) = \sqrt{\pi},\ \text{and}\ \Gamma(0) = 1..$$

Therefore, the field-emitted current is given by

$$I = \frac{2g_v e k_B T}{h} \sum_{n_x=1}^{n_{x\max}} \sum_{n_y=1}^{n_{y\max}} F_0(\eta_{11}) \exp(-\beta_{11}). \tag{1.12}$$

1.2.2 *The Field Emission from Quantum Wires of III–V Semiconductors*

The dispersion relation of the conduction electrons of III–V compounds are described by the models of Kane (both three and two bands) [103, 104], Stillman et al. [105], Newson and Kurobe [106], and Palik et al. [107], respectively. For the purpose of complete and coherent presentation, the FNFE from QWs of III–V semiconductors have also been investigated in accordance with the aforementioned different dispersion relations for indicating the relative comparison as follows:

1.2.2.1 The Three-Band Model of Kane

Under the conditions, $\delta = 0$, $\Delta_\parallel = \Delta_\perp = \Delta$ (isotropic spin–orbit splitting constant) and $m_\parallel^* = m_\perp^* = m_c$ (isotropic effective electron mass at the edge of the conduction band), (1.2) gets simplified into the form

$$\frac{\hbar^2 k^2}{2m_c} = I_{11}(E), \quad I_{11}(E) \equiv \frac{E\left(E+E_{g_0}\right)\left(E+E_{g_0}+\Delta\right)\left(E_{g_0}+\frac{2}{3}\Delta\right)}{E_{g_0}(E_{g_0}+\Delta)\left(E+E_{g_0}+\frac{2}{3}\Delta\right)}, \tag{1.13}$$

which is known as the three-band model of Kane [103, 104] and is often used to study the electronic properties of III–V, ternary, and quaternary semiconductors.

The 1D E–k_z relation can be expressed as

$$k_z^2 = A_{12}(E, n_x, n_y), \tag{1.14}$$

where $A_{12}(E, n_x, n_y) = \frac{2m_c}{\hbar^2} I_{11}(E) - \phi_1(n_x, n_y)$.

The quantized subband energy (E_{12}) is given by

$$I_{11}(E_{12}) = \left[\frac{\hbar^2}{2m_c}\right]\phi_1(n_x, n_y). \tag{1.15}$$

The electron concentration per unit length can be written as

$$n_0 = \frac{2g_v}{\pi}\sum_{n_x=1}^{n_{x\max}}\sum_{n_y=1}^{n_{y\max}}\left[B_{13}\left(E_{\mathrm{F1D}}, n_x, n_y\right) + B_{14}\left(E_{\mathrm{F1D}}, n_x, n_y\right)\right], \tag{1.16}$$

where $B_{13}\left(E_{\mathrm{F1D}}, n_x, n_y\right) = \left[A_{12}\left(E_{\mathrm{F1D}}, n_x, n_y\right)\right]^{1/2}$ and $B_{14}(E_{\mathrm{F1D}}, n_x, n_y) = \sum_{r=1}^{r_0} Z_{\mathrm{1D}}(r)[B_{13}(E_{\mathrm{F1D}}, n_x, n_y)]$.

The field-emitted current can be expressed as

$$I = \frac{2g_v e k_B T}{h}\sum_{n_x=1}^{n_{x\max}}\sum_{n_y=1}^{n_{y\max}} F_0(\eta_{12})\, t_{12}, \tag{1.17}$$

where $\eta_{12} = (E_{\mathrm{F1D}} - E_{12})/k_B T$, $t_{12} = \exp(-\beta_{12})$, $\beta_{12} = \frac{4}{3}[A_{12}(V_0, n_x, n_y)]^{3/2} \cdot [eF_{sz}A'_{12}(V_0)]^{-1}$,

$$A_{12}(V_0, n_x, n_y) = \left[\frac{2m_c}{\hbar^2} I_{11}(V_0) - \phi_1(n_x, n_y)\right],$$

$$I_{11}(V_0) \equiv \frac{V_0(V_0+E_{g_0})(V_0+E_{g_0}+\Delta)\left(E_{g_0}+\frac{2}{3}\Delta\right)}{E_{g_0}(E_{g_0}+\Delta)\left(V_0+E_{g_0}+\frac{2}{3}\Delta\right)} \quad \text{and}$$

$$A'_{12}(V_0) = \left[\frac{2m_c}{\hbar^2} I_{11}(V_0)\left[\frac{1}{V_0} + \frac{1}{V_0+E_{g0}} + \frac{1}{V_0+E_{g0}+\Delta} - \frac{1}{V_0+E_{g0}+(2/3)\Delta}\right]\right].$$

Therefore, the field-emitted current is given by

$$I = \frac{2g_v e k_B T}{h} \sum_{n_x=1}^{n_{x\max}} \sum_{n_y=1}^{n_{y\max}} F_0(\eta_{12}) \exp(-\beta_{12}). \tag{1.18}$$

1.2.2.2 Two-Band Model of Kane

Under the inequalities $\Delta \gg E_{g_0}$ or $\Delta \ll E_{g_0}$, (1.13) assumes the form

$$E(1 + \alpha E) = (\hbar^2 k^2 / 2m_c), \quad \alpha \equiv 1/E_{g_0}. \tag{1.19}$$

Equation (1.19) is known as the two-band model of Kane where α is known as band nonparabolicity parameter and should be as such for studying the electronic properties of the semiconductors whose band structures obey the above inequalities [103, 104].

The 1D E–k_z relation can be expressed as

$$k_z^2 = A_{13}(E, n_x, n_y), \tag{1.20}$$

where $A_{13}(E, n_x, n_y) = \left[\frac{2m_c}{\hbar^2}\{E(1+\alpha E)\} - \phi_1(n_x, n_y)\right]$.

The quantized subband energy (E_{12}) is given by

$$E_{13}(1 + \alpha E_{13}) = \left[\frac{\hbar^2}{2m_c}\right] \phi_1(n_x, n_y). \tag{1.21}$$

The electron concentration per unit length can be written as

$$n_0 = \frac{2g_v}{\pi} \sum_{n_x=1}^{n_{x\max}} \sum_{n_y=1}^{n_{y\max}} \left[B_{15}\left(E_{F1D}, n_x, n_y\right) + B_{16}\left(E_{F1D}, n_x, n_y\right)\right], \tag{1.22}$$

where $B_{15}\left(E_{F1D}, n_x, n_y\right) = \left[A_{13}(E_{F1D}, n_x, n_y)\right]^{1/2}$ and

$$B_{16}(E_{F1D} n_x, n_y) = \sum_{r=1}^{r_0} Z_{1D}(r)\left[B_{15}\left(E_{F1D}, n_x, n_y\right)\right].$$

The field-emitted current can be expressed as

$$I = \frac{2g_v e k_B T}{h} \sum_{n_x=1}^{n_{x\max}} \sum_{n_y=1}^{n_{y\max}} F_0(\eta_{13}) t_{13}, \tag{1.23}$$

where $\eta_{13} = (E_{F1D} - E_{13})/k_B T$, $t_{13} = \exp(-\beta_{13})$, $\beta_{13} = \frac{4}{3}\left[A_{13}\left(V_0, n_x, n_y\right)\right]^{3/2} \cdot \left[eF_{sz} A'_{13}(V_0)\right]^{-1}$, and $A'_{13}(V_0) = \frac{2m_c}{\hbar^2}(1 + 2\alpha V_0)$.

Therefore, the field-emitted current assumes the from

$$I = \frac{2g_v e k_B T}{h} \sum_{n_x=1}^{n_{x\max}} \sum_{n_y=1}^{n_{y\max}} F_0(\eta_{13}) \exp(-\beta_{13}). \tag{1.24}$$

1.2.2.3 Parabolic Energy Bands

The expressions for the electron concentration per unit length and the FNFE from QWs having parabolic energy bands can, respectively, be written as

$$n_0 = \frac{2g_v\sqrt{2\pi m_c k_B T}}{h} \sum_{n_x=1}^{n_{x\max}} \sum_{n_y=1}^{n_{y\max}} F_{-1/2}(\eta_{14}), \tag{1.25}$$

$$I = \frac{2g_v e k_B T}{h} \sum_{n_x=1}^{n_{x\max}} \sum_{n_y=1}^{n_{y\max}} F_0(\eta_{14}) \exp\left[-\frac{4}{3eF_{sz}}\left(\frac{2m_c}{\hbar^2}\right)^{1/2} V_0^{3/2}\left\{1-\frac{\hbar^2\phi_1(n_x,n_y)}{2m_c V_0}\right\}^{3/2}\right], \tag{1.26}$$

where $\eta_{14} = \left[E_{F1D} - \frac{\hbar^2\phi_1(n_x,n_y)}{2m_c}\right][k_B T]^{-1}$.

Converting the summations over n_x and n_y to the corresponding integrations, (1.26) gets transformed as

$$J = \frac{g_v e^2 F_{sz}^2}{8\pi h \phi_w} \exp\left\{-\frac{4}{3eF_{sz}}\left(\frac{2m_c}{\hbar^2}\right)^{1/2} \phi_w^{3/2}\right\}. \tag{1.27}$$

Equation (1.27) is the well-known expression of the FNFE from bulk semiconductors having parabolic energy bands [205, 206].

1.2.2.4 The Model of Stillman et al.

In accordance with the model of Stillman et al. [105], the electron dispersion law of III–V materials assumes the form

$$E = \bar{t}_{11}k^2 - \bar{t}_{12}k^4, \tag{1.28}$$

where $\bar{t}_{11} \equiv \frac{\hbar^2}{2m_c}$, $\bar{t}_{12} \equiv \left(1-\frac{m_c}{m_0}\right)^2\left(\frac{\hbar^2}{2m_c}\right)^2\left[\left(3E_{g0}+4\Delta+\frac{2\Delta^2}{E_{g0}}\right)\cdot\{(E_{g0}+\Delta)(2\Delta+3E_{g0})\}^{-1}\right]$, and m_0 is the free electron mass.

Equation (1.28) can be expressed as

$$\frac{\hbar^2 k^2}{2m_c} = I_{12}(E), \tag{1.29}$$

where $I_{12}(E) \equiv a_{11}[1-(1-a_{12}E)^{1/2}]$ and $a_{11} \equiv \left(\frac{\hbar^2 \bar{t}_{11}}{4m_c \bar{t}_{12}}\right)$, and $a_{12} \equiv \frac{4\bar{t}_{12}}{\bar{t}_{11}^2}$.

The 1D E–k_z relation can be written as

$$k_z^2 = A_{14}\left(E, n_x, n_y\right), \tag{1.30}$$

where $A_{14}(E, n_x, n_y) = \left[\frac{2m_c}{\hbar^2}\{I_{12}(E)\} - \phi_1(n_x, n_y)\right]$.

The quantized subband energy (E_{14}) is given by

$$I_{12}(E_{14}) = \left[\frac{\hbar^2}{2m_c}\right]\phi_1\left(n_x, n_y\right). \tag{1.31}$$

The electron concentration per unit length can be expressed as

$$n_0 = \frac{2g_v}{\pi}\sum_{n_x=1}^{n_{x\max}}\sum_{n_y=1}^{n_{y\max}}\left[B_{17}\left(E_{\mathrm{F1D}}, n_x, n_y\right) + B_{18}\left(E_{\mathrm{F1D}}, n_x, n_y\right)\right], \tag{1.32}$$

where $B_{17}\left(E_{\mathrm{F1D}}, n_x, n_y\right) = \left[A_{14}\left(E_{\mathrm{F1D}}, n_x, n_y\right)\right]^{1/2}$ and $B_{18}\left(E_{\mathrm{F1D}}, n_x, n_y\right) = \sum_{r=1}^{r_0} Z_{\mathrm{1D}}(r)\left[B_{17}\left(E_{\mathrm{F1D}}, n_x, n_y\right)\right]$.

The field-emitted current can be written as

$$I = \frac{2g_v e k_B T}{h}\sum_{n_x=1}^{n_{x\max}}\sum_{n_y=1}^{n_{y\max}} F_0(\eta_{15})\exp(-\beta_{15}), \tag{1.33}$$

where $\eta_{15} = (E_{\mathrm{F1D}} - E_{14})/k_B T$, $\beta_{15} = \frac{4}{3}\left[A_{14}\left(V_0, n_x, n_y\right)\right]^{3/2} \cdot [eF_{sz}A'14(V_0)]^{-1}$, $A_{14}(V_0, n_x, n_y) = \left[\frac{2m_c}{\hbar^2}\{I_{12}(V_0)\} - \phi_1(n_x, n_y)\right]$, $A'_{14}(V_0) = \frac{2m_c}{\hbar^2}I'_{12}(V_0)$, and $I'_{12}(V_0) = \left(\frac{a_{11}a_{12}}{2}\right)(1-a_{12}V_0)^{-1/2}$.

1.2.2.5 The Model of Newson and Kurobe

In accordance with the model of Newson and Kurobe [106], the electron dispersion law in this case assumes the form

$$E = a_{13}k_z^4 + \left[\frac{\hbar^2}{2m_c} + a_{14}k_s^2\right]k_z^2 + \frac{\hbar^2}{2m_c}k_s^2 + a_{14}k_x^2k_y^2 + a_{13}\left(k_x^4 + k_y^4\right), \tag{1.34}$$

where a_{13} is the nonparabolicity constant, $a_{14}(\equiv 2a_{13} + a_{15})$, and a_{15} is known as the warping constant.

The 1D E–k_z relation can be expressed as

$$k_z^2 = A_{15}(E, n_x, n_y), \tag{1.35}$$

where $A_{15}(E, n_x, n_y) = (2a_{13})^{-1}[-\overline{L_1}(n_x, n_y) + [\{\overline{L_1}(n_x, n_y)\}^2 - 4a_{13}[\overline{L_2}(n_x, n_y) - E)]^{1/2}]$, $\overline{L_1}(n_x, n_y) = \frac{\hbar^2}{2m_c} + a_{14}\left[\left(\frac{n_x\pi}{d_x}\right) + \left(\frac{n_y\pi}{d_y}\right)^2\right]$, and $\overline{L_2}(n_x, n_y) = \left[\frac{\hbar^2}{2m_c}\phi_1(n_x, n_y) + a_{14}\left(\frac{n_x\pi}{d_x} \cdot \frac{n_y\pi}{d_y}\right)^2 + a_{13}\left[\left(\frac{n_x\pi}{d_x}\right)^4 + \left(\frac{n_y\pi}{d_y}\right)^4\right]\right]$.

The quantized subband energy (E_{16}) is given by

$$E_{16} = \overline{L_2}(n_x, n_y). \tag{1.36}$$

The electron concentration per unit length can be written as

$$n_0 = \frac{2g_v}{\pi}\sum_{n_x=1}^{n_{x\max}}\sum_{n_y=1}^{n_{y\max}}\left[B_{19}\left(E_{F1D}, n_x, n_y\right) + B_{20}\left(E_{F1D}, n_x, n_y\right)\right], \tag{1.37}$$

where $B_{19}\left(E_{F1D}, n_x, n_y\right) = \left[A_{15}\left(E_{F1D}, n_x, n_y\right)\right]^{1/2}$ and

$$B_{20}(E_{F1D}, n_x, n_y) = \sum_{r=1}^{r_0} Z_{1D}(r)[B_{19}(E_{F1D}, n_x, n_y)]. \tag{1.38}$$

The field-emitted current assumes the form

$$I = \frac{2g_v e k_B T}{h}\sum_{n_x=1}^{n_{x\max}}\sum_{n_y=1}^{n_{y\max}} F_0(\eta_{16})\exp(-\beta_{16}), \tag{1.39}$$

where $\eta_{16} = (E_{F1D} - E_{16})/k_BT$, $\beta_{16} = \frac{4}{3}\left[A_{15}(V_0, n_x, n_y)\right]^{3/2} \cdot \left[eF_{sz}A'_{15}(V_0, n_x, n_y)\right]^{-1}$, and $A'_{15}(V_0, n_x, n_y) = [\{\overline{L_1}(n_x, n_y)\}^2 - 4a_{13}[\overline{L_2}(n_x, n_y) - V_0)]^{-1/2}]$.

1.2.2.6 Model of Palik et al.

The energy spectrum of the conduction electrons in III–V semiconductors up to the fourth order in effective mass theory, taking into account the interactions of heavy hole, light hole, and the split-off holes can be expressed in accordance with the model of Palik et al. [107] as

$$E = \frac{\hbar^2 k^2}{2m_c} - \overline{B}_{11} k^4, \tag{1.40}$$

where $\overline{B}_{11} = \left[\frac{\hbar^4}{4E_{g0}(m_c)^2}\right]\left[\frac{1+\frac{x_{11}^2}{2}}{1+\frac{x_{11}}{2}}\right](1-y_{11})^2, x_{11} = \left[1+\left(\frac{\Delta}{E_{g0}}\right)\right],^{-1}$ and $y_{11} = \frac{m_c}{m_o}$.

Equation (1.40) gets simplified as

$$\frac{\hbar^2 k^2}{2m_c} = I_{13}(E), \tag{1.41}$$

where $I_{13}(E) = \bar{b}_{12}\left[\bar{a}_{12} - \left((\bar{a}_{12})^2 - 4E\overline{B}_{11}\right)^{1/2}\right]$ and $\bar{a}_{12} = \left(\frac{\hbar^2}{2m_c}\right)$, and $\bar{b}_{12} = \left[\frac{\bar{a}_{12}}{2\overline{B}_{11}}\right]$.

The 1D E–k_z relation can be written as

$$k_z^2 = A_{16}(E, n_x, n_y), \tag{1.42}$$

where $A_{16}(E, n_x, n_y) = \left[\frac{2m_c}{\hbar^2}\{I_{13}(E)\} - \phi_1(n_x, n_y)\right]$.

The electron concentration per unit length can be expressed as

$$n_0 = \frac{2g_v}{\pi}\sum_{n_x=1}^{n_{x\max}}\sum_{n_y=1}^{n_{y\max}}\left[B_{21}\left(E_{\mathrm{F1D}}, n_x, n_y\right) + B_{22}\left(E_{\mathrm{F1D}}, n_x, n_y\right)\right], \tag{1.43}$$

where $B_{21}\left(E_{\mathrm{F1D}}, n_x, n_y\right) = \left[A_{16}\left(E_{\mathrm{F1D}}, n_x, n_y\right)\right]^{1/2}$ and $B_{22}\left(E_{\mathrm{F1D}}, n_x, n_y\right) = \sum_{r=1}^{r_0} Z_{\mathrm{1D}}(r)\left[B_{21}\left(E_{\mathrm{F1D}}, n_x, n_y\right)\right]$.

The field-emitted current can be written as

$$I = \frac{2g_v e k_B T}{h}\sum_{n_x=1}^{n_{x\max}}\sum_{n_y=1}^{n_{y\max}} F_0\left(\eta_{16}\right)\exp(-\beta_{16}), \tag{1.44}$$

where $\eta_{16} = (E_{\mathrm{F1D}} - E_{17})/k_B T$ and E_{17} is the root of

$$\frac{\hbar^2}{2m_c}\phi_1(n_x, n_y) = I_{13}(E_{17}), \tag{1.45}$$

where $\beta_{16} = \frac{4}{3}\left[A_{16}\left(V_0, n_x, n_y\right)\right]^{3/2} \cdot \left[eF_{sz}A'_{16}(V_0)\right]^{-1}$, $A'_{16}(V_0) = \frac{2m_c}{\hbar^2}I'_{13}(V_0)$, and $I'_{13}(V_0) = 2\bar{b}_{12}\overline{B}_{11}\left[(\bar{a}_{12})^2 - 4V_0\overline{B}_{11}\right]^{-1/2}$.

1.2.3 *The Field Emission from Quantum Wires of II–VI Semiconductors*

The carrier energy spectra in bulk specimens of II–VI compounds in accordance with Hopfield model [124] can be written as

$$E = A_0k_s^2 + B_0k_z^2 \pm C_0k_s, \tag{1.46}$$

where $A_0 \equiv \hbar^2/2m_\perp^*$, $B_0 \equiv \hbar^2/2m_\parallel^*$, and C_0 represents the splitting of the two-spin states by the spin–orbit coupling and the crystalline field.

The 1D dispersion relation for quantum wires of II–VI semiconductors can be expressed as

$$E = B_0k_z^2 + G_{3,\pm}(n_x, n_y), \tag{1.47}$$

where $G_{3,\pm}\left(n_x, n_y\right) \equiv \left[A_0\left\{\left(\frac{\pi n_x}{d_x}\right)^2 + \left(\frac{\pi n_y}{d_y}\right)^2\right\} \pm C_0\left\{\left(\frac{\pi n_x}{d_x}\right)^2 + \left(\frac{\pi n_y}{d_y}\right)^2\right\}^{1/2}\right]$.

The FNFE in this case is given by

$$I = \frac{eg_vk_BT}{h}\sum_{n_x=1}^{n_{x\max}}\sum_{n_y=1}^{n_{y\max}}\left[\left\{F_0\left\{(k_BT)^{-1}\left[E_{F1D} - \left[G_{3,+}\left(n_x, n_y\right)\right]\right]\right\}\right.\right.$$
$$\left.\left. \times \exp\left(-\overline{\beta}_{016,+}\right)\right\} + F_0\left\{(k_BT)^{-1}\left[E_{F1D} - \left[G_{3,-}\left(n_x, n_y\right)\right]\right]\right\}\exp\left(-\overline{\beta}_{016,-}\right)\right] \tag{1.48}$$

and $\overline{\beta}_{016,\pm} = \frac{4}{3}\left[\left(V_0 - G_{3,\pm}\left(n_x, n_y\right)\right)\right]^{3/2} \cdot \left[eF_{sz}\hbar\right]^{-1}\sqrt{2m_\parallel^*}$.

The 1D electron statistics can be written as

$$n_{1D} = \frac{g_v}{\pi\sqrt{B_0}}\sum_{n_x=1}^{n_{x\max}}\sum_{n_y=1}^{n_{y\max}}\left[t_7\left(E_{F1D}, n_x, n_y\right) + t_8\left(E_{F1D}, n_x, n_y\right)\right], \tag{1.49}$$

where $t_7\left(E_{F1D}, n_x, n_y\right) \equiv \left[E_{F1D} - \left[G_{3,+}\left(n_x, n_y\right)\right]\right]^{1/2} + \left[E_{F1D} - \left[G_{3,-}\left(n_x, n_y\right)\right]\right]^{1/2}$ and $t_8\left(E_{F1D}, n_x, n_y\right) = \sum_{1}^{0} Z_{1D}(r)\left[t_7(E_{F1D}, n_x, n_y)\right]$

1.2.4 *The Field Emission from Quantum Wires of Bismuth*

1.2.4.1 The McClure and Choi Model

The dispersion relation of the carriers in Bi can be written, following the McClure and Choi [135], as

$$E(1+\alpha E) = \frac{p_x^2}{2m_1} + \frac{p_y^2}{2m_2} + \frac{p_z^2}{2m_3} + \frac{p_y^2}{2m_2}\alpha E\left\{1 - \left(\frac{m_2}{m_2'}\right)\right\} + \frac{p_y^4\alpha}{4m_2m_2'} - \frac{\alpha p_x^2 p_y^2}{4m_1m_2} - \frac{\alpha p_y^2 p_z^2}{4m_2m_3}, \tag{1.50}$$

where $p_i \equiv \hbar k_i$, $i = x, y, z$, m_1, m_2, and m_3 are the effective carrier masses at the band edge along the x, y, and z directions, respectively, and m_2' is the effective mass tensor component at the top of the valence band (for electrons) or at the bottom of the conduction band (for holes).

The 1D dispersion relation of the carriers in Bi in this case assumes the form

$$E(1+\alpha E) = \left\{\frac{\hbar^2 k_x^2}{2m_1}\left[1 - \frac{\alpha\hbar^2}{2m_2}\left(\frac{\pi n_y}{d_y}\right)^2\right] + G_{12} + \frac{\hbar^2}{2m_2}\alpha E\left\{1 - \left(\frac{m_2}{m_2'}\right)\right\}\left(\frac{\pi n_y}{d_y}\right)^2\right\}, \tag{1.51}$$

where $n_z = 1, 2, 3, \ldots$ is the size quantum number along the z direction, d_z is the nanothickness along the z direction, and

$$G_{12} \equiv \left\{\frac{\hbar^2}{2m_2}\left(\frac{\pi n_y}{d_y}\right)^2 + \frac{\hbar^2}{2m_3}\left(\frac{\pi n_z}{d_z}\right)^2 + \frac{\alpha\hbar^4}{4m_2m_2'}\left(\frac{\pi n_y}{d_y}\right)^4 - \frac{\alpha}{4m_2m_3}\left(\frac{\hbar^2 n_y n_z \pi^2}{d_y d_z}\right)^2\right\}.$$

Using (1.51), the 1D electron statistics can be expressed as

$$n_{1D} = \frac{2g_v}{\pi}\frac{\sqrt{2m_1}}{\hbar}\sum_{n_y=1}^{n_{y\max}}\sum_{n_z=1}^{n_{z\max}}\left[t_{27}\left(E_{F1D}, n_y, n_z\right) + t_{28}\left(E_{F1D}, n_y, n_z\right)\right], \tag{1.52}$$

where

$$t_{27}\left(E_{F1D}, n_y, n_z\right) \equiv \left\{\left[1 - \frac{\alpha\hbar^2}{2m_2}\left(\frac{\pi n_y}{d_y}\right)^2\right]^{-1/2}\left[E_{F1D}\left(1+\alpha E_{F1D}\right) - G_{12} - \frac{\hbar^2}{2m_2}\alpha E_{F1D}\left\{1 - \left(\frac{m_2}{m_2'}\right)\right\}\left(\frac{\pi\hbar n_y}{d_y}\right)^2\right]^{1/2}\right\}$$

and

$$t_{28}\left(E_{\mathrm{F1D}},n_y,n_z\right)\equiv\sum_{r=1}^{s_0}Z_{\mathrm{1D}}(r)\left[t_{27}\left(E_{\mathrm{F1D}},n_y,n_z\right)\right].$$

The field-emitted current is given by

$$I=\frac{2g_v e k_B T}{h}\sum_{n_z=1}^{n_{z\max}}\sum_{n_y=1}^{n_{y\max}}F_0(\eta_{17})\exp(-\beta_{17}), \tag{1.53}$$

where $\eta_{17}=\dfrac{E_{\mathrm{F1D}}-E_{19}}{k_B T}$ and E_{19} is the root of

$$E_{19}\left(1+\alpha E_{19}\right)=\left\{G_{12}+\frac{\hbar^2}{2m_2}\alpha E_{19}\left\{1-\left(\frac{m_2}{m_2'}\right)\right\}\left(\frac{\pi n_y}{d_y}\right)^2\right\}, \tag{1.54}$$

$$\beta_{17}=\frac{4}{3}[A_{17}(V_0,n_z,n_y)]^{3/2}\cdot[eF_{sz}A'_{17}(V_0,n_z,n_y)]^{-1},$$

$$\left[1-\frac{\alpha\hbar^2}{2m_2}\left(\frac{\pi n_y}{d_y}\right)^2\right]^{-1}\left[V_0\left(1+\alpha V_0\right)-G_{12}-\left\{\frac{\hbar^2}{2m_2}\alpha V_0\left\{1-\left(\frac{m_2}{m_2'}\right)\right\}\right.\right.$$
$$\left.\left.\times\left(\frac{\pi n_y}{d_y}\right)^2\right\}\right]\left(\frac{\hbar^2}{2m_1}\right)^{-1}=A_{17}\left(V_0,n_y,n_z\right),$$

and

$$\left[1-\frac{\alpha\hbar^2}{2m_2}\left(\frac{\pi n_y}{d_y}\right)^2\right]^{-1}\left[\left(1+2\alpha V_0\right)-\left\{\frac{\hbar^2}{2m_2}\alpha\left\{1-\left(\frac{m_2}{m_2'}\right)\right\}\left(\frac{\pi n_y}{d_y}\right)^2\right\}\right]$$
$$\times\left(\frac{\hbar^2}{2m_1}\right)^{-1}=A'_{17}\left(V_0,n_y,n_z\right).$$

1.2.4.2 The Hybrid Model

The dispersion relation of the carriers in bulk specimens of Bi in accordance with the Hybrid model can be written as [134]

$$E\left(1+\alpha E\right)=\frac{\theta_0\left(E\right)\left(\hbar k_y\right)^2}{2M_2}+\frac{\alpha\gamma_0\hbar^4k_y^4}{4M_2^2}+\frac{\hbar^2k_x^2}{2m_1}+\frac{\hbar^2k_z^2}{2m_3}, \tag{1.55}$$

where $\theta_0(E)\equiv[1+\alpha E(1-\gamma_0)+\overline{\delta}_0]$, $\gamma_0\equiv\frac{M_2}{m_2}$, $\overline{\delta}_0\equiv\frac{M_2}{M_2'}$, and the other notations are defined in [134].

The 1D dispersion relation in this case assumes the form

$$E\left(1+\alpha E\right)=\frac{\hbar^2 k_x^2}{2m_1}+G_{14}+\frac{\hbar^2}{2M_2}\left(\frac{\pi n_y}{d_y}\right)^2\alpha E\left(1-\gamma_0\right), \tag{1.56}$$

where $G_{14}=\left[\frac{\hbar^2}{2m_3}\left(\frac{\pi n_z}{d_z}\right)^2+\frac{\hbar^2}{2M_2}\left(\frac{\pi n_y}{d_y}\right)^2\left(1+\bar{\delta}_0\right)+\frac{\alpha\gamma_0\hbar^4}{4M_2^2}\left(\frac{\pi n_y}{d_y}\right)^4\right]$.

The use of (1.56) leads to the expression for the electron concentration per unit length as

$$n_{D1}=\frac{2g_v}{\pi}\frac{\sqrt{2m_1}}{\hbar}\sum_{n_y=1}^{n_{y\max}}\sum_{n_z=1}^{n_{z\max}}\left[t_{31}\left(E_{F1D},n_y,n_z\right)+t_{32}\left(E_{F1D},n_y,n_z\right)\right], \tag{1.57}$$

where

$$t_{31}\left(E_{F1D},n_y,n_z\right)\equiv\left[E_{F1D}\left(1+\alpha E_{1DF}\right)-G_{14}-\frac{\hbar^2}{2M_2}\left(\frac{\pi n_y}{d_y}\right)^2\alpha E_{F1D}\left(1-\gamma_0\right)\right]^{1/2},$$

and $t_{32}\left(E_{F1D},n_y,n_z\right)\equiv\sum_{r=1}^{s_0}Z_{1D}(r)\left[t_{31}\left(E_{F1D},n_y,n_z\right)\right]$.

The field-emitted current is given by

$$I=\frac{2g_v e k_B T}{h}\sum_{n_z=1}^{n_{z\max}}\sum_{n_y=1}^{n_{y\max}}F_0(\eta_{18})\exp(-\beta_{18}), \tag{1.58}$$

where $\eta_{18}=\dfrac{E_{F1D}-E_{20}}{k_B T}$.

E_{20} is the root of

$$E_{20}\left(1+\alpha E_{20}\right)=G_{14}+\frac{\hbar^2}{2M_2}\left(\frac{\pi n_y}{d_y}\right)^2\alpha E_{20}\left(1-\gamma_0\right), \tag{1.59}$$

$$\beta_{18}=\frac{4}{3}[A_{18}(V_0,n_z,n_y)]^{3/2}\cdot[eF_{sz}A'_{18}(V_0,n_z,n_y)]^{-1},$$

$$\left[V_0\left(1+\alpha V_0\right)-G_{14}-\left\{\frac{\hbar^2}{2M_2}\alpha V_0\left\{1-\gamma_0\right\}\left(\frac{\pi n_y}{d_y}\right)^2\right\}\right]\left(\frac{\hbar^2}{2m_1}\right)^{-1}$$

$$=A_{18}\left(V_0,n_y,n_z\right),$$

and $\left[\left(1+2\alpha V_0\right)-\left\{\frac{\hbar^2}{2M_2}\alpha\left\{1-\gamma_0\right\}\left(\frac{\pi n_y}{d_y}\right)^2\right\}\right]\left(\frac{\hbar^2}{2m_1}\right)^{-1}=A'_{18}\left(V_0,n_y,n_z\right).$

1.2.4.3 The Cohen Model

In accordance with the Cohen model [133], the dispersion law of the carriers in Bi is given by

$$E\,(1+\alpha E)=\frac{p_x^2}{2m_1}+\frac{p_z^2}{2m_3}-\frac{\alpha E p_y^2}{2m_2'}+\frac{p_y^2(1+\alpha E)}{2m_2}+\frac{\alpha p_y^4}{4m_2m_2'}. \tag{1.60}$$

The 1D carrier dispersion law in this case can be written as

$$\alpha E^2+El_7-G_{15}=\frac{\hbar^2k_x^2}{2m_1}, \tag{1.61}$$

where $l_7=\left[1-\frac{\alpha\hbar^2}{2m_2}\left(\frac{\pi n_y}{d_y}\right)^2+\frac{\alpha\hbar^2}{2m_2'}\left(\frac{\pi n_y}{d_y}\right)^2\right]$

and $G_{15}=\left[\frac{\hbar^2}{2m_3}\left(\frac{\pi n_z}{d_z}\right)^2+\frac{\hbar^2}{2m_2}\left(\frac{\pi n_y}{d_y}\right)^2+\frac{\alpha\hbar^4}{4m_2m_2'}\left(\frac{\pi n_y}{d_y}\right)^4\right]$.

The 1D electron concentration per unit length assumes the form

$$n_{1D}=\frac{2g_v}{\pi}\frac{\sqrt{2m_1}}{\hbar}\sum_{n_y=1}^{n_{y\max}}\sum_{n_z=1}^{n_{z\max}}\left[t_{35}\left(E_{F1D},n_y,n_z\right)+t_{36}\left(E_{F1D},n_y,n_z\right)\right], \tag{1.62}$$

where $t_{35}\left(E_{F1D},n_y,n_z\right)=\left[\alpha E_{F1D}^2+E_{F1D}l_7-G_{15}\right]^{1/2}$ and $t_{36}\left(E_{F1D},n_y,n_z\right)=\sum_{r=1}^{s_o}Z_{1D}\left(r\right)\left[t_{35}\left(E_{F1D},n_y,n_z\right)\right]$.

The field-emitted current is given by

$$I=\frac{2g_v e k_B T}{h}\sum_{n_z=1}^{n_{z\max}}\sum_{n_y=1}^{n_{y\max}}F_0(\eta_{19})\exp(-\beta_{19}), \tag{1.63}$$

where $\eta_{19}=\frac{E_{F1D}-E_{21}}{k_BT}$.

E_{21} is the root of

$$E_{21}\left(l_7+\alpha E_{21}\right)=G_{15}, \tag{1.64}$$

$$\beta_{19}=\tfrac{4}{3}[A_{21}(V_0,n_z,n_y)]^{3/2}\cdot[eF_{sz}A_{21}'(V_0,n_z,n_y)]^{-1},$$

$$[V_0(l_7+\alpha V_0)-G_{15}]\left(\frac{\hbar^2}{2m_1}\right)^{-1}=A_{21}(V_0,n_y,n_z),\quad\text{and}$$

$$[(l_7+2\alpha V_0)]\left(\frac{\hbar^2}{2m_1}\right)^{-1}=A_{21}'(V_0,n_y,n_z).$$

1.2.4.4 The Lax Model

The electron energy spectra in bulk specimens of Bi in accordance with the Lax model can be written as [129]

$$E(1+\alpha E)=\frac{p_x^2}{2m_1}+\frac{p_y^2}{2m_2}+\frac{p_z^2}{2m_3}. \quad (1.65)$$

The 1D dispersion relation in this case can be expressed as

$$E(1+\alpha E)=\frac{\hbar^2k_x^2}{2m_1}+G_{16}, \quad (1.66)$$

where $G_{16}=\frac{\hbar^2}{2m_2}\left(\frac{\pi n_y}{d_y}\right)^2+\frac{\hbar^2}{2m_3}\left(\frac{\pi n_z}{d_z}\right)^2$.

The 1D electron statistics is given by

$$n_{1D}=\frac{2g_v\sqrt{2m_1}}{\pi\hbar}\sum_{n_y=1}^{n_{y\max}}\sum_{n_z=1}^{n_{z\max}}\left[t_{37}\left(E_{F1D},n_y,n_z\right)+t_{38}\left(E_{F1D},n_y,n_z\right)\right], \quad (1.67)$$

where $t_{37}\left(E_{F1D},n_y,n_z\right)=\left[E_{F1D}\left(1+\alpha E_{F1D}\right)-G_{16}\right]^{1/2}$ and $t_{38}\left(E_{F1D},n_y,n_z\right)=\sum_{r=1}^{s_o}Z_{1D}(r)\left[t_{37}\left(E_{F1D},n_y,n_z\right)\right]$.

The field-emitted current in this case assumes the form

$$I=\frac{2g_vek_BT}{h}\sum_{n_z=1}^{n_{z\max}}\sum_{n_y=1}^{n_{y\max}}F_0(\eta_{20})\exp(-\beta_{20}), \quad (1.68)$$

where $\eta_{20}=\frac{E_{F1D}-E_{22}}{k_BT}$.

E_{22} is the root of

$$E_{22}\left(l_7+\alpha E_{22}\right)=G_{16}, \quad (1.69)$$

$$\beta_{20}=\tfrac{4}{3}\left[A_{22}(V_0,n_z,n_y)\right]^{3/2}\cdot\left[eF_{sz}A'_{22}(V_0)\right]^{-1},$$

$$\left[V_0\left(1+\alpha V_0\right)-G_{16}\right]\left(\frac{\hbar^2}{2m_1}\right)^{-1}=A_{22}\left(V_0,n_y,n_z\right),\quad \text{and}$$

$$\left[\left(1+2\alpha V_0\right)\right]\left(\frac{\hbar^2}{2m_1}\right)^{-1}=A'_{22}\left(V_0\right).$$

It may be noted that under the conditions $\alpha\to 0$, $M_2'\to\infty$ and isotropic effective electron mass at the edge of the conduction band, all models of Bismuth convert into isotropic parabolic energy bands. Thus under the aforementioned conditions and the conversion of the summations over the quantum numbers to the corresponding integrations, all the equations for FNFE for Bismuth lead to the well-known expression of the FNFE as given by (1.27).

1.2.5 The Field Emission from Quantum Wires of IV–VI Semiconductors

The dispersion relation of the conduction electrons in IV–VI semiconductors can be expressed in accordance with Dimmock [207] as

$$\left[\bar{\varepsilon} - \frac{E_{g_0}}{2} - \frac{\hbar^2 k_s^2}{2m_t^-} - \frac{\hbar^2 k_z^2}{2m_l^-}\right]\left[\bar{\varepsilon} + \frac{E_{g_0}}{2} + \frac{\hbar^2 k_s^2}{2m_t^-} + \frac{\hbar^2 k_z^2}{2m_l^-}\right] = P_\perp^2 k_s^2 + P_\parallel^2 k_z^2, \tag{1.70}$$

where $\bar{\varepsilon}$ is the energy as measured from the center of the band gap E_{g_0}, and $m_t^\pm$ and $m_l^\pm$ represent the contributions to the transverse and longitudinal effective masses of the external L_6^+ and L_6^- bands arising from the $\vec{k}.\vec{p}$ perturbations with the other bands taken to the second order.

Using $\varepsilon = E + (E_{g_0}/2)$, $P_\perp^2 = \dfrac{\hbar^2 E_{g_0}}{2m_t^*}$, and $P_\parallel^2 = \dfrac{\hbar^2 E_{g_0}}{2m_l^*}$ (m_t^* and m_l^* are the transverse and longitudinal effective electron masses at $k = 0$) in (1.70), we can write

$$\left[E - \frac{\hbar^2 k_s^2}{2m_t^-} - \frac{\hbar^2 k_z^2}{2m_l^-}\right]\left[1 + \alpha E + \alpha\frac{\hbar^2 k_s^2}{2m_t^+} + \alpha\frac{\hbar^2 k_z^2}{2m_l^+}\right] = \frac{\hbar^2 k_s^2}{2m_t^*} + \frac{\hbar^2 k_z^2}{2m_l^*}. \tag{1.71}$$

The 1D dispersion relation can be expressed from (1.71) as

$$k_z^2 = A_{23}(E, n_x, n_y), \tag{1.72}$$

where $A_{23}(E, n_x, n_y) = (2h_4)^{-1}\left[h_6(E, n_x, n_y) - \left[h_6^2(E, n_x, n_y) + 4h_4 h_7 (E, n_x, n_y)\right]^{1/2}\right]$, $h_4 = \left[\dfrac{\alpha\hbar^4}{4x_3 x_6}\right]$,

$$x_3 = \frac{3m_t^- m_l^-}{2m_l^- + m_t^-}, \quad x_6 = \frac{3m_t^+ m_l^+}{2m_l^+ + m_t^+},$$

$$h_6(E, n_x, n_y) = \left[\frac{\alpha E\hbar^2}{2x_6} - \frac{\alpha\hbar^2}{2x_6}\left[\left(\frac{\pi n_x}{d_x}\right)^2\frac{\hbar^2}{2x_1} + \left(\frac{\pi n_y}{d_y}\right)^2\frac{\hbar^2}{2x_2}\right] - \frac{\alpha\hbar^2}{2x_3}\left[\left(\frac{\pi n_x}{d_x}\right)^2\frac{\hbar^2}{2x_4} + \left(\frac{\pi n_y}{d_y}\right)^2\frac{\hbar^2}{2x_5}\right] - \frac{\hbar^2}{2m_3} - \frac{(1+\alpha E)\hbar^2}{2x_3}\right],$$

$$x_1 = m_t^-, \; x_2 = \frac{m_t^- + 2m_l^-}{3}, \; x_4 = m_t^+, \; x_5 = \frac{m_t^+ + 2m_l^+}{3}, \; m_3 = \frac{3m_t^* m_l^*}{m_t^* + 2m_l^*},$$

$$h_7(E,n_x,n_y) = \left[E(1+\alpha E) + \alpha E\left[\left(\frac{\pi n_x}{d_x}\right)^2 \frac{\hbar^2}{2x_4} + \left(\frac{\pi n_y}{d_y}\right)^2 \frac{\hbar^2}{2x_5}\right]\right.$$
$$- (1+\alpha E)\left[\left(\frac{\pi n_x}{d_x}\right)^2 \frac{\hbar^2}{2x_1} + \left(\frac{\pi n_y}{d_y}\right)^2 \frac{\hbar^2}{2x_2}\right] - \alpha\left[\left(\frac{\pi n_x}{d_x}\right)^2 \frac{\hbar^2}{2x_1}\right.$$
$$\left. + \left(\frac{\pi n_y}{d_y}\right)^2 \frac{\hbar^2}{2x_2}\right]\left[\left(\frac{\pi n_x}{d_x}\right)^2 \frac{\hbar^2}{2x_4} + \left(\frac{\pi n_y}{d_y}\right)^2 \frac{\hbar^2}{2x_5}\right]$$
$$\left. - \left[\left(\frac{\pi n_x}{d_x}\right)^2 \frac{\hbar^2}{2m_1} + \left(\frac{\pi n_y}{d_y}\right)^2 \frac{\hbar^2}{2m_2}\right]\right],$$
$$\times m_1 = m_t^*, \quad \text{and} \quad m_2 = \frac{m_t^* + 2m_l^*}{3}.$$

The electron concentration is given by

$$n_0 = \frac{2g_v}{\pi} \sum_{n_x=1}^{n_{x\max}} \sum_{n_y=1}^{n_{y\max}} [B_{32}(E_{F1D}, n_x, n_y) + B_{33}(E_{F1D}, n_x, n_y)], \tag{1.73}$$

where $B_{32}(E_{F1D}, n_x, n_y) = [A_{23}(E_{F1D}, n_x, n_y)]^{1/2}$ and $B_{33}(E_{F1D}, n_x, n_y) = \sum_{r=1}^{r_0} Z_{1D}(r)[B_{32}(E_{F1D}, n_x, n_y)]$.

The field-emitted current can be written as

$$I = \frac{2g_v e k_B T}{h} \sum_{n_x=1}^{n_{x\max}} \sum_{n_y=1}^{n_{y\max}} F_0(\eta_{22}) \exp(-\beta_{22}), \tag{1.74}$$

where $\eta_{22} = \dfrac{E_{F1D} - \overline{E_{22}}}{k_B T}$.

The subband energy $\overline{E_{22}}$ assumes the form

$$\overline{E_{22}} = (2\alpha)^{-1}\left[-\zeta_{70}(n_x, n_y) + [\zeta_{70}^2(n_x, n_y) + 4\alpha\zeta_{71}(n_x, n_y)]^{1/2}\right] \tag{1.75}$$

in which

$$\zeta_{70}(n_x,n_y) = \left[1 + \alpha\left[\left(\frac{\pi n_x}{d_x}\right)^2 \frac{\hbar^2}{2x_4} + \left(\frac{\pi n_y}{d_y}\right)^2 \frac{\hbar^2}{2x_5}\right]\right.$$
$$\left. - \alpha\left[\left(\frac{\pi n_x}{d_x}\right)^2 \frac{\hbar^2}{2x_1} + \left(\frac{\pi n_y}{d_y}\right)^2 \frac{\hbar^2}{2x_2}\right]\right],$$

$$\zeta_{71}\left(n_x,n_y\right)=\left[\left[\left(\frac{\pi n_x}{d_x}\right)^2\frac{\hbar^2}{2x_1}+\left(\frac{\pi n_y}{d_y}\right)^2\frac{\hbar^2}{2x_2}\right]+\alpha\left[\left(\frac{\pi n_x}{d_x}\right)^2\frac{\hbar^2}{2x_1}\right.\right.$$
$$\left.+\left(\frac{\pi n_y}{d_y}\right)^2\frac{\hbar^2}{2x_2}\right]\left[\left(\frac{\pi n_x}{d_x}\right)^2\frac{\hbar^2}{2x_4}+\left(\frac{\pi n_y}{d_y}\right)^2\frac{\hbar^2}{2x_5}\right]$$
$$\left.+\left[\left(\frac{\pi n_x}{d_x}\right)^2\frac{\hbar^2}{2m_1}+\left(\frac{\pi n_y}{d_y}\right)^2\frac{\hbar^2}{2m_2}\right]\right],$$

$$\beta_{22}=\frac{4}{3}\left[A_{23}(V_0,n_x,n_y)\right]^{3/2}\cdot\left[eF_{sz}A'_{23}(V_0,n_x,n_y)\right]^{-1},$$

$$A'_{23}(V_0,n_x,n_y)=(2h_4)^{-1}\left[h'_6-\frac{h_6\left(V_0,n_x,n_y\right)h'_6+2h_4h'_7\left(V_0,n_x,n_y\right)}{\left[h_6^2\left(V_0,n_x,n_y\right)+4h_4h_7\left(V_0,n_x,n_y\right)\right]^{1/2}}\right],$$

$$h'_6=\frac{\alpha\hbar^2}{2}\left(\frac{1}{x_6}-\frac{1}{x_3}\right),$$

and

$$h'_7(V_0,n_x,n_y)=\left[1+2\alpha V_0+\alpha\left[\frac{\hbar^2}{2x_4}\left(\frac{\pi n_x}{d_x}\right)^2+\left(\frac{\pi n_y}{d_y}\right)^2\frac{\hbar^2}{2x_5}\right]\right.$$
$$\left.-\alpha\left[\frac{\hbar^2}{2x_1}\left(\frac{\pi n_x}{d_x}\right)^2+\left(\frac{\pi n_y}{d_y}\right)^2\frac{\hbar^2}{2x_2}\right]\right].$$

1.2.6 The Field Emission from Quantum Wires of Stressed Semiconductors

The dispersion relation of the conduction electrons in bulk specimens of stressed semiconductors can be written as [208–211]

$$\frac{k_x^2}{[a^*(E)]^2}+\frac{k_y^2}{[b^*(E)]^2}+\frac{k_z^2}{[c^*(E)]^2}=1, \tag{1.76}$$

where

$$\left[a^*\left(E\right)\right]^2=\frac{K_0\left(E\right)}{M_0\left(E\right)+\frac{1}{2}N_0\left(E\right)},$$

$$K_0\left(E\right)=\left[E-C_1\varepsilon-\frac{2C_2^2\varepsilon_{xy}^2}{3E'_{g0}\left(E\right)}\right]\left(\frac{3E'_{g0}\left(E\right)}{2B_2^2}\right).$$

C_1 is the conduction band deformation potential constant, ε is the trace of the strain tensor $\hat{\varepsilon}$ which can be written as $\hat{\varepsilon} = \begin{bmatrix} \varepsilon_{xx} & \varepsilon_{xy} & 0 \\ \varepsilon_{xy} & \varepsilon_{yy} & 0 \\ 0 & 0 & \varepsilon_{zz} \end{bmatrix}$, C_2 is a constant which describes the strain interaction between the conduction and valance bands, $E'_{g_0}(E) = E_{g_0} + E - C_1\varepsilon$, E_{g_0} is the band gap in the absence of stress, B_2 is the momentum matrix element,

$$M_0(E) = \left[1 - \frac{(\bar{a}_0 + C_1)\,\varepsilon}{E'_{g_0}(E)} + \frac{3\bar{b}_0\varepsilon_{xx}}{2E'_{g_0}(E)} - \frac{\bar{b}_0\varepsilon}{2E'_{g_0}(E)}\right], \quad \bar{a}_0 = -\frac{1}{3}\left(\bar{b}_0 + 2\bar{m}\right),$$

$\bar{a}_0 = -\frac{1}{3}\left(\bar{l} + 2\bar{m}\right)$ $\bar{b}_0 = \frac{1}{3}\left(\bar{l} - \bar{m}\right)$, and $\bar{d}_0 = \frac{2\bar{n}}{\sqrt{3}}$, $\bar{l}, \bar{m}, \bar{n}$ are the matrix elements of the strain perturbation operator, $N_0(E) = \left(\bar{d}_0\sqrt{3}\right)\frac{\varepsilon_{xy}}{E'_{g_0}(E)}$,

$$\left[b^*(E)\right]^2 = \frac{K_0(E)}{M_0(E) - \frac{1}{2}N_0(E)}, \quad \left[c^*(E)\right]^2 = \frac{K_0(E)}{L_0(E)}, \quad \text{and}$$

$$L_0(E) = \left[1 - \frac{(\bar{a}_0 + C_1)\,\varepsilon}{E'_{g_0}(E)} + \frac{3\bar{b}_0\varepsilon_{zz}}{2E'_{g_0}(E)} - \frac{\bar{b}_0\varepsilon}{2E'_{g_0}(E)}\right].$$

The 1D dispersion relation of the carriers in stressed materials in this case can be written as

$$k_z^2 = A_{24}(E, n_x, n_y), \tag{1.77}$$

where $A_{24}(E, n_x, n_y) = [c^*(E)]^2\left[1 - \left(\frac{\pi n_x}{d_x}\right)^2 [a^*(E)]^{-2} - \left(\frac{\pi n_y}{d_y}\right)^2 [b^*(E)]^{-2}\right]$.

The subband energy E_{23} assumes the form

$$\left[\left(\frac{\pi n_x}{d_x}\right)^2 \left[a^*(E)\right]^{-2} + \left(\frac{\pi n_y}{d_y}\right)^2 \left[b^*(E)\right]^{-2}\right] = 1. \tag{1.78}$$

Using (1.77), the 1D electron statistics can be expressed as

$$n_{1D} = \frac{2g_v}{\pi}\sum_{n_y=1}^{n_{y\max}}\sum_{n_z=1}^{n_{z\max}}\left[B_{34}\left(E_{F1D}, n_y, n_z\right) + B_{35}\left(E_{F1D}, n_y, n_z\right)\right], \tag{1.79}$$

where $B_{34}\left(E_{F1D}, n_y, n_z\right) = \sqrt{A_{24}(E_{F1D}, n_x, n_y)}$ and $B_{35}\left(E_{F1D}, n_x, n_y\right) = \sum_{r=1}^{s_o} Z_{1D}(r)\left[B_{34}\left(E_{F1D}, n_y, n_z\right)\right]$.

The field-emitted current is given by

$$I = \frac{2g_v e k_B T}{h} \sum_{n_x=1}^{n_{x\max}} \sum_{n_y=1}^{n_{y\max}} F_0(\eta_{23}) \exp(-\beta_{23}), \tag{1.80}$$

where

$$\eta_{23} = \frac{E_{F1D} - E_{23}}{k_B T}, \quad \beta_{23} = \frac{4}{3}\left[A_{24}(V_0, n_x, n_y)\right]^{3/2} \cdot \left[eF_{sz} A'_{24}(V_0, n_x, n_y)\right]^{-1},$$

$$A'_{24}(V_0, n_x, n_y) = \left[\frac{K'_0(V_0)}{L_0(V_0)} - \frac{K_0(V_0)L'_0(V_0)}{L_0^2(V_0)} - \left(\frac{n_x \pi}{d_x}\right)^2 \frac{\left[M'_0(V_0) + \frac{1}{2}N'_0(V_0)\right]}{L_0(V_0)}\right.$$
$$+ \left(\frac{n_x \pi}{d_x}\right)^2 \cdot \frac{L'_0(V_0)}{L_0^2(V_0)}\left[M_0(V_0) + \frac{1}{2}N_0(V_0)\right] + \left(\frac{n_y \pi}{d_y}\right)^2 \cdot \frac{L'_0(V_0)}{L_0^2(V_0)}$$
$$\left. \times \left[M_0(V_0) - \frac{1}{2}N_0(V_0)\right] - \left(\frac{n_y \pi}{d_y}\right)^2 \cdot \frac{1}{L_0(V_0)}\left[M'_0(V_0) - \frac{1}{2}N'_0(V_0)\right]\right],$$

$$K'_0(V_0) = \left[\left[1 + \frac{2C_2^2 \varepsilon_{xy}^2}{3\{E'_{g_0}(V_0)\}^2}\right]\left(\frac{3E'_{g_0}(V_0)}{2B_2^2}\right)\right.$$
$$\left. + \left[V_0 - C_1\varepsilon - \frac{2C_2^2 \varepsilon_{xy}^2}{3E'_{g_0}(V_0)}\right]\left(\frac{3}{2B_2^2}\right)\right],$$

$$M'_0(V_0) = \left[\frac{(\bar{a}_0 + C_1)\varepsilon}{(E'_{g_0}(V_0))^2} - \frac{3\bar{b}_0 \varepsilon_{xx}}{2(E'_{g_0}(V_0))^2} + \frac{\bar{b}_0 \varepsilon}{2(E'_{g_0}(V_0))^2}\right], \quad \text{and}$$

$$N'_0(V_0) = -\left(\bar{d}_0\sqrt{3}\right)\frac{\varepsilon_{xy}}{(E'_{g_0}(V_0))^2}.$$

1.2.7 The Field Emission from Quantum Wires of Tellurium

The dispersion relation of the conduction electrons in Te can be expressed as [212]

$$E = \psi_1 k_z^2 + \psi_2 k_s^2 \pm [\psi_3^2 k_z^2 + \psi_4^2 k_s^2]^{1/2}, \tag{1.81}$$

where ψ_1, ψ_2, ψ_3, and ψ_4 are system constants.

From (1.81), the1D dispersion relation can be written as

$$k_z^2 = A_{25,\pm}(E, n_x, n_y), \tag{1.82}$$

where

$$A_{25,\pm}(E,n_x,n_y) = \left[(\psi_5(E) - \psi_6 k_s^2 \pm \psi_7\left[\psi_8(E) - k_s^2\right]^{1/2}\right],\ k_s^2 = \phi_1(n_x,n_y),$$

$$\psi_5(E) = \left[\frac{E}{\psi_1} + \frac{E}{2\psi_1^2}\right],\ \psi_6 = \frac{\psi_2}{\psi_1},\ \psi_7 = (2\psi_1^2)^{-1}\left[4\psi_3^2\psi_1\psi_2 - 4\psi_1^2\psi_4^2\right]^{1/2},\ \text{and}$$

$$\psi_8(E) = \left[\frac{\psi_4^4 + 4E\psi_3^2\psi_1}{4\psi_3^2\psi_1\psi_2 - 4\psi_1^2\psi_4^2}\right].$$

The subband energies are given by

$$E_{26,\pm} = \psi_2\phi_1\left(n_x,n_y\right) \pm \psi_4\left(\phi_1\left(n_x,n_y\right)\right)^{1/2}. \tag{1.83}$$

The electron concentration per unit length can be expressed as

$$n_0 = \frac{g_v}{\pi}\sum_{n_x=1}^{n_{x\max}}\sum_{n_y=1}^{n_{y\max}}\left[B_{36,\pm}\left(E_{F1D},n_x,n_y\right) + B_{37,\pm}\left(E_{F1D},n_x,n_y\right)\right], \tag{1.84}$$

where $B_{36,\pm}\left(E_{F1D},n_x,n_y\right) = \sqrt{A_{25,+}(E_{F1D},n_x,n_y)} + \sqrt{A_{25,-}(E_{F1D},n_x,n_y)}$ and

$$B_{37,\pm}\left(E_{F1D},n_x,n_y\right) = \sum_{r=1}^{s_o} Z_{1D}(r)\left[B_{36,+}\left(E_{F1D},n_x,n_y\right) + B_{36,-}\left(E_{F1D},n_x,n_y\right)\right].$$

The field-emitted current assumes the form

$$I = \frac{g_v e k_B T}{h}\sum_{n_x=1}^{n_{x\max}}\sum_{n_y=1}^{n_{y\max}}\left[F_0(\eta_{24,+})\exp(-\beta_{24,+}) + F_0(\eta_{24,-})\exp(-\beta_{24,-})\right], \tag{1.85}$$

where $\eta_{24,\pm} = \dfrac{E_{F1D} - E_{26,\pm}}{k_B T}$, $\beta_{24,\pm} = \dfrac{4}{3}\left[A_{25,\pm}\left(V_0,n_x,n_y\right)\right]^{3/2}\cdot\left[eF_{sz}A'_{25,\pm},\right.$ $\left.(V_0 n_x,n_y)\,\right]^{-1}$, $A'_{25,\pm}\left(V_0,n_x,n_y\right) = \left[\frac{1}{\psi_1} \pm \dfrac{\psi_7}{2}\left[\psi_8(V_0) - \phi_1\left(n_x,n_y\right)\right]^{-1/2}\psi'_8(V_0)\right]$, and $\psi'_8(V_0) = \left[\frac{4\psi_3^2\psi_1}{4\psi_3^2\psi_1\psi_2 - 4\psi_1^2\psi_4^2}\right]$

1.2.8 *The Field Emission from Quantum Wires of Gallium Phosphide*

The energy spectrum of the conduction electrons in n-GaP can be written as [213]

$$E = \frac{\hbar^2 k_s^2}{2m_\perp^*} + \frac{\hbar^2}{2m_\parallel^*}\left[\overline{A}' k_s^2 + k_z^2\right] - \left[\frac{\hbar^4 k_0^2}{m_\parallel^{*2}}\left(k_s^2 + k_z^2\right) + |V_G|^2\right]^{1/2} + |V_G|, \tag{1.86}$$

where k_0 and $|V_G|$ are constants of the energy spectrum and $\overline{A}' = 1$. The 1D dispersion relation assumes the form

$$k_z^2 = A_{26}(E, n_x, n_y), \tag{1.87}$$

where

$$A_{26}\left(E, n_x, n_y\right) = \left[t_1 E + t_2 - t_3\phi_1\left(n_x, n_y\right) - \left[t_4\phi_1\left(n_x, n_y\right) + t_6 E + t_5\right]^{1/2}\right],$$

$$t_1 = (b_0)^{-1}$$

$$b_0 = \left(\hbar^2/2m_{\parallel}^*\right),\ t_2 = [-2b_0 D_0 + C]\left[2b_0^2\right]^{-1},\ D_0 = \left|V_g\right|,\ C = \left(\hbar^2 k_0/m_{\parallel}^*\right)^2,$$

$$t_3 = (a_0/b_0),\ a_0 = \left[\left(\hbar^2/2m_{\perp}^*\right) + \left(A\hbar^2/2m_{\parallel}^*\right)\right],\ t_4 = \left(\bar{a}_1/\left(4b_0^4\right)\right),$$

$$\bar{a} = 4b_0 C\left(b_0 - C\right),$$

$$t_6 = \left(\bar{a}_3/\left(4b_0^4\right)\right),\ \bar{a}_3 = 4Cb_0,\ t_5 = \left(\bar{a}_2/\left(4b_0^4\right)\right), \text{ and}$$

$$\bar{a}_2 = \left[C^2 + 4b_0^2 D_0^2 - 4b_0 C D_0\right].$$

The subband energy E_{27} can be written as

$$E_{27} = a_0\phi_1(n_x, n_y) - (C\phi_1(n_x, n_y) + D_0^2)^{1/2} + D_0. \tag{1.88}$$

The electron concentration per unit length can be expressed as

$$n_0 = \frac{2g_v}{\pi}\sum_{n_x=1}^{n_{x\max}}\sum_{n_y=1}^{n_{y\max}}\left[B_{38}\left(E_{F1D}, n_x, n_y\right) + B_{39}\left(E_{F1D}, n_x, n_y\right)\right], \tag{1.89}$$

where $B_{38}\left(E_{F1D}, n_x, n_y\right) = \sqrt{A_{26}(E_{F1D}, n_x, n_y)}$ and $B_{39}\left(E_{F1D}, n_x, n_y\right) = \sum_{r=1}^{s_o} Z_{1D}(r)\left[B_{38}\left(E_{F1D}, n_x, n_y\right)\right]$.

The field-emitted current assumes the form

$$I = \frac{2g_v e k_B T}{h}\sum_{n_x=1}^{n_{x\max}}\sum_{n_y=1}^{n_{y\max}}\left[F_0(\eta_{26})\exp(-\beta_{26})\right], \tag{1.90}$$

where $\eta_{26} = \dfrac{E_{F1D} - E_{27}}{k_B T}$, $\beta_{26} = \dfrac{4}{3}\left[A_{26}\left(V_0, n_x, n_y\right)\right]^{3/2}\cdot\left[eF_{sz}A_{26}'\left(V_0, n_x, n_y\right)\right]^{-1}$, and $A_{26}'\left(V_0, n_x, n_y\right) = \left[t_1 - \dfrac{1}{2}t_6\left[t_4\phi_1\left(n_x, n_y\right) + t_6 V_0 + t_5\right]^{-1/2}\right]$.

1.2.9 The Field Emission from Quantum Wires of Platinum Antimonide

The dispersion relation for $n - PtSb_2$ is given by [214]

$$\left[E + \overline{\lambda_0}\frac{(\overline{a})^2}{4}k^2 - \overline{l}\frac{(\overline{a})^2}{4}k_s^2\right]\left[E + \overline{\delta_0} - \overline{\nu}\frac{(\overline{a})^2}{4}k^2 - \overline{n}\frac{(\overline{a})^2}{4}k_s^2\right] = \frac{I\,(\overline{a})^4}{16}k^4, \tag{1.91}$$

where $\overline{\lambda_0}, \overline{a}, \overline{l}, \overline{\delta_0}, \overline{\nu}, \overline{n}$, and I are system constants.

The 1D dispersion relation can be written as

$$k_x^2 = \left[2\overline{A_9}\right]^{-1}\left[-\overline{A_{10}}\,(E,n_z) + \left[\overline{A_{10}}^2\,(E,n_z) + 4\left(\overline{A_9}\right)\overline{A_{11}}\,(E,n_z)\right]\right]^{1/2} - \left(\frac{n_y\pi}{d_Y}\right)^2, \tag{1.92}$$

where

$$\overline{A_9} = (I_1 + \omega_1\omega_3), \quad I_1 = I\frac{(\overline{a})^4}{16},$$

$$\overline{A_{10}}\,(E,n_z) = \left[\omega_3 E + \omega_1\left[E + \overline{\delta_0} - \omega_4\left(\frac{n_z\pi}{d_z}\right)^2\right] + \omega_2\omega_3\left(\frac{n_z\pi}{d_z}\right)^2 + 2I_1\left(\frac{n_z\pi}{d_z}\right)^2\right],$$

$$\omega_1 = \frac{(\overline{a})^2}{4}\left[\left(\overline{\lambda_0}\right) - \overline{l}\right], \quad \omega_2 = \left(\overline{\lambda_0}\frac{(\overline{a})^2}{4}\right), \quad \omega_3 = \frac{(\overline{a})^2}{4}\,(\overline{n} + \overline{\nu}), \quad \omega_4 = \frac{(\overline{a})^2}{4}\overline{\nu}, \text{ and}$$

$$\overline{A_{11}}\,(E,n_z) = \left[E\left[E + \overline{\delta_0} - \omega_4\left(\frac{n_z\pi}{d_z}\right)^2\right] + \omega_2\left(\frac{n_z\pi}{d_z}\right)^2\left[E + \overline{\delta_0} - \omega_4\left(\frac{n_z\pi}{d_z}\right)^2\right] - I_1\left(\frac{n_z\pi}{d_z}\right)^4\right].$$

Therefore, (1.92) can be expressed as

$$k_x^2 = A_{27}\left(E,n_y,n_z\right), \tag{1.93}$$

where

$$A_{27}\left(E,n_y,n_z\right) = \left[2\overline{A_9}\right]^{-1}\left[-\overline{A_{10}}\,(E,n_z) + \left[\overline{A_{10}}^2\,(E,n_z) + 4\left(\overline{A_9}\right)\overline{A_{11}}\,(E,n_z)\right]\right]^{1/2} - \left(\frac{n_y\pi}{d_y}\right)^2.$$

The electron concentration per unit length assumes the form

$$n_0 = \frac{2g_v}{\pi} \sum_{n_x=1}^{n_{x\max}} \sum_{n_y=1}^{n_{y\max}} \left[B_{40}\left(E_{F1D}, n_x, n_y\right) + B_{41}\left(E_{F1D}, n_x, n_y\right)\right], \tag{1.94}$$

where $B_{40}\left(E_{F1D}, n_x, n_y\right) = \sqrt{A_{27}(E_{F1D}, n_x, n_y)}$ and $B_{41}\left(E_{F1D}, n_x, n_y\right) = \sum_{r=1}^{s_o} Z_{1D}(r)\left[B_{40}\left(E_{F1D}, n_x, n_y\right)\right]$.

The field-emitted current can be written as

$$I = \frac{2g_v e k_B T}{h} \sum_{n_x=1}^{n_{x\max}} \sum_{n_y=1}^{n_{y\max}} [F_0(\eta_{27}) \exp(-\beta_{27})], \tag{1.95}$$

where $\eta_{27} = \dfrac{E_{F1D} - E_{28}}{k_B T}$. E_{28} is the root of the equation

$$\left[2\overline{A_9}\right]^{-1}\left[-\overline{A_{10}}(E_{28}, n_z) + \left[\overline{A_{10}}^2(E_{28}, n_z) + 4\left(\overline{A_9}\right)\overline{A_{11}}(E_{28}, n_z)\right]\right]^{1/2} - \left(\frac{n_y \pi}{d_Y}\right)^2 = 0, \tag{1.96}$$

$$\beta_{27} = \frac{4}{3}[A_{27}(V_0, n_x, n_y)]^{3/2} \cdot [eF_{sz} A_2' 7(V_0, n_x, n_y)]^{-1},$$

$$A_{27}'\left(E, n_y, n_z\right) = \left[2\overline{A_9}\right]^{-1}\left[-(\overline{A_{10}})' + \left[\overline{A_{10}}^2(V_0, n_z) + 4\left(\overline{A_9}\right)\overline{A_{11}}(V_0, n_z)\right]^{1/2} \cdot \left[\overline{A_{10}}(V_0, n_z)\left(\overline{A_{10}}\right)' + 2\left(\overline{A_9}\right)\left(\overline{A_{11}}(V_0, n_z)\right)'\right]\right],$$

$$A_{10}' = (\omega_1 + \omega_3),$$

and

$$(\overline{A_{11}}(V_0, n_z))' = \left[2V_0 + \overline{\delta_0} - \omega_4\left(\frac{n_z \pi}{d_z}\right)^2 + \omega_2\left(\frac{n_z \pi}{d_z}\right)^2\right].$$

1.2.10 The Field Emission from Quantum Wires of Bismuth Telluride

The dispersion relation of the conduction electrons in Bi_2Te_3 can be written as [215–217]

$$E(1 + \alpha E) = \bar{\omega}_1 k_x^2 + \bar{\omega}_2 k_y^2 + \bar{\omega}_3 k_z^2 + \bar{\omega}_4 k_z k_y, \tag{1.97}$$

where $\bar{\omega}_1 = \frac{\hbar^2}{2m_0}\bar{\alpha}_{11}, \bar{\omega}_2 = \frac{\hbar^2}{2m_0}\bar{\alpha}_{22}, \bar{\omega}_3 = \frac{\hbar^2}{2m_0}\bar{\alpha}_{33}$, and $\bar{\omega}_4 = \frac{\hbar^2}{2m_0}\bar{\alpha}_{23}$ in which $\bar{\alpha}_{11}, \bar{\alpha}_{22}, \bar{\alpha}_{33}$, and $\bar{\alpha}_{23}$ are system constants.

The 1D electron energy spectrum assumes the form

$$k_x^2 = A_{28}(E, n_y, n_z), \tag{1.98}$$

where $A_{28}(E, n_y, n_z) = \left[E\,(1+\alpha E) - \bar{\omega}_2\left(\frac{n_y\pi}{d_y}\right)^2 - \bar{\omega}_3\left(\frac{n_z\pi}{d_z}\right)^2 - \bar{\omega}_4\left(\frac{n_y\pi}{d_y}\right)\left(\frac{n_z\pi}{d_z}\right)\right](\bar{\omega}_1)^{-1}$.

The subband energy (E_{30}) can be expressed as

$$E_{30} = (2\alpha)^{-1}\left[-1 + \sqrt{1 + 4\alpha\theta_{32}(n_y, n_z)}\right], \tag{1.99}$$

where $\theta_{32}(n_y, n_z) = \left[\bar{\omega}_2\left(\frac{n_y\pi}{d_y}\right)^2 + \bar{\omega}_3\left(\frac{n_z\pi}{d_z}\right)^2 + \bar{\omega}_4\left(\frac{n_y\pi}{d_y}\right)\left(\frac{n_z\pi}{d_z}\right)\right]$.

The electron concentration per unit length is given by

$$n_0 = \frac{2g_{\mathrm{v}}}{\pi}\sum_{n_z=1}^{n_{z\max}}\sum_{n_y=1}^{n_{y\max}}\left[B_{42}\left(E_{\mathrm{F1D}}, n_z, n_y\right) + B_{43}\left(E_{\mathrm{F1D}}, n_z, n_y\right)\right], \tag{1.100}$$

where $B_{42}\left(E_{\mathrm{F1D}}, n_z, n_y\right) = \sqrt{A_{28}(E_{\mathrm{F1D}}, n_y, n_z)}$ and $B_{43}\left(E_{\mathrm{F1D}}, n_z, n_y\right) = \sum_{r=1}^{s_o} Z_{\mathrm{1D}}(r)\left[B_{42}B\left(E_{\mathrm{F1D}}, n_z, n_y\right)\right]$.

The field-emitted current in this case can be written as

$$I = \frac{2g_{\mathrm{v}}ek_{\mathrm{B}}T}{h}\sum_{n_z=1}^{n_{z\max}}\sum_{n_y=1}^{n_{y\max}}\left[F_0(\eta_{28})\exp(-\beta_{28})\right], \tag{1.101}$$

where

$$\eta_{28} = \frac{E_{\mathrm{F1D}} - E_{30}}{k_{\mathrm{B}}T}, \quad \beta_{28} = \frac{4}{3}\left[A_{28}\left(V_0, n_y, n_z\right)\right]^{3/2} \cdot \left[eF_{sz}A_2'8(V_0)\right]^{-1}, \quad \text{and}$$

$A_{28}'(V_0) = (\bar{\omega}_1)^{-1}\left[1 + 2\alpha V_0\right]$.

1.2.11 The Field Emission from Quantum Wires of Germanium

It is well known that the conduction electrons of n-Ge obey two different types of dispersion laws since band nonparabolicity has been included in two different ways as given in the literature [218–220].

(a) The energy spectrum of the conduction electrons in bulk specimens of n-Ge can be expressed in accordance with Cardona et al. [218] as

$$E = -\frac{E_{g0}}{2} + \frac{\hbar^2 k_z^2}{2m_{\parallel}^*} + \left[\frac{E_{g0}^2}{4} + E_{g0} k_s^2 \left(\frac{\hbar^2}{2m_{\perp}^*}\right)\right]^{1/2}, \tag{1.102}$$

where in this case $m_{\parallel}^*$ and $m_{\perp}^*$ are the longitudinal and transverse effective masses along the $\langle 111 \rangle$ direction at the edge of the conduction band, respectively.

The 1D electron energy spectrum assumes the form

$$k_y^2 = A_{29}(E, n_x, n_z), \tag{1.103}$$

where

$$A_{29}(E, n_x, n_z) = \left[\left[\gamma_{15}(E, n_z) - \left(\frac{\hbar^2}{2m_1^*}\right)\left(\frac{n_x \pi}{d_x}\right)^2\right](2m_2^*/\hbar^2)\right],$$

$$\gamma_{15}(E, n_z) = \left[E(1+\alpha E) - (1+2\alpha E)\left(\frac{\hbar^2}{2m_3^*}\right)\left(\frac{n_z \pi}{d_z}\right)^2 + \alpha\left[\left(\frac{\hbar^2}{2m_3^*}\right)\left(\frac{n_z \pi}{d_z}\right)^2\right]^2\right],$$

$$m_1^* = m_{\perp}^*, \quad m_2^* = \frac{m_{\perp}^* + 2m_{\parallel}^*}{3} \quad \text{and} \quad m_3^* = \frac{3m_{\perp}^* m_{\parallel}^*}{m_{\perp}^* + 2m_{\parallel}^*}.$$

The quantized energy levels (E_{31}) can be expressed through the equation

$$E_{31} = (2\alpha)^{-1}\left[-\rho_{91}(n_z) + \sqrt{\rho_{91}(n_z)^2 - 4\alpha\rho_{92}(n_z)}\right], \tag{1.104}$$

where

$$\rho_{91}(n_z) = \left[1 - 2\alpha\frac{\hbar^2}{2m_3^*}\left(\frac{n_z \pi}{d_z}\right)^2\right] \quad \text{and}$$

$$\rho_{92}(n_z) = \left[\frac{\hbar^2}{2m_3^*}\left(\frac{n_z \pi}{d_z}\right)^2 - \alpha\left[\frac{\hbar^2}{2m_3^*}\left(\frac{n_z \pi}{d_z}\right)^2\right]^2\right].$$

The electron concentration per unit length is given by

$$n_0 = \frac{2g_v}{\pi} \sum_{n_x=1}^{n_{x\max}} \sum_{n_z=1}^{n_{z\max}} \left[B_{44}\left(E_{F1D}, n_x, n_z\right) + B_{45}\left(E_{F1D}, n_x, n_z\right) \right], \tag{1.105}$$

where $B_{44}\left(E_{F1D}, n_x, n_z\right) = \sqrt{A_{29}(E_{F1D}, n_x, n_z)}$ and $B_{45}\left(E_{F1D}, n_x, n_z\right) = \sum_{r=1}^{s_o} Z_{1D}(r)\left[B_{44}\ \left(E_{F1D}, n_x, n_z\right)\right]$.

The field-emitted current assumes the form

$$I = \frac{2g_v e k_B T}{h} \sum_{n_x=1}^{n_{x\max}} \sum_{n_z=1}^{n_{z\max}} \left[F_0(\eta_{29}) \exp(-\beta_{29}) \right], \tag{1.106}$$

where $\eta_{29} = \dfrac{E_{F1D} - E_{31}}{k_B T}$,

$\beta_{29} = \frac{4}{3}\left[A_{29}(V_0, n_y, n_z)\right]^{3/2} \cdot \left[eF_{sz}A_2'9(V_0)\right]^{-1}$, and

$$A_{29}'\left(V_0, n_x, n_z\right) = \left[\left[1 + 2\alpha V_0 - \left(\alpha \frac{\hbar^2}{m_3^*}\right)\left(\frac{n_z \pi}{d_z}\right)^2\right]\left(2m_2^*/\hbar^2\right)\right].$$

(b) The dispersion relation of the conduction electron in bulk specimens of n-Ge can be expressed in accordance with the model of Wang and Ressler [220] and can be written as

$$E = \frac{\hbar^2 k_z^2}{2m_{\|}^*} + \frac{\hbar^2 k_s^2}{2m_{\perp}^*} - \bar{c}_1 \left(\frac{\hbar^2 k_s^2}{2m_{\perp}^*}\right)^2 - \bar{d}_1 \left(\frac{\hbar^2 k_s^2}{2m_{\perp}^*}\right)\left(\frac{\hbar^2 k_z^2}{2m_{\|}^*}\right) - \bar{e}_1 \left(\frac{\hbar^2 k_z^2}{2m_{\|}^*}\right)^2, \tag{1.107}$$

where $\bar{c}_1 = \overline{C}\left(2m_{\perp}^*/\hbar^2\right)^2$, $\overline{C} = 1.4\overline{A}$, $\overline{A} = \frac{1}{4}\left(\hbar^4/E_{g0}m_{\perp}^{*2}\right)\left(1 - \frac{m_{\perp}^*}{m_0}\right)^2$, $\bar{d}_1 = \bar{d}\left(\frac{4m_{\perp}^* m_{\|}^*}{\hbar^4}\right)$, $\bar{d} = 0.8\overline{A}$, $\bar{e}_1 = \bar{e}_0\left(2m_{\|}^*/\hbar^2\right)^2$, and $\bar{e}_0 = 0.005\bar{A}$.

The 1D electron energy spectrum assumes the form

$$k_y^2 = A_{30}(E, n_x, n_z), \tag{1.108}$$

where $A_{30}(E, n_x, n_z) = \left[\left[I_{29}(E, n_z) - \left(\frac{\hbar^2}{2m_1^*}\right)\left(\frac{n_x \pi}{d_x}\right)^2\right]\left(2m_2^*/\hbar^2\right)\right]$,

$$I_{29}(E,n_z) = \left[2\overline{C_1}\right]^{-1}\left[\overline{A_6}(n_z) + \left[\overline{A_6}^2(n_z) - 4\overline{C}_1 E + 4\left(\overline{C_1}\right)\overline{A_5}(n_z)\right]^{\frac{1}{2}}\right],$$

$$\overline{A_5}(n_z) = \left(\frac{\hbar^2}{2m_3^*}\right)\left(\frac{n_z\pi}{d_z}\right)^2\left[1 - \bar{e}_1\left(\frac{\hbar^2}{2m_3^*}\right)\left(\frac{n_z\pi}{d_z}\right)^2\right]\sum_{i=1}^{n} X_i^2, \quad \text{and}$$

$$\overline{A_6}(n_z) = \left[1 - \bar{d}_1\left(\frac{\hbar^2}{2m_3^*}\right)\left(\frac{n_z\pi}{d_z}\right)^2\right].$$

The quantized energy levels (E_{32}) can be expressed through the equation

$$E_{32} = \overline{A}_5(n_z) + \left(\frac{1}{4\overline{C}_1}\right)\left[\left[\frac{\overline{C}_1\hbar^2}{m_1^*}\left(\frac{n_x\pi}{d_x}\right)^2\right]^2 - 2\overline{A}_6(n_z)\frac{\bar{C}_1\hbar^2}{m_1^*}\left(\frac{n_x\pi}{d_x}\right)^2\right]. \tag{1.109}$$

The electron concentration per unit length is given by

$$n_0 = \frac{2g_v}{\pi}\sum_{n_x=1}^{n_{x\max}}\sum_{n_z=1}^{n_{z\max}}\left[B_{46}(E_{F1D}, n_x, n_z) + B_{47}(E_{F1D}, n_x, n_z)\right], \tag{1.110}$$

where

$$B_{46}(E_{F1D}, n_x, n_z) = \sqrt{A_{30}(E_{F1D}, n_x, n_z)} \quad \text{and}$$

$$B_{47}(E_{F1D}, n_x, n_z) = \sum_{r=1}^{s_o} Z_{1D}(r)\left[B_{46}(E_{F1D}, n_x, n_z)\right].$$

The field-emitted current assumes the form

$$I = \frac{2g_v e k_B T}{h}\sum_{n_x=1}^{n_{x\max}}\sum_{n_z=1}^{n_{z\max}}\left[F_0(\eta_{30})\exp(-\beta_{30})\right], \tag{1.111}$$

where $\eta_{30} = \dfrac{E_{F1D} - E_{32}}{k_B T}$, $\beta_{30} = \dfrac{4}{3}\left[A_{30}(V_0, n_y, n_z)\right]^{3/2} \cdot \left[eF_{sz}A'_{30}(V_0)\right]^{-1}$,

$$A'_{30}(V_0, n_x, n_z) = (2m_2^*/\hbar^2)I'_{29}(V_0, n_z), \quad \text{and}$$

$$I'_{29}(V_0, n_z) = \left[\left[\overline{A_6}^2(n_z) - 4\overline{C}_1 V_0 + 4\left(\overline{C_1}\right)\overline{A_5}(n_z)\right]^{-1/2}\right].$$

1.2.12 The Field Emission from Quantum Wires of Gallium Antimonide

The dispersion relation of the conduction electrons in n-GaSb can be written as [221]

$$E = \frac{\hbar^2 k^2}{2m_0} - \frac{\overline{E'}_{go}}{2} + \frac{\bar{E'}_{go}}{2}\left[1 + \frac{2\hbar^2 k^2}{\overline{E'}_{go}}\left(\frac{1}{m_c} - \frac{1}{m_0}\right)\right]^{\frac{1}{2}}, \tag{1.112}$$

where $\overline{E'}_{go} = \left[E_{go} + \dfrac{5.10^{-5}T^2}{2(112+T)}\right]$ eV.

Equation (1.112) can be expressed as

$$\frac{\hbar^2 k^2}{2m_c} = I_{16}(E), \tag{1.113}$$

where

$$\begin{aligned} I_{16}(E) = {} & [E + \overline{E'}_{g0} - (m_c/m_0)(\overline{E'}_{g0}/2) - [(\bar{E'}_{g0}/2)^2 \\ & + [((\overline{E'}_{g0})^2/2)(1 - (m_c/m_0))] + [(\overline{E'}_{g0}/2)(1 - (m_c/m_0))]^2 \\ & + E\overline{E'}_{g0}(1 - (m_c/m_0))]^{1/2}]. \end{aligned}$$

The 1D electron energy spectrum assumes the form

$$k_z^2 = A_{31}(E, n_x, n_y), \tag{1.114}$$

where $A_{31}(E, n_x, n_y) = [I_{16}(E)(2m_c/\hbar^2) - \phi_1(n_x, n_y)]$.

The quantized energy levels (E_{33}) can be expressed through the equation

$$I_{16}(E_{33}) = \left(\frac{\hbar^2}{2m_c}\right)\phi_1(n_x, n_y). \tag{1.115}$$

The electron concentration per unit length is given by

$$n_0 = \frac{2g_v}{\pi}\sum_{n_x=1}^{n_{x\max}}\sum_{n_y=1}^{n_{y\max}}\left[B_{48}\left(E_{F1D}, n_x, n_y\right) + B_{49}\left(E_{F1D}, n_x, n_y\right)\right], \tag{1.116}$$

where

$$B_{48}\left(E_{F1D}, n_x, n_y\right) = \sqrt{A_{31}(E_{F1D}, n_x, n_y)} \quad \text{and}$$

$$B_{49}\left(E_{F1D}, n_x, n_y\right) = \sum_{r=1}^{s_o} Z_{1D}(r)\left[B_{48}\left(E_{F1D}, n_x, n_y\right)\right].$$

The field-emitted current can be written as

$$I = \frac{2g_v e k_B T}{h} \sum_{n_x=1}^{n_{x\max}} \sum_{n_y=1}^{n_{y\max}} [F_0(\eta_{31}) \exp(-\beta_{31})], \tag{1.117}$$

where $\eta_{31} = \dfrac{E_{F1D} - E_{33}}{k_B T}$, $\beta_{31} = \dfrac{4}{3}\left[A_{31}\left(V_0, n_x, n_y\right)\right]^{3/2} \cdot \left[eF_{sz} A'_{31}(V_0)\right]^{-1}$, and $A'_{31}(V_0) = [1 - (m_c/m_0)(\overline{E'}_{g0}/2)[(\overline{E'}_{g0}/2)^2 + [((\overline{E'}_{g0})^2/2)(1 - (m_c/m_0))] + [((\overline{E'}_{g0})^2/2)(1-(m_c/m_0))] + [(\overline{E'}_{g0}/2)(1-(m_c/m_0))]^2 + V_0 \overline{E'}_{g0}(1-(m_c/m_0))] - 1/2](2m_c/\hbar^2)$.

1.2.13 *The Field Emission from Quantum Wires of II–V Materials*

The dispersion relation of the holes are given by [222–224]

$$\begin{aligned} E &= \theta_1 k_x^2 + \theta_2 k_y^2 + \theta_3 k_z^2 + \delta_4 k_x \\ &\mp \left[\left\{\theta_5 k_x^2 + \theta_6 k_y^2 + \theta_7 k_z^2 + \delta_5 k_x\right\}^2 + G_3^2 k_y^2 + \Delta_3^2\right]^{\frac{1}{2}} \pm \Delta_3, \end{aligned} \tag{1.118}$$

where k_x, k_y, and k_z are expressed in the units of 10^{10}m^{-1},

$$\theta_1 = \frac{1}{2}(a_1 + b_1),\ \theta_2 = \frac{1}{2}(a_2 + b_2),\ \theta_3 = \frac{1}{2}(a_3 + b_3),\ \delta_4 = \frac{1}{2}(A + B),$$

$$\theta_5 = \frac{1}{2}(a_1 - b_1),\ \theta_6 = \frac{1}{2}(a_2 - b_2),\ \theta_7 = \frac{1}{2}(a_3 - b_3),\ \delta_5 = \frac{1}{2}(A - B),\ \text{and}$$

$a_i (i = 1, 2, 3, 4), b_i, A, B, G_3$, and Δ_3 are system constants.

The 1D electron energy spectrum assumes the form

$$k_z^2 = A_{32,\pm}(E, n_x, n_y), \tag{1.119}$$

where $A_{32,\pm}(E, n_x, n_y) = \alpha_{4,\pm}(n_x, n_y) + \beta_4 E \pm [\beta_5 E^2 + E\beta_{6,\pm}(n_x, n_y) + \beta_{7,\pm}(n_x, n_y)]^{\frac{1}{2}}$,

$$\alpha_{4,\pm}(n_x, n_y) = \left[2\left(\theta_3^2 - \theta_7^2\right)\right]^{-1}\left[2\theta_7 \alpha_2(n_x, n_y) - 2\alpha_{1,\mp}(n_x, n_y)\theta_3\right],$$

$$\alpha_{1,\mp}(n_x, n_y) = \theta_1\left(\frac{n_x \pi}{d_x}\right)^2 + \theta_2\left(\frac{n_y \pi}{d_y}\right)^2 + \delta_4\left(\frac{n_x \pi}{d_x}\right) \mp \Delta_3,$$

$$\alpha_2(n_x, n_y) = \left[\theta_5\left(\frac{n_x \pi}{d_x}\right)^2 + \theta_6\left(\frac{n_y \pi}{d_y}\right)^2 + \delta_5\left(\frac{n_x \pi}{d_x}\right)\right],$$

$$\beta_4 = 2\theta_3\left[2\left(\theta_3^2-\theta_7^2\right)\right]^{-1}, \quad \beta_5 = \left[2\left(\theta_3^2-\theta_7^2\right)\right]^{-2}\left[4\theta_7^2\right],$$

$$\beta_{6,\pm}(n_x,n_y) = \left[8\theta_3\theta_7\alpha_2(n_x,n_y) - 8\theta_7^2\alpha_{1,\mp}(n_x,n_y)\right]\left[2\left(\theta_3^2-\theta_7^2\right)\right]^{-1},$$

$$\beta_{7,\pm}(n_x,n_y) = \left[2(\theta_3^2-\theta_7^2)\right]^{-2}\left[4\theta_7^2\alpha_2^2(n_x,n_y) - 8\theta_7\theta_3\alpha_2(n_x,n_y)\alpha_{1,\mp}(n_x,n_y) + 4\theta_3^2\alpha_3(n_y) + 4\theta_7^2\alpha_{1,\mp}(n_x,n_y) - 4\alpha_3(n_y)\theta_7^2\right], \quad \text{and}$$

$$\alpha_3(n_y) = G_3^2\left(\frac{n_y\pi}{d_y}\right)^2 + \Delta_3^2.$$

The quantized energy levels $(E_{34,\pm})$ can be expressed through the equation

$$E_{34,\pm} = \alpha_{1,\mp}(n_x,n_y) \pm [\alpha_2^2(n_x,n_y) + \alpha_3(n_y)]^{\frac{1}{2}}. \tag{1.120}$$

The electron concentration per unit length is given by

$$n_0 = \frac{g_v}{\pi}\sum_{n_x=1}^{n_{x\max}}\sum_{n_y=1}^{n_{y\max}}\left[B_{49}\left(E_{F1D},n_x,n_y\right) + B_{50}\left(E_{F1D},n_x,n_y\right)\right], \tag{1.121}$$

where $B_{49}\left(E_{F1D},n_x,n_y\right) = \left[\sqrt{A_{32,+}(E_{F1D},n_x,n_y)} + \sqrt{A_{32,-}(E_{F1D},n_x,n_y)}\right]$ and $B_{50}\left(E_{F1D},n_x,n_y\right) = \sum_{r=1}^{s_o} Z_{1D}(r)\left[B_{49}\left(E_{F1D},n_x,n_y\right)\right]$.

The field-emitted current assumes the form

$$I = \frac{g_v e k_B T}{h}\sum_{n_x=1}^{n_{x\max}}\sum_{n_y=1}^{n_{y\max}}[F_0(\eta_{32,+})\exp(-\beta_{32,+}) + F_0(\eta_{32,-})\exp(-\beta_{32,-})], \tag{1.122}$$

where $\eta_{32,\pm} = \dfrac{E_{F1D} - E_{34,\pm}}{k_B T}$, $\beta_{32,\pm} = \dfrac{4}{3}\left[A_{32,\pm}(V_0,n_x,n_y)\right]^{3/2}\cdot\left[eF_{sz}A'_{32,\pm}(V_0, n_x,n_y)\right]^{-1}$, and $A'_{32,\pm}(V_0,n_x,n_y) = \left[\beta_4 \pm \dfrac{1}{2}\left[2\beta_5V_0 + \beta_{6,\pm}(n_x,n_y)\right].\left[\beta_5V_0^2 + \beta_{6,\pm}(n_x,n_y)V_0 + \beta_{7,\pm}(n_x,n_y)\right]^{-\frac{1}{2}}\right]$.

1.3 Result and Discussions

Using (1.5) and (1.12) and taking the energy band constants as given in the Table 1.1, we have plotted the FNFE current from QWs of $CdGeAs_2$ (an example of nonlinear optical materials) as a function of d_y as shown by the dotted plot of Fig. 1.1, in which the plot corresponds to the solid line represents the same for the two-band model of Kane. Figure 1.2 exhibits the plot FNFE current from QWs of n-InSb as a function of film thickness in accordance with the three- and two-band models of

Kane together with the models of Stillman et al., Newson et al., and Palik et al., respectively for the purpose of assessing the influence of dispersion relations on the FNFE current from QWs of III–V semiconductors. Figure 1.3 exhibits the variation of the field-emitted current with electric field which is normal to the two directions of quantizations in the wavevector space of the material for all cases Fig. 1.2. Figure 1.4 presents the FNFE current as function of film thickness for two different

Table 1.1 The numerical values of the energy band constants of few materials

	Materials	Numerical values of the energy band constants
1	(a) The conduction electrons of n-Cadmium Germanium Arsenide can be described by three types of band models	1. The values of the energy band constants in accordance with the generalized electron dispersion relation of nonlinear optical materials (as given by (1.2)) are as follows: $E_{g_0} = 0.57\,\text{eV}$, $\Delta_{\parallel} = 0.30\,\text{eV}$, $\Delta_{\perp} = 0.36\,\text{eV}$, $m_{\parallel}^* = 0.034m_0$, $m_{\perp}^* = 0.039m_0$, $T = 4\,\text{K}$, $\delta = -0.21$ eV, $g_v = 1$ [47,225], $\varepsilon_{sc} = 18.4\varepsilon_0$[226] ($\varepsilon_{sc}$ and ε_0 are the permittivity of the semiconductor material and free space, respectively), and W(electron affinity) = 4eV [227–229]
		2. In accordance with the three-band model of Kane (as given by (1.13)), the spectrum constants are given by $\Delta = (\Delta_{\parallel} + \Delta_{\perp})/2 = 0.33\text{eV}$, $E_{g_0} = 0.57\text{eV}$, $m_c = (m_{\parallel}^* + m_{\perp}^*)/2 = 0.0365m_0$, and $\delta = 0\text{eV}$
		3. In accordance with two-band model of Kane (as given by (1.19)), the spectrum constants are given by $E_{g_0} = 0.57\,\text{eV}$ and $m_c = 0.0365m_0$
	(b) The conduction electrons of n-Cadmium Arsenide can be described by three types of band models	1. The values of the energy band constants in accordance with the generalized electron dispersion relation of nonlinear optical materials (as given by (1.2)) are as follows: $\lvert E_{g_0}\rvert = 0.095\,\text{eV}$, $\Delta_{\parallel} = 0.27\,\text{eV}$, $\Delta_{\perp} = 0.25\,\text{eV}$, $m_{\parallel}^* = 0.00697m_0$, $m_{\perp}^* = 0.013933m_0$, $T = 4\,\text{K}$, $\delta = 0.085\,\text{eV}$, $g_v = 1$ [47,225], and $\varepsilon_{sc} = 16\varepsilon_0$ [227–229]
		2. In accordance with the three-band model of Kane (as given by (1.13)), the spectrum constants are given by $\Delta = \left(\Delta_{\parallel} + \Delta_{\perp}\right)/2 = 0.26\,\text{eV}$, $\left\lvert E_{g_0}\right\rvert = 0.095\,\text{eV}$, $m_c = \left(m_{\parallel}^* + m_{\perp}^*\right)/2 = 0.020903m_0$, and $\delta = 0\,\text{eV}$
		3. In accordance with two-band model of Kane (as given by (1.19)), the spectrum constants are given by $\left\lvert E_{g_0}\right\rvert = 0.095\,\text{eV}$ and $m_c = 0.020903m_0$
2	n-Indium Arsenide	The values $E_{g_0} = 0.36\,\text{eV}$, $\Delta = 0.43\,\text{eV}$, $m_c = 0.026m_0$, $g_v = 1$, $\varepsilon_{sc} = 12.25\varepsilon_0$ [103,104], and $W = 5.06\,\text{eV}$ [230] are valid for three-band model of Kane as given by (1.13)
3	n-Gallium Arsenide	The values $E_{g_0} = 1.55\,\text{eV}$, $\Delta = 0.35\,\text{eV}$, $m_c = 0.07m_0$, $g_v = 1$, $\varepsilon_{sc} = 12.9\varepsilon_0$ [103,104], and $W = 4.07\,\text{eV}$ [231] are valid for three-band model of Kane as given by (1.13). The values $a_{13} = -1.97 \times 10^{-37}\,\text{eVm}^4$ and $a_{15} = -2.3 \times 10^{-34}\,\text{eVm}^4$ [106] are valid for the Newson and Kurobe model [106] as given by (1.34)

(continued)

Table 1.1 (continued)

	Materials	Numerical values of the energy band constants
4	n-Gallium Aluminium Arsenide	$E_{g_0} = (1.424 + 1.266x + 0.26x^2)$eV, $\Delta = (0.34 - 0.5x)$eV, $g_v = 1$, $m_c = [0.066 + 0.088x]\, m_0$, $\varepsilon_{sc} = [13.18 - 3.12x]\,\varepsilon_0$ [232], and $W = (3.64 - 0.14x)$eV [233]
5	n-Mercury Cadmium Telluride	$E_{g_0} = (-0.302 + 1.93x + 5.35 \times 10^{-4}(1-2x)T - 0.810x^2 + 0.832x^3)$eV, $\Delta = \left(0.63 + 0.24x - 0.27x^2\right)$ eV, $m_c = 0.1m_0 E_{g_0}(\text{eV})^{-1}$, $g_v = 1$, $\varepsilon_{sc} = \left[20.262 - 14.812x + 5.22795x^2\right]\varepsilon_0$ [234], and $W = \left(4.23 - 0.813\left(E_{g_0} - 0.083\right)\right)$ eV [235]
6	n-Indium Gallium Arsenide Phosphide lattice matched to Indium Phosphide	$E_{g_0} = \left(1.337 - 0.73y + 0.13y^2\right)$ eV, $\Delta = \left(0.114 + 0.26y - 0.22y^2\right)$ eV, $y = (0.1896 - 0.4052x)/(0.1896 - 0.0123x)$, $m_c = (0.08 - 0.039y)\, m_0$, $g_v = 1$, $\varepsilon_{sc} = [10.65 + 0.1320y]\,\varepsilon_0$, and [231] $W(x, y) = [5.06\,(1-x)\,y + 4.38(1-x)(1-y) + 3.64xy + 3.75\{x(1-y)\}]$ eV
7	n-Indium Antimonide	$E_{g_0} = 0.2352$ eV, $\Delta = 0.81$ eV, $m_c = 0.01359m_0$, $g_v = 1$, $\varepsilon_{sc} = 15.56\varepsilon_0$ [103, 104], and $W = 4.72$ eV [230]
8	n-Gallium Antimonide	The values of $E_{g_0} = 0.81$ eV, $\Delta = 0.80$ eV, $P = 9.48 \times 10^{-10}$ eVm, $\bar{\zeta}_0 = -2.1$, $\bar{v}_0 = -1.49$, $\bar{\omega}_0 = 0.42$, $g_v = 1$ [238], and $\varepsilon_{sc} = 15.85\varepsilon_0$ [239] are valid for the model of Seiler et al. [238] as given by (R1.5).)
9	n-Cadmium Sulphide	$m_{\parallel}^* = 0.7m_0$, $m_{\perp}^* = 1.5m_0$, $C_0 = 1.4 \times 10^{-8}$ eVm, $g_v = 1$ [103, 104], $\varepsilon_{sc} = 15.5\varepsilon_0$ [240], and $W = 4.5$ eV [230]
10	n-Lead Telluride	The values $m_t^- = 0.070m_0$, $m_l^- = 0.54m_0$, $m_t^+ = 0.010m_0$, $m_l^+ = 1.4m_0$, $P_{\parallel} = 141$ meVnm, $P_{\perp} = 486$ meVnm, $E_{g_0} = 190$ meV, $g_v = 4$ [103, 104], $\varepsilon_{sc} = 33\varepsilon_0$ [103, 104, 241], and $W = 4.6$ eV [242, 244] are valid for the Dimmock model [207] as given by (1.70) The values $m_1 = 0.0239m_0$, $m_2 = 0.024m_0$, $m_2' = 0.31m_0$, and $m_3 = 0.24m_0$ [243] are valid for the Cohen model [133] as given by (1.60)
11	Stressed n-Indium Antimonide	The values $m_c = 0.048m_o$, $E_{g_0} = 0.081$ eV, $B_2 = 9 \times 10^{-10}$ eVm, $C_1 = 3$ eV, $C_2 = 2$eV, $\bar{a}_0 = -10$ eV, $\bar{b}_0 = -1.7$ eV, $\bar{d}_0 = -4.4$ eV, $S_{xx} = 0.6 \times 10^{-3}(\text{kbar})^{-1}$, $S_{yy} = 0.42 \times 10^{-3}(\text{kbar})^{-1}$, $S_{zz} = 0.39 \times 10^{-3}(\text{kbar})^{-1}$, $S_{xy} = 0.5 \times 10^{-3}(\text{kbar})^{-1}$, $\varepsilon_{xx} = \sigma S_{xx}$, $\varepsilon_{yy} = \sigma S_{yy}$, $\varepsilon_{zz} = \sigma S_{zz}$, $\varepsilon_{xy} = \sigma S_{xy}$, σ is the stress in kilobar, and $g_v = 1$ [208–211] are valid for the model of Seiler et al. [208–211] as given by (1.76)
12	Bismuth	$E_{g_0} = 0.0153$ eV, $m_1 = 0.00194m_0$, $m_2 = 0.313m_0$, $m_3 = 0.00246m_0$, $m_2' = 0.36m_0$, $g_v = 3$ [245, 246], $M_2 = 1.25m_0$, $M_2' = 0.36m_0$ [247], and $W = 4.34$ eV
13	Mercury Telluride	$m_v^* = 0.028m_0$, $g_v = 1$, $\varepsilon_{\infty} = 15.2\varepsilon_0$ [248], and $W = 5.5$ eV [249]

(continued)

Table 1.1 (continued)

	Materials	Numerical values of the energy band constants
14	Platinum Antimonide	For valence bands, along the ⟨100⟩ direction, $\bar{\lambda}_0 = (0.02/4)\text{eV}, \bar{l} = (-0.32/4)\text{eV}, \bar{\nu} = (0.39/4)\text{eV}$, $\bar{n} = (-0.65/4)\text{eV}, \bar{a} = 0.643\,\text{nm}, I = 0.30\,(\text{eV})^2$, $\bar{\delta}_0 = 0.02\,\text{eV}$, $g_v = 6$ [214], $\varepsilon_{sc} = 30\varepsilon_0$ [250], and $\phi_w \approx 3.0\,\text{eV}$ [214, 251] For conduction bands, along the ⟨111⟩ direction, $g_v = 8$ [214, 251], $\bar{\lambda}_0 = (0.33/4)\text{eV}, \bar{l} = (1.09/4)\text{eV}, \bar{\nu} = (0.17/4)\text{eV}$, and $\bar{n} = (0.22/4)\text{eV}$
15	n-Gallium Phosphide	$m^*_{\|\|} = 0.92m_0, \quad m^*_{\perp} = 0.25m_0$, $k_0 = 1.7 \times 10^{15}\,\text{m}^{-1}, \quad \|V_G\| = 0.21\,\text{eV}, \quad g_v = 6$ [213], and $W = 3.75\,\text{eV}$ [230]
16	Germanium	$E_{g0} = 0.785\,\text{eV}, \quad m^*_{\|\|} = 1.57m_0, \quad m^*_{\perp} = 0.0807m_0$ [230], $W = 4.14\,\text{eV}$ [231], and $g_v = 4$
17	Tellurium	The values $\psi_1 = 6.7 \times 10^{-16}\,\text{meVm}^2$, $\psi_2 = 4.2 \times 10^{-16}\,\text{meVm}^2$, $\psi_3 = 6 \times 10^{-8}\,\text{meVm}$, and $\psi_4 = (3.6 \times 10^{-8}\,\text{meVm})$ [212] are valid for the model of Bouat et al. [212] as given by (1.81)
18	Graphite	The values $\bar{\Delta} = -0.0002\,\text{eV}, \bar{\gamma}_1 = 0.392\,\text{eV}$, $\bar{\gamma}_5 = 0.194\,\text{eV}, c_6 = 0.674\,\text{nm}, \bar{\gamma}_2 = -0.019\,\text{eV}$, $a_6 = 0.246\,\text{nm}, \bar{\gamma}_0 = 3\,\text{eV}, \bar{\gamma}_4 = 0.193\,\text{eV}, \overline{\gamma_3} = 0.21\,\text{eV}$ [252], and $W = 4.6\,\text{eV}$ [253] are valid for the model of Brandt et al. [252] as given by (R.1.12)
19	Lead Germanium Telluride	The values $g_v = 4$ [254] and $\phi_w \approx 6\,\text{eV}$ [255] are valid for the model of Vassilev [254] as given by (R.1.10)
20	Cadmium Antimonide	The values $a_1 = -32.3 \times 10^{-20}\,\text{eVm}^2$, $b_1 = -60.7 \times 10^{-20}\,\text{eVm}^2, a_2 = -16.3 \times 10^{-20}\,\text{eVm}^2$, $b_2 = -24.4 \times 10^{-20}\,\text{eVm}^2, a_3 = -91.9 \times 10^{-20}\,\text{eVm}^2$, $b_3 = -105 \times 10^{-20}\,\text{eVm}^2, A = 2.92 \times 10^{-10}\,\text{eVm}$, $B = -3.47 \times 10^{-10}\,\text{eVm}, G_3 = 1.3 \times 10^{-10}\,\text{eVm}$, $\Delta_3 = 0.070\,\text{eV}$ [222], and $\phi_w \approx 2\,\text{eV}$ [256]
21	Cadmium Diphosphide	The values $\beta_1 = 8.6 \times 10^{-21}\,\text{eVm}^2$, $\beta_2 = 1.8 \times 10^{-21}\,(\text{eVm})^2, \beta_4 = 0.0825\,\text{eV}$, $\beta_5 = -1.9 \times 10^{-19}\,\text{eVm}^2$ [257], and $\phi_w \approx 5\,\text{eV}$ [258] are valid for the model of Chuiko [257] and is given by (R.1.20)
22	Zinc Diphosphide	The values $\beta_1 = 8.7 \times 10^{-21}\,\text{eVm}^2$, $\beta_2 = 1.9 \times 10^{-21}\,(\text{eVm})^2, \beta_4 = 0.0875\,\text{eV}$, $\beta_5 = -1.9 \times 10^{-19}\,\text{eVm}^2$ [257], and $W \approx 3.9\,\text{eV}$ [258] are valid for the model of Chuiko [257] and is given by (R.1.20)
23	Bismuth Telluride	The values $E_{g0} = 0.145\,\text{eV}, \bar{\alpha}_{11} = 4.9, \bar{\alpha}_{22} = 5.92, \bar{\alpha}_{33} = 9.5$, $\bar{\alpha}_{23} = 4.22$, $g_v = 6$ [215–217], and $\phi_w = 5.3\,\text{eV}$ [259] are valid for the model of Stordeur et al. [215–217] using (1.97)
24	Carbon Nanotube	The values $a_c = 0.144\,\text{nm}$ [260], $t_c = 2.7\,\text{eV}$ [261], $\bar{r}_0 = 0.7\,\text{nm}$ [262, 263], and $W = 3.2\,\text{eV}$ [265] are valid for graphene band structure realization of carbon nanotube [262, 263]

(continued)

Table 1.1 (continued)

	Materials	Numerical values of the energy band constants
25	Antimony	The values $\alpha_{11} = 16.7, \alpha_{22} = 5.98, \alpha_{33} = 11.61, \alpha_{23} = 7.54$ [265], and $W = 4.63\,\mathrm{eV}$ are valid for the model of Ketterson [265] and are given by (R.1.13) and (R.1.14), respectively
26	Zinc Selenide	$m_{c2} = 0.16m_0,\ \Delta_2 = 0.42\,\mathrm{eV},\ E_{g02} = 2.82\,\mathrm{eV}$ [231], and $W = 3.2\,\mathrm{eV}$ [266]
27	Lead Selenide	$m_t^- = 0.23m_0,\ m_l^- = 0.32m_0,\ m_t^+ = 0.115m_0,$ $m_l^+ = 0.303m_0,\ P_{\|} \approx 138\,\mathrm{meVnm},$ $P_\perp = 471\,\mathrm{meVnm},\ E_{g_0} = 0.28\,\mathrm{eV}$ [267], $\varepsilon_{sc} = 21.0\varepsilon_0$ [231], and $W = 4.2\,\mathrm{eV}$ [268]

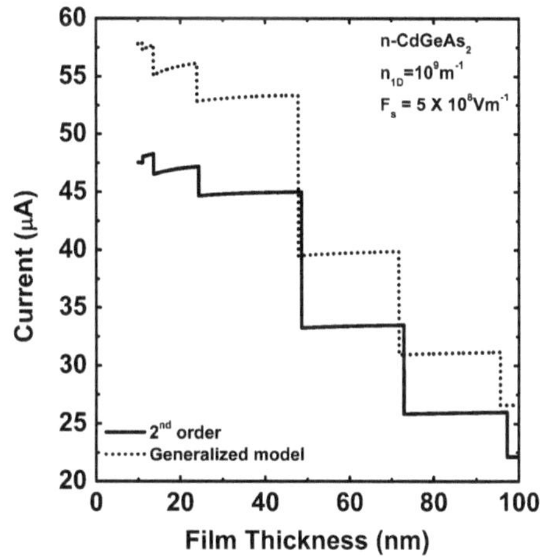

Fig. 1.1 Plot of the FN field emission current as a function of film thickness d_y for QWs of $\mathrm{n-CdGeAs_2}$. The *dotted and solid curves* correspond to the generalized and the two-band models of Kane, respectively, where $d_z = 30\,\mathrm{nm}$

values of alloy composition for QWs of $\mathrm{n-Hg_{1-x}Cd_xTe}$. Figure 1.5 shows the carrier concentration dependence of FNFE current from QWs of $\mathrm{n-Hg_{1-x}Cd_xTe}$, n-InSb, n-InAs, and n-GaAs, respectively, for the purpose of assessing the influence of different energy band constants on the field-emitted current from QWs of III–V materials. In Fig. 1.6, exhibits the film thickness dependence of FNFE current from QWs of $\mathrm{n-In_{1-x}Ga_xAs_yP_{1-y}}$ in accordance with the three- and two-band models of Kane together with the models of Stillman et al., Newson et al., and Palik et al., respectively. Figure 1.7 shows the dependence of FNFE current on alloy composition from QWs of ternary and quaternary materials in accordance with the two-band model of Kane. Figurc 1.8 exhibits the film thickness dependence of FNFE current from QWs of II–VI materials taking p-CdS as an example. Figure 1.9 shows the FNFE current as a function of carrier concentration for the case of Fig. 1.8. In Fig. 1.10, we have plotted the FNFE current from QWs of Bismuth as a function of film thickness in accordance with the models of McClure and Choi, Hybrid, Cohen, and Lax, respectively. Figure 1.11 exhibits the variation of the FNFE current as a function of film thickness for QWs of stressed materials taking stressed n-InSb

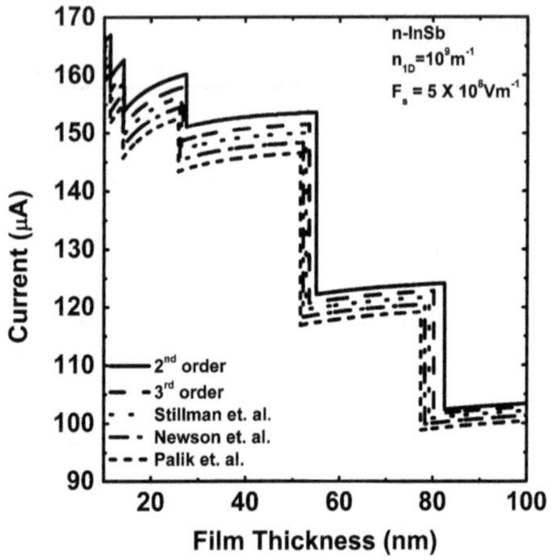

Fig. 1.2 Plot of the FN field emission current as a function of film thickness d_y for QWs of n-InSb in accordance with the three and two-band models of Kane together with the models of Stillman et al., Newson et al., and Palik et al., respectively, where $d_z = 30\,\text{nm}$

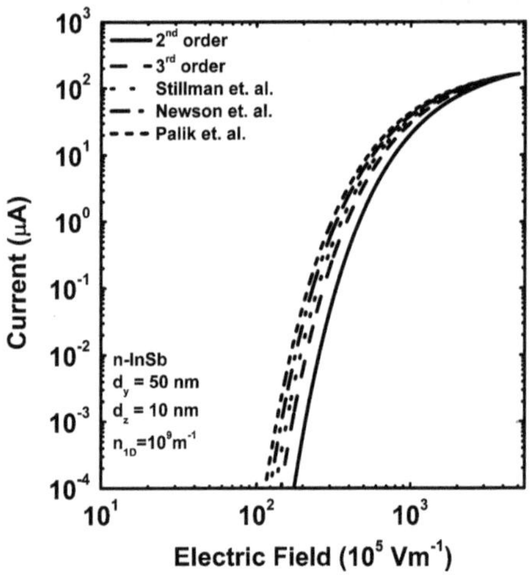

Fig. 1.3 Plot of the FN field emission current as a function of electric field for QWs of n-InSb in accordance with all the cases of Fig. 1.2

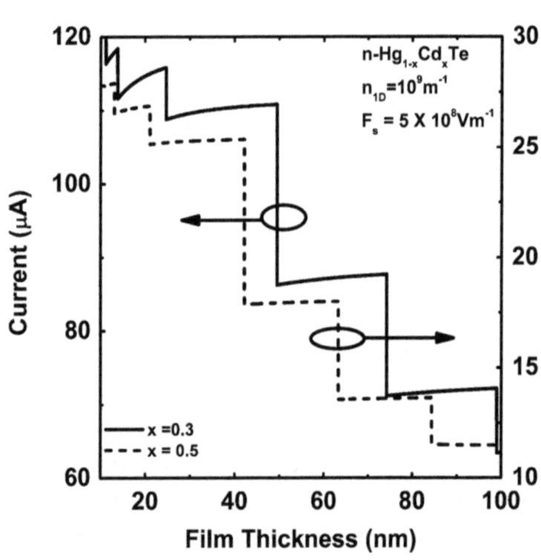

Fig. 1.4 Plot of the FN field emission current as a function of film thickness d_y for QWs of $n-Hg_{1-x}Cd_xTe$ in accordance with the two-band model of Kane for two different values of alloy composition

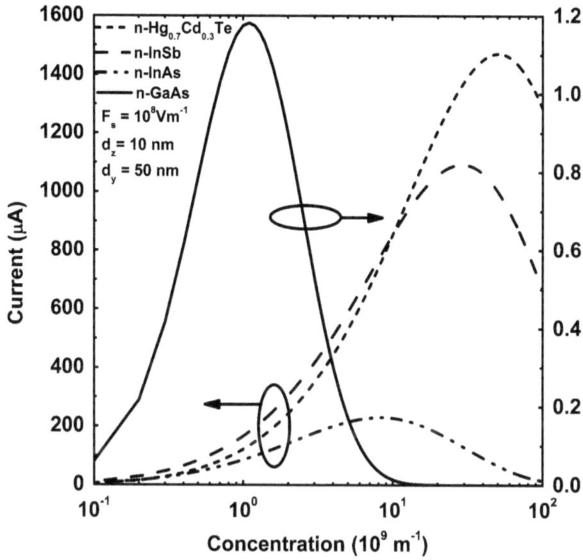

Fig. 1.5 Plot of the FN field emission current as a function of carrier concentration for QWs of $n-Hg_{0.7}Cd_{0.3}Te$, n-InSb, n-InAs, and n-GaAs in accordance with the two-band model of Kane at the lowest subband

Fig. 1.6 Plot of the field current as a function of film thickness d_y for quantum wires of $n-In_{1-x}Ga_xAs_yP_{1-y}$ in accordance with the three- and two-band models of Kane together with the models of Stillman et al., Newson et al., and Palik et al. respectively, where $d_z = 30\,\text{nm}$

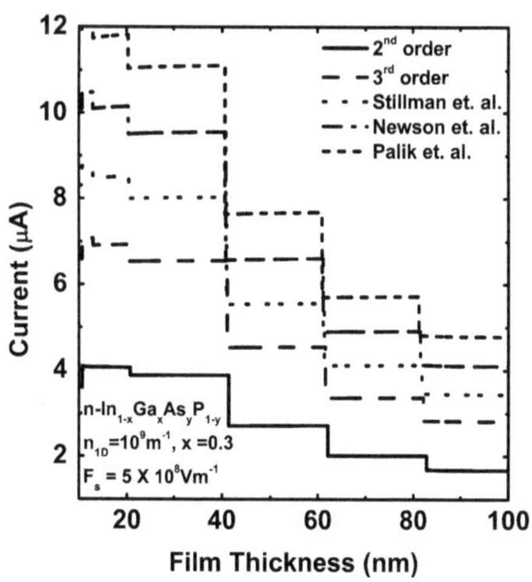

as an example. Figures 1.12 and 1.13 explore the stress dependence of the FNFE current from QWs of stressed n-InSb for different values of doping for the purpose of assessing the influence of carrier concentration on the field-emitted current in this case. In Fig. 1.14, the field-emitted current as a function of film thickness has been plotted for QWs of n-Ge (in accordance with both types of band models of n-Ge), n-GaP,Te, n-PbTe, and $p-Bi_2Te_3$, respectively. The plot (a) of Fig. 1.15 shows the

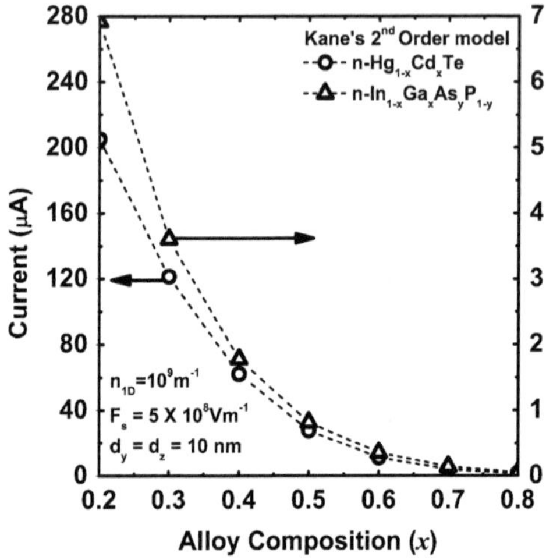

Fig. 1.7 Plot of the field-emitted current as a function of alloy composition for QWs of $n-Hg_{1-x}Cd_xTe$ and $In_{1-x}Ga_xAs_yP_{1-y}$ in accordance with the two-band model of Kane

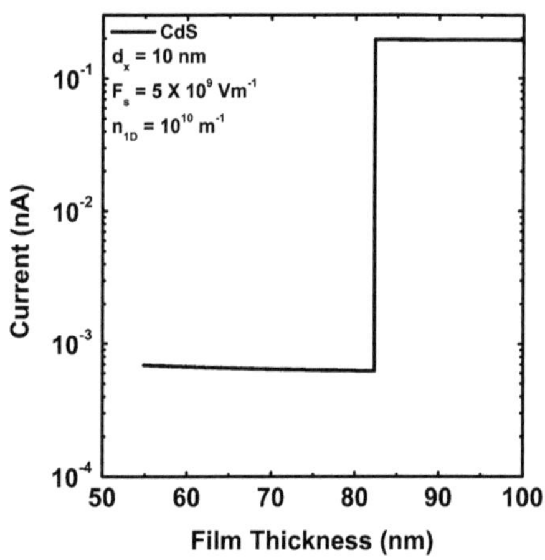

Fig. 1.8 Plot of the field-emitted current as a function of film thickness d_y for QWs of CdS

variation of FNFE current on the electric field for QWs of PbTe and Te, while (b) exhibits the same for QWs of Ge, GaP, and Bi_2Te_3, respectively.

The salient features of the above figures are described as follows:

From Fig. 1.1, we observe that the field emission current exhibits a step-functional decreasing dependence with increase in film thickness for QWs of $n-CdGeAs_2$. The combined influence of the anisotropies of the energy band constants and the crystal field splitting is to enhance the field-emitted current as compared with the same as obtained on the basis of two-band model of Kane in the whole range of thicknesses as considered in Fig. 1.1. The periodicity with respect to the film thickness is same in both the cases and is invariant of the energy band constants. It should be noted that the field-emitted current in general, is a product of two quantities inside the summation signs. One of them is $F_0(\eta_{ij})$, ($i = 1, 2, 3\ldots$ and $j = 1, 2, 3, \ldots$) and the other one is $\exp(-\beta_{ij})$, where both of them are functions

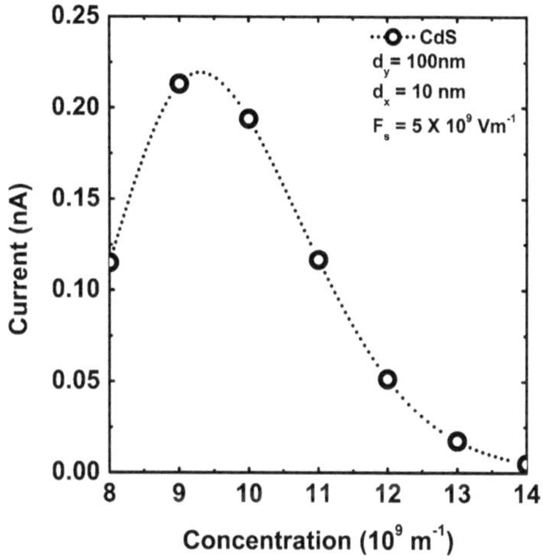

Fig. 1.9 Plot of the field-emitted current as a function of concentration for QWs of CdS

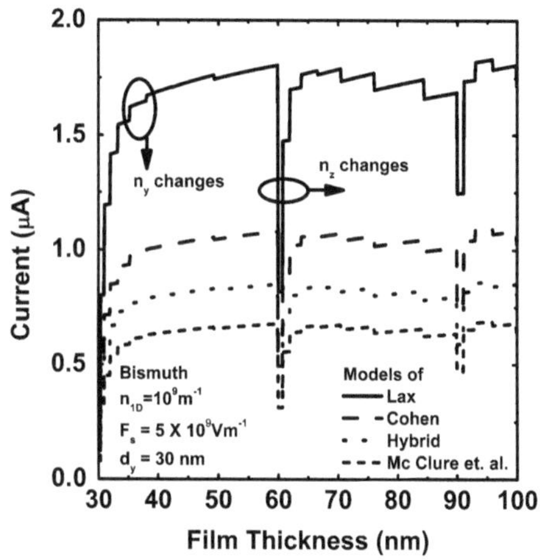

Fig. 1.10 Plot of the field-emitted current as a function of film thickness for QWs of Bi in accordance with the models of Mc Clure and Choi, Hybrid, Cohen, and Lax, respectively

of Fermi energy, effective mass, and various parameters of the system in a complex way. Thus, we observe that field-emitted current depends totally on the product of these two functions within the summation signs and a prefactor outside the summation signs. The product of these two functions ultimately determines the behavior of the field-emitted current. Although we know [247] that the Fermi energy of low-dimensional systems decreases with increasing size, one cannot be always certain that I will decrease with increasing film thickness due to the particular form of field-emitted current in case of QWs. If the rate of increase of $F_0(\eta_{ij})$ overcomes the rate of change of $\exp(-\beta_{ij})$, I will increase, whereas, for the opposite case, I will decrease. This important physical fact determines the magnitude of the field-emitted current and its dependence with respect to any other physical variable.

From Fig. 1.2, we observe that I decrease with increasing film thickness for the three- and two-band models of Kane together with the models of Stillman et al.,

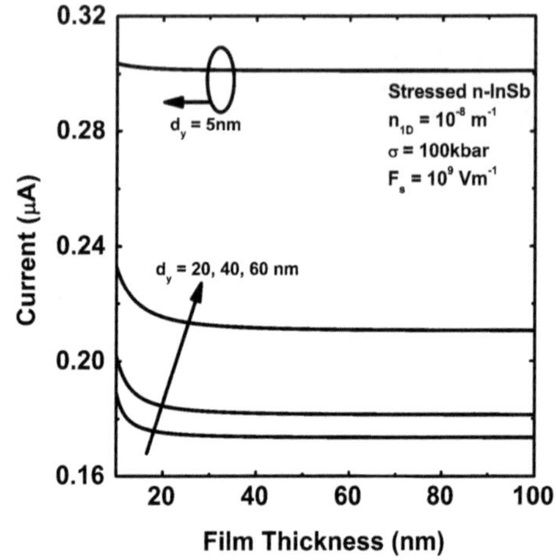

Fig. 1.11 Plot of the field-emitted current as a function of film thickness for QWs of stressed n-InSb for different values of d_y

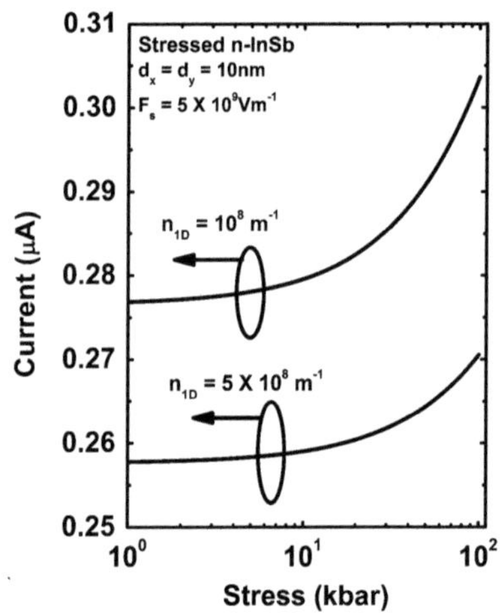

Fig. 1.12 Plot of the field-emitted current as a function of stress for QWs of stressed n-InSb for two different values of carrier concentration as shown

Newson et al., and Palik et al., respectively. The I exhibits different magnitudes which is the direct signature of the dispersion relation on the field-emitted current. Numerical computations reflect the fact that the field-emitted current for a $30 \times 10\,\mathrm{nm}^2$ size quantum wire can reach nearly $160\,\mu\mathrm{A}$ for n-InSb at low temperatures and at a field strength of $5 \times 10^8\,\mathrm{V\,m^{-1}}$ with carrier concentration of $10^9\,\mathrm{m^{-1}}$. From Fig. 1.3, it can be stated that field strength of nearly $10^7\,\mathrm{V\,m^{-1}}$ is sufficient to produce a tenth of microampers. Incidentally, due to the velocity saturation phenomena, we observe that beyond $10^8\,\mathrm{V\,m^{-1}}$, I saturates converging to a unique value and becomes invariant of dispersion relations. The field-emitted current from QWs of $\mathrm{n-Hg_{1-x}Cd_x}$ Te as function of film thickness has been exhibited in Fig. 1.4 and can be compared with the corresponding cases of earlier figures. It appears that for lower values of film thickness, I increases for a particular subband. As thickness increases, generation of different subbands occurs which

Fig. 1.13 Plot of the field-emitted current as a function of stress for QWs of stressed n-InSb for two different values of carrier concentration as shown

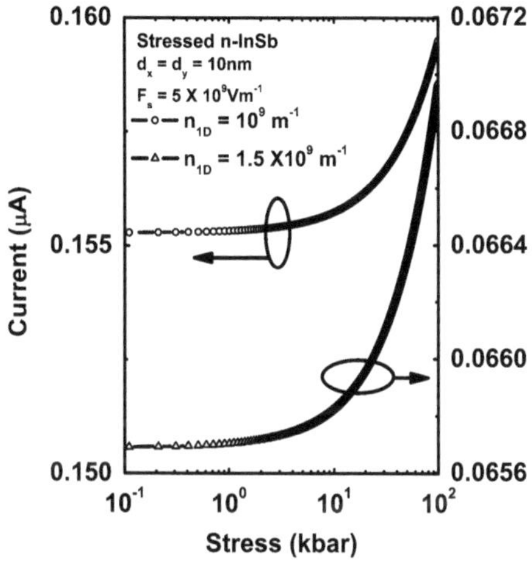

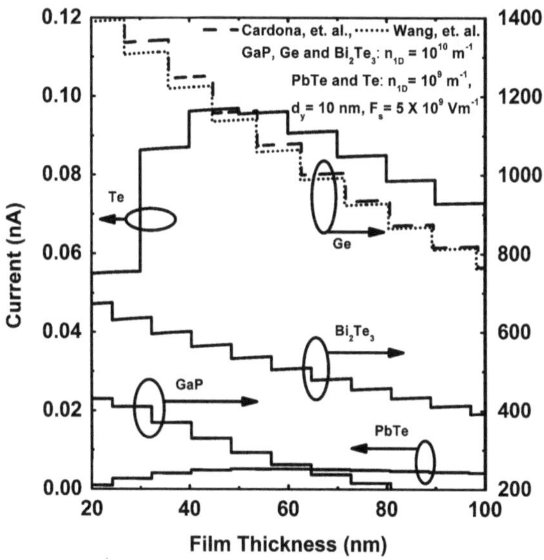

Fig. 1.14 Plot of the field-emitted current as a function of film thickness for QWs of Ge (in accordance with both types of band models), GaP, Bi_2Te_3, PbTe, and Te, respectively

leads to the overall decrease in I. The variation of I over a large range of carrier concentration has been plotted in Fig. 1.5 for the quantum limit case. It appears that I initially increases due to low value of Fermi energy, and as the concentration increases, the magnitude of the current decreases sharply exhibiting a peak. The amount of broadening is highly sensitive to the spectrum parameters of a particular semiconductor. It appears that for n-GaAs, the broadening is more as compared with others as shown in the figure. Figure. 1.6 exhibits the variation of I as function of

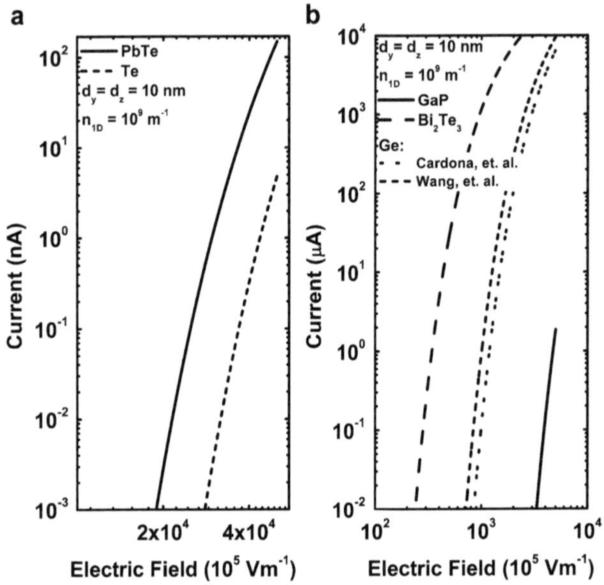

Fig. 1.15 Plot of the field current as a function of electric field for quantum wires of (**a**) PbTe and Te and (b) GaP, Bi_2Te_3, and Ge, respectively

film thickness for QWs of $n - In_{1-x}Ga_xAsP_{1-y}$ in accordance with the three- and two-band models of Kane together with the models of Stillman et al., Newson et al., and Palik et al. respectively. Figure. 1.6 shows almost a constant step-functional dependence of the current for a particular regime of film thickness for all types of band models. This implies that the difference in the Fermi energy and the quantized subband energy is almost invariant of film thickness. From Fig. 1.7, one observes that as alloy composition increases, the field-emitted current decreases for QWs of both ternary and quaternary compounds. For lower values of alloy composition, I from QWs of quaternary semiconductor is more than that of the corresponding ternary one, whereas for higher value of alloy composition, it converges. Step-functional dependence of field-emitted current with film thickness for QWs of CdS has been observed in Fig. 1.8. We note a nearly constant field-emitted current per subband, until a new subband is being occupied. Comparing with the earlier figures on the thickness dependence, we observe in this case an opposite trend because of the fact of overriding of the term $F_0(\eta_{ij})$ by $\exp(-\beta_{ij})$. Figure. 1.9 exhibits the fact that for p-CdS, the field current is of the order of a few nano amperes even at field strength of $5 \times 10^9\,\mathrm{V\,m^{-1}}$.

Composite oscillations in the field-emitted current from QWs of Bi with film thickness has been exhibited in Fig. 1.10 for the models of McClure and Choi, Hybrid, Cohen and Lax, respectively. We observe that for $n_z = 1$, electrons will be populated in the various other subbands corresponding to n_y. After a certain value of film thickness, when $n_z = 2$, the redistribution of the electrons in the quantized energy levels is being repeated leading to a composite oscillations. It should be noted that this behavior is a general feature in thickness dependence of field emission from quantum wires. The energy band parameters determine the composite periodicity. The effect of film thickness on the field current in stressed InSb has been plotted in Fig. 1.11 in the quantum limit case, where the current decreases with increase in d_z for different values of d_y. For $d_y = 5$ nm, the field-emitted current is highest although is an approximately a constant quantity with respect to d_z. We observe here that the product of the two terms namely $F_0(\eta_{ij})$ and $\exp(-\beta_{ij})$ becomes independent of d_z for large values.

The effect of stress on the field-emitted current has been exhibited in Figs. 1.12 and 1.13 for different values of carrier concentration. From Fig. 1.12, it appears that with the increase in stress, the current increases having a magnitude of few tenths of microamperes. With the increase in carrier degeneracy, the current almost reduces to about 10 times. From Fig. 1.14, we observe that the field current from QWs of Ge (in accordance with the models of Cardona et al. and Wang et al. respectively), Bi_2Te_3, and GaP decreases as film thickness increases because of the fact that the term $\exp(-\beta_{ij})$ dominates over the term $F_0(\eta_{ij})$ in the whole range of thickness as considered here. For QWs of PbTe and Te, the current initially increases since the term $\exp(-\beta_{\mathrm{ij}})$ dominates over the term $F_0(\eta_{ij})$ and then decreases exhibiting the fact that the opposite dominancy exists. For QWs of Ge and Bi_2Te_3, the field current has been observed to be in theorder of hundreds of microampere, where as for other 1D systems as considered in this figures the currents are in the order of few nanoamperes exhibiting the influence of the carrier dispersion relation of a particular semiconductor. In Fig. 1.15 (a), the cut-in electric field for the field currents are in the order of $10^9\,\mathrm{V\,m^{-1}}$ for QWs of PbTe and Te, while about 10^7–$10^8\,\mathrm{V\,m^{-1}}$ field strength is sufficient for field current to reach few microamperes.

The influence of quantum confinement is immediately apparent from Figs. 1.1, 1.2, 1.4, 1.6, 1.8, 1.10, 1.11, and 1.14 since the field-emitted current depends strongly on the thickness of the quantum-confined materials in contrast with the corresponding bulk specimens. The current changes with increasing film thickness in an oscillatory way with different numerical magnitudes. It appears from the aforementioned figures that the FNFE exhibits spikes for particular values of film thickness which, in turn, depends on the particular band structure of the specific semiconductor. Moreover, the FNFE from QWs of different compounds can be smaller than of bulk specimens of the same materials, which is also a direct signature of quantum confinement. This oscillatory dependence will be less and less prominent with increasing film thickness. For bulk specimens of the same material, the FNFE will be found to increase continuously with increasing electron degeneracy in a non-oscillatory manner. The appearance of the discrete jumps in the respective figures is due to the redistribution of the electrons among the quantized energy levels when the size quantum number corresponding to the highest occupied level changes from one fixed value to the others.

With varying electron degeneracy, a change is reflected in the FNFE through the redistribution of the electrons among the size-quantized levels. It may be noted that at the transition zone from one subband to another, the height of the peaks between any two subbands decreases with the increasing in the degree of quantum confinement and is clearly shown in all the curves. It should be noted that although, the FNFE changes in various manners with all the variables as evident from all the figures, the rates of variations are totally band structure dependent.

It is imperative to state that the present investigation excludes the many-body, hot electron, broadening, and the allied effects in the simplified theoretical formalism due to the absence of proper analytical techniques for including them for generalized systems as considered here. We have also approximated the variation of value of the work function from its bulk value in the present system. Our simplified approach will be appropriate for the purpose of comparisons when the methods of tackling the formidable problems after inclusion of the said effects for the generalized systems emerge. The results of this simplified approach get transformed to the well-known formulation of the FNFE for wide gap materials having parabolic energy bands. This indirect test not only exhibits the mathematical compatibility of the formulation but also shows the fact that this simple analysis is more generalized one, since one can obtain the corresponding results for materials having parabolic energy bands under certain limiting conditions from the present derivation. For the purpose of computer simulations for obtaining the plots of FNFE versus various external variables, we have taken very low temperatures since the quantization effects are basically low-temperature phenomena together with the fact that the temperature dependence of all the energy band constants of all the semiconductors and their nanostructures as considered in this chapter are not available in the literature. Our results as formulated in this chapter are valid for finite temperatures and are useful in comparing the results for temperature variations of FNFE after the availability of the temperature dependences of such constants of various dispersion relations in this context. The experimental results for the verification of theoretical formulations of FNFE are still not available in the literature. It is worth noting that the nature of the curves of field-emitted current with various physical variables based on our simplified formulations as presented here would be useful to analyze the experimental results when they materialize. The inclusion of the said effects would certainly increase the accuracy of the results although the qualitative features of FNFE would not change in the presence of the aforementioned effects.

It can be noted that on the basis of the dispersion relations of the various quantized structures as discussed above the effective electron mass,the Debye screening length,the plasma frequency, the activity coefficient, the carrier contribution to the elastic constants, the diffusion coefficient of minority carriers, the third-order nonlinear optical susceptibility, the heat capacity, the dia- and paramagnetic susceptibilities, and the various important dc/ac transport coefficients can be probed for all types of QWs as considered here. Thus, our theoretical formulation comprises the dispersion relation dependent properties of various technologically important quantum-confined semiconductors having different band structures. We have not considered other types of compounds in order to keep the presentation concise and succinct. With different sets of energy band parameters, we shall get different numerical values of the FNFE. The nature of variations of the FNFE as shown here would be similar for the other types of materials and the simplified analysis of this chapter exhibits the basic qualitative features of the FNFE. It may be noted that the basic aim of this chapter is not solely to demonstrate the influence of quantum confinement on the FNFE for a wide class of quantized materials but also to formulate the appropriate carrier statistics in the most generalized form, since the transport and other phenomena in modern nanostructured devices having different band structures and the derivation of the expressions of many important carrier properties are based on the temperature-dependent carrier statistics in such systems. For the purpose of condensed presentation, the carrier statistics and the FNFE from different QWs as considered in this chapter have been presented in Table 1.2.

1.4 Open Research Problems

The problems under these sections of this monograph are by far the most important part and few open research problems from this chapter till end are being presented. The numerical values of the energy band constants for various semiconductors are given in Table 1.1 for the related computer simulations.

(R.1.1) Investigate the FNFE from all the bulk semiconductors whose respective dispersion relations of the carriers are given in this chapter by converting the summations over the quantum numbers to the corresponding integrations by including the uniqueness conditions in the appropriate cases and considering the effect of image force in the subsequent study in each case.

(R.1.2) Repeat R.1.1 for the bulk semiconductors whose respective dispersion relations of the carriers in the absence of any field are given below:

(a) The electron dispersion law in n-GaP can be written as [269]

$$E = \frac{\hbar^2 k_s^2}{2m_{\|}^*} + \frac{\hbar^2 k_s^2}{2m_{\perp}^*} \mp \frac{\bar{\Delta}}{2} \pm \left[\left(\frac{\bar{\Delta}}{2} \right)^2 + P_1 k_z^2 + D_1 k_x^2 k_y^2 \right]^{1/2}, \tag{R1.1}$$

where $\bar{\Delta} = 335\,\text{meV}$, $P_1 = 2\times10^{-10}\text{eVm}$, $D_1 = P_1 a_1$, and $a_1 = 5.4\times10^{-10}\text{m}$.

(b) In addition to the Cohen model, the dispersion relation for the conduction electrons for IV–VI semiconductors can also be described by the models of Bangert et al. [270] and Foley et al. [271], respectively.

1. In accordance with Bangert et al. [270], the dispersion relation is given by

$$\Gamma(E) = F_1(E) k_s^2 + F_2(E) k_z^2, \tag{R1.2}$$

where $\Gamma(E) \equiv 2E$, $F_1(E) \equiv \dfrac{R_1^2}{E + E_g} + \dfrac{S_1^2}{E + \Delta_c'} + \dfrac{Q_1^2}{E + E_g}$,

Table 1.2 The carrier statistics and the Fowler–Nordheim field emission current from quantum wires of nonlinear optical, III–V, II–VI, Bismuth, IV–VI, stressed materials, Te, n-GaP, $PtSb_2$, Bi_2Te_3, n-Ge, GaSb, and II–V Compounds

Type of materials	The carrier statistics	The Fowler—Nordheim field emission current
1. Nonlinear optical materials	In accordance with the generalized electron dispersion relation as given by (1.2), $n_0 = \frac{2g_v}{\pi}\sum_{n_x=1}^{n_{x\max}}\sum_{n_y=1}^{n_{y\max}}[B_{11}(E_{F1D}, n_x, n_y) + B_{12}(E_{F1D}, n_x, n_y)]$ (1.5)	On the basis of (1.5), $I = \frac{2g_v e k_B T}{h}\sum_{n_x=1}^{n_{x\max}}\sum_{n_y=1}^{n_{y\max}} F_0(\eta_{11})\exp(-\beta_{11})$ (1.12)
2. III–V materials, the conduction electrons of which can be defined by five types of energy wave vector dispersion relations as described in the column of carrier statistics	(a) Three-band model of Kane: In accordance with the three-band model of Kane (1.13), which is the special case of (1.2) $n_0 = \frac{2g_v}{\pi}\sum_{n_x=1}^{n_{x\max}}\sum_{n_y=1}^{n_{y\max}}[B_{13}(E_{F1D}, n_x, n_y) + B_{14}(E_{F1D}, n_x, n_y)]$ (1.16)	On the basis of (1.16), $I = \frac{2g_v e k_B T}{h}\sum_{n_x=1}^{n_{x\max}}\sum_{n_y=1}^{n_{y\max}} F_0(\eta_{12})\exp(-\beta_{12})$ (1.18)
	(b) Two-band model of Kane: In accordance with the two-band model of Kane (1.19), $n_0 = \frac{2g_v}{\pi}\sum_{n_x=1}^{n_{x\max}}\sum_{n_y=1}^{n_{y\max}}[B_{15}(E_{F1D}, n_x, n_y) + B_{16}(E_{F1D}, n_x, n_y)]$ (1.22)	On the basis of (1.22), $I = \frac{2g_v e k_B T}{h}\sum_{n_x=1}^{n_{x\max}}\sum_{n_y=1}^{n_{y\max}} F_0(\eta_{13})\exp(-\beta_{13})$ (1.24)
	(c) The model of Stillman et al.: In accordance with the model of Stillman et al. (1.28), $n_0 = \frac{2g_v}{\pi}\sum_{n_x=1}^{n_{x\max}}\sum_{n_y=1}^{n_{y\max}}[B_{17}(E_{F1D}, n_x, n_y) + B_{18}(E_{F1D}, n_x, n_y)]$ (1.32)	On the basis of (1.32), $I = \frac{2g_v e k_B T}{h}\sum_{n_x=1}^{n_{x\max}}\sum_{n_y=1}^{n_{y\max}} F_0(\eta_{15})\exp(-\beta_{15})$ (1.33)
	(d) The model of Newson and Kurobe: In accordance with the model of Newson and Kurobe (1.34), $n_0 = \frac{2g_v}{\pi}\sum_{n_x=1}^{n_{x\max}}\sum_{n_y=1}^{n_{y\max}}[B_{19}(E_{F1D}, n_x, n_y) + B_{20}(E_{F1D}, n_x, n_y)]$ (1.37)	On the basis of (1.37), $I = \frac{2g_v e k_B T}{h}\sum_{n_x=1}^{n_{x\max}}\sum_{n_y=1}^{n_{y\max}} F_0(\eta_{16})\exp(-\beta_{16})$ (1.39)

	(e) The model of Palik et al.: In accordance with the model of Palik et al. (1.40), $n_0 = \frac{2g_v}{\pi}\sum_{n_x=1}^{n_{x\max}}\sum_{n_y=1}^{n_{y\max}}[B_{21}(E_{F1D}, n_x, n_y) + B_{22}(E_{F1D}, n_x, n_y)]$ (1.43)	On the basis of (1.43), $I = \frac{2g_v e k_B T}{h}\sum_{n_x=1}^{n_{x\max}}\sum_{n_y=1}^{n_{y\max}} F_0(\eta_{16})\exp(-\beta_{16})$ (1.44)
3. II–VI materials as described by the Hopfield model	In accordance with (1.46), $n_{1D} = \frac{g_v}{\pi\sqrt{B_0}}\sum_{n_x=1}^{n_{x\max}}\sum_{n_y=1}^{n_{y\max}}[t_7(E_{F1D}, n_x, n_y) + t_8(E_{F1D}, n_x, n_y)]$ (1.49)	On the basis of (1.49), $I = \frac{e g_v k_B T}{h}\sum_{n_x=1}^{n_{x\max}}\sum_{n_y=1}^{n_{y\max}}\Big[\Big\{F_0\Big\{(k_B T)^{-1}[E_{F1D} - [G_{3,+}(n_x, n_y)]]\Big\}\exp(-\overline{\beta}_{016,+})\Big\} + F_0\Big\{(k_B T)^{-1}[E_{F1D} - [G_{3,-}(n_x, n_y)]]\Big\} \times \exp(-\overline{\beta}_{016,-})\Big]$ (1.48)
4. Bismuth, the carriers of which can be defined by four types of energy band models as described in the column of carrier statistics	(a) The McClure and Choi model: In accordance with (1.50), $n_{1D} = \frac{2g_v}{\pi}\frac{\sqrt{2m_1}}{\hbar}\sum_{n_y=1}^{n_{y\max}}\sum_{n_z=1}^{n_{z\max}}[t_{27}(E_{F1D}, n_y, n_z) + t_{28}(E_{F1D}, n_y, n_z)]$ (1.52)	On the basis of (1.52), $I = \frac{2g_v e k_B T}{h}\sum_{n_z=1}^{n_{z\max}}\sum_{n_y=1}^{n_{y\max}} F_0(\eta_{17})\exp(-\beta_{17})$ (1.53)
	(b) The Hybrid model: In accordance with (1.55), $n_{1D} = \frac{2g_v}{\pi}\frac{\sqrt{2m_1}}{\hbar}\sum_{n_y=1}^{n_{y\max}}\sum_{n_z=1}^{n_{z\max}}\left[t_{31}(E_{F1D}, n_y, n_z) + t_{32}(E_{F1D}, n_y, n_z)\right]$ (1.57)	On the basis of (1.57), $I = \frac{2g_v e k_B T}{h}\sum_{n_z=1}^{n_{z\max}}\sum_{n_y=1}^{n_{y\max}} F_0(\eta_{18})\exp(-\beta_{18})$ (1.58)
	(c) The Cohen model: In accordance with (1.60), $n_{1D} = \frac{2g_v}{\pi}\frac{\sqrt{2m_1}}{\hbar}\sum_{n_y=1}^{n_{y\max}}\sum_{n_z=1}^{n_{z\max}}\left[t_{35}(E_{F1D}, n_y, n_z) + t_{36}(E_{F1D}, n_y, n_z)\right]$ (1.62)	On the basis of (1.62), $I = \frac{2g_v e k_B T}{h}\sum_{n_z=1}^{n_{z\max}}\sum_{n_y=1}^{n_{y\max}} F_0(\eta_{19})\exp(-\beta_{19})$ (1.63)
	(d) The Lax model: In accordance with (1.65), $n_{1D} = \frac{2g_v}{\pi}\frac{\sqrt{2m_1}}{\hbar}\sum_{n_y=1}^{n_{y\max}}\sum_{n_z=1}^{n_{z\max}}\left[t_{37}(E_{F1D}, n_y, n_z) + t_{38}(E_{F1D}, n_y, n_z)\right]$ (1.67)	On the basis of (1.67), $I = \frac{2g_v e k_B T}{h}\sum_{n_z=1}^{n_{z\max}}\sum_{n_y=1}^{n_{y\max}} F_0(\eta_{20})\exp(-\beta_{20})$ (1.68)

(continued)

Table 1.2 (Continued)

Type of materials	The carrier statistics	The Fowler—Nordheim field emission current
5. IV–VI materials, the carriers of which can be defined by the model of Dimmock	In accordance with the model of Dimmock (1.70), $n_0 = \frac{2g_v}{\pi} \sum_{n_x=1}^{n_{x\max}} \sum_{n_y=1}^{n_{y\max}} [B_{32}(E_{F1D}, n_x, n_y) + B_{33}(E_{F1D}, n_x, n_y)]$ (1.73)	On the basis of (1.73), $I = \frac{2g_v e k_B T}{h} \sum_{n_x=1}^{n_{x\max}} \sum_{n_y=1}^{n_{y\max}} F_0(\eta_{22}) \exp(-\beta_{22})$ (1.74)
6. Stressed materials, as defined by the model of Seiler et al.	In accordance with (1.76), $n_{1D} = \frac{2g_v}{\pi} \sum_{n_y=1}^{n_{y\max}} \sum_{n_z=1}^{n_{z\max}} [B_{34}(E_{F1D}, n_y, n_z) + B_{35}(E_{F1D}, n_y, n_z)]$ (1.79)	On the basis of (1.80), $I = \frac{2g_v e k_B T}{h} \sum_{n_x=1}^{n_{x\max}} \sum_{n_y=1}^{n_{y\max}} F_0(\eta_{23}) \exp(-\beta_{23})$ (1.80)
7. Tellurium the conduction electrons of which can be defined by the model of Bouat et al	In accordance with (1.81), $n_0 = \frac{g_v}{\pi} \sum_{n_x=1}^{n_{x\max}} \sum_{n_y=1}^{n_{y\max}} [B_{36,\pm}(E_{F1D}, n_x, n_y) + B_{37,\pm}(E_{F1D}, n_x, n_y)]$ (1.84)	On the basis of (1.84), $I = \frac{g_v e k_B T}{h} \sum_{n_x=1}^{n_{x\max}} \sum_{n_y=1}^{n_{y\max}} [F_0(\eta_{24,+}) \exp(-\beta_{24,+}) + F_0(\eta_{24,-}) \exp(-\beta_{24,-})]$ (1.85)
8. n-GaP as described by the Rees model	In accordance with (1.86), $n_0 = \frac{2g_v}{\pi} \sum_{n_x=1}^{n_{x\max}} \sum_{n_y=1}^{n_{y\max}} [B_{38}(E_{F1D}, n_x, n_y) + B_{39}(E_{F1D}, n_x, n_y)]$ (1.89)	On the basis of (1.89), $I = \frac{2g_v e k_B T}{h} \sum_{n_x=1}^{n_{x\max}} \sum_{n_y=1}^{n_{y\max}} [F_0(\eta_{26}) \exp(-\beta_{26})]$ (1.89)
9. $PtSb_2$, as defined by the Emtage model	In accordance with (1.91), $n_0 = \frac{2g_v}{\pi} \sum_{n_x=1}^{n_{x\max}} \sum_{n_y=1}^{n_{y\max}} [B_{40}(E_{F1D}, n_x, n_y) + B_{41}(E_{F1D}, n_x, n_y)]$ (1.94)	On the basis of (1.94), $I = \frac{2g_v e k_B T}{h} \sum_{n_x=1}^{n_{x\max}} \sum_{n_y=1}^{n_{y\max}} [F_0(\eta_{27}) \exp(-\beta_{27})]$ (1.95)
10. Bi_2Te_3, which follows the model of Stordeur et al.	In accordance with (1.97), $n_0 = \frac{2g_v}{\pi} \sum_{n_z=1}^{n_{z\max}} \sum_{n_y=1}^{n_{y\max}} [B_{42}(E_{F1D}, n_z, n_y) + B_{43}(E_{F1D}, n_z, n_y)]$ (1.100)	On the basis of (1.100), $I = \frac{2g_v e k_B T}{h} \sum_{n_z=1}^{n_{z\max}} \sum_{n_y=1}^{n_{y\max}} [F_0(\eta_{28}) \exp(-\beta_{28})]$ (1.101)

11. n-Ge, the conduction electrons of which can be defined by two types of energy band models as described in the column of carrier statistics	(a) In accordance with the model of Cardona et al. (1.102), $$n_0 = \frac{2g_v}{\pi} \sum_{n_x=1}^{n_{x\max}} \sum_{n_z=1}^{n_{z\max}} [B_{44}(E_{F1D}, n_x, n_z) + B_{45}(E_{F1D}, n_x, n_z)] \quad (1.105)$$	On the basis of (1.105), $$I = \frac{2g_v e k_B T}{h} \sum_{n_x=1}^{n_{x\max}} \sum_{n_z=1}^{n_{z\max}} [F_0(\eta_{29}) \exp(-\beta_{29})] \quad (1.106)$$
	(b) In accordance with the model of Wang and Ressler (1.107), $$n_0 = \frac{2g_v}{\pi} \sum_{n_x=1}^{n_{x\max}} \sum_{n_z=1}^{n_{z\max}} [B_{46}(E_{F1D}, n_x, n_z) + B_{47}(E_{F1D}, n_x, n_z)] \quad (1.110)$$	On the basis of (1.110), $$I = \frac{2g_v e k_B T}{h} \sum_{n_x=1}^{n_{x\max}} \sum_{n_z=1}^{n_{z\max}} [F_0(\eta_{30}) \exp(-\beta_{30})] \quad (1.111)$$
12. Gallium Antimonide, the carriers of which can be defined by the model of Mathur et al.	In accordance with (1.112), $$n_0 = \frac{2g_v}{\pi} \sum_{n_x=1}^{n_{x\max}} \sum_{n_y=1}^{n_{y\max}} [B_{48}(E_{F1D}, n_x, n_y) + B_{49}(E_{F1D}, n_x, n_y)]$$	On the basis of (1.116), $$I = \frac{2g_v e k_B T}{h} \sum_{n_x=1}^{n_{x\max}} \sum_{n_y=1}^{n_{y\max}} [F_0(\eta_{31}) \exp(-\beta_{31})] \quad (1.117)$$
13. II–V materials, as defined by the model of Yamada	In accordance with (1.118), $$n_0 = \frac{g_v}{\pi} \sum_{n_x=1}^{n_{x\max}} \sum_{n_y=1}^{n_{y\max}} [B_{49}(E_{F1D}, n_x, n_y) + B_{50}(E_{F1D}, n_x, n_y)] \quad (1.121)$$	On the basis of (1.121), $$I = \frac{g_v e k_B T}{h} \sum_{n_x=1}^{n_{x\max}} \sum_{n_y=1}^{n_{y\max}} [F_0(\eta_{32,+}) \exp(-\beta_{32,+}) + F_0(\eta_{32,-}) \exp(-\beta_{32,-})] \quad (1.122)$$

$$F_2(E) \equiv \frac{2C_5^2}{E+E_g} + \frac{(S_1+Q_1)^2}{E+\Delta_c''},$$

$R_1^2 = 2.3 \times 10^{-19}\,(\text{eVm})^2$, $C_5^2 = 0.83 \times 10^{-19}\,(\text{eVm})^2$, $Q_1^2 = 1.3R_1^2$, $S_1^2 = 4.6R_1^2$,

$\Delta_c' = 3.07$ eV, $\Delta_c'' = 3.28$ eV, and $g_v = 4$. It may be noted that under the substitution $S_1 = 0$, $Q_1 = 0$, $R_1^2 \equiv \dfrac{\hbar^2 E_g}{m_\perp^*}$, $C_5^2 \equiv \dfrac{\hbar^2 E_g}{2m_{\parallel}^*}$, (R1.2) assumes the form $E(1+\alpha E) = \dfrac{\hbar^2 k_s^2}{2m_\perp^*} + \dfrac{\hbar^2 k_z^2}{2m_{\parallel}^*}$, which is the simplified Lax model.

2. The carrier energy spectrum of IV–VI semiconductors in accordance with Foley et al. [271] can be written as

$$E + \frac{E_g}{2} = E_-(k) + \left[\left[E_+(k) + \frac{E_g}{2}\right]^2 + P_\perp^2 k_s^2 + P_{\parallel}^2 k_z^2\right]^{1/2}, \tag{R1.3}$$

where $E_+(k) = \dfrac{\hbar^2 k_s^2}{2m_\perp^+} + \dfrac{\hbar^2 k_z^2}{2m_{\parallel}^+}$, $E_-(k) = \dfrac{\hbar^2 k_s^2}{2m_\perp^-} + \dfrac{\hbar^2 k_z^2}{2m_{\parallel}^-}$ represents the contribution from the interaction of the conduction and the valance band edge states with the more distant bands and the free electron term, $1/m_\perp^\pm = \dfrac{1}{2}[1/m_{tc} \pm 1/m_{tv}]$, $\dfrac{1}{m_{\parallel}^\pm} = \dfrac{1}{2}\left[\dfrac{1}{m_{1c}} \pm \dfrac{1}{m_{1v}}\right]$. For n-PbTe, $P_\perp = 4.61 \times 10^{-10}$ eVm, $P_{\parallel} = 1.48 \times 10^{-10}$ eVm, $m_0/m_{tv} = 10.36$, $m_0/m_{1v} = 0.75$, $m_0/m_{tc} = 11.36$, $m_0/m_{1c} = 1.20$, and $g_v = 4$.

(c) The hole energy spectrum of p-type zero-gap semiconductors (e.g., HgTe) is given by [272]

$$E = \frac{\hbar^2 k^2}{2m_v^*} + \frac{3e^2}{128\varepsilon_\infty} k - \left(\frac{2E_B}{\pi}\right) \ln\left|\frac{k}{k_0}\right|, \tag{R1.4}$$

where m_v^* is the effective mass of the hole at the top of the valence band, ε_∞ is the semiconductor permittivity in the high-frequency limit, $E_B \equiv m_0 e^2/2\hbar^2\varepsilon_\infty^2$, and $k_0 \equiv m_0 e^2/\hbar^2 \varepsilon_\infty$.

(d) The conduction electrons of n-GaSb obey the following two dispersion relations:

1. In accordance with the model of Seiler et al. [238]

$$E = \left[-\frac{E_g}{2} + \frac{E_g}{2}\left[1+\alpha_4 k^2\right]^{1/2} + \frac{\bar{\varsigma}_0 \hbar^2 k^2}{2m_o} + \frac{\bar{v}_0 f_1(k)\hbar^2}{2m_o} \pm \frac{\bar{\omega}_0 f_2(k)\hbar^2}{2m_o}\right], \tag{R1.5}$$

where $\alpha_4 \equiv 4P^2\left(E_g + \frac{2}{3}\Delta\right)\left[E_g^2\left(E_g + \Delta\right)\right]^{-1}$, P is the isotropic momentum matrix element, $f_1(k) \equiv k^{-2}\left[k_x^2k_y^2 + k_y^2k_z^2 + k_z^2k_x^2\right]$ represents the warping of the Fermi surface, $f_2(k) \equiv \left[\left\{k^2\left(k_x^2k_y^2 + k_y^2k_z^2 + k_z^2k_x^2\right) - 9k_x^2k_y^2k_z^2\right\}^{1/2} k^{-1}\right]$ represents the inversion asymmetry splitting of the conduction band, and $\bar{\varsigma}_0, \bar{v}_0$, and $\bar{\omega}_0$ represent the constants of the electron spectrum in this case.

2. In accordance with the model of Zhang et al. [273]

$$E = \left[E_2^{(1)} + E_2^{(2)}K_{4,1}\right]k^2 + \left[E_4^{(1)} + E_4^{(2)}K_{4,1}\right]k^4 + k^6\left[E_6^{(1)} + E_6^{(2)}K_{4,1} + E_6^{(3)}K_{6,1}\right], \tag{R1.6}$$

where $K_{4,1} \equiv \frac{5}{4}\sqrt{21}\left[\frac{k_x^4 + k_y^4 + k_z^4}{k^4} - \frac{3}{5}\right]$,

$K_{6,1} \equiv \sqrt{\frac{639639}{32}}\left[\frac{k_x^2k_y^2k_z^2}{k^6} + \frac{1}{22}\left(\frac{k_x^4 + k_y^4 + k_z^4}{k^4} - \frac{3}{5}\right) - \frac{1}{105}\right]$, the coefficients are in eV, the values of k are $10\left(\frac{a}{2\pi}\right)$times those of k in atomic units (a is the lattice constant), $E_2^{(1)} = 1.0239620$, $E_2^{(2)} = 0$, $E_4^{(1)} = -1.1320772$, $E_4^{(2)} = 0.05658$, $E_6^{(1)} = 1.1072073$, $E_6^{(2)} = -0.1134024$, and $E_6^{(3)} = -0.0072275$.

(e) In addition to the well-known band models as discussed in this monograph, the conduction electrons of III–V semiconductors obey the following three dispersion relations:

1. In accordance with the model of Rossler [274]

$$E = \frac{\hbar^2k^2}{2m^*} + \bar{\alpha}_{10}k^4 + \bar{\beta}_{10}\left[k_x^2k_y^2 + k_y^2k_z^2 + k_z^2k_x^2\right] \pm \bar{\gamma}_{10}\left[k^2\left(k_x^2k_y^2 + k_y^2k_z^2 + k_z^2k_x^2\right) - 9k_x^2k_y^2k_z^2\right]^{1/2}, \tag{R1.7}$$

where $\bar{\alpha}_{10} = \bar{\alpha}_{11} + \bar{\alpha}_{12}k$, $\bar{\beta}_{10} = \bar{\beta}_{11} + \bar{\beta}_{12}k$, and $\bar{\gamma}_{10} = \bar{\gamma}_{11} + \bar{\gamma}_{12}k$, in which $\bar{\alpha}_{11} = -2132 \times 10^{-40}\,\text{eVm}^4$, $\bar{\alpha}_{12} = 9030 \times 10^{-50}\,\text{eVm}^5$, $\bar{\beta}_{11} = -2493 \times 10^{-40}\,\text{eVm}^4$, $\bar{\beta}_{12} = 12594 \times 10^{-50}\,\text{eVm}^5$, $\bar{\gamma}_{11} = 30 \times 10^{-30}\,\text{eVm}^3$, and $\bar{\gamma}_{12} = -154 \times 10^{-42}\,\text{eVm}^4$.

2. In accordance with Johnson and Dickey [275], the electron energy spectrum assumes the form

$$E = -\frac{E_g}{2} + \frac{\hbar^2 k^2}{2}\left[\frac{1}{m_0} + \frac{1}{m_{\gamma b}}\right] + \frac{E_g}{2}\left[1 + 4\frac{\hbar^2 k^2}{2m_c'}\frac{\bar{f}_1(E)}{E_g}\right]^{1/2}, \quad \text{(R1.8)}$$

where $\frac{m_0}{m_c'} \equiv P^2\left[\frac{\left(E_g + \frac{2\Delta}{3}\right)}{E_g(E_g+\Delta)}\right]$, $\bar{f}_1(E) \equiv \frac{(E_g+\Delta)\left(E+E_g+\frac{2\Delta}{3}\right)}{\left(E_g+\frac{2\Delta}{3}\right)(E+E_g+\Delta)}$,
$m_c' = 0.139 m_0$, and $m_{\gamma b} = \left[\frac{1}{m_c'} - \frac{2}{m_0}\right]^{-1}$.

3. In accordance with Agafonov et al. [276], the electron energy spectrum can be written as

$$E = \frac{\bar{\eta} - E_g}{2}\left[1 - \frac{\hbar^2 k^2}{2\bar{\eta} m^*}\left\{\frac{D\sqrt{3} - 3\overline{B}}{2\left(\frac{\hbar^2}{2m^*}\right)}\right\}\left[\frac{k_x^4 + k_y^4 + k_z^4}{k^4}\right]\right], \quad \text{(R1.9)}$$

where $\bar{\eta} \equiv (E_g^2 + (8/3)P^2k^2)^{1/2}$, $\overline{B} \equiv -21\frac{\hbar^2}{2m_0}$, and $D \equiv -40\left(\frac{\hbar^2}{2m_0}\right)$.

(f) The dispersion relation of the carriers in n-type $Pb_{1-x}Ga_xTe$ with $x = 0.01$ can be written following Vassilev [254] as

$$\left[E - 0.606k_s^2 - 0.0722k_z^2\right]\left[E + \overline{E}_g + 0.411k_s^2 + 0.0377k_z^2\right]$$
$$= 0.23k_s^2 + 0.02k_z^2 \pm \left[0.06\overline{E}_g + 0.061k_s^2 + 0.0066k_z^2\right]k_s \quad \text{(R1.10)}$$

where $\overline{E}_g (= 0.21\,\text{eV})$ is the energy gap for the transition point, the zero of the energy E is at the edge of the conduction band of the Γ point of the Brillouin zone and is measured positively upwards, k_x, k_y, and k_z are in the units of 10^9m^{-1}.

(g) The energy spectrum of the carriers in the two higher valance bands and the single lower valance band of Te can, respectively, be expressed as [277]

$$\overline{E} = A_{10}k_z^2 + B_{10}k_s^2 \pm \left[\Delta_{10}^2 + (\beta_{10}k_z)^2\right]^{1/2} \text{ and } \overline{E} = \Delta_{||} + A_{10}k_z^2 + B_{10}k_s^2 \pm \beta_{10}k_z, \quad \text{(R1.11)}$$

where $\overline{E}$ is the energy of the hole as measured from the top of the valance and within it, $A_{10} = 3.77 \times 10^{-19}\,\text{eVm}^2$, $B_{10} = 3.57 \times 10^{-19}\,\text{eVm}^2$, $\Delta_{10} = 0.628\,\text{eV}$, $(\beta_{10})^2 = 6 \times 10^{-20}\,(\text{eVm})^2$, and $\Delta_{||} = 1004 \times 10^{-5}\,\text{eV}$ are the spectrum constants.

(h) The dispersion relation for the electrons in graphite can be written following Brandt [252] as

$$E = \frac{1}{2}[E_2 + E_3] \pm \left[\frac{1}{4}(E_2 - E_3)^2 + \eta_2^2 k^2\right]^{1/2}, \quad \text{(R1.12)}$$

where $E_2 \equiv \bar{\Delta} - 2\bar{\gamma}_1 \cos\phi_0 + 2\bar{\gamma}_5 \cos^2\phi_0, \phi_0 \equiv c_6 k_z/2, E_3 \equiv 2\bar{\gamma}_2 \cos^2\phi_0$, and $\eta_2 \equiv \left(\frac{\sqrt{3}}{2}\right) a_6(\bar{\gamma}_0 + 2\bar{\gamma}_4 \cos\phi_0)$in which the band constants are $\bar{\Delta}$, $\bar{\gamma}_0$, $\bar{\gamma}_1, \bar{\gamma}_2, \bar{\gamma}_4, \bar{\gamma}_5, a_6$, and c_6, respectively.

(i) The dispersion relation of the conduction electrons in Antimony (Sb) in accordance with Ketterson [265] can be written as

$$2m_0 E = \alpha_{11} p_x^2 + \alpha_{22} p_y^2 + \alpha_{33} p_z^2 + 2\alpha_{23} p_y p_z \quad \text{(R1.13)}$$

and

$$2m_0 E = a_1 p_x^2 + a_2 p_y^2 + a_3 p_z^2 + a_4 p_y p_z \pm a_5 p_x p_z \pm a_6 p_x p_y, \quad \text{(R1.14)}$$

where $a_1 = \frac{1}{4}(\alpha_{11} + 3\alpha_{22}), a_2 = \frac{1}{4}(\alpha_{22} + 3\alpha_{11}\alpha_{22} + 3\alpha_{11}), a_3 = \alpha_{33}, a_4 = \alpha_{33}$, $a_5 = \sqrt{3}$, and $a_6 = \sqrt{3}(\alpha_{22} - \alpha_{11})$in which $\alpha_{11}, \alpha_{22}, \alpha_{33}$, and α_{23} are the system constants.

(j) The dispersion relation of the holes in p-InSb can be written in accordance with Cunningham [278] as

$$\overline{E} = c_4(1 + \gamma_4 f_4)k^2 \pm \frac{1}{3}[2\sqrt{2}\sqrt{c_4}\sqrt{16 + 5\gamma_4}\sqrt{E_4} g_4 k], \quad \text{(R1.15)}$$

where $c_4 \equiv \hbar^2/2m_0 + \theta_4, \theta_4 \equiv 4.7(\hbar^2/2m_0), \gamma_4 \equiv b_4/c_4, b_4 \equiv [3/2(b_5) + 2\theta_4]$, $b_5 \equiv 2.4(\hbar^2/2m_0)$, $f_4 \equiv \frac{1}{4}\left[\sin^2 2\theta + \sin^4\theta \sin^2 2\phi\right]$, θ is measured from the positive z-axis, ϕ is measured from positive x-axis, $g_4 \equiv \sin\theta \left[\cos^2\theta + \frac{1}{4} \sin^4\theta \sin^2 2\phi\right]$, and $E_4 = 5 \times 10^{-4}$ eV.

(k) The energy spectrum of the valance bands of CuCl in accordance with Yekimov et al. [279] can be written as

$$E_h = (\gamma_6 - 2\gamma_7)\frac{\hbar^2 k^2}{2m_0} \quad \text{(R1.16)}$$

and

$$E_{l,s} = (\gamma_6 + \gamma_7)\frac{\hbar^2 k^2}{2m_0} - \frac{\Delta_1}{2} \pm \left[\frac{\Delta_1^2}{4} + \gamma_7 \Delta_1 \frac{\hbar^2 k^2}{2m_0} + 9\left(\frac{\gamma_7 \hbar^2 k^2}{2m_0}\right)^2\right]^{1/2}, \quad \text{(R1.17)}$$

where $\gamma_6 = 0.53, \gamma_7 = 0.07$, and $\Delta_1 = 70$ meV.

(l) In the presence of stress, χ_6 along the $\langle 001\rangle$ and $\langle 111\rangle$ directions, the energy spectra of the holes in semiconductors having diamond structure valance bands can be respectively expressed following Roman [280] et al. as

$$E = A_6 k^2 \pm \left[\overline{B}_7^2 k^4 + \delta_6^2 + B_7\delta_6\left(2k_z^2 - k_s^2\right)\right]^{1/2} \tag{R1.18}$$

and

$$E = A_6 k^2 \pm \left[\overline{B}_7^2 k^4 + \delta_7^2 + \frac{D_6}{\sqrt{3}}\delta_7(2k_z^2 - k_s^2)\right]^{1/2}, \tag{R1.19}$$

where A_6, B_7, D_6, and C_6 are inverse mass band parameters in which $\delta_6 \equiv l_7\left(\overline{S}_{11} - \bar{S}_{12}\right)\chi_6$, $\overline{S}_{ij}$ are the usual elastic compliance constants, $\overline{B}_7^2 \equiv (B_7^2 + c_6^2/5)$, and $\delta_7 \equiv (d_8 S_{44}/2\sqrt{3})\chi_6$. For gray tin, $d_8 = -4.1\,\text{eV}$, $l_7 = -2.3\,\text{eV}$, $A_6 = 19.2(\hbar^2/2m_0)$, $B_7 = 26.3(\hbar^2/2m_0)$, $D_6 = 31(\hbar^2/2m_0)$, and $c_6^2 = -1112(\hbar^2/2m_0)$.

(m) The dispersion relation of the carriers of cadmium and zinc diphosphides are given by [257]

$$E = \left[\beta_1 + \frac{\beta_2\beta_3(k)}{8\beta_4}\right]k^2 \pm \left\{\left[\beta_4\beta_3(k)\left(\beta_5 - \frac{\beta_2\beta_3(k)}{8\beta_4}\right)k^2\right] + 8\beta_4^2\left(1 - \frac{\beta_3^2(k)}{4}\right) - \beta_2\left(1 - \frac{\beta_3^2(k)}{4}\right)k^2\right\}^{1/2} \tag{R1.20}$$

where β_1, β_2, β_4, and β_5 are system constants and $\beta_3(k) = k_x^2 + k_y^2 - 2k_z^2/k^2$.

(R1.3) Investigate the FNFE from quantum wells, wires, and dots of all the semiconductors as considered in R1.1 and R1.2, respectively.

(R1.4) Investigate the FNFE from bulk specimens of heavily doped semiconductors in the presence of Gaussian, exponential, Kane, Halperian, Lax, and Bonch-Burevich types of band tails [103, 104] for all systems whose unperturbed carrier energy spectra are defined in R1.1 and R1.2, respectively.

(R1.5) Investigate the FNFE from quantum wells, wires, and dots of all the heavily doped semiconductors as considered in R1.4.

(R1.6) Investigate the FNFE from bulk specimens of the negative refractive index, organic, magnetic, and other advanced optical materials in the presence of an arbitrarily oriented alternating electric field.

(R1.7) Investigate the FNFE from quantum wells, wires, and dots of the negative refractive index, organic, magnetic, and other advanced optical materials in the presence of an arbitrarily oriented alternating electric field.

(R1.8) Investigate the FNFE from the multiple quantum wells, wires, and dots of semiconductors whose unperturbed carrier energy spectra are defined in R1.1, R1.2, and heavily doped semiconductors in the presences of Gaussian, exponential, Kane, Halperian, Lax, and Bonch-Burevich types of band tails [103, 104] for all systems whose unperturbed carrier energy spectra are defined in the same problems respectively.

(R1.9) Investigate the FNFE from all the appropriate low-dimensional systems of this chapter in the presence of finite potential wells.

(R1.10) Investigate the FNFE from all the appropriate low-dimensional systems of this chapter in the presence of parabolic potential wells.

(R1.11) Investigate the FNFE from all the appropriate systems of this chapter forming quantum rings.

(R1.12) Investigate the FNFE from all the above appropriate problems in the presence of elliptical Hill and quantum square rings.

(R1.13) Investigate the FNFE for the appropriate accumulation layers for all the materials whose unperturbed carrier energy spectra are defined in R1.1and R1.2, respectively.

(R1.14) Investigate the FNFE from wedge shaped, cylindrical, ellipsoidal, conical, triangular, circular, parabolic rotational, and parabolic cylindrical quantum dots in the presence of an arbitrarily oriented alternating electric field for all the materials whose unperturbed carrier energy spectra are defined in R1.1 and R1.2, respectively.

(R1.15) Investigate the FNFE from wedge shaped, cylindrical, ellipsoidal, conical, triangular, circular, parabolic rotational, and parabolic cylindrical quantum dots of the negative refractive index, organic, magnetic, and other advanced optical materials in the presence of an arbitrarily oriented alternating electric field.

(R1.16) Formulate the time delay for all the appropriate systems of this chapter.

(R1.17) Formulate the reflection time for all the appropriate systems of this chapter.

(R1.18) Formulate the minimum tunneling, Dwell and Phase tunneling, Buttiker and Landauer, and intrinsic times for all the appropriate systems of this chapter.

(R1.19) Investigate all the appropriate problems of this chapter for a Dirac electron.

(R1.20) Investigate all the appropriate problems of this chapter by including the many body, image force, broadening, and hot carrier effects, respectively.

(R1.21) Investigate all the appropriate problems of this chapter by removing all the mathematical approximations and establishing the respective appropriate uniqueness conditions.

References

1. R.H. Fowler, L. Nordheim, Proc. Roy. Soc. London A **119**, 173 (1928)
2. A. Van Der Ziel, *Solid State Physical Electronics* (Prentice-Hall, Englewood Cliffs, 1957), p.176
3. B. Mitra, K.P. Ghatak, Phys. Lett. A **357**, 146 (1990)
4. B. Mitra, K.P. Ghatak, Phys. Lett. A **142A**, 401 (1989)
5. K.P. Ghatak, M. Mondal, J. Mag. Mag. Mat. **74**, 203 (1988)
6. K.P. Ghatak, B. Mitra, Phys. Lett.**156A**, 233 (1991)
7. K.P. Ghatak, A. Ghosal, S.N. Biswas, M. Mondal, Proc. SPIE USA **1308**, 356 (1990)
8. V.T. Binh, Ch. Adessi, Phys. Rev. Lett. **85**, 864 (2000)
9. R.G. Forbes, Ultramicroscopy **79**, 11 (1999)

10. J.W. Gadzuk, E.W. Plummer, Rev. Mod. Phys **45**, 487 (1973)
11. J.M. Beebe, B. Kim, J.W. Gadzuk, C.D. Frisbie, J.G. Kushmerick, Phys. Rev. Lett **97**, 026801 (1999)
12. Y. Feng, J.P. Verboncoeur, Phys. Plasmas **12,** 103301 (2005)
13. W.S. Koh, L.K. Ang, Nanotechnology **19**, 235402 (2008)
14. M. Razavy, *Quantum Theory of Tunneling* (World Scientific, Singapore, 2003)
15. S.I. Baranchuk, N.V. Mileshkina, Sov. Phys. Solid State **23**, 1715 (1981)
16. P.G. Borzyak, A.A. dadykin, Sov. Phys. Docklady **27**, 335(1982)
17. S. Bono, R.H. Good, Jr., Surface Sci. **134**, 272 (1983)
18. S.M. Lyth, S.R.P. Silva, Appl. Phys. Lett. **90**, 173124 (2007)
19. C. Xu, X. Sun, Int. J. Nanotechnol. **1**, 452 (2004)
20. S.D. Liang, L. Chen, Phys. Rev. Lett. **101**, 027602 (2008)
21. E.C. Heeres, E.P.A.M. Bakkers, A.L. Roest, M. Kaiser, T.H. Oosterkamp, N. de Jonge, Nano Lett. **7**, 536 (2007)
22. L. Dong, J. Jiao, D.W. Tuggle, J.M. Petty, S.A. Elliff, M. Coulter, Appl. Phys. Lett. **82**, 1096 (2003)
23. S.Y. Li, P. Lin, C.Y. Lee, T.Y. Tseng, J. Appl. Phys. **95**, 3711(2004)
24. N.N. Kulkarni, J. Bae, C.K. Shih, S.K. Stanley, S.S. Coffee, J.G. Ekerdt, Appl. Phys. Lett. **87**, 213115 (2005)
25. K Senthil, K. Yong, Mat. Chem. Phys. **112**, 88 (2008)
26. R. Zhou, H.C. Chang, V. Protasenko, M. Kuno, A.K. Singh, D. Jena, H. Xing, J. Appl. Phys. **101**, 073704 (2007)
27. K.S. Yeong, J.T.L. Thong, J. Appl. Phys. **100**, 114325 (2006)
28. C.H. Oon, S.H. Khong, C.B. Boothroyd, J.T.L. Thong, J. Appl. Phys. **99**, 064309 (2006)
29. . B.H. Kim, M.S. Kim, K.T. Park, J.K. Lee, D.H. Park, J. Joo, S.G. Yu, S.H. Lee, Appl. Phys. Lett. **83**, 539 (2003)
30. Z.S. Wu, S.Z. Deng, N.S. Xu, J. Chen, J. Zhou, J. Chen, Appl. Phys. Lett. **80**, 3829 (2002)
31. Y.W. Zhu, T. Yu, F.C. Cheong, X.J. Xu, C.T. Lim, V.B.C. Tan, J.T.L. Thong, C.H. Sow, Nanotechnology **16**, 88 (2005)
32. Y.W. Zhu, H.Z. Zhang, X.C. Sun, S.Q. Feng, J. Xu, Q. Zhao, B. Xiang, R.M. Wang, D.P. Yu, Appl. Phys. Lett. **83**, 144(2003)
33. S. Bhattacharjee, T. Chowdhury, Appl. Phys. Lett. **95**, 061501 (2009)
34. S. Kher, A. Dixit, D.N. Rawat, M.S. Sodha, Appl. Phys. Lett. **96**, 044101 (2010)
35. I. Shigeo, W. Teruo, O. Kazuyoshi, T. Masateru, U. Satoshi, N. Norio, J. Vac. Sci. Technol. B **13**, 487(2009)
36. C.A. Spindt, I. Brodie, L. Humphrey, E.R. Westerberg, J. Appl. Phys. **47**, 5248 (2009)
37. Q. Fu, A.V. Nurmikko, L.A. Kolodziejski, R.L. Gunshor, J.W. Wu, Appl. Phys. Lett. **51**, 578 (2009)
38. P.M. Petroff, A.C. Gossard, W. Wiegmann, Appl. Phys. Lett. **45**, 620 (1984)
39. J.M. Gaines, P.M. Petroff, H. Kroemar, R.J. Simes, R.S. Geels, J.H. English, J. Vac. Sci. Technol. **B 6**, 1378 (1988)
40. J. Cilbert, P.M. Petroff, G.J. Dolan, S.J. Pearton, A.C. Gossard, J.H. English, Appl. Phys. Lett. **49**, 1275 (1986)
41. T. Fujui, H. Saito, Appl. Phys. Lett. **50**, 824 (1987)
42. H. Sasaki, Jpn. J. Appl. Phys. **19**, 94 (1980)
43. P.M. Petroff, A.C. Gossard, R.A. Logan, W. Weigmann, Appl. Phys. Lett. **41** 635 (1982)
44. H. Temkin, G.J. Dolan, M.B. Panish, S.N.G. Chu, Appl. Phys. Lett. **50**, 413 (1988)
45. I. Miller, A. Miller, A. Shahar, U. Koren, P.J. Corvini, Appl. Phys. Lett. **54**, 188 (1989)
46. L.L. Chang, H. Esaki, C.A. Chang, L. Esaki, Phys. Rev. Lett. **38**, 1489 (1977)
47. K. Less, M.S. Shur, J.J. Drunnond, H. Morkoc, IEEE Trans. Electron. Devices **ED-30**, 07 (1983)
48. G. Bastard, *Wave Mechanics Applied to Semiconductor Heterostructures (*Halsted, Les Ulis, Les Editions de Physique, New York, 1988)

49. M.J. Kelly, *Low Dimensional Semiconductors: Materials, Physics, Technology, Devices* (Oxford University Press, Oxford, 1995)
50. C. Weisbuch, B. Vinter, *Quantum Semiconductor Structures* (Boston Academic Press, Boston, 1991)
51. N.T. Linch, Festkorperprobleme,**23**, 27 (1985)
52. D.R. Sciferes, C. Lindstrom, R.D. Burnham, W. Streifer, T.L. Paoli, Electron. Lett. **19**, 170 (1983)
53. P.M. Solomon, Proc. IEEE, **70**, 489 (1982)
54. T.E. Schlesinger, T. Kuech, Appl. Phys. Lett. **49**, 519 (1986)
55. D. Kasemet, C.S. Hong, N.B. Patel, P.D. Dapkus, Appl. Phys. Lett. **41**, 912 (1982)
56. K. Woodbridge, P. Blood, E.D. Pletcher, P.J. Hulyer, Appl. Phys. Lett. **45**, 16 (1984)
57. S. Tarucha, H.O. Okamoto, Appl. Phys. Lett. **45**, 16 (1984)
58. H. Heiblum, D.C. Thomas, C.M. Knoedler, M.I. Nathan, Appl. Phys. Lett. **47**, 1105 (1985)
59. O. Aina, M. Mattingly, F.Y. Juan, P.K. Bhattacharyya, Appl. Phys. Lett. **50**, 43 (1987)
60. I. Suemune, L.A. Coldren, IEEE J. Quant. Electronic. **24**, 1178 (1988)
61. D.A.B. Miller, D.S. Chemla, T.C. Damen, J.H. Wood, A.C. Burrus, A.C. Gossard, W. Weigmann, IEEE J. Quant. Electron. **21**, 1462 (1985)
62. P. Harrison, *Quantum Wells, Wires and Dots* (John Wiley and Sons, Ltd, 2002)
63. B.K. Ridley, *Electrons and Phonons in Semiconductors Multilayers* (Cambridge University Press, Cambridge, 1997)
64. V.V. Martin, A.A. Kochelap, M.A. Stroscio, *Quantum Heterostructures* (Cambridge University Press, Cambridge, 1999)
65. C.S. Lent, D.J. Kirkner, J. Appl. Phys. **67**, 6353 (1990)
66. F. Sols, M. Macucci, U. Ravaioli, K. Hess, Appl. Phys. Lett. **54**, 350 (1980)
67. C.S. Kim, A.M. Satanin, Y.S. Joe, R.M. Cosby, Phys. Rev. B, **60**, 10962 (1999)
68. S. Midgley, J.B. Wang, Phys. Rev. B **64**, 153304 (2001)
69. T. Sugaya, J.P. Bird, M. Ogura, Y. Sugiyama, D.K. Ferry, K.Y. Jang, Appl. Phys. Lett. **80**, 434 (2002)
70. B. Kane, G. Facer, A. Dzurak, N. Lumpkin, R. Clark, L. PfeiKer, K. West, Appl. Phys. Lett. **72**, 3506 (1998)
71. C. Dekker, Phys. Today **52**, 22 (1999)
72. A. Yacoby, H.L. Stormer, N.S. Wingreen, L.N. Pfeiffer, K.W. Baldwin, K.W. West, Phys. Rev. Lett. **77**, 4612 (1996)
73. Y. Hayamizu, M. Yoshita, S. Watanabe, H.A.L. PfeiKer, K. West, Appl. Phys. Lett. **81**, 4937 (2002)
74. S. Frank, P. Poncharal, Z.L. Wang, W.A. Heer, Science **280**, 1744 (1998)
75. I. Kamiya, I.I. Tanaka, K. Tanaka, F. Yamada, Y. Shinozuka, H. Sakaki, Physica E **13**, 131 (2002)
76. A.K. Geim, P.C. Main, N. LaScala, L. Eaves, T.J. Foster, P.H. Beton, J.W. Sakai, F.W. Sheard, M. Henini, G. Hill et al., Phys. Rev. Lett. **72**, 2061 (1994)
77. A.S. Melinkov, V.M. Vinokur, Nature **415**, 60 (2002)
78. K. Schwab, E.A. Henriksen, J.M. Worlock, M.L. Roukes, Nature **404**, 974 (2000)
79. L. Kouwenhoven, Nature **403**, 374 (2000)
80. S. Komiyama, O. Astafiev, V. Antonov, H. Hirai, Nature **403**, 405 (2000)
81. E. Paspalakis, Z. Kis, E. Voutsinas, A.F. Terziz, Phys. Rev. B **69**, 155316 (2004)
82. J.H. Jefferson, M. Fearn, D.L.J. Tipton, T.P. Spiller, Phys. Rev. A **66**, 042328 (2002)
83. J. Appenzeller, C. Schroer, T. Schapers, A. Hart, A. Froster, B. Lengeler, H. Luth, Phys. Rev. B **53**, 9959 (1996)
84. J. Appenzeller, C. Schroer, J. Appl. Phys. **87**, 31659 (2002)
85. P. Debray, O.E. Raichev, M. Rahman, R. Akis, W.C. Mitchel, Appl. Phys. Lett. **74**, 768 (1999)
86. D. Kasemet, C.S. Hong, N.B. Patel, P.D. Dapkus, Appl. Phys. Lett.**41**, 912 (1982)
87. K. Woodbridge, P. Blood, E.D. Pletcher, P.J. Hulyer, Appl. Phys. Lett. **45**, 16 (1984)
88. S. Tarucha, H.O. Okamoto, Appl. Phys. Lett. **45**, 16 (1984)
89. M.I. Nathan, Appl. Phys. Lett. **47**, 1105 (1985)

90. I. Suemune, L.A. Coldren, IEEE J. Quant. Electronic. **24**, 1178 (1988)
91. J.L. Shay, J.W. Wernick, *Ternary Chalcopyrite Semiconductors – Growth, Electronic Properties and Applications* (Pergamon, London, 1975)
92. J.W. Rowe, J.L. Shay, Phys. Rev. B, **3,** 451 (1973)
93. H. Kildal, Phys. Rev. B, **10,** 5082 (1974)
94. J. Bodnar, in Proc. Int. Conf. of the Physics of Narrow-gap Semiconductors (Polish Science Publishers, Warsaw, 1978)
95. G.P. Chuiko, N.N. Chuiko, Sov. Phys. Semicond. **15**, 739 (1981)
96. K.P. Ghatak, S.N. Biswas, Proc. of SPIE, **1484**, 149 (1991)
97. A. Rogalski, J. Alloys Comp. **371**, 53 (2004)
98. A. Baumgartner, A. Chaggar, A. Patanè, L. Eaves, M. Henini, Appl. Phys. Lett. **92**, 091121 (2008)
99. J. Devenson, R. Teissier, O. Cathabard, A.N. Baranov, Proc. SPIE **6909**, 69090U (2008)
100. B.S. Passmore, J. Wu, M.O. Manasreh, G.J. Salamo, Appl. Phys. Lett. **91**, 233508 (2007)
101. M. Mikhailova, N. Stoyanov, I. Andreev, B. Zhurtanov, S. Kizhaev, E. Kunitsyna, K. Salikhov, Y. Yakovlev, Proc. SPIE **6585**, 658526 (2007)
102. W. Kruppa, J.B. Boos, B.R. Bennett, N.A. Papanicolaou, D. Park, R. Bass, Electron. Lett. **42**, 688 (2006)
103. B.R. Nag, *Electron Transport in Compound Semiconductors* (Springer Verlag, 1980)
104. E.O. Kane, in *Semiconductors and Semimetals*, Vol. 1, Ed. By R.K. Willardson, A.C. Beer (Academic Press, New York, 1966), p. 75
105. G.E. Stillman, C.M. Wolfe, J.O. Dimmock in *Semiconductors and Semimetals*, Ed. R.K. Willardon, A.C. Beer 12 (Academic, New York, 1977), p. 169
106. D.J. Newson, A. Karobe, Semiconductor. Sci. Technol. **3**, 786 (1988)
107. E.D. Palik, G.S. Picus, S. Teither, R.E. Wallis, Phys. Rev. **122**, 475 (1961)
108. P.Y. Lu, C.H. Wung, C.M. Williams, S.N.G. Chu, C.M. Stiles, Appl. Phys. Lett. **49** (1986) 1372
109. N.R. Taskar, I.B. Bhat, K.K. Prat, D. Terry, H. Ehasani, S.K. Ghandhi, J. Vac. Sci. Technol. **7A** (1989) 281
110. F. Koch, *Springer Series in Solid States Sciences*, vol. 53 (Springer-Verlag, Berlin, 1984), pp. 20
111. L.R. Tomasetta, H.D. Law, R.C. Eden, I. Reyhimy, K. Nakano, IEEE J. Quant. Electron. **14** (1978) 800
112. T. Yamato, K. Sakai, S. Akiba, Y. Suematsu, IEEE J. Quantum Electronics **14** (1978) 95
113. T.P. Pearsall, B.I. Miller, R.J. Capik, Appl. Phys. Lett. **28** (1976) 499
114. M.A. Washington, R.E. Nahory, M.A. Pollack, E.D. Beeke, Appl. Phys. Lett. **33** (1978) 854
115. M.I. Timmons, S.M. Bedair, R.J. Markunas, J.A. Hutchby, *Proceedings of the 16th IEEE Photovoltaic Specialist Conference* (IEEE, San Diego, California 666, 1982)
116. J.A. Zapien, Y.K. Liu, Y.Y. Shan, H. Tang, C.S. Lee, S.T. Lee, Appl. Phys. Lett. **90**, 213114 (2007)
117. M. Park, Proc. SPIE **2524**, 142 (1995)
118. S.-G. Hur, E.T. -Kim, J.H. -Lee, G.H. -Kim, S.G. -Yoon, Electrochem. Solid-State Lett. **11**, H176 (2008)
119. H. Kroemer, Rev. of Mod. Phys. **73**, 783 (2001)
120. T. Nguyen Duy, J. Meslage, G. Pichard, J. Crys. Growth, **72**, 490 (1985)
121. T. Aramoto, F. Adurodija, Y. Nishiyama, T. Arita, A. Hanafusa, K. Omura, A. Morita, Solar Energy Mat. Solar Cells, **75**, 211 (2003)
122. H.B. Barber, J. Elect. Mat. **25**, 1232 (1996)
123. S. Taniguchi, T. Hino, S. Itoh, K. Nakano, N. Nakayama, A. Ishibashi, M. Ikeda, Elect. Lett. **32**, 552 (1996)
124. J.J. Hopfield, J. Appl. Phys. **32,** 2277 (1961)
125. R.V. Belosludov, A.A. Farajian, H. Mizuseki, K. Miki, Y. Kawazoe, Phys. Rev. B, **75**, 113411 (2007)

126. J. Heremans, C.M. Thrush, Y.-M. Lin, S. Cronin, Z. Zhang, M.S. Dresselhaus, J.F. Mansfield, Phys. Rev. B **61**, 2921 (2000)
127. D. Shoenberg, Proc. Roy. Soc. (London) **170**, 341 (1939)
128. B. Abeles, S. Meiboom, Phys. Rev. **101**, 544 (1956)
129. B. Lax, J.G. Mavroides, H.J. Zieger, R.J. Keyes, Phys. Rev. Lett. **5**, 241 (1960)
130. Y.-H. Kao, Phys. Rev. **129,** 1122 (1963)
131. R.J. Dinger, A.W. Lawson, Phys. Rev. B, **3**, 253 (1971)
132. J.F. Koch, J.D. Jensen, Phys. Rev. **184**, 643 (1969)
133. M.H. Cohen Phys. Rev. **121**, 387 (1961)
134. S. Takaoka, H. Kawamura, K. Murase, S. Takano, Phys. Rev. B **13**, 1428 (1976)
135. J.W. McClure, K.H. Choi, Solid State Commun. **21**, 1015 (1977)
136. G.P. Agrawal, N.K. Dutta, *Semiconductor Lasers* (Van Nostrand Reinhold, New York, 1993), p. 547
137. S. Chatterjee, U. Pal, Opt. Eng., Bellingham, **32**, 2923 (1993)
138. T.K. Chaudhuri, Int. J. Energy Res. **16**, 481 (1992)
139. J.H. Dughaish, Physica B **322,** 205 (2002)
140. C. Wood, Rep. Prog. Phys. **51**, 459 (1988)
141. K.-F. Hsu, S. Loo, F. Guo, W. Chen, J.S. Dyck, C. Uher, T. Hogan, E.K. Polychroniadis, M.G. Kanatzidis, Science **303**, 818 (2004)
142. J. Androulakis, K.F. Hsu, R. Pcionek, H. Kong, C. Uher, J.J. D'Angelo, A. Downey, T. Hogan, M.G. Kanatzidis, Adv. Mater., Weinheim, Ger. **18**, 1170 (2006)
143. P.F.P. Poudeu, J. D'Angelo, A.D. Downey, J.L. Short, T.P. Hogan, M.G. Kanatzidis, Angew. Chem., Int. Ed. **45**, 3835 (2006)
144. P.F. Poudeu, J. D'Angelo, H. Kong, A. Downey, J.L. Short, R. Pcionek, T.P. Hogan, C. Uher, M.G. Kanatzidis, J. Am. Chem. Soc. **128**, 14347 (2006)
145. J.R. Sootsman, R.J. Pcionek, H. Kong, C. Uher, M.G. Kanatzidis, Chem. Mater. **18**, 4993 (2006)
146. A.J. Mountvala, G. Abowitz, J. Am. Ceram. Soc**. 48**, 651 (1965)
147. E.I. Rogacheva, I.M. Krivulkin, O.N. Nashchekina, A. Yu. Sipatov, V.A. Volobuev, M.S. Dresselhaus, Appl. Phys. Lett. **78**, 3238 (2001)
148. H.S. Lee, B. Cheong, T.S. Lee, K.S. Lee, W.M. Kim, J.W. Lee, S.H. Cho, J.Y. Huh, Appl. Phys. Lett. **85**, 2782 (2004)
149. K. Kishimoto, M. Tsukamoto, T. Koyanagi, J. Appl. Phys. **92**, 5331 (2002)
150. E.I. Rogacheva, O.N. Nashchekina, S.N. Grigorov, M.A. Us, M.S. Dresselhaus, S.B. Cronin, Nanotechnology **14**, 53 (2003)
151. E.I. Rogacheva, O.N. Nashchekina, A.V. Meriuts, S.G. Lyubchenko, M.S. Dresselhaus, G. Dresselhaus, Appl. Phys. Lett. **86**, 063103 (2005)
152. E.I. Rogacheva, S.N. Grigorov, O.N. Nashchekina, T.V. Tavrina, S.G. Lyubchenko, A.Yu. Sipatov, V.V. Volobuev, A.G. Fedorov, M.S. Dresselhaus, Thin Solid Films **493**, 41 (2005).
153. X. Qiu, Y. Lou, A.C.S. Samia, A. Devadoss, J.D. Burgess, S. Dayal, C. Burda, Angew. Chem., Int. Ed**. 44,** 5855 (2005)
154. C. Wang, G. Zhang, S. Fan, Y. Li, J. Phys. Chem. Solids **62**, 1957 (2001)
155. B. Poudel, W.Z. Wang, D.Z. Wang, J.Y. Huang, Z.F. Ren, J. Nanosci. Nanotechnol. **6**, 1050 (2006)
156. B. Zhang, J. He, T.M. Tritt, Appl. Phys. Lett. **88**, 043119 (2006)
157. W. Heiss, H. Groiss, E. Kaufmann, G. Hesser, M. Böberl, G. Springholz, F. Schäffler, K. Koike, H. Harada, M. Yano, Appl. Phys. Lett.**88**, 192109 (2006)
158. B.A. Akimov, V.A. Bogoyavlenskiy, L.I. Ryabova, V.N. Vasil'kov, Phys. Rev. B **61**, 16045 (2000)
159. Ya. A. Ugai, A.M. Samoilov, M.K. Sharov, O.B. Yatsenko, B.A. Akimov, Inorg. Mater. **38**, 12 (2002)
160. Ya. A. Ugai, A.M. Samoilov, S.A. Buchnev, Yu.V. Synorov, M.K. Sharov, Inorg. Mater. **38**, 450 (2002)

161. A.M. Samoilov, S.A. Buchnev, Yu.V. Synorov, B.L. Agapov, A.M. Khoviv, Inorg. Mater. **39**, 1132 (2003)
162. A.M. Samoilov, S.A. Buchnev, E.A. Dolgopolova, Yu.V. Synorov, A.M. Khoviv, Inorg. Mater. **40**, 349 (2004)
163. H. Murakami, W. Hattori, R. Aoki, Phys. C **269**, 83 (1996)
164. H. Murakami, W. Hattori, Y. Mizomata, R. Aoki, Phys. C **273**, 41 (1996)
165. H. Murakami, R. Aoki, K. Sakai, Thin Solid Films, **27**, 343 (1999)
166. B.A. Volkov, L.I. Ryabova, D.R. Khokhlov, Phys. Usp. **45**, 819 (2002), and references therein
167. F. Hüe, M. Hÿtch, H. Bender, F. Houdellier, A. Claverie, Phys. Rev. Lett. **100**, 156602 (2008)
168. S. Banerjee, K.A. Shore, C.J. Mitchell, J.L. Sly, M. Missous, IEEE Proc., Circuits Devices Syst. **152**, 497 (2005)
169. M. Razeghi, A. Evans, S. Slivken, J.S. Yu, J.G. Zheng, V.P. Dravid, Proc. SPIE, **5840**, 54 (2005)
170. R.A. Stradling, Semicond. Sci. Technol. **6**,*C*52 (1991)
171. P.K. Weimer, Proc. IEEE, **52**, 608 (1964)
172. G. Ribakovs, A.A. Gundjian, IEEE, J. Quant. Electron. **QE-14**, 42 (1978)
173. S.K. Dey, J. Vac. Sc. Tech. **10**, 227 (1973)
174. S.J. Lynch, Thin Solid Films, **102**, 47 (1983)
175. V.V. Kudzin, V.S. Kulakov, D.R. Pape', S.V. Kulakov, V.V. Molotok, IEEE, Ultrasonics Symposium, **1**, 749 (1997)
176. F. Hatami, V. Lordi, J.S. Harris, H. Kostial, W.T. Masselink, J. Appl. Phys. **97**, 096106 (2005)
177. B.W. Wessels, J. Electrochem. Soc. **122**, 402 (1975)
178. D.W.L. Tolfree, J. Sci. Instrum. **41**, 788, (1964)
179. P.B. Hart, Proc. the IEEE, **61**, 880, 1973
180. M.A. Hines, G.D. Scholes, Adv. Mater. **15**, 1844 (2003)
181. C.A. Wang, R.K. Huang, D.A. Shiau, M.K. Connors, P.G. Murphy, P.W. O'Brien, A.C. Anderson, D.M. DePoy, G. Nichols, M.N. Palmisiano, Appl. Phys. Lett, **83,** 1286 (2003)
182. C.W. Hitchcock, R.J. Gutmann, J.M. Borrego, I.B. Bhat, G.W. Charache, IEEE Trans. Electron. Devices, **46**, 2154 (1999)
183. H.J. Goldsmid, R.W. Douglas, Brit. J. Appl. Phys. **5**, 386 (1954)
184. F.D. Rosi, B. Abeles, R.V. Jensen, J. Phys. Chem. Sol. **10**, 191 (1959)
185. T.M. Tritt (ed.), *Semiconductors and Semimetals, Vols. 69, 70 and 71: Recent Trends in Thermoelectric Materials Research I, II and III* (Academic, New York, 2000)
186. D.M. Rowe (Ed.), *CRC Handbook of Thermoelectrics* (CRC, New York, 1995)
187. D.M. Rowe, C.M. Bhandari, *Modern Thermoelectrics* (Reston, Virginia, 1983)
188. D.M. Rowe (ed.), *Thermoelectrics Handbook: Macro to Nano* (CRC, New York, 2006)
189. H. Choi, M. Chang, M. Jo, S.J. Jung, H. Hwang, Electrochem. Solid-State Lett. **11**, H154 (2008)
190. S. Cova, M. Ghioni, A. Lacaita, C. Samori, F. Zappa, Appl. Opt. **35**, 1956 (1996)
191. H.W.H. Lee, B.R. Taylor, S.M. Kauzlarich, *Nonlinear Optics: Materials, Fundamentals, and Applications* (Technical Digest, 12, 2000)
192. E. Brundermann, U. Heugen, A. Bergner, R. Schiwon, G.W. Schwaab, S. Ebbinghaus, D.R. Chamberlin, E.E. Haller, M. Havenith, *29th International Conference on Infrared and Millimeter Waves and 12th International Conference on Terahertz Electronics*, 283 (2004)
193. A.N. Baranov, T.I. Voronina, N.S. Zimogorova, L.M. Kauskaya, Y.P. Yakoviev, Sov. Phys. Semicond. **19**, 1676 (1985)
194. M. Yano, Y. Suzuki, T. Ishii, Y. Matsushima, M. Kimata, Jpn. J. Appl. Phys. **17**, 2091 (1978)
195. F.S. Yuang, Y.K. Su, N.Y. Li, Jpn. J. Appl. Phys. **30**, 207 (1991)
196. F.S. Yuang, Y.K. Su, N.Y. Li, K.J. Gan, J. Appl. Phys. **68**, 6383 (1990)
197. Y.K. Su, S.M. Chen, J. Appl. Phys. **73**, 8349 (1993)
198. S.K. Haywood, A.B. Henriques, N.J. Mason, R.J. Nicholas, P.J. Walker, Semicond. Sci. Technol. **3**, 315 (1988)
199. C. Young, W.W. Anderson, L.B. Anderson, Trans. Electron Dev., IEEE, **24**, 492 (1977)
200. R.L. Gordon, V.I. Neeley, H.R. Curtin, Proc. IEEE, **54**, 2014 (1966)

201. P.K. Weimer, Proc. IRE, **50**, 1462 (1962)
202. M.J. Lee, S.W. Wright, C.P. Judge, P.Y. Cheung, Display Research Conference, International Conference Record, 211 (1991)
203. M. Abramowitz, I.A. Stegun, *Handbook of Mathematical Functions* (Dover, New York, 1965)
204. A. Haung, *Theoretical Solid State Physics*, vol. 1 (Pergamon, Oxford)
205. S. Flugge, *Practical Quantum Mechanics* (Springer Verlag, Berlin,1974)
206. M. Razavy, *Quantum Theory of Tunneling* (World Scientific, Singapore 2003)
207. J.O. Dimmock in *The Physics of Semimetals and Narrow Gap Semiconductors* Ed. by D.L. Carter, R.T. Bates (Pergamon, Oxford, 1971)
208. D.G. Seiler, B.D. Bajaj, A.E. Stephens, Phys. Rev. B **16**, 2822 (1977)
209. A.V. Germaneko, G.M. Minkov, Phys. Stat. Sol. (b) **184**, 9 (1994)
210. G.L. Bir, G.E. Pikus, *Symmetry and Strain – Induced Effects in Semiconductors* (Nauka, 1972) (in Russian)
211. M. Mondal, K.P. Ghatak, Phys. Stat. Sol. (b) **135**, K21 (1986)
212. J. Bouat, J.C. Thuillier, Surface Science, **73**, 528 (1978)
213. G.J. Rees, Phys. of Compounds, Proc. of the 13th Inter. Nat. Conf. Ed. F.G. Fumi, pp. 1166, North-Holland Company (1976)
214. P.R Emtage, Phys. Rev. **138**, A246 (1965)
215. M. Stordeur, W. Kuhnberger, Phys. Stat. sol. (b), **69**, 377(1975)
216. D.R. Lovett, *Semimetals and Narrow-Bandgap Semiconductor* (Pion, London, 1977)
217. H. Kohler, Phys. Stat. Sol.(b), **74**, 591(1976)
218. M. Cardona, W. Paul, H. Brooks Helv, Acta Phys. **33**, 329 (1960)
219. A.F. Gibson in *"Proceeding of International school of physics "ENRICO FERMI" course XIII*, Ed. R.A Smith, Academic Press, 171 (1963)
220. C.C. Wang, N.W. Ressler, Phys. Rev. **2**, 1827 (1970)
221. P.C Mathur, S. Jain, Phys. Rev. **19**, 1359(1979)
222. M. Singh, P.R. Wallace, S.D. Jog, E. Arushanov, J. Phys. Chem. Solids, **45**, 409 (1984)
223. Y. Yamada, J. Phys. Soc. Japan, **35**, 1600(1973), **37**, 606(1974)
224. J. Cisowski, J.C Portal, E.K Erushanov, J.M Broto, S. Huant, L.C Brumel, Phys.Stat.Sol(b), **121**, 289 (1984)
225. E.A. Arushanov, A.A. Kaynzev, A.N. Natepov, S.I. Radautsan, Sov. Phys. Semicond. **15**, 828 (1981)
226. K.S. Hong, R.F. Speyer, R.A. Condrate, J. Phys. Chem. Solids, **51**, 969 (1990)
227. M. Mondal, S. Banik, K. P. Ghatak, J. Low. Temp. Phys**. 74**, 423 (1989)
228. M.J. Gelten, C.V.M. VanEs, F.A.P. Blom, J.W.F. Jongencelen, Solid State Commun. **33**, 833 (1980)
229. A.A. El-Shazly, H.S. Soliman, H.E.A. El-Sayed, D.A.A. El-Hady, J. Vac. **47**, 53 (1996)
230. S. Adachi, *Properties of Group-IV, III–V and II–VI Semiconductors* (Wiley, New York, 2005)
231. O. Madelung, *Semiconductors: Data Handbook*, 3rd ed. (Springer-Verlag, Berlin, 2003)
232. S. Adachi, J. Appl. Phys. **58**, R1 (1985)
233. S. Adachi, *GaAs and Related Materials: Bulk Semiconductors and Superlattice Properties* (World Scientific, New York, 1994)
234. G.L. Hansen, J.L. Schmit, T.N. Casselman, J. Appl. Phys. **63**, 7079 (1982)
235. J. Wenus, J. Rutkowski, A. Rogalski, IEEE Trans. Elect. Dev. **48**, 1326 (2001)
236. S. Adachi, J. Appl. Phys**. 53**, 8775 (1982)
237. S.K. Sutradhar, D. Chattopadhyay, B.R. Nag, Phys. Rev. (b) **25**, 4069(1982)
238. D.G. Seiler, W.M. Beeker, L.M. Roth, Phys. Rev. **1**, 764 (1970)
239. I.V. Skryabinskii, Yu. I. Ukhanov, Sov. Phys. Solid State, **14**, 2838 (1973)
240. S. Tiwari, S. Tiwari, Cryst. Res. Technol. **41**, 78 (2006)
241. J.R. Lowney, S.D. Senturia, J. Appl. Phys. **47**, 1771 (1976)
242. W.E. Spicer, G.J. Lapeyre Phys. Rev. **139**, A565 (1965)
243. G.M.T. Foley, P.N. Langenberg, Phys. Rev. B, **15**, 4850 (1977)
244. D.R. Lovett, In: *Semimetals and Narrow Band Gap Semiconductors* (Pion, London, 1977)
245. C.C. Wu, C.J. Lin, J. Low. Temp. Phys. **57**, 469 (1984)

246. J. Rose, R. Schuchardt, Phys. Stat. Sol.(b), **117**, 213 (1983)
247. K.P Ghatak, S. Bhattachaya, D. De, *Einstein Relation in Compound Semiconductors and Nanostructures* (Springer Series in Material Sciences, **116**, 2009)
248. V.I. Ivanov-Omskii, A.Sh. Mekhtisev, S.A. Rustambekova, E.N. Ukraintsev, Phys. Stat. Sol. (b) **119**, 159 (1983)
249. H. Kim, K. Cho, H. Song, B. Min, J. Lee, G. Kim, S. Kim, S.H. Kim, T. Noh, Appl. Phys. Lett. **83**, 4619 (2003)
250. R.A. Reynolds, M.J. Brau, R.A. Chapman, J. Phys. Chem. Solids **29**, 755 (1968)
251. J. O'Shaughnessy, C. Smith, Solid State Commun. **8**, 481 (1970)
252. N.B. Brandt, V.N. Davydov, V.A. Kulbachinskii, O.M. Nikitina, Sov. Phys. Sol. Stat. **29**, 1014 (1987)
253. L.M. Viculis, J.J. Mack, O.M. Mayer, H.T. Hahn, R.B. Kaner, J. Mater. Chem. **15**, 974 (2005)
254. L.A. Vassilev, Phys. Stat. Sol. (b), **121**, 203 (1984)
255. D.L. Partin, Superlattices and Microstructures, **1**, 131 (1985)
256. W.J. Turner, A.S. Fischler, W.E. Reese, Phys. Rev. **121**, 759 (1961)
257. G.P. Chuiko, Sov. Phys. Semicond. **19(12)**, 1381 (1985)252
258. W.E. Swank, P.G. Le Comber, Phys. Rev. **153**, 844 (1967)
259. D. Haneman, J. Phys. Chem. Solids, **11**, 205 (1959)
260. N. Wei, G. Wu, J. Dong, Phys. Lett. A **325**, 403 (2004)
261. S. Reich, J. Maultzsch, C. Thomsen, P. Ordejo'n, Phys. Rev. B, **66**, 035412 (2006)
262. M.S. Lundstrom, J. Guo, *Nanoscale Transistors: Device Physics, Modeling and Simulation* (Springer, New York, 2006)
263. J.W. Mintmire, C.T. White, Phys. Rev. Lett. **81**, 2506 (1998)
264. F. Buonocore, F. Trani, D. Ninno, A. Di Matteo, G. Cantele, G. Iadonisi, Nanotechnology, **19**, 025711 (2008)
265. J.B. Ketterson, Phys. Rev. **129**, 18 (1963)
266. R.C. Vilão, J.M. Gil, A. Weidinger, H.V. Alberto, J. Piroto Duarte, N.A. de Campos, R.L. Lichti, K.H. Chow, S.P. Cottrell, S.F.J. Cox, Phys. Rev. B, **77**, 235212 (2008)
267. I. Kang, F.W. Wise, Phys. Rev. B, J. Opt. Soc. Am. B, **14**, 1632 (1997)
268. D. Cui, J. Xu,S.-Y. Xu, G. Paradee, B.A. Lewis, M.D. Gerhold, IEEE Trans. Elect. Dev. **5**, 362 (2006)
269. E.L. Ivchenko, G.E. Pikus. Sov. Phys. Semicond. **13**, 579 (1979)
270. E. Bangert, P. Kastner, Phys. Stat. Sol (b) **61**, 503(1974)
271. G.M.T. Foley, P.N Langenberg, Phys. Rev. B, **15B**, 4850 (1977)
272. V.I. Ivanov-Omskii, A.Sh. Mekhtisev, S.A. Rustambekova, E.N. Ukraintsev, Phys. Stat. Sol. (b) **119**, 159 (1983)
273. H.I. Zhang, Phys. Rev. B, **1**, 3450 (1970)
274. U. Rossler, Solid State Commun. **49**, 943 (1984)
275. J. Johnson, D.H. Dickey, Phys. Rev. **1**, 2676 (1970)
276. V.G. Agafonov, P.M. Valov, B.S. Ryvkin, I.D. Yarashetskin, Sov. Phys. Semiconduct. **12**, 1182 (1978)
277. N.S. Averkiev, V.M. Asnin, A.A. Bakun, A.M. Danishevskii, E.L. Ivchenko, G.E. Pikus, A.A. Rogachev, Sov. Phys. Semicond. **18**, 379 and 402 (1984)
278. R.W. Cunningham,Phys. Rev. **167**, 761(1968)
279. A.I. Yekimov, A.A. Onushchenko, A.G. Plyukhin, AI.L. Efros, J. Expt. Theor. Phys. **88**, 1490 (1985)
280. B.J. Roman, A.W. Ewald, Phys. Rev. B. **5**, 3914 (1972)

Chapter 2
Field Emission from Quantum Wire Superlattices of Non-parabolic Semiconductors

2.1 Introduction

In recent years, modern fabrication techniques have generated altogether a new dimension in the arena of quantum effect devices through the experimental realization of an important artificial structure known as semiconductor superlattice (SL) by growing two similar but different semiconducting materials in alternate layers with finite thicknesses. The materials forming the alternate layers have the same kind of band structure but different energy gaps. The concept of SL was developed for the first time by Keldysh [1] and was successfully fabricated by Esaki and Tsu [2–5]. The SLs are being extensively used in thermal sensors [6,7], quantum cascade lasers [8–10], photodetectors [11, 12], light-emitting diodes [13–16], multiplication [17], frequency multiplication [18], photocathodes [19,20], thin-film transistor [21], solar cells [22,23], infrared imaging [24], thermal imaging [25,26], infrared sensing [27], and also in other microelectronic devices.

The most extensively studied III–V SL is the one consisting of alternate layers of GaAs and $Ga_{1-x}Al_x$. As owing to the relative easiness of fabrication. The GaAs and $Ga_{1-x}Al_x$. As layers form the quantum wells and the potential barriers, respectively. The III–V SLs are attractive for the realization of high-speed electronic and optoelectronic devices [28]. In addition to SLs with usual structure, other types of SLs such as II–VI [29], IV–VI [30], and HgTe/CdTe [31] have also been investigated in the literature. The IV–VI SLs exhibit quite different properties as compared to the III–V SL due to the specific band structure of the constituent materials [32]. The epitaxial growth of II–VI SL is a relatively recent development and the primary motivation for studying the mentioned SLs made of materials with the large band gap is in their potential for optoelectronic operation in the blue [32]. HgTe/CdTe SLs have raised a great deal of attention since 1979 as promising new materials for long wavelength infrared detectors and other electro-optical applications [33]. Interest in Hg-based SLs has been further increased as new properties with potential device applications were revealed [33, 34]. These features arise from the unique zero-band gap material HgTe [35] and the direct band gap semiconductor CdTe,

S. Bhattacharya and K.P. Ghatak, *Fowler–Nordheim Field Emission*, Springer Series in Solid-State Sciences 170, DOI 10.1007/978-3-642-20493-7_2,

which can be described by the three-band mode of Kane [36]. The combination of the aforementioned materials with specified dispersion relation makes HgTe/CdTe SL very attractive, especially because of the tailoring of the material properties for various applications by varying the energy band constants of the SLs.

We note that all the aforementioned SLs have been proposed with the assumption that the interfaces between the layers are sharply defined, of zero thickness, i.e., devoid of any interface effects. The SL potential distribution may be then considered as a one-dimensional array of rectangular potential wells. The aforementioned advanced experimental techniques may produce SLs with physical interfaces between the two materials crystallographically abrupt; adjoining their interface will change at least on an atomic scale. As the potential form changes from a well (barrier) to a barrier (well), an intermediate potential region exists for the electrons. The influence of finite thickness of the interfaces on the electron dispersion law is very important, since the electron energy spectrum governs the electron transport in SLs. In addition to it, for effective mass SLs, the electronic subbands appear continually in real space [37].

In this chapter, we shall study the FNFE from III–V, II–VI, IV–VI, and HgTe/CdTe quantum wire SLs with graded interfaces in Sects. 2.2.1 to 2.2.4, respectively. From Sects. 2.2.5 to 2.2.8, we shall investigate the same from III–V, II–VI, IV–VI, and HgTe/CdTe effective mass quantum wire SLs. In Sect. 2.3, the result and discussions have been written with respect to the dependences of the FNFE as functions of various variables by taking GaAs/$Ga_{1-x}Al_xAs$, CdS/CdTe, PbTe/PbSnTe, and HgTe/CdTe quantum wire SLs with graded interfaces and the corresponding effective mass SLs as examples. Section 2.4 contains open research problems.

2.2 Theoretical Background

2.2.1 The Field Emission from III–V Quantum Wire Superlattices with Graded Interfaces

The energy spectrum of the conduction electrons in bulk specimens of the constituent materials of III–V SLs whose energy band structures are defined by three-band model of Kane can be written as

$$\frac{\hbar^2 k^2}{2m_{ci}} = EG(E, E_{gi}, \Delta_i), \tag{2.1}$$

m_{ci} is the effective electron mass at the edge of the conduction band, $i = 1$ and 2 and

$$G(E, E_{gi}, \Delta_i) \equiv ([E_{gi} + (2/3)\Delta_i](E + E_{gi} + \Delta_i)(E + E_{gi}))/ \\ (E_{gi}(E_{gi} + \Delta_i)[E + E_{gi} + (2/3)\Delta_i]).$$

Therefore, the dispersion law of the electrons of III–V SLs with graded interfaces can be expressed, following Jiang and Lin [38], as

$$\cos(L_o k) = \frac{1}{2}\Phi_{21}(E, k_s), \tag{2.2}$$

where $L_0(\equiv a_0 + b_0)$ is the period length, a_0 and b_0 are the widths of the barrier and the well, respectively,

$$\begin{aligned}
&\Phi_{21}(E, k_s)\\
&\equiv \Bigg[2\cosh\{\beta_{21}(E, k_s)\}\cos\{\gamma_{21}(E, k_s)\} + \varepsilon_{21}(E, k_s)\\
&\times \sinh\{\beta_{21}(E, k_s)\}\sin\{\gamma_{21}(E, k_s)\}\\
&+ \Delta_{21}\left[\left(\frac{K_{21}^2(E, k_s)}{K_{22}(E, k_s)} - 3K_{22}(E, k_s)\right)\cosh\{\beta_{21}(E, k_s)\}\sin\{\gamma_{21}(E, k_s)\}\right.\\
&\left.+ \left(3K_{21}(E, k_s) - \frac{\{K_{22}(E, k_s)\}^2}{K_{21}(E, k_s)}\right)\sinh\{\beta_{21}(E, k_s)\}\cos\{\gamma_{21}(E, k_s)\}\right]\\
&+ \Delta_{21}\Bigg[2\left(\{K_{21}(E, k_s)\} - \{K_{22}(E, k_s)\}^2\right)\cosh\{\beta_{21}(E, k_s)\}\\
&\times \cos\{\gamma_{21}(E, k_s)\}\\
&+ \frac{1}{12}\left[\frac{5\{K_{22}(E, k_s)\}^3}{K_{21}(E, k_s)} + \frac{5\{K_{21}(E, k_s)\}^3}{K_{22}(E, k_s)} - 34K_{22}(E, k_s)K_{21}(E, k_s)\right]\\
&\times\ \sinh\{\beta_{21}(E, k_s)\}\sin\{\gamma_{21}(E, k_s)\}\Bigg]\Bigg],
\end{aligned}$$

$\beta_{21}(E, k_s) \equiv K_{21}(E, k_s)[a_0 - \Delta_{21}]$, Δ_{21} is the interface width, $K_{21}(E, k_s) \equiv \left[\frac{2m_{c2}E'}{\hbar^2}G(E - \overline{V}_o, \alpha_2, \Delta_2) + k_s^2\right]^{1/2}$, $E' \equiv (\overline{V}_0 - E)$, $\overline{V}_0$ is the potential barrier encountered by the electron ($\overline{V}_0 \equiv |E_{g_2} - E_{g_1}|$), $\alpha_i \equiv 1/E_{gi}$, $\gamma_{21}(E, k_s) \equiv K_{22}(E, k_s)[b_0 - \Delta_{21}]$, $K_{22}(E, k_s) \equiv \left[\frac{2m_{c1}E}{\hbar^2}G(E, \alpha_1, \Delta_1) - k_s^2\right]^{1/2}$, and $k_s^2 = k_x^2 + k_y^2$.

The electron dispersion law in III–V quantum wire superlattices (QWSLs) can be written following (2.2) as

$$k_z^2 = \left[\frac{1}{L_0^2}\left\{\cos^{-1}\left[\frac{1}{2}f_{21}(n_x, n_y, E)\right]\right\}^2 - \phi_1(n_x, n_y)\right], \tag{2.3}$$

where the function $\phi_1(n_x, n_y)$ has already been defined in Chap. 1,

$$\begin{aligned}
&f_{21}(n_x, n_y, E)\\
&= \Big[2\cosh\{g_{21}(n_x, n_y, E)\}\cos\{g_{22}(n_x, n_y, E)\}
\end{aligned}$$

$$+ Q_{21}(n_x, n_y, E)\sinh\{g_{21}(n_x, n_y, E)\}\cdot\sin\{g_{22}(n_x, n_y, E)\}$$

$$+ \Delta_{21}\left[\left(\frac{\{J_{21}(n_x, n_y, E)\}^2}{J_{22}(n_x, n_y, E)} - 3J_{22}(n_x, n_y, E)\right)\right.$$

$$\times\cosh\{g_{21}(n_x, n_y, E)\}\cdot\sin\{g_{22}(n_x, n_y, E)\}$$

$$+\left(3J_{21}(n_x, n_y, E) - \frac{\{J_{22}(n_x, n_y, E)\}^2}{J_{21}(n_x, n_y, E)}\right)$$

$$\left.\cdot\sinh\{g_{21}(n_x, n_y, E)\}\cos\{g_{22}(n_x, n_y, E)\}\right]$$

$$+ \Delta_{21}\left[2\left(\{J_{21}(n_x, n_y, E)\}^2 - \{J_{22}(n_x, n_y, E)\}^2\right)\right.$$

$$\times\cosh\{g_{21}(n_x, n_y, E)\}\cos\{g_{22}(n_x, n_y, E)\}$$

$$+\frac{1}{12}\left(\frac{5\{J_{22}(n_x, n_y, E)\}^3}{J_{21}(n_x, n_y, E)} + \frac{5\{J_{21}(n_x, n_y, E)\}^3}{J_{22}(n_x, n_y, E)}\right.$$

$$\left. - \{34J_{21}(n_x, n_y, E)J_{22}(n_x, n_y, E)\}\right)$$

$$\left.\left.\times\ \sinh\{g_{21}(n_x, n_y, E)\}\sin\{g_{22}(n_x, n_y, E)\}\right]\right],$$

$$g_{21}(n_x, n_y, E) \equiv J_{21}(n_x, n_y, E)[a_0 - \Delta_{21}],$$

$$J_{21}(n_x, n_y, E) \equiv \left[\frac{2m_{c2}E'}{\hbar^2}G(E - \overline{V}_0, \alpha_2, \Delta_2) + \phi_1(n_x, n_y)\right]^{1/2},$$

$$g_{22}(n_x, n_y, E) = J_{22}(n_x, n_y, E)[b_0 - \Delta_{21}],$$

$$J_{22}(n_x, n_y, E) \equiv \left[\frac{2m_{c1}E}{\hbar^2}G(E, \alpha_1, \Delta_1) - \phi_1(n_x, n_y)\right]^{1/2}, \quad \text{and}$$

$$Q_{21}(n_x, n_y, E) \equiv \left[\frac{J_{21}(n_x, n_y, E)}{J_{22}(n_x, n_y, E)} - \frac{J_{22}(n_x, n_y, E)}{J_{21}(n_x, n_y, E)}\right]$$

Therefore, (2.3) can be expressed as $k_z^2 = L_{21}(E, n_x, n_y)$, where $L_{21}(E, n_x, n_y) = \left[\frac{1}{L_0^2}\left\{\cos^{-1}\left[\frac{1}{2}f_{21}(n_x, n_y, E)\right]\right\}^2 - \phi_1(n_x, n_y)\right]$.

The electron concentration per unit length is given by

$$n_0 = \frac{2g_v}{\pi}\sum_{n_x=1}^{n_{x\max}}\sum_{n_y=1}^{n_{y\max}}\left[D_{21}(E_{F1D}, n_x, n_y) + D_{22}(E_{F1D}, n_x, n_y)\right], \qquad (2.4)$$

where $D_{21}(E_{F1D}, n_x, n_y) = \sqrt{L_{21}(E_{FID}, n_x, n_y)}$ and $D_{22}(E_{F1D}, n_x, n_y) = \sum_{r=1}^{s_o} Z_{1D}(r)[D_{21}(E_{F1D}, n_x, n_y)]$.

The field-emitted current assumes the form

$$I = \frac{2g_v e k_B T}{h} \sum_{n_x=1}^{n_{x\max}} \sum_{n_y=1}^{n_{y\max}} \left[F_0(\bar{\eta}_{21}) \exp(-\bar{\beta}_{21})\right], \quad (2.5)$$

where $\bar{\eta}_{21} = (E_{F1D} - \overline{E}_{21})/k_B T$, and $\overline{E}_{21}$ is the lowest positive root of the equation.

$$f_{21}(\overline{E}_{21}, n_x, n_y) = 2\cos\left[L_0(\phi_1(n_x, n_y))^{\frac{1}{2}}\right], \quad (2.6)$$

$$\bar{\beta}_{21} = \frac{4}{3}[L_{21}(V_0, n_x, n_y)]^{3/2} \cdot [eF_{sz}\overline{L}_{21}(V_0, n_x, n_y)]^{-1},$$

$$\overline{L}_{21}(V_0, n_x, n_y) = \left(\frac{1}{L_0^2}\left\{\cos^{-1}\left[\frac{1}{2}f_{21}(n_x, n_y, V_0)\right]\right\}\left\{1 - \frac{1}{4}f_{21}^2(n_x, n_y, V_0)\right\}^{-\frac{1}{2}} \cdot \bar{f}_{21}(n_x, n_y, V_0)\right)$$

$$\begin{aligned}
&\bar{f}_{21}(n_x, n_y, V_0) \\
&= \Big[2g'_{21}(n_x, n_y, V_0)\sinh\{g_{21}(n_x, n_y, V_0)\}\cos\{g_{22}(n_x, n_y, V_0)\} \\
&\quad - 2g'_{22}(n_x, n_y, V_0)\sin\{g_{22}(n_x, n_y, V_0)\} \cdot \cosh\{g_{21}(n_x, n_y, V_0)\} \\
&\quad + Q'_{21}(n_x, n_y, V_0)\sinh\{g_{21}(n_x, n_y, V_0)\}\sin\{g_{22}(n_x, n_y, V_0)\} \\
&\quad + Q_{21}(n_x, n_y, V_0)\left[g'_{21}(n_x, n_y, V_0) \cdot \cosh\{g_{21}(n_x, n_y, V_0)\}\right. \\
&\quad \times \sin\{g_{22}(n_x, n_y, V_0)\} + g'_{22}(n_x, n_y, V_0)\cos\{g_{22}(n_x, n_y, V_0)\} \\
&\quad \left.\times \sinh\{g_{21}(n_x, n_y, V_0)\}\right]\Big] \\
&\quad + \Delta_{21} \cdot \left[\left\{\frac{2J_{21}(n_x, n_y, V_0)J'_{21}(n_x, n_y, V_0)}{J_{22}(n_x, n_y, V_0)} - \frac{J_{21}^2(n_x, n_y, V_0)J'_{22}(n_x, n_y, V_0)}{J_{22}^2(n_x, n_y, V_0)}\right.\right. \\
&\quad \left. - 3J'_{22}(n_x, n_y, V_0)\right\}\cosh\{g_{21}(n_x, n_y, V_0)\}\sin\{g_{22}(n_x, n_y, V_0)\} \\
&\quad + \left\{\frac{J_{21}^2(n_x, n_y, V_0)}{J_{22}(n_x, n_y, V_0)} - 3J_{22}(n_x, n_y, V_0)\right\}\{g'_{21}(n_x, n_y, V_0)\sinh\{g_{21}(n_x, n_y, V_0)\} \\
&\quad \cdot \sin\{g_{22}(n_x, n_y, V_0)\} + g'_{22}(n_x, n_y, V_0) \cdot \cos\{g_{22}(n_x, n_y, V_0)\}
\end{aligned}$$

$$\times \cosh\left\{g_{21}\left(n_x,n_y,V_0\right)\right\}\} + \Delta_{21}\Bigg[\left[2J_{21}^2(n_x,n_y,V_0) - 2J_{22}^2(n_x,n_y,V_0)\right]$$

$$\times \left\{g'_{21}\left(n_x,n_y,V_0\right)\sinh\left\{g_{21}\left(n_x,n_y,V_0\right)\right\}\cdot\cos\left\{g_{22}\left(n_x,n_y,V_0\right)\right\}\right.$$

$$\left.- g'_{22}\left(n_x,n_y,V_0\right)\sin\left\{g_{22}\left(n_x,n_y,V_0\right)\right\}\cosh\left\{g_{21}\left(n_x,n_y,V_0\right)\right\}\right.$$

$$+ \left\{4J_{21}(n_x,n_y,V_0)J'_{21}(n_x,n_y,V_0) - 4J_{22}(n_x,n_y,V_0)J'_{22}(n_x,n_y,V_0)\right\}$$

$$\times \left\{\cosh\left\{g_{21}\left(n_x,n_y,V_0\right)\right\}\cos\left\{g_{22}\left(n_x,n_y,V_0\right)\right\}\right\}$$

$$+ \frac{1}{12}\left[\frac{15J_{22}^2(n_x,n_y,V_0)J'_{22}(n_x,n_y,V_0)}{J_{21}(n_x,n_y,V_0)} - \frac{5J_{22}^3(n_x,n_y,V_0)J'_{21}(n_x,n_y,V_0)}{J_{21}^2(n_x,n_y,V_0)}\right.$$

$$- \frac{5J_{21}^3(n_x,n_y,V_0)J'_{22}(n_x,n_y,V_0)}{J_{22}^2(n_x,n_y,V_0)} + \frac{15J_{21}^2(n_x,n_y,V_0)J'_{21}(n_x,n_y,V_0)}{J_{22}(n_x,n_y,V_0)}$$

$$\left.- 34J'_{21}(n_x,n_y,V_0)\cdot J_{22}(n_x,n_y,V_0) - 34J_{21}(n_x,n_y,V_0)J'_{22}(n_x,n_y,V_0)\right]$$

$$\cdot\left\{\sinh\left\{g_{21}\left(n_x,n_y,V_0\right)\right\}\sin\left\{g_{22}\left(n_x,n_y,V_0\right)\right\}\right\}$$

$$+ \left\{\frac{5J_{22}^3(n_x,n_y,V_0)}{J_{21}^2(n_x,n_y,V_0)} + \frac{1}{12}\frac{5J_{21}^3(n_x,n_y,V_0)}{J_{22}^2(n_x,n_y,V_0)} - 34J_{21}(n_x,n_y,V_0)\right.$$

$$\left.\cdot J_{22}(n_x,n_y,V_0)\right\}\left\{g'_{21}(n_x,n_y,V_0)\cosh\{g_{21}(n_x,n_y,V_0)\}\sin\{g_{22}(n_x,n_y,V_0)\}\right.$$

$$\left.+ g'_{22}\left(n_x,n_y,V_0\right)\cos\left\{g_{22}\left(n_x,n_y,V_0\right)\right\}\sinh\left\{g_{21}\left(n_x,n_y,V_0\right)\right\}\right\}\Bigg]\Bigg],$$

$$g'_{21}\left(n_x,n_y,V_0\right) = J'_{21}\left(n_x,n_y,V_0\right)\left[a_0 - \Delta_{21}\right],$$

$$J'_{21}\left(n_x,n_y,V_0\right) = \left[J_{21}\left(n_x,n_y,V_0\right)\right]^{-1}\left[-\frac{m_{c2}}{\hbar^2}G(V_0-\overline{V}_0,\alpha_2,\Delta_2)\right.$$

$$\left.+ \frac{m_{c2}}{\hbar^2}(\overline{V}_0 - V_0)G'(V_0-\overline{V}_0,\alpha_2,\Delta_2)\right],$$

$$G'(V_0-\overline{V}_0,\alpha_2,\Delta_2) = G(V_0-\overline{V}_0,\alpha_2,\Delta_2)\cdot\left[\frac{1}{(V_0-\overline{V}_0+E_{g2}+\Delta_2)}\right.$$

$$\left.+ \frac{1}{(V_0-\overline{V}_0+E_{g2})} - \frac{1}{(V_0-\overline{V}_0+E_{g2}+\frac{2}{3}\Delta_2)}\right],$$

$$g'_{22}\left(n_x,n_y,V_0\right) = J'_{22}\left(n_x,n_y,V_0\right)\left[b_0 - \Delta_{21}\right],$$

$$J'_{22}(n_x,n_y,V_0) = \left[J_{22}(n_x,n_y,V_0)\right]^{-1}\left[\frac{m_{c1}}{\hbar^2}G(V_0,\alpha_1,\Delta_1) + \frac{m_{c1}}{\hbar^2}V_0G'(V_0,\alpha_1,\Delta_1)\right]$$

$$G(V_0-\overline{V}_0,\alpha_2,\Delta_2) = G(E,\alpha_2,\Delta_2)\big|_{E\to(V_0-\overline{V}_0)},$$

and

$$Q'_{21}(n_x,n_y,V_0) \equiv \left[\frac{J'_{21}(n_x,n_y,V_0)}{J_{22}(V_{0,}n_x,n_y)} - \frac{J'_{22}(V_{0,}n_x,n_y)}{J_{21}(n_x,n_y,V_0)} - \frac{J_{21}(n_x,n_y,V_0)\cdot J'_{22}(V_{0,}n_x,n_y)}{J^2_{22}(V_{0,}n_x,n_y)} + \frac{J'_{21}(n_x,n_y,V_0)\cdot J_{22}(V_{0,}n_x,n_y)}{J^2_{21}(V_{0,}n_x,n_y)}\right].$$

2.2.2 The Field Emission from II–VI Quantum Wire Superlattices with Graded Interfaces

The energy spectrum of the conduction electrons of the constituent materials of II–VI SLs is given by

$$E = \frac{\hbar^2 k_s^2}{2m^*_{\perp,1}} + \frac{\hbar^2 k_z^2}{2m^*_{\parallel,1}} \pm C_0 k_s \quad \text{and} \quad \frac{\hbar^2 k^2}{2m_{c2}} = EG(E, E_{g2}, \Delta_2). \tag{2.8}$$

The electron dispersion law in II–VI SLs with graded interfaces can be expressed as

$$\cos(L_o k) = \frac{1}{2}\Phi_{121}(E, k_s), \tag{2.9}$$

where

$$\begin{aligned}
&\Phi_{121}(E,k_s) \\
&\equiv \Bigg[2\cosh\{\beta_{121}(E,k_s)\}\cos\{\gamma_{121}(E,k_s)\} + \varepsilon_{121}(E,k_s) \\
&\quad\times \sinh\{\beta_{121}(E,k_s)\}\sin\{\gamma_{121}(E,k_s)\} \\
&\quad+ \Delta_{21}\Bigg[\left(\frac{K^2_{321}(E,k_s)}{K_{421}(E,k_s)} - 3K_{421}(E,k_s)\right)\cosh\{\beta_{121}(E,k_s)\}\sin\{\gamma_{121}(E,k_s)\} \\
&\quad+ \left(3K_{321}(E,k_s) - \frac{\{K_{421}(E,k_s)\}^2}{K_{321}(E,k_s)}\right)\sinh\{\beta_{121}(E,k_s)\}\cos\{\gamma_{121}(E,k_s)\}\Bigg] \\
&\quad+ \Delta_{21}\Bigg[2(\{K_{321}(E,k_s)\}^2 - \{K_{421}(E,k_s)\}^2)\cosh\{\beta_{121}(E,k_s)\}\cos\{\gamma_{121}(E,k_s)\} \\
&\quad+ \frac{1}{12}\left[\frac{5\{K_{321}(E,k_s)\}^3}{K_{421}(E,k_s)} + \frac{5\{K_{421}(E,k_s)\}^3}{K_{321}(E,k_s)} - 34K_{421}(E,k_s)K_{321}(E,k_s)\right] \\
&\quad\times \sinh\,\{\beta_{121}(E,k_s)\}\sin\{\gamma_{121}(E,k_s)\}\Bigg]\Bigg],
\end{aligned}$$

$$\beta_{121}(E,k_s) \equiv K_{321}(E,k_s)[a_0 - \Delta_{21}],$$

$$K_{321}(E,k_s) \equiv \left[\frac{2m_{c2}E\prime}{\hbar^2}G(E - \overline{V}_o, \alpha_2, \Delta_2) + k_s^2\right]^{1/2}, \quad \alpha_i \equiv 1/E_{gi},$$

$$\gamma_{121}(E,k_s) \equiv K_{421}(E,k_s)[b_0 - \Delta_{21}],$$

$$K_{421}(E,k_s) \equiv \left[\frac{2m^*_{\parallel,1}}{\hbar^2}\left[E - \frac{\hbar^2 k_s^2}{2m^*_{\perp,1}} \mp C_0 k_s\right]\right]^{1/2}.$$

The electron dispersion law in II–VI QWSLs can be written as

$$k_z^2 = \left[\frac{1}{L_0^2}\left\{\cos^{-1}\left[\frac{1}{2}f_{212}(n_x,n_y,E)\right]\right\}^2 - \phi_1(n_x,n_y)\right] \tag{2.10}$$

where

$$\begin{aligned}
&f_{212}(n_x,n_y,E)\\
&= \Bigg[2\cosh\{g_{212}(n_x,n_y,E)\}\cos\{g_{222}(n_x,n_y,E)\}\\
&+ Q_{212}(n_x,n_y,E)\sinh\{g_{212}(n_x,n_y,E)\}\cdot\sin\{g_{222}(n_x,n_y,E)\}\\
&+ \Delta_{212}\Bigg[\left(\frac{\{J_{212}(n_x,n_y,E)\}^2}{J_{222}(n_x,n_y,E)} - 3J_{222}(n_x,n_y,E)\right)\\
&\times\cosh\{g_{212}(n_x,n_y,E)\}\cdot\sin\{g_{222}(n_x,n_y,E)\}\\
&+ \left(3J_{212}(n_x,n_y,E) - \frac{\{J_{222}(n_x,n_y,E)\}^2}{J_{212}(n_x,n_y,E)}\right)\\
&\times\sinh\{g_{212}(n_x,n_y,E)\}\cos\{g_{222}(n_x,n_y,E)\}\Bigg]\\
&+ \Delta_{212}\Bigg[2\left(\{J_{212}(n_x,n_y,E)\}^2 - \{J_{222}(n_x,n_y,E)\}^2\right)\cosh\{g_{212}(n_x,n_y,E)\}\\
&\times\cos\{g_{222}(n_x,n_y,E)\} + \frac{1}{12}\left(\frac{5\{J_{222}(n_x,n_y,E)\}^3}{J_{212}(n_x,n_y,E)} + \frac{5\{J_{212}(n_x,n_y,E)\}^3}{J_{222}(n_x,n_y,E)}\right.\\
&\left. - \{34J_{212}(n_x,n_y,E)J_{222}(n_x,n_y,E)\}\right)\\
&\times\sinh\{g_{212}(n_x,n_y,E)\}\sin\{g_{222}(n_x,n_y,E)\}\Bigg]\Bigg]
\end{aligned}$$

$$g_{212}\left(n_x,n_y,E\right) \equiv J_{212}\left(n_x,n_y,E\right)\left[a_0-\Delta_{212}\right],$$

$$J_{212}\left(n_x,n_y,E\right) \equiv \left[\frac{2m_{c2}E\prime}{\hbar^2}G(E-\overline{V}_0,\alpha_2,\Delta_2)+\phi_1\left(n_x,n_y\right)\right]^{1/2},$$

$$g_{222}\left(n_x,n_y,E\right) = J_{222}\left(n_x,n_y,E\right)\left[b_0-\Delta_{212}\right],$$

$$J_{222}(n_x,n_y,E) \equiv \left[\frac{2m^*_{\parallel,1}}{\hbar^2}\left[E-\frac{\hbar^2}{2m^*_{\perp,1}}\phi_1(n_x,n_y)\mp C_0\sqrt{\phi_1(n_x,n_y)}\right]\right]^{1/2}, \text{ and}$$

$$Q_{212}(n_x,n_y,E) \equiv \left[\frac{J_{212}(n_x,n_y,E)}{J_{222}(n_x,n_y,E)}-\frac{J_{222}(n_x,n_y,E)}{J_{212}(n_x,n_y,E)}\right].$$

Therefore, (2.10) can be expressed as

$$k_z^2 = L_{212}(E,n_x,n_y), \tag{2.11}$$

where $L_{212}(E,n_x,n_y) = \left[\frac{1}{L_0^2}\left\{\cos^{-1}\left[\frac{1}{2}f_{212}\left(n_x,n_y,E\right)\right]\right\}^2-\phi_1(n_x,n_y)\right]$.

The electron concentration per unit length is given by

$$n_0 = \frac{g_v}{\pi}\sum_{n_x=1}^{n_{x\max}}\sum_{n_y=1}^{n_{y\max}}\left[D_{212}(E_{F1D},n_x,n_y)+D_{222}(E_{F1D},n_x,n_y)\right], \tag{2.12}$$

where $D_{212}(E_{F1D},n_x,n_y) = \sqrt{L_{212}(E_{F1D},n_x,n_y)}$ and $D_{222}(E_{F1D},n_x,n_y) = \sum_{r=1}^{s_o} Z_{1D}(r)[D_{212}(E_{F1D},n_x,n_y)]$.

The field-emitted current assumes the form

$$I = \frac{g_v e k_B T}{h}\sum_{n_x=1}^{n_{x\max}}\sum_{n_y=1}^{n_{y\max}}\left[F_0(\bar{\eta}_{22})\exp(-\bar{\beta}_{22})\right] \tag{2.13}$$

where $\bar{\eta}_{22} = (E_{F1D}-\overline{E}_{22})/k_B T$, and $\overline{E}_{22}$ is the lowest positive root of the equation.

$$f_{212}(\overline{E}_{22},n_x,n_y) = 2\cos[L_0(\phi_1(n_x,n_y))^{\frac{1}{2}}], \tag{2.14}$$

$$\bar{\beta}_{22} = \frac{4}{3}[L_{212}(V_0,n_x,n_y)]^{3/2}\cdot[3eF_{sz}\overline{L}_{212}(V_0,n_x,n_y)]^{-1},$$

$$\overline{L}_{212}(V_0,n_x,n_y) = \left(\frac{1}{L_0^2}\left\{\cos^{-1}\left[\frac{1}{2}f_{212}(n_x,n_y,V_0)\right]\right\}\right.$$
$$\left.\times\left\{1-\frac{1}{4}f_{212}^2(n_x,n_y,V_0)\right\}^{-1/2}\cdot\bar{f}_{212}(n_x,n_y,V_0)\right)$$

$$
\begin{aligned}
&\bar{f}_{212}(n_x, n_y, V_0) \\
&= \Big[2g'_{212}(n_x, n_y, V_0) \sinh\{g_{212}(n_x, n_y, V_0)\} \cos\{g_{222}(n_x, n_y, V_0)\} \\
&\quad - 2g'_{222}(n_x, n_y, V_0) \sin\{g_{222}(n_x, n_y, V_0)\} \cdot \cosh\{g_{212}(n_x, n_y, V_0)\} \\
&\quad + Q'_{212}(n_x, n_y, V_0) \sinh\{g_{212}(n_x, n_y, V_0)\} \sin\{g_{222}(n_x, n_y, V_0)\} \\
&\quad + Q_{212}(N_x, n_y, V_0) \big[g'_{212}(n_x, n_y, V_0) \cdot \cosh\{g_{212}(n_x, n_y, V_0)\} \sin\{g_{222}(n_x, n_y, V_0)\} \\
&\quad + g'_{222}(n_x, n_y, V_0). \cos\{g_{222}(n_x, n_y, V_0)\} \sinh\{g_{212}(n_x, n_y, V_0)\}\big] \\
&\quad + \Delta_{212} \cdot \Bigg[\Bigg\{ \frac{2J_{212}(n_x, n_y, V_0) J'_{212}(n_x, n_y, V_0)}{J_{222}(n_x, n_y, V_0)} - \frac{J^2_{212}(n_x, n_y, V_0) J'_{222}(n_x, n_y, V_0)}{J^2_{222}(n_x, n_y, V_0)} \\
&\quad - 3J'_{222}(n_x, n_y, V_0)\Bigg\} \cosh\{g_{212}(n_x, n_y, V_0)\} \sin\{g_{222}(n_x, n_y, V_0)\} \\
&\quad + \left\{ \frac{J^2_{212}(n_x, n_y, V_0)}{J_{222}(n_x, n_y, V_0)} - 3J_{222}(n_x, n_y, V_0) \right\} \{g'_{212}(n_x, n_y, V_0) \\
&\quad \cdot \sinh\{g_{212}(n_x, n_y, V_0)\} \sin\{g_{222}(n_x, n_y, V_0)\} + g'_{222}(n_x, n_y, V_0) \\
&\quad \cdot \cos\{g_{222}(n_x, n_y, V_0)\} \cosh\{g_{212}(n_x, n_y, V_0)\}\} \\
&\quad + \Bigg\{ 3J'_{212}(n_x, n_y, V_0) + \frac{J^2_{222}(n_x, n_y, V_0)}{J^2_{212}(n_x, n_y, V_0)} \cdot J'_{212}(n_x, n_y, V_0) \\
&\quad - \{\{J_{212}(n_x, n_y, V_0)\}^{-1} \cdot \{2J'_{222}(n_x, n_y, V_0) \cdot J_{222}(n_x, n_y, V_0)\}\} \\
&\quad \cdot \{\sinh[g_{212}(n_x, n_y, V_0)] \cdot \cos\{g_{222}(n_x, n_y, V_0)\}\}] \Bigg\} + \{g'_{212}(n_x, n_y, V_0) \\
&\quad \cdot \cosh\{g_{212}(n_x, n_y, V_0)\} \cdot \cos\left\{g_{222}\left(n_x, n_y, V_0\right)\right\} \\
&\quad - g'_{222}\left(n_x, n_y, V_0\right) \sin\left\{g_{222}\left(n_x, n_y, V_0\right)\right\} \\
&\quad \cdot \sinh[g_{212}\left(n_x, n_y, V_0\right)]\}\{3J_{212}\left(n_x, n_y, V_0\right) - \frac{J^2_{222}(n_x, n_y, V_0)}{J_{212}(n_x, n_y, V_0)} \Bigg\} \Bigg] \\
&\quad + \Delta_{212} \Big[\Big[4J_{212}(n_x, n_y, V_0) \cdot J'_{212}(n_x, n_y, V_0) - 4J_{222}(n_x, n_y, V_0) J'_{222}(n_x, n_y, V_0)\} \\
&\quad \times \left\{\cosh\left\{g_{212}\left(n_x, n_y, V_0\right)\right\} \cos\left\{g_{222}\left(n_x, n_y, V_0\right)\right\}\right\} \\
&\quad + \{2J^2_{212}(n_x, n_y, V_0) - 2J^2_{222}(n_x, n_y, V_0)\}\{g'_{212}\left(n_x, n_y, V_0\right) \\
&\quad \times \sinh[g_{212}(n_x, n_y, V_0)] \cos\left\{g_{222}(n_x, n_y, V_0)\right\} \\
&\quad - g'_{222}(n_x, n_y, V_0) \sin\left\{g_{222}(n_x, n_y, V_0)\right\} \\
&\quad \cdot \cosh[g_{212}(n_x, n_y, V_0)] + \frac{1}{12} \Bigg[\frac{15J^2_{222}(n_x, n_y, V_0) J'_{222}(n_x, n_y, V_0)}{J_{212}(n_x, n_y, V_0)}
\end{aligned}
$$

$$- \frac{5J_{222}^3(n_x, n_y, V_0)J_{212}'(n_x, n_y, V_0)}{J_{212}^2(n_x, n_y, V_0)} - \frac{5J_{212}^3(n_x, n_y, V_0)J_{222}'(n_x, n_y, V_0)}{J_{222}^2(n_x, n_y, V_0)}$$

$$+ \frac{15J_{212}^2(n_x, n_y, V_0)J_{212}'(n_x, n_y, V_0)}{J_{212}'(n_x, n_y, V_0)} - 34J_{212}'(n_x, n_y, V_0) \cdot J_{222}(n_x, n_y, V_0)$$

$$- 34J_{212}(n_x, n_y, V_0)J_{222}'(n_x, n_y, V_0) \left\{\{\sinh\{g_{212}(n_x, n_y, V_0)\} \ \sin\{g_{222}(n_x, n_y, V_0)\}\right\}$$

$$+ \frac{1}{12}\left\{\frac{5J_{222}^3(n_x, n_y, V_0)}{J_{212}(n_x, n_y, V_0)} + \frac{5J_{212}^3(n_x, n_y, V_0)}{J_{222}(n_x, n_y, V_0)} - 34J_{222}(n_x, n_y, V_0)J_{212}(n_x, n_y, V_0)\right\}$$

$$\cdot g_{212}'\left(n_x, n_y, V_0\right)\cosh\left\{g_{212}\left(n_x, n_y, V_0\right)\right\}\sin\left\{g_{222}\left(n_x, n_y, V_0\right)\right\}$$

$$+ g_{222}'\left(n_x, n_y, V_0\right)\cos\left\{g_{222}\left(n_x, n_y, V_0\right)\right\}\sinh\left\{g_{212}\left(n_x, n_y, V_0\right)\right\}$$

$$g_{212}'\left(n_x, n_y, V_0\right) = J_{212}'\left(n_x, n_y, V_0\right)[a_0 - \Delta_{21}]$$

$$J_{212}'\left(n_x, n_y, V_0\right) = [J_{212}(n_x, n_y, V_0)]^{-1}\left[-\frac{m_{c2}}{\hbar^2}G(V_0 - \overline{V}_0, \alpha_2, \Delta_2) + \frac{m_{c2}}{\hbar^2}(\overline{V}_0 - V_0)G'(V_0 - \overline{V}_0, \alpha_2, \Delta_2)\right]$$

$$G'(V_0 - \overline{V}_0, \alpha_2, \Delta_2) = G(V_0 - \overline{V}_0, \alpha_2, \Delta_2)^{-1} \cdot \left[\frac{1}{(V_0 - \overline{V}_0 + E_{g2} + \Delta_2)} + \frac{1}{(V_0 - \overline{V}_0 + E_{g2})} - \frac{1}{(V_0 - \overline{V}_0 + E_{g2} + \frac{2}{3}\Delta_2)}\right]$$

$$g_{222}'\left(n_x, n_y, V_0\right) = J_{222}'\left(n_x, n_y, V_0\right)[b_0 - \Delta_{21}],$$

$$J_{222}'\left(n_x, n_y, V_0\right) = \left[J_{222}\left(n_x, n_y, V_0\right)\right]^{-1}\left[\frac{m_{\parallel,1}^*}{\hbar^2}\right]$$

$$G(V_0 - \overline{V}_0, \alpha_2, \Delta_2) = G(E, \alpha_2, \Delta_2)|_{E \to (V_0 - \overline{V}_0)} \quad \text{and}$$

$$Q_{212}'(n_x, n_y, V_0) \equiv \left[\frac{J_{212}'(n_x, n_y, V_0)}{J_{222}(n_x, n_y, V_0)} - \frac{J_{222}'(n_x, n_y, V_0)}{J_{212}(n_x, n_y, V_0)} - \frac{J_{212}(n_x, n_y, V_0) \cdot J_{222}'(n_x, n_y, V_0)}{J_{222}^2(n_x, n_y, V_0)} - \frac{J_{222}(n_x, n_y, V_0) \cdot J_{212}'(n_x, n_y, V_0)}{J_{212}^2(n_x, n_y, V_0)}\right]$$

2.2.3 The Field Emission from IV–VI Quantum Wire Superlattices with Graded Interfaces

The **E–k** dispersion relation of the conduction electrons of the constituent materials of the IV–VI SLs can be expressed as [39]

$$E = a_i k_s^2 + b_i k_z^2 + \left[\left[c_i k_s^2 + d_i k_z^2\right] + \left(e_i k_s^2 + f_i k_y^2 + \frac{E_{g_i}}{2}\right)^2\right]^{1/2} - \frac{E_{g_i}}{2}, \tag{2.15}$$

where $a_i \equiv \left[\frac{\hbar^2}{2m_{\perp,i}^-}\right]$, $b_i \equiv \left(\frac{\hbar^2}{2m_{\|,i}^-}\right)^2$, $c_i \equiv P_{\perp,i}^2$, $d_i \equiv P_{\perp,i}^2$, $e_i \equiv \left[\frac{\hbar^2}{2m_{\perp,i}^+}\right]$ and $f_i \equiv \left(\frac{\hbar^2}{2m_{\|,i}^+}\right)^2$.

The electron dispersion law in IV–VI SLs with graded interfaces can be expressed as

$$\cos(L_o k) = \frac{1}{2}\overline{\Phi}_2(E, k_s), \tag{2.16}$$

where

$$\begin{aligned}
&\overline{\Phi}_2(E, k_s) \\
&\quad \equiv \left[2\cosh\left\{\bar{\beta}_2(E, k_s)\right\}\cos\left\{\bar{\gamma}_2(E, k_s)\right\} + \bar{\varepsilon}_2(E, k_s)\right. \\
&\quad \times \sinh\left\{\bar{\beta}_2(E, k_s)\right\}\sin\left\{\bar{\gamma}_2(E, k_s)\right\} \\
&\quad + \Delta_{21}\left[\left(\frac{\left\{\overline{K}_5(E, k_s)\right\}^2}{\overline{K}_6(E, k_s)} - 3\overline{K}_6(E, k_s)\right)\cosh\left\{\bar{\beta}_2(E, k_s)\right\}\sin\left\{\bar{\gamma}_2(E, k_s)\right\}\right. \\
&\quad \left. + \left(3\overline{K}_5(E, k_s) - \frac{\left\{\overline{K}_6(E, k_s)\right\}^2}{\overline{K}_5(E, k_s)}\right)\sinh\left\{\bar{\beta}_2(E, k_s)\right\}\cos\left\{\bar{\gamma}_2(E, k_s)\right\}\right] \\
&\quad + \Delta_{21}\left[2\left(\left\{\overline{K}_5(E, k_s)\right\}^2 - \left\{\overline{K}_6(E, k_s)\right\}^2\right)\cosh\left\{\bar{\beta}_2(E, k_s)\right\}\cos\left\{\bar{\gamma}_2(E, k_s)\right\}\right. \\
&\quad + \frac{1}{12}\left[\frac{5\left\{\overline{K}_5(E, k_s)\right\}^3}{\overline{K}_6(E, k_s)} + \frac{5\left\{\overline{K}_6(E, k_s)\right\}^3}{\overline{K}_5(E, k_s)} - 34\overline{K}_6(E, k_s)\,\overline{K}_5(E, k_s)\right] \\
&\quad \left.\left. \times \sinh\left\{\bar{\beta}_2(E, k_s)\right\}\sin\left\{\bar{\gamma}_2(E, k_s)\right\}\right]\right],
\end{aligned}$$

$$\bar{\varepsilon}_2(E, k_s) \equiv \left[\frac{\overline{K}_5(E, k_s)}{\overline{K}_6(E, k_s)} - \frac{\overline{K}_6(E, k_s)}{\overline{K}_5(E, k_s)}\right], \bar{\beta}_2(E, k_s) \equiv \overline{K}_5(E, k_s)\left[a_0 - \Delta_{21}\right],$$

$$
\begin{aligned}
\overline{K}_5\left(E,k_x,k_y\right) &\equiv \left[-\left[(E-\overline{V}_0)H_{12}+H_{22}\left(k_x,k_y\right)\right]+\left[H_{32}(E-\overline{V}_0)^2\right.\right.\\
&\quad\left.\left.+(E-\overline{V}_0)H_{42}\left(k_x,k_y\right)+H_{52}\left(k_x,k_y\right)\right]^{1/2}\right]^{1/2}\\
H_{1i} &= b_i.\left(b_i^2-f_i^2\right)^{-1},\\
i &= 1 \text{ and } 2,\; H_{2i}\left(k_x,k_y\right)=\left[2(b_i^2-f_i^2)\right]^{-1}\left[E_{g_i}b_i+d_i+f_iE_{g_i}\right.\\
&\quad\left.+2\left(c_if_i-a_ib_i\right)\left(k_x^2+k_y^2\right)\right],\quad H_{3i}=\frac{f_i^2}{\left(b_i^2-f_i^2\right)^2},\\
H_{4i}\left(k_x,k_y\right) &= \left[4(b_i^2-f_i^2)^2\right]^{-1}\left[4b_i^2E_{g_i}+4b_id_i+4b_if_iE_{g_i}+4f_i^2E_{g_i}\right.\\
&\quad\left.+8\left(k_x^2+k_y^2\right)\left[b_i^2a_i+C_if_ib_i-a_i^2b_i\right]\right]\\
H_{5i}\left(k_x,k_y\right) &\equiv \left[4(b_i^2-f_i^2)^2\right]^{-1}\left[\left(k_x^2+k_y^2\right)^2\left[-8a_ib_iC_if_i+4b_i^2C_i^2+4f_i^2a_i^2\right.\right.\\
&\quad\left.-4f_i^2C_i^2\right]+\left(k_x^2+k_y^2\right)\left[8d_i\,C_if_i-4a_ib_id_i-4a_ib_if_iE_{gi}\right.\\
&\quad\left.\left.+4b_i^2C_i+4b_i^2e_iE_{gi}-4a_if_i^2E_{g_i}-4f_i^2e_iE_{g_i}\right]\right]\\
&\quad+\left[E_{g_i}^2b_i^2+d_i^2+f_i^2E_{g_i}^2+2E_{g_i}f_id_i\right]\Big],\\
\bar{\gamma}_2\left(E,k_s\right) &= \overline{K}_6\left(E,k_s\right)\left[b_0-\Delta_{21}\right],\overline{K}_6\left(E,k_x,k_y\right)\\
&\equiv \left[\left[EH_{11}+H_{21}\left(k_x,k_y\right)\right]-\left[H_{31}E^2+EH_{41}\left(k_x,k_y\right)\right.\right.\\
&\quad\left.\left.+H_{51}\left(k_x,k_y\right)\right]^{1/2}\right]^{1/2} \text{ and}\\
\overline{K}_6\left(E,k_x,k_y\right) &\equiv \left[\left[(E-V_0)^2H_{32}+(E-V_0)H_{42}\left(k_x,k_y\right)+H_{52}\left(k_x,k_y\right)\right]^{1/2}\right.\\
&\quad\left.-\left[(E-V_0)H_{12}+H_{22}\left(k_x,k_y\right)\right]\right]^{1/2}.
\end{aligned}
$$

The electron dispersion law in IV–VI QWSLs can be written from (2.16) as

$$
k_z^2=\left[\frac{1}{L_0^2}\left\{\cos^{-1}\left[\frac{1}{2}I_{23}\left(n_x,n_y,E\right)\right\}^2-\phi_1\left(n_x,n_y\right)\right], \tag{2.17}
$$

where

$$
\begin{aligned}
I_{23}\left(n_x,n_y,E\right) &= \left[2\cosh\{\bar{\beta}_{23}\left(n_x,n_y,E\right)\}\cos\{\bar{\gamma}_{23}\left(n_x,n_y,E\right)\}+\bar{\varepsilon}_{23}\left(n_x,n_y,E\right)\right.\\
&\quad\times\sinh\{\bar{\beta}_{23}\left(n_x,n_y,E\right)\}\sin\{\bar{\gamma}_{23}\left(n_x,n_y,E\right)\}
\end{aligned}
$$

$$+\Delta_{21}\left[\left(\frac{\{\overline{K}_{53}(n_x,n_y,E)\}^2}{\overline{K}_{65}(n_x,n_y,E)}-3\overline{K}_{63}(n_x,n_y,E)\right)\cosh\{\bar{\beta}_{23}(n_x,n_y,E)\}\right.$$

$$\times\sin\{\bar{\gamma}_{23}(n_x,n_y,E)\}+\left(3\overline{K}_{53}(n_x,n_y,E)-\frac{\{\overline{K}_{63}(n_x,n_y,E)\}^2}{\overline{K}_{53}(n_x,n_y,E)}\right)$$

$$\times\sinh\{\bar{\beta}_{23}(n_x,n_y,E)\}\cos\{\bar{\gamma}_{23}(n_x,n_y,E)\}]$$

$$+\Delta_{21}\left[2\left(\{\overline{K}_{53}(n_x,n_y,E)\}^2-\{\overline{K}_{63}(n_x,n_y,E)\}^2\right)\cosh\{\bar{\beta}_{23}(n_x,n_y,E)\}\right.$$

$$\times\cos\{\bar{\gamma}_{23}(n_x,n_y,E)\}+\frac{1}{12}\left(\frac{5\{\overline{K}_{53}(n_x,n_y,E)\}^3}{\overline{K}_{63}(n_x,n_y,E)}+\frac{5\{\overline{K}_{63}(n_x,n_y,E)\}^3}{\overline{K}_{53}(n_x,n_y,E)}\right.$$

$$-\{34\overline{K}_{53}(n_x,n_y,E)\overline{K}_{63}(n_x,n_y,E)\})$$

$$\times\sinh\{\bar{\beta}_{23}(n_x,n_y,E)\}\sin\{\bar{\gamma}_{23}(n_x,n_y,E)\}]]$$

$$\bar{\varepsilon}_{23}(n_x,n_y,E)\equiv\left[\frac{\overline{K}_{53}(n_x,n_y,E)}{\overline{K}_{63}(n_x,n_y,E)}-\frac{\overline{K}_{63}(n_x,n_y,E)}{\overline{K}_{53}(n_x,n_y,E)}\right],$$

$$\bar{\beta}_{23}(n_x,n_y,E)\equiv\overline{K}_{53}(n_x,n_y,E)[a_0-\Delta_{21}],$$

$$\overline{K}_{53}(n_x,n_y,E)\equiv[-[(E-\overline{V}_0)H_{12}+H_{22}(n_x,n_y)]+[(E-\overline{V}_0)^2H_{32}+(E-\overline{V}_0)$$

$$H_{42}(n_x,n_y)+H_{52}(n_x,n_y)]^{1/2}\Big]^{1/2},\ H_{2i}(n_x,n_y)$$

$$=\left[2(b_i^2-f_i^2)\right]^{-1}\left[E_{g_i}b_i+d_i+f_iE_{g_i}+2(C_if_i-a_ib_i)\phi_1(n_x,n_y)\right]$$

$$H_{4i}(n_x,n_y)=\left[4(b_i^2-f_i^2)^2\right]^{-1}\left[4b_i^2E_{g_i}+4b_id_i+4b_if_iE_{g_i}+4f_i^2E_{g_i}\right.$$

$$\left.+8\phi_1(n_x,n_y)\left[b_i^2a_i+C_if_ib_i-a_i^2b_i\right]\right]$$

$$H_{5i}(n_x,n_y)\equiv\left[4(b_i^2-f_i^2)^2\right]^{-1}\left[\phi_1^2(n_x,n_y)\left[-8a_ib_iC_if_i+4b_i^2C_i^2\right.\right.$$

$$\left.+4f_i^2a_i^2-4f_i^2C_i^2\right]+\phi_1(n_x,n_y)[8d_i\,C_if_i-4a_ib_id_i$$

$$-4a_ib_if_iE_{g_i}+4b_i^2C_i+4b_i^2e_iE_{g_i}-4a_if_i^2E_{g_i}-4f_i^2e_iE_{g_i}]]$$

$$\left.+\left[E_{g_i}^2b_i^2+d_i^2+f_i^2E_{g_i}^2+2E_{g_i}f_id_i\right]\right],\quad\bar{\gamma}_{23}(n_x,n_y,E)$$

$$\equiv\overline{K}_{63}(n_x,n_y,E)[b_0-\Delta_{21}]$$

$$\overline{K}_{63}(n_x,n_y,E)\equiv\Big[-\left[E^2H_{31}+EH_{41}(n_x,n_y)+H_{51}(n_x,n_y)\right]^{1/2}$$

$$+\left[EH_{11}+H_{21}(n_x,n_y)\right]\Big]^{1/2}.$$

Therefore, (2.17) can be expressed as

$$k_z^2 = \overline{A}_3(E, n_x, n_y), \tag{2.18}$$

where $\overline{A}_3(E, n_x, n_y) = \left[\frac{1}{L_0^2}\left\{\cos^{-1}\left[\frac{1}{2}I_{23}(n_x, n_y, E)\right]\right\}^2 - \phi_1(n_x, n_y)\right]$.

The electron concentration per unit length is given by

$$n_0 = \frac{2g_v}{\pi}\sum_{n_x=1}^{n_{x\max}}\sum_{n_y=1}^{n_{y\max}}\left[\overline{D}_3(E_{F1D}, n_x, n_y) + \overline{D}_4(E_{F1D}, n_x, n_y)\right], \tag{2.19}$$

where $\overline{D}_3(E_{F1D}, n_x, n_y) = \sqrt{\overline{A}_3(E_{FID}, n_x, n_y)}$ and $\overline{D}_4(E_{F1D}, n_x, n_y) = \sum_{r=1}^{s_o} Z_{1D}(r)$ $[\overline{D}_3(E_{F1D}, n_x, n_y)]$. The field-emitted current assumes the form

$$I = \frac{2g_v e k_B T}{h}\sum_{n_x=1}^{n_{x\max}}\sum_{n_y=1}^{n_{y\max}}\left[F_0(\eta_{23})\exp(-\beta_{23})\right], \tag{2.20}$$

where $\eta_{23} = (E_{F1D} - \overline{E}_{23})/k_B T$, and $\overline{E}_{23}$ is the lowest positive root of the equation.

$$I_{23}(\overline{E}_{23}, n_x, n_y) = 2\cos[L_0(\phi_1(n_x, n_y))^{\frac{1}{2}}], \tag{2.21}$$

$$\beta_{23} = \frac{4}{3}[\overline{A}_3(V_0, n_x, n_y)]^{3/2} \cdot [3eF_{sz}\overline{A}_4(V_0, n_x, n_y)]^{-1},$$

$$\overline{A}_4(V_0, n_x, n_y) = \left(\frac{1}{L_0^2}\left\{\cos^{-1}\left[\frac{1}{2}I_{23}(n_x, n_y, V_0)\right]\right\}\right.$$
$$\left.\left\{1 - \frac{1}{4}I_{23}{}^2(n_x, n_y, V_0)\right\}^{-1/2}\overline{I}_{23}(n_x, n_y, V_0)\right)$$

$$\begin{aligned}\overline{I}_{23}(n_x, n_y, V_0) = {}& \left[2\bar{\beta}_{02}(n_x, n_y, V_0)\sinh\{\bar{\beta}_{23}(n_x, n_y, V_0)\}\cos\{\bar{\gamma}_{23}(n_x, n_y, V_0)\}\right.\\
&- 2g'_{222}(n_x, n_y, V_0)\sin\{g_{222}(n_x, n_y, V_0)\}\cdot\cosh\{g_{212}(n_x, n_y, V_0)\}\\
&+ Q'_{212}(n_x, n_y, V_0)\sinh\{g_{212}(n_x, n_y, V_0)\}\sin\{g_{222}(n_x, n_y, V_0)\}\\
&+ Q_{212}(n_x, n_y, V_0)[g'_{212}(n_x, n_y, V_0)\cdot\cosh\{g_{212}(n_x, n_y, V_0)\}\\
&\times\sin\{g_{222}(n_x, n_y, V_0)\}\\
&+ g'_{222}(n_x, n_y, V_0).\cos\{g_{222}(n_x, n_y, V_0)\}\sinh\{g_{212}(n_x, n_y, V_0)\}]\\
&+ \Delta_{212}\cdot\left[\left\{\frac{2J_{212}(n_x, n_y, V_0)J'_{212}(n_x, n_y, V_0)}{J_{222}(n_x, n_y, V_0)}\right.\right.\end{aligned}$$

$$-\frac{J_{212}^2(n_x,n_y,V_0)J_{222}'(n_x,n_y,V_0)}{J_{222}^2(n_x,n_y,V_0)}$$

$$-3J_{222}'(n_x,n_y,V_0)\}\cosh\{g_{212}(n_x,n_y,V_0)\}\sin\{g_{222}(n_x,n_y,V_0)\}$$

$$+\left\{\frac{J_{212}^2(n_x,n_y,V_0)}{J_{222}(n_x,n_y,V_0)}-3J_{222}(n_x,n_y,V_0)\right\}\{g_{212}'(n_x,n_y,V_0).$$

$$\sinh\{g_{212}(n_x,n_y,V_0)\}\sin\{g_{222}(n_x,n_y,V_0)\}$$

$$+g_{222}'(n_x,n_y,V_0)\cdot\cos\{g_{222}(n_x,n_y,V_0)\}\cosh\{g_{212}(n_x,n_y,V_0)\}\}$$

$$+\{3J_{212}'(n_x,n_y,V_0)+\frac{J_{222}^2(n_x,n_y,V_0)}{J_{212}^2(n_x,n_y,V_0)}.J_{212}'(n_x,n_y,V_0)$$

$$-\{\{J_{212}(n_x,n_y,V_0)\}^{-1}.\{2J_{222}'(n_x,n_y,V_0).J_{222}(n_x,n_y,V_0)\}\}$$

$$.\{\sinh[g_{212}(n_x,n_y,V_0)].\cos\{g_{222}(n_x,n_y,V_0)\}]\}$$

$$+\{g_{212}'(n_x,n_y,V_0).\cosh\{g_{212}(n_x,n_y,V_0)\}.\cos\left\{g_{222}\left(n_x,n_y,V_0\right)\right\}-g_{222}'\left(n_x,n_y,V_0\right)$$

$$\left.\times\left\{3J_{212}\left(n_x,n_y,V_0\right)-\frac{J_{222}^2(n_x,n_y,V_0)}{J_{212}(n_x,n_y,V_0)}\right\}\right]\sin\left\{g_{222}\left(n_x,n_y,V_0\right)\right\}.\sinh[g_{212}(n_x,n_y,V_0)]$$

$$+\Delta_{212}[[4J_{212}(n_x,n_y,V_0).J_{212}'(n_x,n_y,V_0)-4J_{222}(n_x,n_y,V_0)$$

$$J_{222}'(n_x,n_y,V_0)\}\{\cosh\left\{g_{212}\left(n_x,n_y,V_0\right)\right\}\cos\left\{g_{222}\left(n_x,n_y,V_0\right)\right\}\}$$

$$+\{2J_{212}^2(n_x,n_y,V_0)-2J_{222}^2(n_x,n_y,V_0)\}\{g_{212}'\left(n_x,n_y,V_0\right)\sinh[g_{212}\left(n_x,n_y,V_0\right)]$$

$$\cos\left\{g_{222}\left(n_x,n_y,V_0\right)\right\}-g_{222}'\left(n_x,n_y,V_0\right)\sin\left\{g_{222}\left(n_x,n_y,V_0\right)\right\}.$$

$$.\cosh[g_{222}\left(n_x,n_y,V_0\right)]+\frac{1}{12}\left[\frac{15J_{222}^2(n_x,n_y,V_0)J_{222}'(n_x,n_y,V_0)}{J_{212}(n_x,n_y,V_0)}\right.$$

$$-\frac{5J_{222}^3(n_x,n_y,V_0)J_{212}'(n_x,n_y,V_0)}{J_{212}^2(n_x,n_y,V_0)}-\frac{5J_{212}^3(n_x,n_y,V_0)J_{222}'(n_x,n_y,V_0)}{J_{222}^2(n_x,n_y,V_0)}$$

$$+\frac{15\,J_{212}^2(n_x,n_y,V_0)J_{212}'(n_x,n_y,V_0)}{J_{222}(n_x,n_y,V_0)}-34J_{212}'(n_x,n_y,V_0)\cdot J_{222}(n_x,n_y,V_0)$$

$$-34J_{212}(n_x,n_y,V_0)J_{222}'(n_x,n_y,V_0)]\{\sinh\{g_{212}(n_x,n_y,V_0)\}\sin\{g_{222}(n_x,n_y,V_0)\}\}$$

$$+\frac{1}{12}\left\{\frac{5J_{222}^3(n_x,n_y,V_0)}{J_{212}(n_x,n_y,V_0)}+\frac{5J_{212}^3(n_x,n_y,V_0)}{J_{222}(n_x,n_y,V_0)}-34J_{222}(n_x,n_y,V_0)J_{212}(n_x,n_y,V_0)\right\}$$

$$.\{g_{212}'\left(n_x,n_y,V_0\right)\cosh\left\{g_{212}\left(n_x,n_y,V_0\right)\right\}\sin\left\{g_{222}(n_x,n_y,V_0)\right\}$$

$$+g_{222}'\left(n_x,n_y,V_0\right).\cos\left\{g_{222}\left(n_x,n_y,V_0\right)\right\}\sinh\left\{g_{212}\left(n_x,n_y,V_0\right)\right\}\}]],$$

$$\bar{\beta}_{02}(n_x,n_y,V_0)=g_{212}'(n_x,n_y,V_0),\quad \bar{\beta}_{23}(n_x,n_y,V_0)=g_{212}(n_x,n_y,V_0),$$

$$\bar{\gamma}_{02}(n_x,n_y,V_0)=g_{222}'(n_x,n_y,V_0),\quad \bar{\gamma}_{23}(n_x,n_y,V_0)=g_{222}(n_x,n_y,V_0),$$

$$\overline{K}_{53}(n_x, n_y, V_0) = J_{212}(n_x, n_y, V_0), \quad \overline{K}_{05}(n_x, n_y, V_0) = J'_{212}(n_x, n_y, V_0),$$

$$\overline{K}_{65}(n_x, n_y, V_0) = J_{222}(n_x, n_y, V_0), \quad \overline{K}_{06}(n_x, n_y, V_0) = J'_{222}(n_x, n_y, V_0)$$

$$g'_{212}(n_x, n_y, V_0) = J'_{212}\left(n_x, n_y, V_0\right)[a_0 - \Delta_{212}], J'_{21}\left(n_x, n_y, V_0\right)$$

$$= [2\overline{K}_{53}\left(n_x, n_y, V_0\right)]^{-1}$$

$$\left[-H_{12} + \frac{(V_0 - \overline{V}_0)H_{32} + [H_{42}(n_x, n_y)/2]}{\left[(V_0 - \overline{V}_0)^2 H_{32} + (V_0 - \overline{V}_0)H_{42}(n_x, n_y) + H_{52}(n_x, n_y)\right]^{1/2}}\right],$$

$$g'_{222}\left(n_x, n_y, V_0\right) = J'_{222}\left(n_x, n_y, V_0\right)[b_0 - \Delta_{21}],$$

$$J'_{222}(n_x, n_y, V_0) = [2\overline{K}_{06}\left(n_x, n_y, V_0\right)]^{-1}$$

$$\left[H_{11} - \frac{(V_0)H_{31} + [H_{41}(n_x, n_y)/2]}{\left[(V_0)^2 H_{31} + (V_0)H_{41}(n_x, n_y) + H_{51}(n_x, n_y)\right]^{1/2}}\right]$$

and

$$Q'_{212}(n_x, n_y, V_0) \equiv \left[\frac{J'_{212}(n_x, n_y, V_0)}{J_{222}(n_x, n_y, V_0)} - \frac{J'_{222}(n_x, n_y, V_0)}{J_{212}(n_x, n_y, V_0)}\right.$$

$$\left. - \frac{J_{212}(n_x, n_y, V_0).J'_{222}(n_x, n_y, V_0)}{J^2_{222}(n_x, n_y, V_0)} - \frac{J_{222}(n_x, n_y, V_0).J'_{212}(n_x, n_y, V_0)}{J^2_{212}(n_x, n_y, V_0)}\right].$$

2.2.4 *The Field Emission from HgTe/CdTe Quantum Wire Superlattices with Graded Interfaces*

The dispersion relation of the conduction electrons of the constituent materials of HgTe/CdTe SLs can be written as

$$E = \frac{\hbar^2 k^2}{2m_{c1}} + \frac{3\,|e|^2\,k}{128\varepsilon_{sc}} \quad \text{and} \quad \frac{\hbar^2 k^2}{2m_{c2}} = EG(E_1 E_{g2}, \Delta_2). \tag{2.22}$$

The electron energy dispersion law in HgTe/CdTe SL is given by

$$\cos(L_o k) = \frac{1}{2}\Phi_3(E, k_s), \tag{2.23}$$

where

$$\begin{aligned}\Phi_3(E,k_s) \equiv\ & [2\cosh\{\beta_3(E,k_s)\}\cos\{\gamma_3(E,k_s)\} + \varepsilon_3(E,k_s)\sinh\{\beta_3(E,k_s)\}\\ & \times\sin\{\gamma_3(E,k_s)\}\\ & + \Delta_{21}\left[\left(\frac{\{K_7(E,k_s)\}^2}{K_8(E,k_s)} - 3K_8(E,k_s)\right)\cosh\{\beta_3(E,k_s)\}\sin\{\gamma_3(E,k_s)\}\right.\\ & \left. + \left(3K_7(E,k_s) - \frac{\{K_8(E,k_s)\}^2}{K_7(E,k_s)}\right)\sinh\{\beta_3(E,k_s)\}\cos\{\gamma_3(E,k_s)\}\right]\\ & + \Delta_{21}\left[2\left(\{K_7(E,k_s)\}^2 - \{K_8(E,k_s)\}^2\right)\cosh\{\beta_3(E,k_s)\}\cos\{\gamma_3(E,k_s)\}\right.\\ & \left.\left. + \frac{1}{12}\left[\frac{5\{K_8(E,k_s)\}^3}{K_7(E,k_s)} + \frac{5\{K_7(E,k_s)\}^3}{K_8(E,k_s)} - 34K_7(E,k_s)\,K_8(E,k_s)\right]\right]\right]\\ & \sinh\{\beta_3(E,k_s)\}\sin\{\gamma_3(E,k_s)\}]],\end{aligned}$$

$$\varepsilon_3(E,k_s) \equiv \left[\frac{K_7(E,k_s)}{K_8(E,k_s)} - \frac{K_8(E,k_s)}{K_7(E,k_s)}\right],$$

$$\beta_3(E,k_s) \equiv K_7(E,k_s)\,[a_0 - \Delta_{21}],$$

$$\gamma_3(E,k_s) = K_8(E,k_s)\,[b_0 - \Delta_{21}],$$

$$K_8(E,k_s) \equiv \left[\frac{B_0^2 + 2A_0E - B_0\sqrt{B_0^2 + 4A_0E}}{2A_0^2} - k_s^2\right]^{1/2},$$

$$B_0 = \frac{3|e|^2}{128\varepsilon_{sc}},$$

$$A = \frac{\hbar^2}{2m_{c1}} \text{ and } K_7(E,k_s) \equiv \left[\frac{2m_{c2}E\prime}{\hbar^2}G(E-V_0, E_{g_2}, \Delta_2) + k_s^2\right]^{1/2}, E' = \overline{V}_0 - E.$$

The electron dispersion law in HgTe/CdTe QWSLs can be expressed as

$$k_z = \left[\frac{1}{L_0^2}\{\rho_{11}(n_x,n_y,E)\} - \{\phi_1(n_x,n_y)\}\right]^{1/2}, \tag{2.24}$$

where

$$\rho_{11}(n_x,n_y,E) = \left[\cos^{-1}\left\{\frac{1}{2}\psi_{11}(n_x,n_y,E)\right\}\right]^2,$$

$$\begin{aligned}\psi_{11}(n_x,n_y,E) = & [2\cosh\{\beta_{11}(n_x,n_y,E)\}\cos\{\gamma_{11}(n_x,n_y,E)\}\\ & + \varepsilon_{11}(n_x,n_y,E)\sinh\{\beta_{11}(n_x,n_y,E)\}\end{aligned}$$

$$\times \sin\left\{\gamma_{11}\left(n_x,n_y,E\right)\right\} + \Delta_{21}\left[\left[\left(\frac{\left\{K_{13}\left(n_x,n_y,E\right)\right\}^2}{K_{14}\left(n_x,n_y,E\right)} - 3K_{14}\left(n_x,n_y,E\right)\right)\right.\right.$$

$$\times \cosh\left\{\beta_{11}\left(n_x,n_y,E\right)\right\}\sin\left\{\gamma_{11}\left(n_x,n_y,E\right)\right\}$$

$$+\left(3K_{13}\left(n_x,n_y,E\right) - \frac{\left\{K_{14}\left(n_x,n_y,E\right)\right\}^2}{K_{13}\left(n_x,n_y,E\right)}\right)$$

$$\times \sinh\left\{\beta_{11}\left(n_x,n_y,E\right)\right\}\cos\left\{\gamma_{11}\left(n_x,n_y,E\right)\right\}$$

$$+\Delta_{21}\left[2\left(\left\{K_{13}\left(n_x,n_y,E\right)\right\}^2 - \left\{K_{14}\left(n_x,n_y,E\right)\right\}^2\right)\cosh\left\{\beta_{11}\left(n_x,n_y,E\right)\right\}\right.$$

$$\times \cos\left\{\gamma_{11}\left(n_x,n_y,E\right)\right\} + \frac{1}{12}\left(\frac{5\left\{K_{13}\left(n_x,n_y,E\right)\right\}^3}{K_{14}\left(n_x,n_y,E\right)} + \frac{5\left\{K_{14}\left(n_x,n_y,E\right)\right\}^3}{K_{13}\left(n_x,n_y,E\right)}\right.$$

$$\left. - \left\{34K_{14}\left(n_x,n_y,E\right)K_{13}\left(n_x,n_y,E\right)\right\}\right)\sinh\left\{\beta_{11}\left(n_x,n_y,E\right)\right\}$$

$$\left.\left.\times \sin\left\{\gamma_{11}\left(n_x,n_y,E\right)\right\}\right]\right]\varepsilon_{11}\left(n_x,n_y,E\right) \equiv \left[\frac{K_{13}\left(n_x,n_y,E\right)}{K_{14}\left(n_x,n_y,E\right)} - \frac{K_{14}\left(n_x,n_y,E\right)}{K_{13}\left(n_x,n_y,E\right)}\right],$$

$$\beta_{11}\left(n_x,n_y,E\right) \equiv K_{13}\left(n_x,n_y,E\right)\left[a_0 - \Delta_{21}\right],$$

$$\gamma_{11}\left(n_x,n_y,E\right) = K_{14}\left(n_x,n_y,E\right)\left[b_0 - \Delta_{21}\right],$$

$$K_{14}\left(n_x,n_y,E\right) \equiv \left[\frac{B_0^2 + 2A_0E - B_0\sqrt{B_0^2 + 4A_0E}}{2A_0^2} - \left\{\phi_1\left(n_x,n_y\right)\right\}\right]^{1/2},$$

$$K_{13}\left(n_x,n_y,E\right) \equiv \left[\left(\frac{2m_{c2}E\prime}{\hbar^2}\right)G\left(E - \overline{V}_0, E_{g2}, \Delta_2\right) + \left\{\phi_1\left(n_x,n_y\right)\right\}\right]^{1/2}.$$

Therefore, (2.24) can be expressed as

$$k_z^2 = \overline{A}_5(E, n_x, n_y), \tag{2.25}$$

where $\overline{A}_5(E, n_x, n_y) = \left[\frac{1}{L_0^2}\left\{\cos^{-1}\left[\frac{1}{2}\psi_{11}(n_x, n_y, E)\right]\right\}^2 - \phi_1(n_x, n_y)\right]$.

The electron concentration per unit length is given by

$$n_0 = \frac{2g_v}{\pi}\sum_{n_x=1}^{n_{x\max}}\sum_{n_y=1}^{n_{y\max}}\left[\overline{D}_5\left(E_{F1D}, n_x, n_y\right) + \overline{D}_6\left(E_{F1D}, n_x, n_y\right)\right], \tag{2.26}$$

where $\overline{D}_5(E_{F1D}, n_x, n_y) = \sqrt{\overline{A}_5(E_{FID}, n_x, n_y)}$ and $\overline{D}_6(E_{F1D}, n_x, n_y) = \sum_{r=1}^{s_o} Z_{1D}(r)$ $[\ \overline{D}_5(E_{F1D}, n_x, n_y)]$.

The field-emitted current assumes the form

$$I = \frac{2g_v e k_B T}{h} \sum_{n_x=1}^{n_{x\max}} \sum_{n_y=1}^{n_{y\max}} [F_0(\eta_{24}) \exp(-\beta_{24})], \tag{2.27}$$

where $\eta_{24} = (E_{F1D} - E_{24})/k_B T$, and E_{24} is the lowest positive root of the equation.

$$\overline{A}_5(E_{24}, n_x, n_y) = 0, \tag{2.28}$$

$$\begin{aligned}
\beta_{24} &= \frac{4}{3}[\overline{A}_5(V_0, n_x, n_y)]^{3/2} \cdot [3eF_{sz}\overline{A}_6(V_0, n_x, n_y)]^{-1}, \\
\overline{A}_6(V_0, n_x, n_y) &= \left(\frac{1}{L_0^2}\left\{\cos^{-1}[\frac{1}{2}\psi_{11}(n_x, n_y, V_0)]\right\}\bar{\psi}_{11}(V_0, n_x, n_y)\right. \\
&\quad \left. \cdot \left\{1 - \frac{1}{4}\psi_{11}^2(n_x, n_y, V_0)\right\}^{-\frac{1}{2}}\right)\bar{\psi}_{11}(n_x, n_y, V_0) \\
&= \left[2\beta_{011}(n_x, n_y, V_0)\right. \\
&\quad \times \sinh\{\beta_{11}(n_x, n_y, V_0)\}\cos\{\gamma_{11}(n_x, n_y, V_0)\} - 2\gamma_{011}(n_x, n_y, V_0) \\
&\quad \times \sin\{\gamma_{11}(n_x, n_y, V_0)\} \cdot \cosh\{\beta_{11}(n_x, n_y, V_0)\} \\
&\quad + \varepsilon_{11}'(n_x, n_y, V_0)\sinh\{\beta_{11}(n_x, n_y, V_0)\}\sin\{\gamma_{11}(n_x, n_y, V_0)\} \\
&\quad + \varepsilon_{11}(n_x, n_y, V_0)[\beta_{011}(n_x, n_y, V_0) \cdot \cosh\{\beta_{11}(n_x, n_y, V_0)\} \\
&\quad \sin\{\gamma_{11}(n_x, n_y, V_0)\} \\
&\quad \left. + \gamma_{011}(n_x, n_y, V_0) . \cos\{\gamma_{11}(n_x, n_y, V_0)\}\sinh\{\beta_{11}(n_x, n_y, V_0)\}\right] \\
&\quad + \Delta_{21} \cdot \left[-\frac{[K_{13}(n_x, n_y, V_0)]^2 K_{014}(n_x, n_y, V_0)}{K_{14}(n_x, n_y, V_0)} - 3K_{014}(n_x, n_y, V_0)\right] \\
&\quad \times \left\{\frac{2K_{13}(n_x, n_y, V_0)K_{013}(n_x, n_y, V_0)}{K_{14}(n_x, n_y, V_0)} - 3K_{014}(n_x, n_y, V_0)\right\}. \\
&\quad \cosh\{\beta_{11}(n_x, n_y, V_0)\}\sin\{\gamma_{11}(n_x, n_y, V_0)\} \\
&\quad + \left\{\frac{K_{13}^2(n_x, n_y, V_0)}{K_{14}(n_x, n_y, V_0)} - 3K_{14}(n_x, n_y, V_0)\right\}\{\beta_{011}(n_x, n_y, V_0). \\
&\quad \sinh\{\beta_{11}(n_x, n_y, V_0)\}\sin\{\gamma_{11}(n_x, n_y, V_0)\} \\
&\quad + \gamma_{01}(n_x, n_y, V_0) \cdot \cos\{\gamma_{11}(n_x, n_y, V_0)\}\cosh\{\beta_{11}(n_x, n_y, V_0)\}\} \\
&\quad + \{3K_{013}(n_x, n_y, V_0) + \frac{K_{14}^2(n_x, n_y, V_0)}{K_{13}^2(n_x, n_y, V_0)} . K_{013}(n_x, n_y, V_0) \\
&\quad - \{\{K_{13}(n_x, n_y, V_0)\}^{-1} . \{2K_{14}(n_x, n_y, V_0) . K_{014}(n_x, n_y, V_0)\}\} \\
&\quad . \{\sinh[\beta_{11}(n_x, n_y, V_0)] . \cos\{\gamma_{11}(n_x, n_y, V_0)\}\}]\} + \{3K_{13}(n_x, n_y, V_0)
\end{aligned}$$

$$
\left.-\frac{K_{14}^{2}(n_x,n_y,V_0)}{K_{13}(n_x,n_y,V_0)}\right\}.[\beta_{011}\left(n_x,n_y,V_0\right)\cos\left\{\gamma_{11}\left(n_x,n_y,V_0\right)\right\}
$$

$$
.\cosh\left\{\beta_{11}\left(n_x,n_y,V_0\right)\right\}-\gamma_{011}\left(n_x,n_y,V_0\right)\sin\left\{\gamma_{11}\left(n_x,n_y,V_0\right)\right\}
$$

$$
.\sinh[\beta_{11}\left(n_x,n_y,V_0\right)]+\Delta_{21}[[4K_{13}(n_x,n_y,V_0).K_{013}(n_x,n_y,V_0)
$$

$$
-4K_{14}(n_x,n_y,V_0)K_{014}(n_x,n_y,V_0)\}\{\cosh\left\{\beta_{11}\left(n_x,n_y,V_0\right)\right\}
$$

$$
\times\cos\left\{\gamma_{11}\left(n_x,n_y,V_0\right)\right\}\}+\{2K_{13}^{2}(n_x,n_y,V_0)-2K_{14}^{2}(n_x,n_y,V_0)\}
$$

$$
\times\{\beta_{011}\left(n_x,n_y,V_0\right)\sinh[\beta_{11}\left(n_x,n_y,V_0\right)]\cos\left\{\gamma_{11}\left(n_x,n_y,V_0\right)\right\}
$$

$$
-\gamma_{011}\left(n_x,n_y,V_0\right)\sin\left\{\gamma_{11}\left(n_x,n_y,V_0\right)\right\}.
$$

$$
.\cosh[\beta_{11}\left(n_x,n_y,V_0\right)]+\frac{1}{12}\left[\frac{15K_{13}^{2}(n_x,n_y,V_0)K_{013}(n_x,n_y,V_0)}{K_{14}(n_x,n_y,V_0)}\right.
$$

$$
-\frac{5K_{13}^{3}(n_x,n_y,V_0)K_{014}(n_x,n_y,V_0)}{K_{14}^{2}(n_x,n_y,V_0)}-\frac{5K_{14}^{3}(n_x,n_y,V_0)K_{013}(n_x,n_y,V_0)}{K_{13}^{2}(n_x,n_y,V_0)}
$$

$$
+\frac{15K_{14}^{2}(n_x,n_y,V_0)K_{014}(n_x,n_y,V_0)}{K_{13}(n_x,n_y,V_0)}-34\,K_{013}(n_x,n_y,V_0)\cdot K_{14}(n_x,n_y,V_0)
$$

$$
-34\,K_{13}(n_x,n_y,V_0)K_{014}(n_x,n_y,V_0)]\{\sinh\left\{\beta_{11}\left(n_x,n_y,V_0\right)\right\}
$$

$$
\times\sin\left\{\gamma_{11}\left(n_x,n_y,V_0\right)\right\}\}+\{\beta_{011}\left(n_x,n_y,V_0\right)\cosh\left\{\beta_{11}\left(n_x,n_y,V_0\right)\right\}
$$

$$
\times\sin\left\{\gamma_{11}\left(n_x,n_y,V_0\right)\right\}+\{\gamma_{011}\left(n_x,n_y,V_0\right)\cos\left\{\gamma_{11}\left(n_x,n_y,V_0\right)\right\}
$$

$$
\times\sinh\left\{\beta_{11}\left(n_x,n_y,V_0\right)\right\}+\frac{1}{12}\left\{\frac{5\,K_{13}^{3}(n_x,n_y,V_0)}{K_{14}(n_x,n_y,V_0)}\right.
$$

$$
+\frac{5\,K_{14}^{3}(n_x,n_y,V_0)}{K_{13}(n_x,n_y,V_0)}-34\,K_{13}(n_x,n_y,V_0)K_{14}(n_x,n_y,V_0)\}]]
$$

$$
\times\beta_{011}\left(n_x,n_y,V_0\right)=K_{013}\left(n_x,n_y,V_0\right)[a_0-\Delta_{21}]
$$

$$
K_{013}(n_x,n_y,V_0)=[K_{13}(n_x,n_y,V_0)]^{-1}\left[-\frac{m_{c2}}{\hbar^2}G(V_0-\overline{V}_0,\alpha_2,\Delta_2)\right.
$$

$$
\left.+\frac{m_{c2}}{\hbar^2}(\overline{V}_0-V_0)G'(V_0-\overline{V}_0,\alpha_2,\Delta_2)\right],G'(V_0-\overline{V}_0,\alpha_2,\Delta_2)
$$

$$
=G(V_0-\overline{V}_0,\alpha_2,\Delta_2)^{\cdot}\left[\frac{1}{(V_0-\overline{V}_0+E_{g2}+\Delta_2)}\right.
$$

$$
\left.+\frac{1}{(V_0-\overline{V}_0+E_{g2})}-\frac{1}{(V_0-\overline{V}_0+E_{g2}+\frac{2}{3}\Delta_2)}\right]
$$

$$
\gamma_{011}\left(n_x,n_y,V_0\right)=K_{014}\left(n_x,n_y,V_0\right)[b_0-\Delta_{21}],\ K_{014}\left(n_x,n_y,V_0\right)
$$

$$
=[K_{14}\left(n_x,n_y,V_0\right)]^{-1}\left[\tfrac{V_0}{A_0}-\tfrac{B_0}{A_0}\{B_0^2+4A_0V_0\}^{-/2}\right],
$$

and

$$\varepsilon'_{11}(n_x,n_y,V_0) \equiv \left[\frac{K_{013}(n_x,n_y,V_0)}{K_{14}(n_x,n_y,V_0)} - \frac{K_{014}(n_x,n_y,V_0)}{K_{13}(n_x,n_y,V_0)} - \frac{K_{13}(n_x,n_y,V_0).K_{014}(n_x,n_y,V_0)}{K_{14}^2(n_x,n_y,V_0)} - \frac{K_{14}(n_x,n_y,V_0).K_{013}(n_x,n_y,V_0)}{K_{13}^2(n_x,n_y,V_0)}\right].$$

2.2.5 *The Field Emission from Quantum Wire III–V Effective Mass Superlattices*

Following Sasaki [37], the electron dispersion law in III–V effective mass superlattices (EMSLs) can be written as

$$k_x^2 = \left[\frac{1}{L_0^2}\left\{\cos^{-1}\left(f_{21}\left(E,k_y,k_z\right)\right)\right\}^2 - k_\perp^2\right], \tag{2.29}$$

in which

$$f_{21}(E,k_y,k_z) = a_1\cos\left[a_0C_{21}(E,k_\perp)+b_0D_{21}(E,k_\perp)\right] - a_2\cos\left[a_0C_{21}(E,k_\perp) - b_0D_{21}(E,k_\perp)\right], \quad k_\perp^2 = k_y^2 + k_z^2,$$

$$a_1 = \left[\sqrt{\frac{m_{c2}}{m_{c1}}}+1\right]^2\left[4\left(\frac{m_{c2}}{m_{c1}}\right)^{1/2}\right]^{-1},$$

$$a_2 = \left[-1+\sqrt{\frac{m_{c2}}{m_{c1}}}\right]^2\left[4\left(\frac{m_{c2}}{m_{c1}}\right)^{1/2}\right]^{-1},$$

$$C_{21}(E,k_\perp) \equiv \left[\left(\frac{Em_{c1}}{\hbar^2}\right)G\left(E,E_{g_1},\Delta_1\right) - k_\perp^2\right]^{1/2} \text{ and}$$

$$D_{21}(E,k_\perp) \equiv \left[\left(\frac{Em_{c2}}{\hbar^2}\right)G\left(E,E_{g_2},\Delta_2\right) - k_\perp^2\right]^{1/2}$$

The electron dispersion law in III–V effective mass quantum wire superlattices (EMQWSLs) can be expressed as

$$k_x^2 = \left[\rho_{21}\left(n_y,n_z,E\right)\right], \tag{2.30}$$

in which

$$\rho_{21}(n_y,n_z,E) = \frac{1}{L_0^2}\left[\cos^{-1}\left(f_{22}\left(n_y,n_z,E\right)\right)\right]^2 - \left\{\phi_2(n_y,n_z)\right\},$$

$$\phi_2(n_y, n_z) = \left\{ \left(\frac{n_y \pi}{d_y}\right)^2 + \left(\frac{n_z \pi}{d_z}\right)^2 \right\},$$

$$f_{22}(n_y, n_z, E) = a_1 \cos\left[a_0 C_{22}(n_y, n_z, E) + b_0 D_{22}(n_y, n_z, E)\right] - a_2 \cos\left[a_0 C_{22}(n_y, n_z, E) - b_0 D_{22}(n_y, n_z, E)\right],$$

$$C_{22}(n_y, n_z, E) \equiv \left[\left(\frac{2m_{c1}E}{\hbar^2}\right) G(E, E_{g_1}, \Delta_1) - \{\phi_2(n_y, n_z)\}\right]^{1/2}$$

and
$$D_{22}(n_y, n_z, E) \equiv \left[\left(\frac{2m_{c2}E}{\hbar^2}\right) G(E, E_{g_1}, \Delta_2) - \{\phi_2(n_y, n_z)\}\right]^{1/2}.$$

The electron concentration per unit length is given by

$$n_0 = \frac{2g_v}{\pi} \sum_{n_z=1}^{n_{z\max}} \sum_{n_y=1}^{n_{y\max}} \left[\overline{D}_7(E_{F1D}, n_z, n_y) + \overline{D}_8(E_{F1D}, n_z, n_y)\right], \tag{2.31}$$

where $\overline{D}_7(E_{F1D}, n_z, n_y) = \sqrt{\rho_{21}(E_{FID}, n_z, n_y)}$ and $\overline{D}_8(E_{F1D}, n_z, n_y) = \sum_{r=1}^{s_0} Z_{1D}(r)\left[\overline{D}_7(E_{F1D}, n_z, n_y)\right]$.

The field-emitted current assumes the form

$$I = \frac{2g_v e k_B T}{h} \sum_{n_z=1}^{n_{z\max}} \sum_{n_y=1}^{n_{y\max}} \left[F_0(\eta_{26}) \exp(-\beta_{26})\right], \tag{2.32}$$

where $\eta_{26} = (E_{F1D} - \bar{E}_{26})/k_B T$, and $\bar{E}_{26}$ is the lowest positive root of the equation.

$$\rho_{21}(E_{26}, n_z, n_y) = 0, \tag{2.33}$$

$$\beta_{26} = \frac{4}{3}[\rho_{21}(V_0, n_z, n_y)]^{3/2} \cdot [3eF_{sz}\bar{\rho}_{21}(V_0, n_z, n_y)]^{-1},$$

$$\bar{\rho}_{21}(V_0, n_z, n_y) = \left(\frac{2}{L_0^2}\{\cos^{-1}[f_{22}(n_z, n_y, V_0)]\}\, \bar{f}_{22}(V_0, n_z, n_y) .\{1 - f_{22}^2(n_x, n_y, V_0)\}^{-1/2}\right)$$

$$\bar{f}_{22}(V_0, n_y, n_z) = [a_1 \sin\{a_0 C_{22}(V_0, n_y, n_z) + b_0 D_{22}(V_0, n_y, n_z)\}\{a_0 \overline{C}_{22}(V_0, n_y, n_z) + b_0 \overline{D}_{22}(V_0, n_y, n_z)\} - a_2 \sin\{a_0 C_{22}(V_0, n_y, n_z) - b_0 D_{22}(V_0, n_y, n_z)\}\{a_0 \overline{C}_{22}(V_0, n_y, n_z) - b_0 \overline{D}_{22}(V_0, n_y, n_z)\}]$$

$$\overline{C}_{22}(V_0,n_z,n_y) = \left(\frac{m_{c1}}{\hbar^2}\right)[C_{22}(V_0,n_z,n_y)]^{-1}[G(V_0,E_{g1},\Delta_1)$$

$$\cdot\left[1+V_0\left[\frac{1}{(V_0+E_{g1}+\Delta_1)}+\frac{1}{(V_0+E_{g1})}-\frac{1}{(V_0+E_{g1}+\frac{2}{3}\Delta_1)}\right]\right]\right]$$

$$\overline{D}_{22}(V_0,n_z,n_y) = \left(\frac{m_{c2}}{\hbar^2}\right)[D_{22}(V_0,n_z,n_y)]^{-1}[G(V_0,E_{g2},\Delta_2)$$

$$\cdot\left[1+V_0\left[\frac{1}{(V_0+E_{g2}+\Delta_2)}+\frac{1}{(V_0+E_{g2})}-\frac{1}{(V_0+E_{g2}+\frac{2}{3}\Delta_2)}\right]\right]\right].$$

2.2.6 The Field Emission from Quantum Wire II–VI Effective Mass Superlattices

Following Sasaki [37], the electron dispersion law in II–VI EMSLs can be written as

$$k_z^2 = \left[\frac{1}{L_0^2}\left\{\cos^{-1}\left(f_{23}\left(E,k_x,k_y\right)\right)\right\}^2 - k_s^2\right], \tag{2.34}$$

in which

$$f_{23}\left(E,k_x,k_y\right) = \{a_3\cos\left[a_0C_{23}\left(E,k_s\right)+b_0D_{23}\left(E,k_s\right)\right]$$
$$-a_4\cos\left[a_0C_{23}\left(E,k_s\right)-b_0D_{23}\left(E,k_s\right)\right]\},\quad k_s^2=k_x^2+k_y^2,$$

$$a_3 = \left[\sqrt{\frac{m_{c2}}{m_{\parallel,1}^*}}+1\right]^2\left[4\left(\frac{m_{c2}}{m_{\parallel,1}^*}\right)^{1/2}\right]^{-1},$$

$$a_4 = \left[-1+\sqrt{\frac{m_{c2}}{m_{\parallel,1}^*}}\right]^2\left[4\left(\frac{m_{c2}}{m_{\parallel,1}^*}\right)^{1/2}\right]^{-1},$$

$$C_{23}\left(E,k_s\right) \equiv \left(\frac{2m_{\parallel,1}^*}{\hbar^2}\right)^{1/2}\left[E-\frac{\hbar^2k_s^2}{2m_{\perp,1}^*}\mp C_0k_s\right]^{1/2} \quad \text{and}$$

$$D_{23}\left(E,k_s\right) \equiv \left[\left(\frac{2m_{c2}}{\hbar^2}\right)\left[EG(E,E_{g2},\Delta_2)\right]-k_s^2\right]^{1/2}.$$

The dispersion law in II–VI, EMQWSLs can be expressed as

$$k_z^2 = \frac{1}{L_0^2}\left[\cos^{-1}\left(f_{24}\left(n_x, n_y, E\right)\right)\right]^2 - \left\{\phi_1\left(n_x, n_y\right)\right\}, \tag{2.35}$$

in which

$$f_{24}\left(n_x, n_y, E\right) = a_3 \cos\left[a_0 C_{24}\left(n_x, n_y, E\right) + b_0 D_{24}\left(n_x, n_y, E\right)\right]$$
$$-a_4 \cos\left[a_0 C_{24}\left(n_x, n_y, E\right) + b_0 D_{24}\left(n_x, n_y, E\right)\right],$$

$$C_{24}\left(n_x, n_y, E\right) \equiv \left(\frac{2m_{\parallel,1}^*}{\hbar^2}\right)^{1/2}\left[E - \left\{\frac{\hbar^2}{2m_{\perp,1}^*}\phi_1\left(n_x, n_y\right)\right\}\right.$$
$$\left. \mp C_0\left\{\phi_1\left(n_x, n_y\right)\right\}^{1/2}\right]^{1/2} \text{ and}$$

$$D_{24}\left(n_x, n_y, E\right) \equiv \left[\left(\frac{2m_{c2}}{\hbar^2}\right) EG\left(E, E_{g_2}, \Delta_2\right) - \phi_1\left(n_x, n_y\right)\right]^{1/2}.$$

Equation (2.35) can be written as

$$k_z^2 = \left[\rho_{22}\left(n_x, n_y, E\right)\right], \tag{2.36}$$

where $\rho_{22}(n_x, n_y, E) = \frac{1}{L_0^2}\left[\cos^{-1}\left(f_{24}(n_x, n_y, E)\right)\right]^2 - \{\phi_1(n_x, n_y)\}$.

The electron concentration per unit length is given by

$$n_0 = \frac{g_v}{\pi}\sum_{n_x=1}^{n_{x\max}}\sum_{n_y=1}^{n_{y\max}}\left[\overline{D}_9\left(E_{F1D}, n_x, n_y\right) + \overline{D}_{10}(E_{F1D}, n_x, n_y)\right], \tag{2.37}$$

where $\overline{D}_9(E_{F1D}, n_x, n_y) = \sqrt{\rho_{22}(E_{F1D}, n_x, n_y)}$ and $\overline{D}_{10}(E_{F1D}, n_x, n_y) = \sum\limits_{r=1}^{s_o} Z_{1D}(r)$ $[\overline{D}_9(E_{F1D}, n_x, n_y)]$.

The field-emitted current assumes the form

$$I = \frac{g_v e k_B T}{h}\sum_{n_x=1}^{n_{x\max}}\sum_{n_y=1}^{n_{y\max}}\left[F_0(\eta_{27})\exp(-\beta_{27})\right], \tag{2.38}$$

where $\eta_{27} = (E_{F1D} - \overline{E}_{27})/k_B T$, and $\overline{E}_{27}$ is the lowest positive root of the equation.

$$\rho_{22}(\overline{E}_{27}, n_x, n_y) = 0, \tag{2.39}$$

$$\beta_{27} = \frac{4}{3}[\rho_{22}(V_0, n_x, n_y)]^{3/2} \cdot [3eF_{sz}\bar{\rho}_{22}(V_0, n_x, n_y)]^{-1},$$

$$\bar{\rho}_{22}(V_0, n_x, n_y) = \left(\frac{2}{L_0^2}\{\cos^{-1}[f_{24}(n_x, n_y, V_0)]\}\,\bar{f}_{24}(V_0, n_x, n_y)\right.$$
$$\left. .\{1 - f_{24}^2(n_x, n_y, V_0)\}^{-1/2}\right)$$

$$\bar{f}_{24}(V_0, n_x, n_y) = [a_3 \sin\{a_0 C_{24}(V_0, n_x, n_y) + b_0 D_{24}(V_0, n_x, n_y)\}$$
$$\{a_0 \overline{C}_{24}(V_0, n_x, n_y) + b_0 \overline{D}_{24}(V_0, n_x, n_y)\}$$
$$- a_4 \sin\{a_0 C_{24}(V_0, n_x, n_y) - b_0 D_{24}(V_0, n_x, n_y)\}\{a_0 \overline{C}_{24}(V_0, n_x, n_y)$$
$$- b_0 \overline{D}_{24}(V_0, n_x, n_y)\}]$$

$$\overline{C}_{24}(V_0, n_x, n_y) = \left(\frac{m_{\parallel,1}^*}{\hbar^2}\right)[C_{24}(V_0, n_x, n_y)]^{-1}$$

$$\overline{D}_{24}(V_0, n_x, n_y) = \left(\frac{m_{c2}}{\hbar^2}\right)[D_{24}(V_0, n_x, n_y)]^{-1}[G(V_0, E_{g2}, \Delta_2)$$
$$\cdot\left[1 + V_0\left[\frac{1}{(V_0+E_{g2}+\Delta_2)} + \frac{1}{(V_0+E_{g2})} - \frac{1}{(V_0+E_{g2}+\frac{2}{3}\Delta_2)}\right]\right]\Bigg]$$

2.2.7 *The Field Emission from Quantum Wire IV–VI Effective Mass Superlattices*

Following Sasaki [37], the electron dispersion law in IV–VI, EMSLs can be written as

$$k_x^2 = \left[\frac{1}{L_0^2}\left\{\cos^{-1}\left(f_{26}\left(E, k_y, k_z\right)\right)\right\}^2 - k_\perp^2\right] \tag{2.40}$$

in which,

$$f_{26}\left(E, k_y, k_z\right) = a_5 \cos\left[a_0 C_{42}\left(E, k_y, k_z\right) + b_0 D_{42}\left(E, k_y, k_z\right)\right]$$
$$- a_6 \cos\left[a_0 C_{42}\left(E, k_y, k_z\right) - b_0 D_{42}\left(E, k_y, k_z\right)\right],$$

$$a_5 = \left[\sqrt{\frac{m_2^*}{m_1^*}} + 1\right]^2 \left[4\left(\frac{m_2^*}{m_1^*}\right)^{1/2}\right]^{-1}, \quad k_\perp^2 = k_y^2 + k_z^2,$$

$$m_i^* = \left[\frac{\hbar^2}{2\{a_i^2 - C_i^2\}}\right]\left[a_i - \left[a_i C_i + a_i e_i E_{g_i} + C_i^2 E_{g_i}\right]\right.$$
$$\left.\left[E_{g_i}^2 a_i^2 + C_i^2 + e_i^2 E_{g_i}^2 + 2C_i a_i E_{g_i} + 2E_{g_i} e_i C_i + 2e_i a_i E_{g_i}^2\right]^{-1/2}\right]$$

$$a_6 = \left[-1+\sqrt{\frac{m_2^*}{m_1^*}}\right]^2 \left[4\left(\frac{m_2^*}{m_1^*}\right)^{1/2}\right]^{-1},$$

$$C_{42}\left(E,k_y,k_z\right) \equiv \left[\left[-k_y^2 + ER_{21} + R_{31}\left(k_z\right)\right]\right.$$
$$\left. -\left[E^2R_{41} + EH_{51}\left(k_z\right) + R_{61}\left(k_z\right)\right]^{1/2}\right]^{1/2}, \; i = 1,2$$

$$R_{2i} = a_i.\left(a_i^2 - C_i^2\right)^{-1},$$

$$R_{3i}\left(k_z\right) = \left[2(a_i^2 - C_i^2)\right]^{-1}\left[E_{g_i}a_i + C_i + e_iE_{g_i} + 2k_z^2\left(C_if_i - a_ib_i\right)\right],$$

$$R_{4i} = \frac{C_i^2}{\left(a_i^2 - C_i^2\right)^2},$$

$$R_{5i}\left(k_z\right) = \left[(a_i^2 - C_i^2)\right]^{-2}\left[k_z^2(2b_iC_i^2 + 2C_if_ia_i) + \left[C_ia_i + e_iE_{g_i}a_i\right]\right]$$

$$R_{6i}\left(k_z\right) \equiv \left[2(a_i^2 - C_i^2)\right]^{-2}\left[(k_z)^4\left[-8a_ib_if_i^2 + 4a_i^2f_i^2 + 4b_i^2C_i^2\right] + \left(k_z^2\right)\right.$$
$$\left[-4a_iC_ib_i - 4a_ib_ie_iE_{g_i} + 4a_iC_if_iE_{g_i} + 4f_i^2C_i^2 + 4C_if_ie_iE_{g_i}\right.$$
$$\left. + 4a_i^2d_i + 4a_i^2f_iE_{g_i} - 4b_iC_i^2E_{g_i} - 4C_i^2d_i + 4C_i^2f_iE_{g_i}\right]$$
$$\left. + \left[E_{g_i}^2a_i^2 + C_i^2 + e_i^2E_{g_i}^2 + 2E_{g_i}a_iC_i + 2E_{g_i}e_iC_i + 2E_{g_i}^2a_ie_i\right]\right],$$

$$D_{42}\left(E,k_y,k_z\right) \equiv \left[\left[-k_y^2 + ER_{22} + R_{32}\left(k_z\right)\right]\right.$$
$$\left. -\left[E^2R_{42} + ER_{52}\left(k_z\right) + R_{62}\left(k_z\right)\right]^{1/2}\right]^{1/2}$$

The dispersion law in IV–VI, EMQWSLs can be expressed as

$$k_x^2 = \left[\rho_{29}\left(n_y,n_z,E\right)\right] \tag{2.41}$$

in which,

$$\rho_{29}\left(n_y,n_z,E\right) = \frac{1}{L_0^2}\left[\cos^{-1}\left(f_{29}\left(n_y,n_z,E\right)\right)\right]^2 - \left\{\phi_2\left(n_y,n_z\right)\right\},$$

$$f_{29}\left(n_y,n_z,E\right) = a_5\cos\left[a_0C_{44}\left(n_y,n_z,E\right) + b_0D_{44}\left(n_y,n_z,E\right)\right]$$
$$- a_6\cos\left[a_0C_{44}\left(n_y,n_z,E\right) - b_0D_{44}\left(n_y,n_z,E\right)\right],$$

$$C_{44}(E,n_y,n_z) = \left[-\left(\frac{n_y\pi}{d_y}\right)^2 + ER_{21} + R_{31}(n_z) - \left[R_{41}E^2\right.\right.$$
$$\left.\left. + R_{51}(n_z)E + R_{61}(n_z)\right]^{1/2}\right]$$

$$R_{3i}(n_z) = \left[2(a_i^2 - C_i^2)\right]^{-1}\left[E_{g_i}a_i + C_i + e_iE_{g_i} + 2\left(\frac{n_z\pi}{d_z}\right)^2(C_if_i - a_ib_i)\right],$$

$$R_{5i}(n_z) = \left[(a_i^2 - C_i^2)\right]^{-2}\left[(n_z\pi d_z)^2(2b_iC_i^2 + 2C_if_ia_i) + \left[C_ia_i + e_iE_{g_i}a_i\right]\right]$$

$$\begin{aligned}R_{6i}(n_z) \equiv{} & \left[2(a_i^2 - C_i^2)\right]^{-2}\left[(n_z\pi d_z)^4\left[-8a_ib_if_i^2 + 4a_i^2f_i^2 + 4b_i^2C_i^2\right]\right. \\ & + (n_z\pi/d_z)^2\left[-4a_i\,C_ib_i - 4a_ib_ie_iE_{g_i} + 4f_i^2C_i^2 + 4C_if_ie_iE_{g_i}\right. \\ & \left. + 4a_i^2d_i + 4a_i^2f_iE_{g_i} - 4b_iC_i^2E_{g_i} - 4C_i^2d_i + 4C_i^2f_iE_{g_i}\right] \\ & \left. + \left[E_{g_i}^2a_i^2 + C_i^2 + e_i^2E_{g_i}^2 + 2E_{g_i}a_iC_i + 2E_{g_i}e_iC_i + 2E_{g_i}^2a_ie_i\right]\right]\end{aligned}$$

$$\begin{aligned}D_{44}(E, n_y, n_z) ={} & [-(n_y\pi/d_y)^2 + ER_{22} + R_{32}(n_z) \\ & -[R_{42}E^2 + R_{52}(n_z)E + R_{62}(n_z)]^{1/2}]^{1/2}\end{aligned}$$

The electron concentration per unit length is given by

$$n_0 = \frac{2g_v}{\pi}\sum_{n_y=1}^{n_{y\max}}\sum_{n_z=1}^{n_{z\max}}\left[D_{11}\left(E_{F1D}, n_y, n_z\right) + D_{12}\left(E_{F1D}, n_y, n_z\right)\right] \tag{2.42}$$

where, $D_{11}(E_{F1D}, n_y, n_z) = \sqrt{\rho_{29}(E_{FID}, n_y, n_z)}$ and $D_{12}(E_{F1D}, n_y, n_z) = \sum_{r=1}^{s_o} Z_{1D}(r)[D_{11}(E_{F1D}, n_y, n_z)]$

The field-emitted current assumes the form

$$I = \frac{2g_vek_BT}{h}\sum_{n_y=1}^{n_{y\max}}\sum_{n_z=1}^{n_{z\max}}[F_0(\eta_{28})\exp(-\beta_{28})] \tag{2.43}$$

$\eta_{28} = \frac{E_{F1D} - \overline{E}_{28}}{k_BT}$, $\overline{E}_{28}$ is the lowest positive root of the equation.

$$\rho_{29}(\overline{E}_{28}, n_y, n_z) = 0, \tag{2.44}$$

$$\beta_{28} = \frac{4}{3}[\rho_{29}(V_0, n_y, n_z)]^{3/2}\cdot[3eF_{sz}\rho_{30}(V_0, n_y, n_z)]^{-1},$$

$$\begin{aligned}\rho_{30}(V_0, n_y, n_z) ={} & \left(\frac{2}{L_0^2}\left\{\cos^{-1}[f_{29}\left(n_y, n_z, V_0\right)]\right\}f_{30}(V_0, n_y, n_z)\right. \\ & \left. .\{1 - f_{29}^2\left(n_y, n_z, V_0\right)\}^{-1/2}\right)\end{aligned}$$

$$\begin{aligned}f_{30}(V_0, n_y, n_z) ={} & [a_5\sin\{a_0C_{44}(V_0, n_y, n_z) \\ & + b_0D_{44}(V_0, n_y, n_z)\}\{a_0\overline{C}_{44}(V_0, n_y, n_z) \\ & + b_0\overline{D}_{44}(V_0, n_y, n_z)\} - a_6\sin\{a_0C_{44}(V_0, n_y, n_z)\end{aligned}$$

$$- b_0 D_{44}(V_0, n_y, n_z)\}\{a_0 \overline{C}_{44}(V_0, n_y, n_z) - b_0 \overline{D}_{44}(V_0, n_y, n_z)\}]$$

$$\overline{C}_{44}(V_0, n_y, n_z) = [2C_{44}(V_0, n_y, n_z)]^{-1} \cdot \left[R_{21} - \frac{\left[R_{41} V_0 + \frac{1}{2} R_{51}(n_z)\right]}{[R_{41} V_0^2 + R_{51}(n_z) V_0 + R_{61}(n_z)]^{1/2}} \right]$$

$$\overline{D}_{44}(V_0, n_y, n_z) = [2D_{44}(V_0, n_y, n_z)]^{-1} \cdot \left[R_{22} - \frac{[R_{42} V_0 + \frac{1}{2} R_{52}(n_z)]}{[R_{42} V_0^2 + R_{52}(n_z) V_0 + R_{62}(n_z)]^{1/2}} \right]$$

2.2.8 *The Field Emission from Quantum Wire HgTe/CdTe Effective Mass Superlattices*

Following Sasaki [37], the electron dispersion law in HgTe/CdTe EMSLs can be written as

$$k_x^2 = \left[\frac{1}{L_0^2} \left\{ \cos^{-1} \left(f_{31} \left(E, k_y, k_z \right) \right) \right\}^2 - k_\perp^2 \right] \tag{2.45}$$

in which,

$$f_{31} \left(E, k_y, k_z \right) = a_7 \cos \left[a_0 C_{47} \left(E, k_\perp \right) + b_0 D_{47} \left(E, k_\perp \right) \right] - a_8 \cos \left[a_0 C_{47} \left(E, k_\perp \right) - b_0 D_{47} \left(E, k_\perp \right) \right],$$

$$a_7 = \left[\sqrt{\frac{m_{c2}}{m_{c1}}} + 1 \right]^2 \left[4 \left(\frac{m_{c2}}{m_{c1}} \right)^{1/2} \right]^{-1},$$

$$a_8 = \left[-1 + \sqrt{\frac{m_{c2}}{m_{c1}}} \right]^2 \left[4 \left(\frac{m_{c2}}{m_{c1}} \right)^{1/2} \right]^{-1}, C_{47} \left(E, k_\perp \right) \equiv \left[\frac{B_0^2 + 2A_0 E - B_0 \sqrt{B_0^2 + 4A_0 E}}{2A_0^2} - k_\perp^2 \right]^{1/2}$$

$$D_{47} \left(E, k_\perp \right) \equiv \left[\left(\frac{2E m_{c2}}{\hbar^2} \right) G \left(E, E_{g_2}, \Delta_2 \right) - k_\perp^2 \right]^{1/2}, \quad k_\perp^2 = k_y^2 + k_z^2$$

The dispersion law in HgTe/CdTe, EMQWSLs can be expressed as

$$k_x^2 = \left[\rho_{30} \left(n_y, n_z, E \right) \right] \tag{2.46}$$

in which,

$$\rho_{30}\left(n_y, n_z, E\right) = \frac{1}{L_0^2}\left[\cos^{-1}\left(f_{32}\left(n_y, n_z, E\right)\right)\right]^2 - \left\{\phi_2\left(n_y, n_z\right)\right\},$$

$$f_{32}\left(n_y, n_z, E\right) = a_7 \cos\left[a_0 C_{48}\left(n_y, n_z, E\right) + b_0 D_{48}\left(n_y, n_z, E\right)\right] - a_8 \cos\left[a_0 C_{48}\left(n_y, n_z, E\right) - b_0 D_{48}\left(n_y, n_z, E\right)\right],$$

$$C_{48}\left(n_y, n_z, E\right) \equiv \left[\frac{B_0^2 + 2A_0 E - B_0\sqrt{B_0^2 + 4A_0 E}}{2A_0^2} - \left\{\phi_2\left(n_y, n_z\right)\right\}\right]^{1/2}$$

and

$$D_{48}\left(n_y, n_z, E\right) \equiv \left[\left(\frac{2Em_{c2}}{\hbar^2}\right) G\left(E, E_{g_2}, \Delta_2\right) - \left\{\phi_2\left(n_y, n_z\right)\right\}\right]^{1/2}.$$

The electron concentration per unit length is given by

$$n_0 = \frac{2g_v}{\pi} \sum_{n_z=1}^{n_{z\max}} \sum_{n_y=1}^{n_{y\max}} \left[D_{15}\left(E_{F1D}, n_x, n_y\right) + D_{16}\left(E_{F1D}, n_z, n_y\right)\right] \tag{2.47}$$

where, $D_{15}(E_{F1D}, n_z, n_y) = \sqrt{\rho_{30}(E_{FID}, n_z, n_y)}$ and $D_{16}(E_{F1D}, n_z, n_y) = \sum\limits_{r=1}^{s_o} Z_{1D}(r)[D_{15}(E_{F1D}, n_z, n_y)]$

The field-emitted current assumes the form

$$I = \frac{2g_v e k_B T}{h} \sum_{n_z=1}^{n_{z\max}} \sum_{n_y=1}^{n_{y\max}} [F_0(\eta_{30}) \exp(-\beta_{30})] \tag{2.48}$$

$\eta_{30} = (E_{F1\text{D}} - \overline{E}_{30})/k_{\text{B}}T$, and $\overline{E}_{30}$ is the lowest positive root of the equation.

$$\rho_{30}(\overline{E}_{30}, n_z, n_y) = 0, \tag{2.49}$$

$$\beta_{30} = \frac{4}{3}[\rho_{30}(V_0, n_z, n_y)]^{3/2} \cdot [3eF_{sz}\rho_{31}(V_0, n_z, n_y)]^{-1},$$

$$\rho_{31}(V_0, n_z, n_y) = \left(\frac{2}{L_0^2}\left\{\cos^{-1}[f_{32}\left(n_z, n_y, V_0\right)]\right\} f_{33}(V_0, n_z, n_y) .\{1 - f_{32}^2\left(n_z, n_y, V_0\right)\}^{-1/2}\right)$$

$$f_{33}(V_0, n_z, n_y) = [a_7 \sin\{a_0 C_{48}(V_0, n_z, n_y) + b_0 D_{48}(V_0, n_z, n_y)\}\{a_0 C_{49}(V_0, n_z, n_y) + b_0 D_{49}(V_0, n_z, n_y)\} - a_8 \sin\{a_0 C_{48}(V_0, n_z, n_y) - b_0 D_{48}(V_0, n_z, n_y)\}\{a_0 C_{49}(V_0, n_z, n_y) - b_0 D_{49}(V_0, n_z, n_y)\}]$$

$$C_{49}(V_0, n_z, n_y) = [2C_{48}(V_0, n_z, n_y)]^{-1} \cdot \left[\frac{1}{A_0} - \frac{B_0}{A_0}(B_0^2 + 4A_0 V_0)^{1/2} \right]$$

$$D_{49}(V_0, n_z, n_y) = [D_{48}(V_0, n_z, n_y)]^{-1} [\left(\frac{m_{c2}}{\hbar^2} \right)$$

$$\cdot \left[1 + V_0 \left[\frac{1}{(V_0 + E_{g2} + \Delta_2)} + \frac{1}{(V_0 + E_{g2})} \right.\right. \tag{2.50}$$

$$\left.\left. - \frac{1}{(V_0 + E_{g2} + \frac{2}{3}\Delta_2)} \right] \right] G(V_0, n_z, n_y)].$$

2.3 Result and Discussions

Using Table 1.1 and $\Delta_{21} = 0.4\,\text{nm}$[9.16] together with the (2.4), (2.5); (2.12), (2.13); (2.19), (2.20); and (2.26), (2.27), we have plotted the field-emitted current in Figs. 2.1–2.3 as functions of film thickness, electron concentration per unit length and electric field for GaAs/AlGaAs, CdS/CdTe, PbTe/PbSe, and HgTe/CdTe quantum wires superlattices with graded interfaces, respectively. Using (2.31), (2.32); (2.37), (2.38), (2.42); (2.43), (2.47) and (2.48) we have plotted the field-emitted current as function of film thickness, electron concentration per unit length, and electric field for GaAs/AlGaAs, CdS/CdTe, PbTe/PbSe, and HgTe/CdTe quantum wire effective mass superlattices in Figs. 2.4–2.6, respectively.

It appears from Fig. 2.1 that the field-emitted current increases with the film thickness for quantum wires of GaAs/AlGaAs superlattices with graded interfaces exhibiting step functional dependence. For PbTe/PbSe quantum wire superlattices with graded interfaces, the said dependence is relatively much less in magnitude as compared with the previous one, whereas for CdS/CdTe and HgTe/CdTe quantum

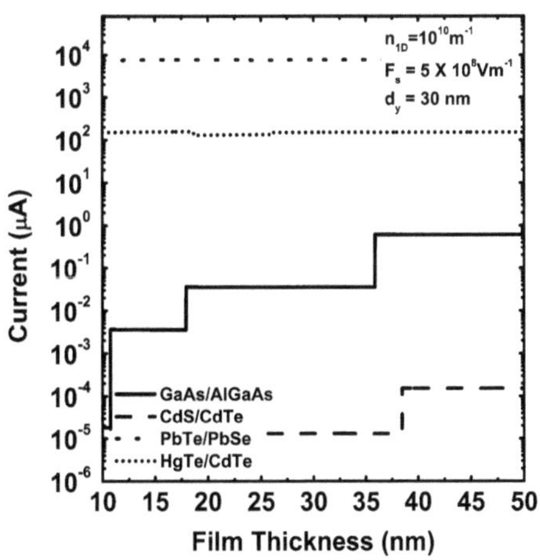

Fig. 2.1 Plot of the field-emitted current as a function of film thickness for quantum wire superlattices of GaAs/AlGaAs, CdS/CdTe, PbTe/PbSe, and HgTe/CdTe with graded interfaces

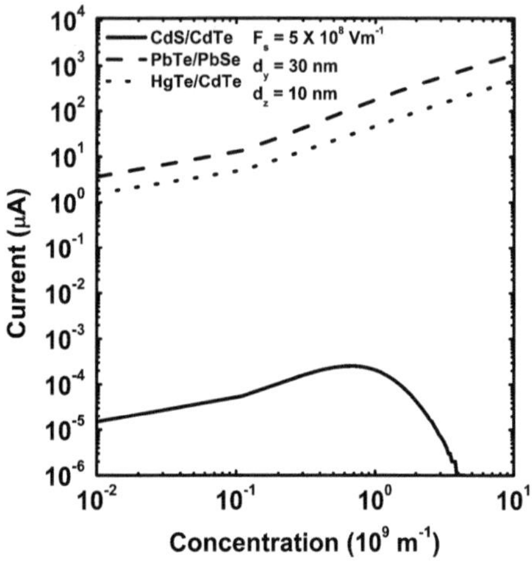

Fig. 2.2 Plot of the field-emitted current as a function of carrier concentration per unit length for quantum wire superlattices of CdS/CdTe, PbTe/PbSe, and HgTe/CdTe with graded interfaces

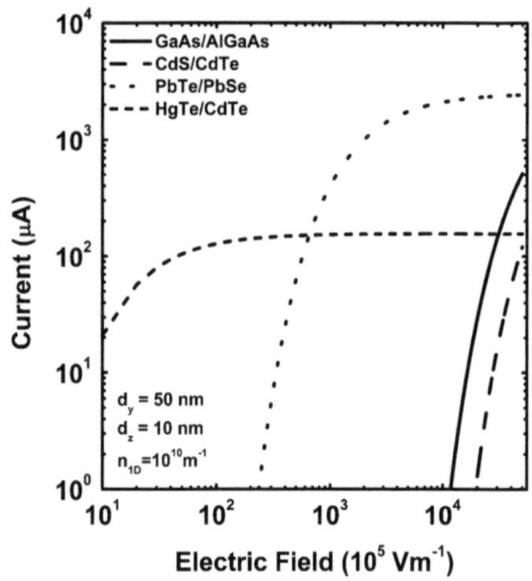

Fig. 2.3 Plot of the field-emitted current as a function of electric field for quantum wire superlattices of GaAs/AlGaAs, CdS/CdTe, PbTe/PbSe, and HgTe/CdTe with graded interfaces

wire superlattices with graded interfaces, the field-emitted current is relatively large and invariant with respect to thickness variation. This behavior is the direct consequence of the carrier dispersion relation in the respective cases. From Fig. 2.2, we observe that the field-emitted current increases with increasing concentration for PbTe/PbSe and HgTe/CdTe quantum wire superlattices with graded interfaces, whereas for CdS/CdTe quantum wire superlattices with graded interfaces, the current increases with increasing concentration and decreases after a carrier degeneracy of about $10^9\,\mathrm{m}^{-1}$. Comparing the individual curves, we note that the numerical values of field current for PbTe/PbSe and HgTe/CdTe quantum wire superlattices with graded interfaces are much greater as compared with CdS/CdTe quantum wire superlattices with graded interfaces.

Figure 2.3 explores the fact that the cut-in value of the electric field for quantum wire superlattices of HgTe/CdTe with graded interfaces is even less than $10^6\mathrm{Vm}^{-1}$

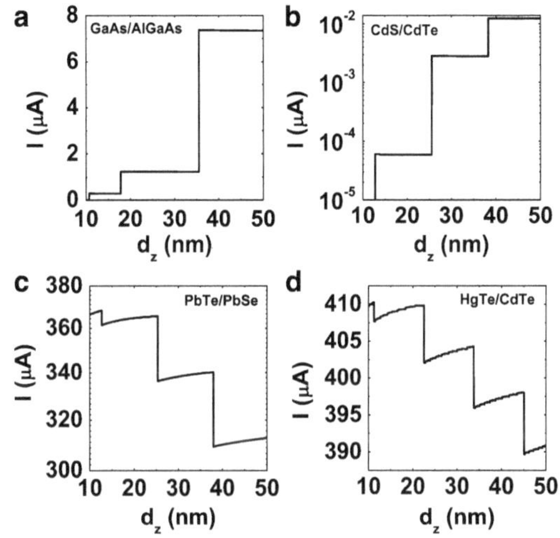

Fig. 2.4 Plot of the field-emitted current as a function of film thickness for quantum wire effective mass superlattices of GaAs/AlGaAs, CdS/CdTe, PbTe/PbSe, and HgTe/CdTe

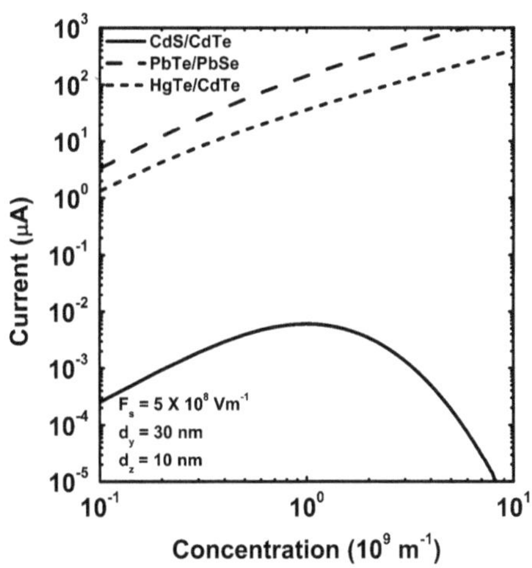

Fig. 2.5 Plot of the field-emitted current as a function of carrier concentration per unit length for quantum wire effective mass superlattices of CdS/CdTe, PbTe/PbSe, and HgTe/CdTe

and saturates beyond $10^7 \mathrm{Vm}^{-1}$. For the other cases, the cut-in values are larger well beyond $10^7 \mathrm{Vm}^{-1}$.

From Fig. 2.4, we observe that the variations of the field-emitted current with film thickness in GaAs/AlGaAs and CdS/CdTe quantum wire effective mass superlattices are in opposite trend with that of the PbTe/PbSe and HgTe/CdTe. The reason for this variation has already been discussed in "Result and Discussions" of Chap. 1. We note that HgTe/CdTe quantum wire effective mass superlattices exhibits a maximum field current as compared with that of the others. Figure 2.5 exhibits the fact that the field-emitted current becomes very small beyond $10^9 \mathrm{m}^{-1}$ for CdS/CdTe quantum wires of effective mass superlattices. The Fig. 2.6 indicates that the cut-in field for quantum wires of effective mass superlattices HgTe/CdTe is nearly $10^6 \mathrm{Vm}^{-1}$ rather than that of CdS/CdTe which is beyond $10^8 \mathrm{Vm}^{-1}$. For the

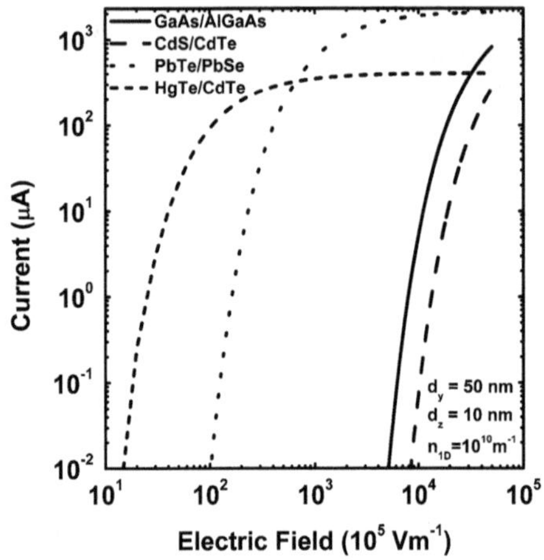

Fig. 2.6 Plot of the field-emitted current as a function of electric field for quantum wire effective mass superlattices of GaAs/AlGaAs, CdS/CdTe, PbTe/PbSe, and HgTe/CdTe

purpose of condensed presentation, the carrier statistics and the FNFE from different quantized materials as considered in this chapter have been presented in Table 2.1.

2.4 Open Research Problems

(R.2.1) [(a)] Investigate the FNFE from all the superlattices whose respective dispersion relations of the carriers are given in this chapter by converting the summations over the quantum numbers to the corresponding integrations by including the uniqueness conditions in the appropriate cases and considering the effect of image force in the subsequent study in each case.

(b) Investigate the FNFE in the presence of an arbitrarily oriented non-quantizing magnetic field for all types of quantum wire superlattices as considered in this chapter by considering the electron spin.

(R.2.2) Investigate the FNFE in the presence of an additional arbitrarily oriented electric field for all types of quantum wire superlattices.

(R.2.3) Investigate the FNFE for all types of quantum wire superlattices as considered in this chapter under arbitrarily oriented crossed electric and magnetic fields.

(R.2.4) Investigate the FNFE in III–V, II–VI, IV–VI, and HgTe/CdTe quantum well superlattices with graded interfaces,

(R.2.5) Investigate the FNFE for all problems of R.2.1 to R.2.3 for III–V, II–VI, IV–VI, and HgTe/CdTe quantum well superlattices with graded interfaces.

(R.2.6) Investigate the FNFE for III–V, II–VI, IV–VI, and HgTe/CdTe quantum well effective mass superlattices.

(R.2.7) Investigate the FNFE for all problems of R.2.1 to R.2.3 for III–V, II–VI, IV–VI, and HgTe/CdTe quantum well effective mass superlattices.

Table 2.1 The carrier statistics and the Fowler–Nordheim field emission current from III–V, II–VI, IV–VI, HgTe/CdTe quantum wire superllatices with graded interfaces, and effective mass quantum wire superlattices of the aforementioned materials

The Carrier Statistics	The Fowler–Nordheim field emission current
III–V superlattices with graded interfaces $n_0=\frac{2g_v}{\pi}\sum_{n_x=1}^{n_{x\max}}\sum_{n_y=1}^{n_{y\max}}\left[D_{21}(E_{F1D},n_x,n_y)\ +D_{22}(E_{F1D},n_x,n_y)\right]$ (2.4)	On the basis of (2.4) $I=\frac{2g_v ek_B T}{h}\sum_{n_x=1}^{n_{x\max}}\sum_{n_y=1}^{n_{y\max}}[F_0(\bar{\eta}_{21})\exp(-\bar{\beta}_{21})]$ (2.5)
II–VI superlattices with graded interfaces $n_0=\frac{g_v}{\pi}\sum_{n_x=1}^{n_{x\max}}\sum_{n_y=1}^{n_{y\max}}\left[D_{212}(E_{F1D},n_x,n_y)+D_{222}(E_{F1D},n_x,n_y)\right]$ (2.12)	On the basis of (2.12) $I=\frac{g_v ek_B T}{h}\sum_{n_x=1}^{n_{x\max}}\sum_{n_y=1}^{n_{y\max}}[F_0(\bar{\eta}_{22})\exp(-\bar{\beta}_{22})]$ (2.13)
IV–VI superlattices with graded interfaces $n_0=\frac{2g_v}{\pi}\sum_{n_x=1}^{n_{x\max}}\sum_{n_y=1}^{n_{y\max}}\left[\overline{D}_3(E_{F1D},n_x,n_y)\ +\overline{D}_4(E_{F1D},n_x,n_y)\right]$ (2.19)	On the basis of (2.19) $I=\frac{2g_v ek_B T}{h}\sum_{n_x=1}^{n_{x\max}}\sum_{n_y=1}^{n_{y\max}}[F_0(\eta_{23})\exp(-\beta_{23})]$ (2.20)
HgTe/CdTe superlattices with graded interfaces $n_0=\frac{2g_v}{\pi}\sum_{n_x=1}^{n_{x\max}}\sum_{n_y=1}^{n_{y\max}}\left[\overline{D}_5(E_{F1D},n_x,n_y)\ +\overline{D}_6(E_{F1D},n_x,n_y)\right]$ (2.26)	On the basis of (2.26) $I=\frac{2g_v ek_B T}{h}\sum_{n_z=1}^{n_{z\max}}\sum_{n_y=1}^{n_{y\max}}[F_0(\eta_{24})\exp(-\beta_{24})]$ (2.27)
III–V effective mass superlattices $n_0=\frac{2g_v}{\pi}\sum_{n_z=1}^{n_{z\max}}\sum_{n_y=1}^{n_{y\max}}\left[\overline{D}_7(E_{F1D},n_z,n_y)+\overline{D}_8(E_{F1D},n_z,n_y)\right]$ (2.31)	On the basis of (2.31) $I=\frac{2g_v ek_B T}{h}\sum_{n_z=1}^{n_{z\max}}\sum_{n_y=1}^{n_{y\max}}[F_0(\eta_{26})\exp(-\beta_{26})]$ (2.32)
II–VI effective mass superlattices $n_0=\frac{g_v}{\pi}\sum_{n_x=1}^{n_{x\max}}\sum_{n_y=1}^{n_{y\max}}\left[\overline{D}_9(E_{F1D},n_x,n_y)+\overline{D}_{10}(E_{F1D},n_x,n_y)\right]$ (2.37)	On the basis of (2.37) $I=\frac{g_v ek_B T}{h}\sum_{n_x=1}^{n_{x\max}}\sum_{n_y=1}^{n_{y\max}}[F_0(\eta_{27})\exp(-\beta_{27})]$ (2.38)
IV–VI effective mass superlattices $n_0=\frac{2g_v}{\pi}\sum_{n_y=1}^{n_{y\max}}\sum_{n_z=1}^{n_{z\max}}\left[D_{11}(E_{F1D},n_y,n_z)+D_{12}(E_{F1D},n_y,n_z)\right]$ (2.42)	On the basis of (2.42) $I=\frac{2g_v ek_B T}{h}\sum_{n_y=1}^{n_{y\max}}\sum_{n_z=1}^{n_{z\max}}[F_0(\eta_{28})\exp(-\beta_{28})]$ (2.43)
HgTe/CdTe effective mass superlattices $n_0=\frac{2g_v}{\pi}\sum_{n_z=1}^{n_{z\max}}\sum_{n_y=1}^{n_{y\max}}\left[D_{15}(E_{F1D},n_x,n_y)+D_{16}(E_{F1D},n_z,n_y)\right]$ (2.47)	On the basis of (2.47) $I=\frac{2g_v ek_B T}{h}\sum_{n_z=1}^{n_{z\max}}\sum_{n_y=1}^{n_{y\max}}[F_0(\eta_{30})\exp(-\beta_{30})]$ (2.48)

(R.2.8) Investigate the FNFE for short period, strained layer, random, and Fibonacci quantum wire superlattices.

(R.2.9) Investigate the FNFE for short period, strained layer, random, and Fibonacci quantum well superlattices in the presence of an arbitrarily oriented magnetic field by considering electron spin and broadening.

(R.2.10) Investigate the FNFE for strained layer, random, Fibonacci, polytype, and sawtooth superlattices in the presence of an arbitrarily oriented electric field.

(R.2.11) Investigate the FNFE for strained layer, random, Fibonacci, polytype, and sawtooth superlattices in the presence of an arbitrarily oriented crossed electric and magnetic field.

(R.2.12) Investigate the FNFE for strained layer, random, Fibonacci, polytype, and sawtooth quantum well and quantum wires superlattices in the presence of an arbitrarily oriented electric field.

(R.2.13) Investigate the FNFE for strained layer, random, Fibonacci, polytype, and sawtooth quantum well and quantum wires superlattices in the presence of arbitrarily oriented crossed electric and quantizing magnetic fields.

(R.2.14) [(a)] Formulate the minimum tunneling, Dwell and phase tunneling, Buttiker and Landauer and intrinsic times for all types of superlattices as discussed in this chapter.

(b) Investigate all the appropriate problems of this chapter for the Dirac electron.

(c) Investigate all the problems of this chapter by removing all the mathematical approximations and establishing the respective appropriate uniqueness conditions.

References

1. L.V. Keldysh, *Sov. Phys. Solid State* **4**, 1658 (1962)
2. L. Esaki, R. Tsu, *IBM J. Res. Develop.* **14**, 61 (1970)
3. G. Bastard, *Wave Mechanics Applied to Heterostructures* (Editions de Physique, Les Ulis, 1990)
4. E.L. Ivchenko, G. Pikus, *Superlattices and Other Heterostructures* (Springer, Berlin, 1995)
5. R. Tsu, *Superlattices to Nanoelectronics* (Elsevier, Amsterdam, 2005)
6. P. Fürjes, Cs. Dücs, M. Ádám, J. Zettner, I. Bársony, *Superlatt. Microstr.* **35**, 455 (2004)
7. T. Borca-Tasciuc, D. Achimov, W.L. Liu, G. Chen, H.-W. Ren, C.-H. Lin, S.S. Pei, *Microscale Thermophys. Eng.* **5**, 225 (2001)
8. B.S. Williams, *Nat. Photonics* **1**, 517 (2007)
9. A. Kosterev, G. Wysocki, Y. Bakhirkin, S. So, R. Lewicki, F. Tittel, R.F. Curl, *Appl. Phys. B* **90**, 165 (2008)
10. M.A. Belkin, F. Capasso, F. Xie, A. Belyanin, M. Fischer, A. Wittmann, J. Faist, *Appl. Phys. Lett.* **92**, 201101 (2008)
11. G.J. Brown, F. Szmulowicz, R. Linville, A. Saxler, K. Mahalingam, C.-H. Lin, C.H. Kuo, W.Y. Hwang, *IEEE Photon. Technol. Lett.* **12**, 684 (2000)
12. H.J. Haugan, G.J. Brown, L. Grazulis, K. Mahalingam, D.H. Tomich, *Phys. E Low Dimens. Syst. Nanostr.* **20**, 527 (2004)

13. S.A. Nikishin, V.V. Kuryatkov, A. Chandolu, B.A. Borisov, G.D. Kipshidze, I. Ahmad, M. Holtz, H. Temkin, *Jpn. J. Appl. Phys.* **42**, L1362 (2003)
14. Y.-K. Su, H.-C. Wang, C.-L. Lin, W.-B. Chen, S.-M. Chen, *Jpn. J. Appl. Phys.* **42**, L751 (2003)
15. C.H. Liu, Y.K. Su, L.W. Wu, S.J. Chang, R.W. Chuang, *Semicond. Sci. Technol.* **18**, 545 (2003)
16. S.-B. Che, I. Nomura, A. Kikuchi, K. Shimomura, K. Kishino, *Phys. Stat. Sol. B* **229**, 1001 (2002).
17. C.P. Endres, F. Lewen, T.F. Giesen, S. Schlemmer, D.G. Paveliev, Y.I. Koschurinov, V.M. Ustinov, A.E. Zhucov, *Rev. Sci. Instrum.* **78**, 043106 (2007)
18. F. Klappenberger, K.F. Renk, P. Renk, B. Rieder, Y.I. Koshurinov, D.G. Pavelev, V. Ustinov, A. Zhukov, N. Maleev, A. Vasilyev, *Appl. Phys. Lett.* **84**, 3924 (2004)
19. X. Jin, Y. Maeda, T. Saka, M. Tanioku, S. Fuchi, T. Ujihara, Y. Takeda, N. Yamamoto, Y. Nakagawa, A. Mano, S. Okumi, M. Yamamoto, T. Nakanishi, H. Horinaka, T. Kato, T. Yasue, T. Koshikawa, *J. Cryst. Growth* **310**, 5039 (2008)
20. X. Jin, N. Yamamoto, Y. Nakagawa, A. Mano, T. Kato, M. Tanioku, T. Ujihara, Y. Takeda, S. Okumi, M. Yamamoto, T. Nakanishi, T. Saka, H. Horinaka, T. Kato, T. Yasue, T. Koshikawa, *Appl. Phys. Express* **1**, 045002 (2008)
21. B.H. Lee, K.H. Lee, S. Im, M.M. Sung, *Org. Electron.* **9**, 1146 (2008)
22. P.-H. Wu, Y.-K. Su, I.-L. Chen, C.-H. Chiou, J.-T. Hsu, W.-R. Chen, *Jpn. J. Appl. Phys.* **45**, L647 (2006)
23. A.C. Varonides, *Renew. Ener.* **33**, 273 (2008)
24. M. Walther, G. Weimann, *Phys. Stat. Sol. B* **203**, 3545 (2006)
25. R. Rehm, M. Walther, J. Schmitz, J. Flei β ner, F. Fuchs, J. Ziegler, W. Cabanski, *Opto-electron. Rev.* **14**, 19 (2006)
26. R. Rehm, M. Walther, J. Scmitz, J. Fleissner, J. Ziegler, W. Cabanski, R. Breiter, *Electron. Lett.* **42**, 577 (2006)
27. G.J. Brown, F. Szmulowicz, H. Haugan, K. Mahalingam, S. Houston,*Microelectron. J.* **36**, 256 (2005)
28. K.V. Vaidyanathan, R.A. Jullens, C.L. Anderson, H.L. Dunlap, *Solid State Electron.***26**, 717 (1983)
29. B.A. Wilson, IEEE *J. Quantum Electron.* **24**, 1763 (1988)
30. M. Krichbaum, P. Kocevar, H. Pascher, G. Bauer, *IEEE J. Quantum Electron.* **24**, 717 (1988)
31. J.N. Schulman, T.C. McGill, *Appl. Phys. Lett.***34**, 663 (1979)
32. H. Kinoshita, T. Sakashita, H. Fajiyasu, *J. Appl. Phys.* **52**, 2869 (1981)
33. L. Ghenin, R.G. Mani, J.R. Anderson, J.T. Cheung, *Phys. Rev. B* **39**, 1419 (1989)
34. C.A. Hoffman, J.R. Mayer, F.J. Bartoli, J.W. Han, J.W. Cook, J.F. Schetzina, J.M. Schubman, *Phys. Rev. B.* **39**, 5208 (1989)
35. V.A. Yakovlev, *Sov. Phys. Semicond.* **13**, 692 (1979)
36. E.O. Kane, *J. Phys. Chem. Solids* **1**, 249 (1957)
37. H. Sasaki, *Phys. Rev. B* **30**, 7016 (1984)
38. H.X. Jiang, J.Y. Lin, *J. Appl. Phys.* **61**, 624 (1987)
39. G.M.T. Foley, P.N. Langenberg, *Phys. Rev. B* **15B**, 4850 (1977)

Chapter 3
Field Emission from Quantum Confined Semiconductors Under Magnetic Quantization

3.1 Introduction

It is well known that the band structure of semiconductors can be dramatically changed by applying the external fields [1–68]. The effects of the quantizing magnetic field on the band structure of compound semiconductors are more striking and can be observed easily in experiments. Under magnetic quantization, the motion of the electron parallel to the magnetic field remains unaltered, while the area of the wave-vector space perpendicular to the direction of the magnetic field gets quantized in accordance with the Landau's rule of area quantization in the wave-vector space [40–68] The energy levels of the carriers in a magnetic field (with the component of the wave-vector parallel to the direction of magnetic field be equated with zero) are termed as the Landau levels and the quantized energies are known as the Landau subbands. It is important to note that the same conclusion may be arrived either by solving the single-particle time-independent Schrödinger differential equation in the presence of a quantizing magnetic field or by using the operator method. The quantizing magnetic field tends to remove the degeneracy and increases the band gap. A semiconductor, placed in a magnetic field B, can absorb radiative energy with the frequency ($\omega_0 = (|e|B/m_c)$). This phenomenon is known as cyclotron or diamagnetic resonance. The effect of energy quantization is experimentally noticeable when the separation between any two consecutive Landau levels is greater than $k_B T$. A number of interesting transport phenomena originate from the change in the basic band structure of the semiconductor in the presence of a quantizing magnetic field. These have been widely investigated and also served as diagnostic tools for characterizing the different materials having various band structures. The discreteness in the Landau levels leads to a whole crop of magneto-oscillatory phenomena, important among which are (1) Shubnikov–de Haas oscillations in magneto-resistance; (2) de Haas–Van Alphen oscillations in magnetic susceptibility; and (3) magneto-phonon oscillations in thermoelectric power, etc.

S. Bhattacharya and K.P. Ghatak, *Fowler–Nordheim Field Emission*, Springer Series in Solid-State Sciences 170, DOI 10.1007/978-3-642-20493-7_3,

In this chapter in Sect. 3.2.1, of the theoretical background, the FNFE has been investigated in nonlinear optical semiconductors in the presence of a quantizing magnetic field. Section 3.2.2 explores the FNFE from III–V, ternary and quaternary compounds under magnetic quantization in accordance with the three- and the two-band models of Kane and forms the special case of Sect. 3.2.1. In the same section, the well-known result of FNFE from semiconductors having parabolic energy bands in the absence of any field has been noted and the magneto-FNFE in accordance with the models of Stillman et al. and Palik et al. have farther been presented for the purpose of relative comparison. Section 3.2.3 contains the study of the magneto-FNFE from II–VI semiconductors. In Sect. 3.2.4, the magneto-FNFE from Bismuth has been investigated in accordance with the models of the McClure and Choi, the Cohen and the Lax nonparabolic ellipsoidal respectively. In Sect. 3.2.5, the FNFE in IV–VI semiconductors under magnetic quantization has been discussed in accordance with the model of Dimmock, Bangert and Kastner, and Foley and Landenberg, respectively. In Sect. 3.2.6, the magneto-FNFE for the stressed Kane-type semiconductors has been investigated. In Sect. 3.2.7, the FNFE in Te has been studied under magnetic quantization. In Sect. 3.2.8, the magneto-FNFE in n-GaP has been studied. In Sect. 3.2.9, the FNFE in $PtSb_2$ has been investigated under magnetic quantization. In Sect. 3.2.10, the magneto-FNFE in Bi_2Te_3 has been studied. Section 3.2.11 contains the formulation of FNFE from Ge under magnetic quantization in accordance with the models of Cardona et al. and Wang et al., respectively. In Sects. 3.2.12 and3.2.13, the magneto-FNFE in n-GaSb and II–V compounds has respectively been investigated. Section 3.3 contains the result and discussions and Sect. 3.4 explores open research problems for this chapter in this context.

3.2 Theoritical Background

3.2.1 The Field Emission from Nonlinear Optical Semiconductors Under Magnetic Quantization

In the presence of a quantizing magnetic field ($\vec{B}$) along the z-direction, the magneto-electron energy spectrum can be expressed as follows [69]:

$$\gamma(E) = f_1(E) \cdot \frac{2eB}{\hbar}\left(n + \frac{1}{2}\right) + f_2(E)k_z^2 \pm \frac{eB\hbar\Delta_\parallel E_{g0}}{6}\left[\frac{E_{g0} + \Delta_\perp}{m_\perp^*\left(E_{g0} + \frac{2}{3}\Delta_\perp\right)}\right] \cdot \left[E + E_{g0} + \delta + \frac{\Delta_\parallel^2 - \Delta_\perp^2}{3\Delta_\parallel}\right], \tag{3.1}$$

where $n(= 0, 1, 2 \ldots.)$ is the Landau magnetic quantum number.

Equation (3.1) can be written as

$$k_z^2 = A_{31,\mp}(E,n), \tag{3.2}$$

where

$$A_{31,\mp}(E,n) = \left[[f_2(E)]^{-1}\left[\gamma(E) - f_1(E)\cdot\frac{2eB}{\hbar}\left(n+\frac{1}{2}\right) \mp \frac{eB\hbar\Delta_{\parallel}E_{g0}}{6}\right.\right.$$
$$\left.\left.\times\left[\frac{E_{g0}+\Delta_{\perp}}{m_{\perp}^*(E_{g0}+\frac{2}{3}\Delta_{\perp})}\right]\cdot\left[E+E_{g0}+\delta+\frac{\Delta_{\parallel}^2-\Delta_{\perp}^2}{3\Delta_{\parallel}}\right]\right]\right].$$

The field-emitted current density under magnetic quantization from nonlinear optical materials is given by

$$J = \frac{1}{2}\sum_{n=0}^{n_{\max}}\int_{E_{n_{31},\pm}}^{\infty} e\cdot\frac{1}{\hbar}\cdot\frac{\partial E}{\partial k_z}\cdot\frac{g_v eB}{2\pi^2\hbar}\frac{\partial k_z}{\partial E}f(E)t_{31}, \tag{3.3}$$

where $E_{n31\pm}$ is the Landau subbands in this case, t_{31} is the transmission coefficient in this case, which can be written as

$$t_{31} = \exp\left[-2\int_0^{z_{t31}}[A_{31,\pm}(V_0,n) - eF_s z B_{31,\pm}(V_0,n)]^{1/2}\,\mathrm{d}z\right], \tag{3.4}$$

where

$$B_{31,\pm}(V_0,n) = \left[\frac{-A_{31,\mp}(V_0,n)f_2'(V_0)}{f_2(V_0)} + \frac{1}{f_2(V_0)}\left[\gamma'(V_0) - f_1'(V_0)\frac{2eB}{\hbar}\left(n+\frac{1}{2}\right)\right.\right.$$
$$\left.\left.\mp\frac{eB\hbar\Delta_{\parallel}E_{g0}}{6}\left[\frac{E_{g0}+\Delta_{\perp}}{m_{\perp}^*\left(E_{g0}+\frac{2}{3}\Delta_{\perp}\right)}\right]\right]\right],$$

$$z_{t,31} = \frac{A_{31,\mp}(V_0,n)}{eF_S B_{31,\mp}(V_0,n)}, \quad \text{and}$$

$$\gamma(E_{n_{31},\pm}) = f_1(E_{n_{31},\pm})\left(\frac{2eB}{\hbar}\left(n+\frac{1}{2}\right)\right) \pm \left(\frac{eB\hbar\Delta_{\parallel}E_{g0}}{6}\right)\left(\frac{E_g+\Delta_{\perp}}{m_{\perp}^*(E_{g0}+\frac{2}{3}\Delta_{\perp})}\right)$$
$$\times\left(E_{n_{31},\pm} + E_{g0} + \delta + \frac{\Delta_{\parallel}^2-\Delta_{\perp}^2}{3\Delta_{\parallel}}\right).$$

Therefore

$$t_{31} = \exp[-\beta_{31,\pm}], \tag{3.5}$$

where $\beta_{31,\pm} = \dfrac{4[A_{31,\mp}(V_0,n)]^{3/2}}{3eF_{sz}B_{31,\pm}(V_0,n)}$.

Therefore

$$J = \frac{e^2 g_v B k_B T}{4\pi^2 \hbar^2} \sum_{n=0}^{n_{max}} [F_0(\eta_{31,+}) \exp[-\beta_{31,+}] + F_0(\eta_{31,-}) \exp[-\beta_{31,-}]], \quad (3.6)$$

where $\eta_{31,\pm} = (k_B T)^{-1}[E_{FB} - E_{n_{31},\pm}]$ and E_{FB} is the Fermi energy in the presence of magnetic quantization as measured from the edge of the conduction band in the absence of any quantization in the vertically upward direction.

The electron concentration can be expressed as

$$n_0 = \frac{eBg_v}{2\pi^2 \hbar} \sum_{n=0}^{n_{max}} [Y_{31}(E_{FB}, n) + Z_{31}(E_{FB}, n)], \quad (3.7)$$

where $Y_{31}(E_{FB}, n) = [\sqrt{A_{31,+}(E_{FB}, n)} + \sqrt{A_{31,-}(E_{FB}, n)}]$ and

$$Z_{31}(E_{FB}, n) = \sum_{r=1}^{s_0} Z_B(r)[Y_{31}(E_{FB}, n)],\ Z_B(r)$$
$$= 2(k_B T)(1 - 2^{1-2r})\zeta(2r)(\partial^{2r}/\partial^{2r} E_{FB}).$$

3.2.2 *The Field Emission from III–V Semiconductors Under Magnetic Quantization*

3.2.2.1 Three-Band Model of Kane

Under the conditions, $m_{\|}^* = m_{\perp}^* = m_c$, $\Delta_{\|} = \Delta_{\perp} = \Delta$, and $\delta = 0$, (3.1) assumes the form [70, 71]

$$I_{11}(E) = \left(n + \frac{1}{2}\right)\hbar\omega_0 + \frac{\hbar^2 k_z^2}{2m_c} \pm \frac{eB\hbar\Delta}{6m_c\left(E + E_{g0} + \frac{2}{3}\Delta\right)}, \quad (3.8)$$

which is the well-known dispersion relation of magneto-three-band Kane model.

From (3.8), we can write

$$k_z^2 = \frac{2m_c}{\hbar^2}\left[I_{11}(E) - \left(n + \frac{1}{2}\right)\hbar\omega_0 \pm \frac{eB\hbar\Delta}{6m_c\left(E + E_{g0} + \frac{2}{3}\Delta\right)}\right].$$

Thus,

$$k_z^2 = \frac{2m_c}{\hbar^2} A_{32,\pm}(E, n), \quad (3.9)$$

where $A_{32,\pm}(E, n) = I_{11}(E) - \left(n + \frac{1}{2}\right)\hbar\omega_0 \pm \frac{eB\hbar\Delta}{6m_c\left(E + E_{g0} + \frac{2}{3}\Delta\right)}$.

Transmission coefficient t_{32} is given by

$$t_{32} = \exp\left\{-2\int_0^{z_{t,32}} [A_{32,\pm}(V_0,n) - eF_{sz}zB_{32,\pm}(E)]^{1/2}\cdot\frac{\sqrt{2m_c}}{\hbar}\mathrm{d}z\right\},$$

where $z_{t,32} = \dfrac{A_{32,\pm}(V_0,n)}{eF_{sz}B_{32,\pm}(V_0)}$ and

$$B_{32,\pm}(V_0) = \left\{\left[I_{11}(E)\left(\frac{1}{E}+\frac{1}{E+E_{g0}}+\frac{1}{E+E_{g0}+\Delta}-\frac{1}{E+E_{g0}+\frac{2}{3}\Delta}\right)\right]\right.$$
$$\left.\mp\frac{(eB\hbar\Delta)}{(6m_c)}\left(E+E_{g0}+\frac{2}{3}\Delta\right)^{-2}\right\}\Bigg|_{E=V_0}.$$

Thus, we can write

$$t_{32} = \exp[-\beta_{32,\pm}], \tag{3.10}$$

where $\beta_{32,\pm} = \dfrac{4\sqrt{2m_c}[A_{32,\mp}(V_0,n)]^{3/2}}{3eF_{sz}\hbar B_{32,\mp}(V_0)}$, $A_{32,\pm}(V_0,n) = I_{11}(V_0) - \left(n+\frac{1}{2}\right)\hbar\omega_0 \pm \dfrac{eB\hbar\Delta}{6m_c(V_0+E_g+\frac{2}{3}\Delta)}$,

$$I_{11}(V_0) = \frac{V_0(V_0+E_{g0})(V_0+E_{g0}+\Delta)(E_{g0}+\frac{2}{3}\Delta)}{E_{g0}(E_{g0}+\Delta)(V_0+E_{g0}+\frac{2}{3}\Delta)}, \quad \text{and}$$

$$B_{32,\mp}(V_0) = \left\{\left[I_{11}(V_0)\left(\frac{1}{V_0}+\frac{1}{V_0+E_{g0}}+\frac{1}{V_0+E_{g0}+\Delta}-\frac{1}{V_0+E_{g0}+\frac{2}{3}\Delta}\right)\right]\right.$$
$$\left.\mp\frac{(eB\hbar\Delta)}{(6m_c)}\left(V_0+E_{g0}+\frac{2}{3}\Delta\right)^{-2}\right\}.$$

The field-emitted current density under magnetic quantization is given by

$$J = \frac{e^2 g_v B k_B T}{4\pi^2\hbar^2}\sum_{n=0}^{n_{\max}}[F_0(\eta_{32,+})\exp[-\beta_{32,+}] + F_0(\eta_{32,-})\exp[-\beta_{32,-}]], \tag{3.11}$$

where $\eta_{32,\pm} = (k_B T)^{-1}[E_{FB} - E_{n_{32},\pm}]$, and $E_{n_{32},\pm}$ is the lowest positive root of the equation.

$$I_{11}(E_{n,32\pm}) = \left(n+\frac{1}{2}\right)\hbar\omega_0 \pm \frac{eB\hbar\Delta}{6m_c\left(E_{n,32\pm}+E_{g0}+\frac{2}{3}\Delta\right)}. \tag{3.12}$$

The electron concentration can be expressed as

$$n_0 = \frac{\sqrt{2m_c}eBg_v}{2\pi^2\hbar^2} \sum_{n=0}^{n_{\max}} [\sqrt{A_{32,+}(E_{FB}, n)} + \sqrt{A_{32,-}(E_{FB}, n)}] + Z_{32}(E_{FB}, n), \tag{3.13}$$

where

$$Z_{32}(E_{FB}, n) = \sum_{r=1}^{s_0} Z_B(r)[\sqrt{A_{32,+}(E_{FB}, n)} + \sqrt{A_{32,-}(E_{FB}, n)}].$$

3.2.2.2 Two-Band Model of Kane

Under the condition $\Delta \to 0$, (3.8) assumes the form

$$E(1 + \alpha E) = \left(n + \frac{1}{2}\right)\hbar\omega_0 + \frac{\hbar^2 k_z^2}{2m_c} \pm \frac{1}{2}\mu_0 g^* B, \tag{3.14}$$

where μ_0 is the Bohr magneton and g^* is the effective g factor as the edge of the conduction band.

From (3.14), we can write

$$k_z^2 = \frac{2m_c}{\hbar^2}\overline{A}_{32,\pm}(E, n), \tag{3.15}$$

where $\overline{A}_{32,\pm}(E, n) = E(1 + \alpha E) - (n + \frac{1}{2})\hbar\omega_0 \mp \frac{1}{2}\mu_0 g^* B$.

Transmission coefficient t_{33} is given by

$$t_{33} = \exp\left[-\beta_{33,\pm}\right], \tag{3.16}$$

where $\beta_{33,\pm} = \dfrac{4\sqrt{2m_c}[\overline{A}_{32,\mp}(V_0, n)]^{3/2}}{3eF_{sz}\hbar(1 + 2\alpha V_0)}$.

The field-emitted current density can be written as

$$J = \frac{e^2 g_v B k_B T}{4\pi^2\hbar^2} \sum_{n=0}^{n_{\max}} [F_0(\eta_{33,+})\exp\left[-\beta_{33,+}\right] + F_0(\eta_{33,-})\exp\left[-\beta_{33,-}\right]], \tag{3.17}$$

where $\eta_{33,\pm} = (k_B T)^{-1}[E_{FB} - E_{n_{33},\pm}]$, and $E_{n_{33},\pm}$ is the lowest positive root of the equation.

$$E_{n_{33},\pm} = (2\alpha)^{-1}\left[-1 + \left[1 + 4\alpha\left\{\left(n + \frac{1}{2}\right)\hbar\omega_0 \pm \frac{1}{2}\mu_0 g^* B\right\}^{1/2}\right]\right]. \tag{3.18}$$

The electron concentration can be expressed as

$$n_0 = \frac{\sqrt{2m_c}eBg_v}{2\pi^2\hbar^2}\sum_{n=0}^{n_{\max}} Y_{32}(E_{FB},n) + Z_{32}(E_{FB},n), \tag{3.19}$$

where $Y_{32}(E_{FB},n) = \left[\sqrt{\bar{A}_{32,+}(E_{FB},n)} + \sqrt{\bar{A}_{32,-}(E_{FB},n)}\right]$ and $Z_{32}(E_{FB},n) = \sum_{r=1}^{s_0} Z_B(r)[Y_{32}(E_{FB},n)]$.

3.2.2.3 Parabolic Energy Bands

For parabolic energy bands the expressions of the electron concentration and the field-emitted current density under magnetic quantization can, respectively, be written as,

$$n_0 = (g_v N_c \theta/2)\sum_{n=0}^{n_{\max}} F_{-1/2}(\eta'_{B,\pm}), \tag{3.20}$$

and

$$J = \frac{e^2 g_v B k_B T}{h^2}\sum_{n=0}^{n_{\max}}\left[F_0(\eta'_{B,+})\exp[-\beta_{34,+}(V_0,n)] + F_0(\eta'_{B,-})\exp[-\beta_{34,-}(V_0,n)]\right], \tag{3.21}$$

where $N_c = 2(2\pi m_c k_B T/h^2)^{3/2}$, $\theta = (\hbar\omega_0/k_B T)$, $\eta'_{B,\pm} = \frac{E_{FB}-(n+\frac{1}{2})\hbar\omega_0 \pm \frac{1}{2}\mu_0 g^* B}{k_B T}$, and

$$\beta_{34,\pm}(V_0,n) = \frac{4\sqrt{2m_c}[V_0 - (n+\frac{1}{2})\hbar\omega_0 \pm \frac{1}{2}\mu_0 g^* B]^{3/2}}{3eF_{sz}\hbar}.$$

In the absence of spin and under the condition of extreme degeneracy, we get

$$J = \frac{2e^2 B g_v}{h^2}\sum_{n=0}^{n_{\max}}\left[\left[E_{FB} - \left(n+\frac{1}{2}\right)\hbar\omega_0\right] \cdot \exp\left[\frac{-4\sqrt{2m_c}}{3eF_{sz}\hbar}\left(V_0 - \left(n+\frac{1}{2}\right)\hbar\omega_0\right)^{3/2}\right]\right] \tag{3.22}$$

and

$$n_0 = \frac{\sqrt{2m_c}eBg_v}{\pi^2\hbar^2}\sum_{n=0}^{n_{\max}}\sqrt{E_{FB} - \left(n+\frac{1}{2}\right)\hbar\omega_0}. \tag{3.23}$$

In the absence of magnetic field $B \to 0$, (3.22) gets simplified into the form as given in (1.27). Besides in the absence of magnetic quantization, (3.23) assumes the form

$$n_0 = \frac{g_v}{3\pi^2}\left(\frac{2m^* E_{F0}}{\hbar^2}\right)^{3/2}. \tag{3.24}$$

It may be noted that (3.24) is the well-known expression of the Fermi energy in bulk semiconductors having parabolic energy bands under the condition of extreme degeneracy.

3.2.2.4 The Model of Stillman et al.

In accordance with model, the electron energy spectrum in III–V semiconductors in the presence of the quantizing magnetic field $\vec{B}$ along the z-direction can be written following (1.29) as

$$k_z^2 = \frac{2m_c}{\hbar^2}\left[I_{12}(E) - \left(n + \frac{1}{2}\right)\hbar\omega_0\right]. \tag{3.25}$$

Therefore,

$$k_z^2 = \frac{2m_c}{\hbar^2} A_{33}(E, n), \tag{3.26}$$

where $A_{33}(E,n) = I_{12}(E) - (n + \frac{1}{2})\hbar\omega_0$.

The field-emitted current density can be written as

$$J = \frac{e^2 g_v B k_B T}{2\pi^2\hbar^2}\sum_{n=0}^{n_{\max}}[F_0(\eta_{34})\exp[-\beta_{34}]], \tag{3.27}$$

where $\eta_{34} = (k_B T)^{-1}[E_{FB} - E_{34}]$.

E_{34} can be expressed as

$$E_{34} = \bar{t}_1\left[\frac{2eB}{\hbar}\left(n + \frac{1}{2}\right)\right] - \bar{t}_2\left[\frac{2eB}{\hbar}\left(n + \frac{1}{2}\right)\right]^2,$$

$$\beta_{34} = \frac{4\sqrt{2m_c}[A_{33}(V_0,n)]^{3/2}}{3eF_{sz}\hbar[I'_{12}(V_0)]}, \quad I'_{12}(V_0) = \left(\frac{\bar{a}_{11}\bar{a}_{12}}{2}\right)(1 - \bar{a}_{12}V_0)^{-1/2},$$

$$A_{33}(V_0, n) = \left[I_{12}(V_0) - \left(n + \frac{1}{2}\right)\hbar\omega_0\right], \tag{3.28}$$

and

$$I_{12}(V_0) = a_{11}\left[1 - (1 - V_0 a_{12})^{1/2}\right].$$

The electron concentration is given by

$$n_0 = \frac{\sqrt{2m_c}eBg_v}{\pi^2\hbar^2}\sum_{n=0}^{n_{\max}}[Y_{33}(E_{FB},n)+Z_{33}(E_{FB},n)], \tag{3.29}$$

where $Y_{33}(E_{FB},n) = [\sqrt{A_{33}(E_{FB},n)}]$ and $Z_{33}(E_{FB},n) = \sum_{r=1}^{s_0} Z_B(r)[Y_{33}(E_{FB},n)]$.

3.2.2.5 The Model of Palik et al.

To the fourth order in effective mass theory and taking into account the interactions of the conduction, light-hole, heavy-hole, and split-off hole bands, the electron energy spectrum in III–V semiconductors in the presence of a quantizing magnetic field $\vec{B}$ can be written in accordance with the present model extending (1.40) as

$$E = J_{31} + \left(n+\frac{1}{2}\right)\hbar\omega_0 + \frac{\hbar^2k_z^2}{2m_c} \pm \frac{1}{4}\left(\frac{m_c}{m_0}\right)\hbar\omega_0 g_0^* \pm k_{30}\alpha\left(n+\frac{1}{2}\right)(\hbar\omega_0)^2$$
$$\pm k_{31}\alpha\hbar\omega_0\left(\frac{\hbar^2k_z^2}{2m_c}\right) + k_{32}\alpha\left[\hbar\omega_0\left(n+\frac{1}{2}\right)+\frac{\hbar^2k_z^2}{2m_c}\right]^2, \tag{3.30}$$

where $J_{31} = -\frac{1}{2}\alpha\hbar\omega_0\left[(1-y_{11})/(2+x_{11})^2\right]\cdot J_{32}$,

$$J_{32} = \left\{\left[\frac{1}{3}(1-x_{11})^2 - (2+x_{11}^2)\right](2+x_{11})\cdot y_{11} + \frac{1}{2}(1-x_{11}^2)(1+x_{11})(1+y_{11})\right\},$$

$$g_0^* = 2\left\{1-\left[\frac{(1-x_{11})}{(2+x_{11})}\right]\left[\frac{(1-y_{11})}{y_{11}}\right]\right\},$$

$$k_{30} = (1-y_{11})(1-x_{11})\left\{\left[\left(2+\frac{3}{2}x_{11}+x_{11}^2\right)\cdot\frac{(1-y_{11})}{(2+x_{11})^2}\right]-\frac{2}{3}y_{11}\right\},$$

$$k_{31} = (1-y_{11})\left[\frac{(1-x_{11})}{(2+x_{11})}\right]\cdot\left\{\left[\left(2+\frac{3}{2}x_{11}+x_{11}^2\right)\cdot\frac{(1-y_{11})}{(2+x_{11})}\right]-\frac{2}{3}(1-x_{11})y_{11}\right\}, \quad \text{and}$$

$$k_{32} = -\left[\left(1+\frac{1}{2}x_{11}^2\right)\Big/\left(1+\frac{1}{2}x_{11}\right)\right](1-y_{11})^2.$$

Equation (3.30) assumes the form

$$J_{34}k_z^4 + J_{35,\pm}(n)k_z^2 + J_{36,\pm}(n) - E = 0 \tag{3.31}$$

where $J_{34} = \alpha k_{32}\left(\hbar^2/2m_c\right)^2$, $J_{35,\pm}(n) = \left[\frac{\hbar^2}{2m_c} \pm \alpha k_{31}\hbar\omega_0 \cdot \frac{\hbar^2}{2m_c} + \alpha k_{32}\hbar\omega_0 \cdot \frac{\hbar^2}{2m_c}\right.$ $\left.(n+\frac{1}{2})\right]$,

$$J_{36,\pm}(n) = \left[J_{31} \pm \frac{1}{4}\left(\frac{m_c}{m_0}\right)\hbar\omega_0 g_0^* \pm k_{30}\alpha(\hbar\omega_0)^2\left(n+\frac{1}{2}\right)\right.$$
$$\left. + k_{32}\alpha\left[(\hbar\omega_0)\left(n+\frac{1}{2}\right)\right]^2\right]$$

From (3.31) we get

$$2J_{34}k_z^2 = -J_{35,\pm} + \sqrt{(J_{35,\pm})^2 - 4J_{34}\left[J_{36,\pm} - E\right]},$$
$$k_z^2 = A_{35,\pm}(E,n), \tag{3.32}$$

where $A_{35,\pm}(E,n) = (2J_{34})^{-1}\left[-J_{35,\pm}(n) + \sqrt{(J_{35,\pm}(n))^2 - 4J_{34}\left[J_{36,\pm}(n) - E\right]}\right]$

The field-emitted current density is given by

$$J = \frac{e^2 g_v B k_B T}{h^2}\sum_{n=0}^{n_{max}}\left[F_0(\eta_{35,+})\exp\left[-\beta_{35,+}\right] + F_0(\eta_{35,-})\exp\left[-\beta_{35,-}\right]\right], \tag{3.33}$$

where $\eta_{35,\pm} = (k_B T)^{-1}[E_{FB} - E_{35,\pm}]$, and $E_{35,\pm}$ is the lowest positive root of the equation.

$$E_{35,\pm} = J_{36,\pm}(n) \tag{3.34}$$

$$\beta_{35,\pm} = \frac{4[A_{35,\pm}(V_0,n)]^{3/2}}{3eF_{sz}[A_{36,\pm}(V_0,n)]}, \quad A_{35,\pm}(V_0,n)$$
$$= (2J_{34})^{-1}\left[-J_{35,\pm}(n) + \sqrt{(J_{35,\pm}(n))^2 - 4J_{34}[J_{36,\pm}(n) - V_0]}\right],$$

$$A_{36,\pm}(V_0,n) = \frac{1}{\sqrt{(J_{35,\pm}(n))^2 - 4J_{34}[J_{36,\pm}(n) - V_0]}}.$$

The electron concentration can be expressed as

$$n_0 = \frac{eBg_v}{2\pi^2 h}\sum_{n=0}^{n_{max}}\left[Y_{34}(E_{FB},n) + Z_{34}(E_{FB},n)\right], \tag{3.35}$$

where $Y_{34}(E_{FB},n) = \left[\sqrt{A_{35,+}(E_{FB},n)} + \sqrt{A_{35,-}(E_{FB},n)}\right]$ and $Z_{34}(E_{FB},n) = \sum_{r=1}^{s_0} Z_B(r)[Y_{34}(E_{FB},n)]$.

3.2.3 *The Field Emission from II–VI Semiconductors Under Magnetic Quantization*

The magneto-dispersion relation assumes the form

$$E = E_{36,\pm}(n) + B_0 k_z^2, \tag{3.36a}$$

where $E_{36,\pm}(n) = \left[A_0 \cdot \frac{2eB}{\hbar}\left(n+\frac{1}{2}\right) \pm \frac{1}{2}\mu_0 g^* B \pm C_0\sqrt{\left(\frac{2eB}{\hbar}\right)\left(n+\frac{1}{2}\right)}\right]$.

Therefore,

$$k_z^2 = \frac{E - E_{36,\pm}(n)}{B_0}. \tag{3.36b}$$

The field-emitted current density is given by

$$J = \frac{e^2 g_v B k_B T}{h^2} \sum_{n=0}^{n_{\max}} [F_0(\eta_{36,+}) \exp[-\beta_{36,+}] + F_0(\eta_{36,-}) \exp[-\beta_{36,-}]], \tag{3.37}$$

where $\eta_{36,\pm} = (k_B T)^{-1}[E_{FB} - E_{36,\pm}]$, and $E_{36,\pm}$ is the lowest positive root of the equation.

$$\beta_{36,\pm} = \frac{4\sqrt{2m_{\parallel}^*}[V_0 - E_{36,\pm}]^{3/2}}{3eF_{sz}\hbar}.$$

The electron concentration can be written as

$$n_0 = \frac{eBg_v\sqrt{2m_{\parallel}^*}}{2\pi^2\hbar^2} \sum_{n=0}^{n_{\max}} [Y_{35}(E_{FB}, n) + Z_{35}(E_{FB}, n)], \tag{3.38}$$

where $Y_{35}(E_{FB}, n) = [\sqrt{E_{FB} - E_{36,+}(n)} + \sqrt{E_{FB} - E_{36,-}(n)}]$ and $Z_{35}(E_{FB}, n) = \sum_{r=1}^{s_0} Z_B(r)[Y_{35}(E_{FB}, n)]$

3.2.4 *The Field Emission from Under Bismuth Magnetic Quantization*

3.2.4.1 The McClure and Choi Model

The electron energy spectrum in Bi in accordance with the model of McClure and Choi under magnetic quantization upto the first order can be expressed following [72, 73] as

$$E(1+\alpha E) = \left(n+\frac{1}{2}\right)\hbar\omega(E) + (n^2+1+n)\left(\frac{\alpha\hbar^2\omega^2(E)}{4}\right) \pm \frac{1}{2}\mu_0 g^* B + \frac{\hbar^2 k_z^2}{2m_3}\left[1 - \frac{\alpha\left(n+\frac{1}{2}\right)\hbar\omega(E)}{2}\right], \tag{3.39}$$

where $\omega(E) = \frac{eB}{\sqrt{m_1 m_2}}\left[1+\alpha E\left(1-\frac{m_2}{m_2'}\right)\right]^{\frac{1}{2}}$.

Therefore,

$$k_z^2 = A_{36,\pm}(E,n), \tag{3.40}$$

where

$$A_{36,\pm}(E,n) = \frac{2m_3}{\hbar^2}\left[1-\frac{\alpha\left(n+\frac{1}{2}\right)\hbar\omega(E)}{2}\right]^{-1}\left[E(1+\alpha E) - \left(n+\frac{1}{2}\right)\hbar\omega(E) - (n^2+1+n)\left(\frac{\alpha\hbar^2\omega^2(E)}{4}\right) \mp \frac{1}{2}\mu_0 g^* B\right].$$

The field-emitted current density is given by

$$J = \frac{e^2 g_v B k_B T}{h^2}\sum_{n=0}^{n_{max}}\left[F_0(\eta_{37,+})\exp\left[-\beta_{37,+}\right] + F_0(\eta_{37,-})\exp\left[-\beta_{37,-}\right]\right], \tag{3.41}$$

where $\eta_{37,\pm} = (k_B T)^{-1}[E_{FB} - E_{37,\pm}(n)]$, and $E_{37,\pm}(n)$ is the lowest positive root of the equation.

$$(E_{37,\pm}(n))[1+\alpha(E_{37,\pm}(n))] = \left(n+\frac{1}{2}\right)\hbar\omega(E_{37,\pm}(n)) + (n^2+1+n) \times \left(\frac{\alpha\hbar^2\omega^2(E_{37,\pm}(n))}{4}\right) \pm \frac{1}{2}\mu_0 g^* B, \tag{3.42a}$$

where $\omega(E_{37,\pm}(n)) = \frac{eB}{\sqrt{m_1 m_2}}\left[1+\alpha(E_{37,\pm}(n))\left(1-\frac{m_2}{m_2'}\right)\right]^{\frac{1}{2}}$,

$$\beta_{37,\pm} = \frac{4[A_{36,\pm}(V_0,n)]^{3/2}}{3eF_{sz}[A_{37,\pm}(V_0,n)]}, \quad A_{36,\pm}(V_0,n) = \frac{2m_3}{\hbar^2}\left[1-\frac{\alpha\left(n+\frac{1}{2}\right)\hbar\omega(V_0)}{2}\right]^{-1}$$

$$\left[V_0(1+\alpha V_0) - \left(n+\frac{1}{2}\right)\hbar\omega(V_0) - (n^2+1+n)\left(\frac{\alpha\hbar^2\omega^2(V_0)}{4}\right) \mp \frac{1}{2}\mu_0 g^* B\right],$$

$$\omega(V_0) = \frac{eB}{\sqrt{m_1 m_2}}\left[1+\alpha(V_0)\left(1-\frac{m_2}{m_2'}\right)\right]^{\frac{1}{2}},$$

$$\omega_1(V_0) = \left[\left(\frac{eB}{\sqrt{m_1 m_2}}\right)\left(\alpha\left(1-\frac{m_2}{m_2'}\right)\right)(2\omega(V_0))^{-1}\right],$$

and

$$A_{37,\pm}(V_0,n)$$
$$= \left(\frac{\alpha\left(n+\frac{1}{2}\right)\hbar\omega_1(V_0)}{2}\right)\left[\frac{A_{36,\pm}(V_0,n)}{1-\frac{\alpha\left(n+\frac{1}{2}\right)\hbar\omega(V_0)}{2}}\right]+\frac{2m_3}{\hbar^2}\left[1-\frac{\alpha\left(n+\frac{1}{2}\right)\hbar\omega(V_0)}{2}\right]^{-1}$$
$$\times\left[(1+2\alpha V_0)-\left(n+\frac{1}{2}\right)\hbar\omega_1(V_0)-(n^2+1+n)\left(\frac{\alpha\hbar^2\omega(V_0)\omega_1(V_0)}{2}\right)\right].$$

The electron concentration can be written as

$$n_0 = \frac{eBg_v\sqrt{2m_3}}{2\pi^2\hbar^2}\sum_{n=0}^{n_{\max}}[Y_{36}(E_{FB},n)+Z_{36}(E_{FB},n)], \tag{3.42b}$$

where $Y_{36}(E_{FB},n) = [\sqrt{A_{38,+}(E_{FB},n)}+\sqrt{A_{38,-}(E_{FB},n)}]$,

$$A_{38,\pm}(E_{FB},n)$$
$$= \left[1-\frac{\alpha\left(n+\frac{1}{2}\right)\hbar\omega(E_{FB})}{2}\right]^{-1}\left[E_{FB}(1+\alpha E_{FB})\right.$$
$$\left.-\left(n+\frac{1}{2}\right)\hbar\omega(E_{FB})-\left(n^2+1+n\right)\left(\frac{\alpha\hbar^2\omega^2(E_{FB})}{4}\right)\pm\frac{1}{2}\mu_0 g^* B\right],$$

and $Z_{36}(E_{FB},n) = \sum_{r=1}^{s_0} Z_B(r)[Y_{36}(E_{FB},n)]$.

3.2.4.2 The Cohen Model

The magneto-dispersion relation for the conduction electrons in Bi in accordance with the Cohen model can be written as [72, 73]

$$E(1+\alpha E) = \left(n+\frac{1}{2}\right)\hbar\omega(E)+\frac{\hbar^2 k_z^2}{2m_3}+\hbar^2\omega^2(E)\left(n^2+\frac{1}{2}+n\right)\left(\frac{3\alpha}{8}\right)\pm\frac{1}{2}\mu_0 g^* B. \tag{3.43}$$

From (3.43), we get,

$$k_z^2 = \frac{2m_3}{\hbar^2}[A_{38,\pm}(E,n)], \tag{3.44}$$

where $A_{38,\pm}(E,n) = \left[E(1+\alpha E) - \left(n+\frac{1}{2}\right)\hbar\omega(E) - \hbar^2\omega^2(E)\left(n^2+\frac{1}{2}+n\right)\left(\frac{3\alpha}{8}\right) \mp \frac{1}{2}\mu_0 g^* B\right]$.

The field-emitted current density under magnetic quantization for this model can be expressed as

$$J = \frac{e^2 g_v B k_B T}{h^2}\sum_{n=0}^{n_{max}}[F_0(\eta_{38,+})\exp[-\beta_{38,+}] + F_0(\eta_{38,-})\exp[-\beta_{38,-}]], \tag{3.45}$$

where $\eta_{38,\pm} = (k_B T)^{-1}[E_{FB} - E_{38,\pm}]$, and $E_{38,\pm}$ is the lowest positive root of the equation.

$$E_{38,\pm}[1+\alpha(E_{38,\pm})] = \left(n+\frac{1}{2}\right)\hbar\omega(E_{38,\pm}) + \hbar^2\omega^2(E_{38,\pm})(n^2+1+n)\left(\frac{3\alpha}{8}\right) \pm \frac{1}{2}\mu_0 g^* B, \tag{3.46}$$

where $\beta_{38\pm} = (4[A_{38,\pm}(V_0,n)]^{3/2}\sqrt{2m_3})/(3eF_{sz}\hbar[A_{39}(V_0,n)])$,

$$\omega(E_{38\pm}) = (eB/\sqrt{m_1 m_2})[1+\alpha(1-(m_2/m_2'))(E_{38\pm})]^{1/2}$$

$$A_{38,\pm}(V_0,n) = \left[V_0(1+\alpha V_0) - \left(n+\frac{1}{2}\right)\hbar\omega(V_0) - \hbar^2\omega^2(V_0)\left(n^2+\frac{1}{2}+n\right)\left(\frac{3\alpha}{8}\right) \mp \frac{1}{2}\mu_0 g^* B\right],$$

$$A_{39}(V_0,n) = \left[1+2\alpha V_0 - \left(n+\frac{1}{2}\right)\hbar\omega_1(V_0) - \hbar^2\omega(V_0)\omega_1(V_0)\left(n^2+\frac{1}{2}+n\right)\left(\frac{3\alpha}{4}\right)\right],$$

$$\omega(V_0) = \frac{eB}{\sqrt{m_1 m_2}}\left[1+\alpha\left(1-\frac{m_2}{m_2'}\right)(V_0)\right]^{\frac{1}{2}}, \quad \text{and}$$

$$\omega_1(V_0) = \left[\left(\frac{eB}{\sqrt{m_1 m_2}}\right)\left(\alpha\left(1-\frac{m_2}{m_2'}\right)\right)(2\omega(V_0))^{-1}\right].$$

The electron concentration is given by

$$n_0 = \frac{eBg_v\sqrt{2m_3}}{2\pi^2\hbar^2}\sum_{n=0}^{n_{max}}[Y_{37}(E_{FB},n) + Z_{37}(E_{FB},n)], \tag{3.47}$$

where $Y_{37}(E_{FB},n) = [\sqrt{A_{38,+}(E_{FB},n)} + \sqrt{A_{38,-}(E_{FB},n)}]$ and $Z_{37}(E_{FB},n) = \sum_{r=1}^{s_0} Z_B(r)[Y_{37}(E_{FB},n)]$.

3.2.4.3 The Lax Model

In accordance with this model, the magneto-dispersion relation assumes the form [72, 73]

$$E(1+\alpha E) = \left(n+\frac{1}{2}\right)\hbar\omega_{03} + \frac{\hbar^2 k_z^2}{2m_3} \pm \frac{1}{2}\mu_0 g^* B, \tag{3.48}$$

where $\omega_{03} = eB/\sqrt{m_1 m_2}$.

Therefore

$$k_z^2 = \frac{2m_3}{\hbar^2}[A_{40,\pm}(E,n)], \tag{3.49}$$

where $A_{40,\pm}(E,n) = E(1+\alpha E) - \left(n+\frac{1}{2}\right)\hbar\omega_{03} \mp \frac{1}{2}\mu_0 g^* B$.

The field-emitted current density is given by

$$J = \frac{e^2 g_v B k_B T}{h^2} \sum_{n=0}^{n_{\max}} [F_0(\eta_{39,+})\exp[-\beta_{39,+}] + F_0(\eta_{39,-})\exp[-\beta_{39,-}]], \tag{3.50}$$

where $\eta_{39,\pm} = (k_B T)^{-1}[E_{FB} - E_{39,\pm}]$, and $E_{39,\pm}$ can be determined from the equation

$$E_{39,\pm} = (2\alpha)^{-1}\left[-1+\left[1+4\alpha\left\{\left(n+\frac{1}{2}\right)\hbar\omega_{03} \pm \frac{1}{2}\mu_0 g^* B\right\}\right]^{1/2}\right],$$

$$\beta_{39,\pm} = \frac{4[A_{40,\pm}(V_0,n)]^{3/2}\sqrt{2m_3}}{3eF_{sz}\hbar[A_{41}(V_0)]}$$

$$A_{40,\pm}(V_0,n) = \left[V_0(1+\alpha V_0) - \left(n+\frac{1}{2}\right)\hbar\omega_{03} \mp \frac{1}{2}\mu_0 g^* B\right], \quad \text{and}$$

$$A_{41}(V_0) = [1+2\alpha V_0]. \tag{3.51}$$

The electron concentration can be expressed as

$$n_0 = \frac{eBg_v\sqrt{2m_3}}{2\pi^2\hbar^2} \sum_{n=0}^{n_{\max}} [Y_{40}(E_{FB},n) + Z_{40}(E_{FB},n)], \tag{3.52}$$

where $Y_{40}(E_{FB},n) = [\sqrt{A_{40,+}(E_{FB},n)} + \sqrt{A_{40,-}(E_{FB},n)}]$ and $Z_{40}(E_{FB},n) = \sum_{r=1}^{s_0} Z_B(r)[Y_{40}(E_{FB},n)]$.

3.2.4.4 Ellipsoidal Parabolic Energy Bands

For $\alpha \to 0$, from (3.48) we get

$$E = \left(n + \frac{1}{2}\right)\hbar\omega_0 + \frac{\hbar^2 k_z^2}{2m_3} \pm \frac{1}{2}\mu_0 g^* B. \tag{3.53}$$

The expressions of J and n_0 for this model are the special cases of the models of McClure and Choi, Cohen and Lax, respectively.

3.2.5 *The Field Emission from IV–VI Semiconductors Under Magnetic Quantization*

3.2.5.1 The Dimmock Model

In accordance with Dimmock model, the electron energy spectrum in IV–VI semiconductors in the presence of a quantizing magnetic field $\vec{B}$ along the z-direction can be written following (1.71) as

$$\left[E - \frac{\hbar^2}{2m_t^-}\cdot\frac{2eB}{\hbar}\left(n+\frac{1}{2}\right) - \frac{\hbar^2 k_z^2}{2m_l^-}\right]\left[1+\alpha E+\alpha\frac{\hbar^2}{2m_t^+}\cdot\frac{2eB}{\hbar}\left(n+\frac{1}{2}\right)+\frac{\alpha\hbar^2 k_z^2}{2m_l^+}\right]$$
$$= \frac{\hbar eB}{m_t^*}\left(n+\frac{1}{2}\right) + \frac{\hbar^2 k_z^2}{2m_l^*}. \tag{3.54}$$

Thus, (3.54) assumes the form

$$k_z^2 = A_{42}(E,n), \tag{3.55}$$

where $A_{42}(E,n) = [2C_{31}]^{-1}\left[-C_{32}(E,n) + \left[C_{32}^2(E,n) - 4C_{31}\{C_{33}(E,n) - E\right.\right.$ $\left.\left.(1+\alpha E)\}\right]^{1/2}\right]$,

$$C_{31} = \frac{\alpha\hbar^4}{4m_l^+ m_l^-},\quad C_{32}(E,n) = \left[\frac{-\alpha E\hbar^2}{2m_l^+} + \frac{\alpha\hbar^3 eB}{2m_l^+ m_t^-}\left(n+\frac{1}{2}\right)\right.$$
$$\left. + \frac{(1+\alpha E)\hbar^2}{2m_l^-} + \frac{\alpha\hbar^3 eB}{2m_l^- m_t^+}\left(n+\frac{1}{2}\right) + \frac{\hbar^2}{2m_l^*}\right],\quad \text{and}$$

$$C_{33}(E,n) = \left[\frac{\hbar eB}{m_t^*}\left(n+\frac{1}{2}\right) - \frac{\alpha E\hbar eB}{m_t^+}\left(n+\frac{1}{2}\right)\right.$$
$$\left. + \frac{(1+\alpha E)\hbar eB}{m_t^-}\left(n+\frac{1}{2}\right) + \frac{\alpha(\hbar eB)^2}{m_t^+ m_t^-}\left(n+\frac{1}{2}\right)^2\right].$$

The field-emitted current density is given by

$$J = \frac{e^2 g_v B k_B T}{2\pi^2\hbar^2} \sum_{n=0}^{n_{\max}} [F_0(\eta_{40}) \exp[-\beta_{40}]], \tag{3.56}$$

where $\eta_{40} = (k_B T)^{-1}[E_{FB} - E_{40}]$,and E_{40} is the root of the following equation:

$$E_{40} = \frac{-C_{34}(n) + \sqrt{C_{34}^2(n) + 4\alpha C_{35}(n)}}{2\alpha}, \tag{3.57}$$

where

$$C_{34}(n) = \left[1 + \frac{\alpha\hbar eB}{m_t^+}\left(n + \frac{1}{2}\right) - \frac{\alpha\hbar eB}{m_t^-}\left(n + \frac{1}{2}\right)\right],$$

$$C_{35}(n) = \left[\frac{\hbar eB}{m_t^*}\left(n + \frac{1}{2}\right) + \frac{\hbar eB}{m_t^-}\left(n + \frac{1}{2}\right) + \frac{\alpha(\hbar eB)^2}{m_t^+ m_t^-}\left(n + \frac{1}{2}\right)^2\right],$$

$$\beta_{40} = \frac{4[A_{42}(V_0, n)]^{3/2}}{3eF_{sz}[A_{43}(V_0, n)]},$$

$$A_{42}(V_0, n) = [2C_{31}]^{-1}\big[-C_{32}(V_0, n) + \big[C_{32}^2(V_0, n) - 4C_{31}\{C_{33}(V_0, n) - V_0(1 + \alpha V_0)\}\big]^{1/2}\big],$$

$$A_{43}(V_0, n) = [2C_{31}]^{-1}\left[-C_{36}(n) + \frac{\{C_{32}(V_0, n)C_{36}(n) - 2C_{31}\{C_{37}(n) - (1 + 2\alpha V_0)\}\}}{\left[C_{32}^2(V_0, n) - 4C_{31}\{C_{33}(V_0, n) - V_0(1 + \alpha V_0)\}\right]^{1/2}}\right],$$

$$C_{36}(n) = \left[\frac{\alpha\hbar^2}{2m_l^-} - \frac{\alpha\hbar^2}{2m_l^+}\right], \quad \text{and} \quad C_{37}(n) = \left[\frac{-\alpha\hbar eB}{m_t^+}\left(n + \frac{1}{2}\right) + \frac{\alpha\hbar eB}{m_t^-}\left(n + \frac{1}{2}\right)\right].$$

The electron concentration can be written as

$$n_0 = \left(\frac{eBg_v}{\pi^2\hbar}\right) \sum_{n=0}^{n_{\max}} [Y_{41}(E_{FB}, n) + Z_{41}(E_{FB}, n)], \tag{3.58}$$

where $Y_{41}(E_{FB}, n) = \left[\sqrt{A_{42}(E_{FB}, n)}\right]$ and $Z_{41}(E_{FB}, n) = \sum_{r=1}^{s_0} Z_B(r)[Y_{41}(E_{FB}, n)]$.

3.2.5.2 The Model of Bangert and Kastner

The electron energy spectrum of IV–VI materials in accordance with the model of Bangert and Kastner can be written as [70]

$$\Gamma(E) = \overline{F}_1(E)k_s^2 + \overline{F}_2(E)k_z^2, \tag{3.59}$$

where $\Gamma(E) = 2E$, $\overline{F}_1(E) = \left[\frac{(\overline{R})^2}{E+E_{g0}} + \frac{(\overline{S})^2}{E+\Delta_l'} + \frac{(\overline{Q})^2}{E+\Delta_l''}\right]$, $\overline{F}_2(E) = \left[\frac{2(\overline{A})^2}{E+E_{g0}} + \frac{(\overline{S}+\overline{Q})^2}{E+\Delta_l''}\right]$, and

$\overline{R}$, $\overline{S}$, $\overline{Q}$, $\overline{A}$, Δ_l', Δ_l'' are the electron energy spectrum constants.

In the presence of a quantizing magnetic field $\vec{B}$ along the z-direction, (3.59) assumes the form

$$\Gamma(E) = \overline{F}_1(E)\frac{2eB}{\hbar}\left(n+\frac{1}{2}\right) + \overline{F}_2(E)k_z^2. \tag{3.60}$$

Therefore,

$$k_z^2 = A_{44}(E,n), \tag{3.61}$$

where $A_{44}(E,n) = \frac{\Gamma(E)-\overline{F}_1(E)2eB\left(n+\frac{1}{2}\right)}{\overline{F}_2(E)}$.

The field-emitted current density is given by

$$J = \frac{e^2 g_v B k_B T}{2\pi^2\hbar^2}\sum_{n=0}^{n_{\max}}[F_0(\eta_{41})\exp[-\beta_{41}]], \tag{3.62}$$

where $\eta_{41} = (k_B T)^{-1}[E_{FB} - E_{41}]$, and E_{41} can be determined from the equation

$$\Gamma(E_{41}) = \overline{F}_1(E_{41})\frac{2eB}{\hbar}\left(n+\frac{1}{2}\right), \tag{3.63}$$

$$\beta_{41} = \frac{4[A_{44}(V_0,n)]^{3/2}}{3eF_{sz}[A_{45}(V_0,n)]}, \quad A_{44}(V_0,n) = \frac{\Gamma(V_0) - \overline{F}_1(V_0)2eB(n+\frac{1}{2})}{\overline{F}_2(V_0)},$$

$$\Gamma(V_0) = 2V_0,$$

$$\overline{F}_1(V_0) = \left[\frac{(\overline{R})^2}{V_0+E_{g0}} + \frac{(\overline{S})^2}{V_0+\Delta_l'} + \frac{(\overline{Q})^2}{V_0+\Delta_l''}\right],$$

$$\overline{F}_2(V_0) = \left[\frac{2(\overline{A})^2}{V_0+E_{g0}} + \frac{(\overline{S}+\overline{Q})^2}{V_0+\Delta_l''}\right], \quad \text{and}$$

$$A_{45}(V_0,n) = \left\{A_{44}(V_0,n)\left[\frac{2(\overline{A})^2}{(V_0+E_{g0})^2}+\frac{(\overline{S}+\overline{Q})^2}{(V_0+\Delta_l'')^2}\right]+\frac{1}{\overline{F}_2(V_0)} \times\left[2+\frac{2eB}{\hbar}\left(n+\frac{1}{2}\right)\left[\frac{(\overline{R})^2}{(V_0+E_{g0})^2}+\frac{(\overline{S})^2}{(V_0+\Delta_l')^2}+\frac{(\overline{Q})^2}{(V_0+\Delta_l'')^2}\right]\right]\right\}.$$

The electron concentration can be expressed as

$$n_0 = \left(\frac{eBg_v}{\pi^2\hbar}\right)\sum_{n=0}^{n_{\max}}[Y_{42}(E_{FB},n)+Z_{42}(E_{FB},n)], \tag{3.64}$$

where $Y_{42}(E_{FB},n) = [\sqrt{A_{44}(E_{FB},n)}]$ and $Z_{42}(E_{FB},n) = \sum_{r=1}^{s_0} Z_B(r)[Y_{42}(E_{FB},n)]$.

3.2.5.3 The Model of Foley and Landenberg

In accordance with the model of Foley and Landenberg, the electron energy spectrum in IV–VI semiconductors can be written as [71]

$$E+\frac{E_{g0}}{2} = \frac{\hbar^2k_s^2}{2m_\perp^-}+\frac{\hbar^2k_z^2}{2m_\parallel^-}+\left[\left[\frac{\hbar^2k_s^2}{2m_\perp^+}+\frac{\hbar^2k_z^2}{2m_\parallel^+}+\frac{E_{g0}}{2}\right]^2+P_\parallel^2k_z^2+P_\perp^2k_s^2\right]^{1/2}, \tag{3.65}$$

where $\frac{1}{m_\perp^\pm} = \frac{1}{2}\left[\frac{1}{m_{tc}}\pm\frac{1}{m_{t2}}\right]$, $\frac{1}{m_\parallel^\pm} = \frac{1}{2}\left[\frac{1}{m_{lc}}\pm\frac{1}{m_{l2}}\right]$, m_{tc} and m_{lc} are the transverse and longitudinal effective electron masses of the conduction electrons at the edge of the conduction band, and m_{t2} and m_{l2} are the transverse and longitudinal effective hole masses at the edge of the valence band. In the presence of magnetic quantization $\vec{B}$ along the z-direction, (3.65) assumes the form

Therefore

$$k_z^2 = A_{46}(E,n), \tag{3.66}$$

where $A_{46}(E,n) = (2D_{31})^{-1}\left[-D_{32}(E,n)+\left[D_{32}^2(E,n)+4[E(E+E_{g0})-D_{33}(E,n)]D_{31}\right]^{\frac{1}{2}}\right]$,

$$D_{31} = \left[\frac{\hbar^4}{4(m_\parallel^+)^2}-\frac{\hbar^4}{4(m_\parallel^-)^2}\right],$$

$$D_{32}(E,n) = \left[\frac{\hbar^2}{2m_\parallel^+}\left\{E_{g0}+\frac{2\hbar eB}{m_\perp^+}\left(n+\frac{1}{2}\right)\right\}+P_\parallel^2-\frac{\hbar^3eB}{m_\perp^-m_\parallel^-}\left(n+\frac{1}{2}\right) + \left(E_{g0}+2E\right)\frac{\hbar^2}{2m_\parallel^-}\right], \quad \text{and}$$

$$D_{33}(E,n) = \left[-\left\{\frac{\hbar eB}{m_\perp^-}\left(n+\frac{1}{2}\right)\right\}^2 + (E_{g0}+2E)\frac{\hbar eB}{m_\perp^-}\left(n+\frac{1}{2}\right) + \left\{\frac{\hbar eB}{m_\perp^+}\left(n+\frac{1}{2}\right)\right\}^2 \right.$$
$$\left. + E_{g0}\frac{\hbar eB}{m_\perp^+}\left(n+\frac{1}{2}\right) + P_\perp^2\frac{2eB}{\hbar}\left(n+\frac{1}{2}\right)\right]$$

The field-emitted current density is given by

$$J = \frac{e^2 g_v B k_B T}{2\pi^2\hbar^2}\sum_{n=0}^{n_{\max}}[F_0(\eta_{42})\exp[-\beta_{42}]], \tag{3.67}$$

where $\eta_{42} = (k_B T)^{-1}[E_{FB} - E_{42}]$, and E_{42} can be determined from the equation

$$E_{42} = \frac{-D_{45}(n) + \sqrt{D_{45}^2(n) + 4D_{44}(n)}}{2}, \tag{3.68}$$

where $D_{45}(n) = \left[E_{g0} - \frac{2\hbar eB}{m_\perp^-}\left(n+\frac{1}{2}\right)\right]$,

$$D_{44}(n) = \left[-\left\{\frac{\hbar eB}{m_\perp^-}\left(n+\frac{1}{2}\right)\right\}^2 + (E_{g0})\frac{\hbar eB}{m_\perp^-}\left(n+\frac{1}{2}\right) + \left\{\frac{\hbar eB}{m_\perp^+}\left(n+\frac{1}{2}\right)\right\}^2 \right.$$
$$\left. + E_{g0}\frac{\hbar eB}{m_\perp^+}\left(n+\frac{1}{2}\right) + P_\perp^2\frac{2eB}{\hbar}\left(n+\frac{1}{2}\right)\right],$$

$$\beta_{42} = \frac{4[A_{46}(V_0,n)]3/2}{3eF_{sz}[A_{47}(V_0,n)]},$$

$$A_{46}(V_0,n) = (2D_{31})^{-1}\left[-D_{32}(V_0,n) + \left[D_{32}^2(V_0,n) + 4D_{31}\left[V_0(1+\alpha V_0)\right.\right.\right.$$
$$\left.\left.\left. -D_{33}(V_0,n)\right]\right]^{\frac{1}{2}}\right],$$

$$A_{47}(V_0,n) = (2D_{31})^{-1}\left[-D_{46}(n) + \frac{[D_{32}(V_0,n)D_{46}(n) + 2D_{31}\left[1+2\alpha V_0 - D_{47}(n)\right]]}{\left[D_{32}^2(V_0,n) + 4D_{31}\left[V_0(1+\alpha V_0) - D_{33}(V_0,n)\right]\right]^{\frac{1}{2}}}\right],$$

$$D_{46}(n) = \frac{\hbar^2}{m_\parallel^-}, \quad \text{and} \quad D_{47}(n) = \frac{2\hbar eB}{m_\perp^-}\left(n+\frac{1}{2}\right).$$

The electron concentration can be expressed as

$$n_0 = \left(\frac{eBg_v}{\pi^2\hbar}\right)\sum_{n=0}^{n_{max}}[Y_{43}(E_{FB},n) + Z_{43}(E_{FB},n)], \tag{3.69}$$

where $Y_{43}(E_{FB},n) = [\sqrt{A_{46}(E_{FB},n)}]$ and $Z_{43}(E_{FB},n) = \sum_{r=1}^{s_0} Z_B(r)[Y_{43}(E_{FB},n)]$.

3.2.6 The Field Emission from Stressed Semiconductors Under Magnetic Quantization

The electron energy spectrum under magnetic quantization can be written following (1.76) as

$$\left(n+\frac{1}{2}\right)\frac{2eB}{\hbar}\left[M_0^2(E) - \frac{1}{4}N_0^2(E)\right]^{1/2} + L_0(E)k_z^2 = k_0(E). \tag{3.70}$$

Therefore, (3.70) can be expressed as

$$k_z^2 = A_{48}(E,n), \tag{3.71}$$

where $A_{48}(E,n) = \dfrac{\left[k_0(E) - \frac{2eB}{\hbar}\left(n+\frac{1}{2}\right)\left[M_0^2(E) - \frac{1}{4}N_0^2(E)\right]1/2\right]}{L_0(E)}$.

The field-emitted current density is given by

$$J = \frac{e^2 Bg_v k_B T}{2\pi^2\hbar^2}\sum_{n=0}^{n_{max}}[F_0(\eta_{43})\exp[-\beta_{43}]], \tag{3.72}$$

where $\eta_{43} = (k_B T)^{-1}[E_{FB} - E_{43}]$, and E_{43} is the root of the equation.

$$k_0(E_{43}) = \frac{2eB}{\hbar}\left(n+\frac{1}{2}\right)\left[M_0^2(E_{43}) - \frac{1}{4}N_0^2(E_{43})\right]^{1/2}, \tag{3.73}$$

$$\beta_{43} = \frac{4[A_{48}(V_0,n)]^{3/2}}{3eF_{sz}[A_{49}(V_0,n)]},$$

$$A_{48}(V_0,n) = \frac{\left[k_0(V_0) - \frac{2eB}{\hbar}\left(n+\frac{1}{2}\right)\left[M_0^2(V_0) - \frac{1}{4}N_0^2(V_0)\right]^{1/2}\right]}{L_0(V_0)} \quad \text{and}$$

$$A_{49}(V_0,n) = \left[\frac{-A_{48}(V_0,n)}{L_0(V_0)}L_0'(V_0) + \frac{1}{L_0(V_0)}\left[k_0'(V_0) - \frac{eB}{\hbar}\left(n+\frac{1}{2}\right)\right.\right.$$
$$\left.\left.\times\left[M_0^2(V_0) - \frac{1}{4}N_0^2(V_0)\right]^{-1/2}\left[2M_0(V_0)M_0'(V_0) - \frac{1}{2}N_0(V_0)N_0'(V_0)\right]\right]\right].$$

The electron concentration can be expressed as

$$n_0 = \left(\frac{eBg_v}{\pi^2\hbar}\right)\sum_{n=0}^{n_{\max}}[Y_{44}(E_{FB},n) + Z_{44}(E_{FB},n)], \tag{3.74}$$

where $Y_{44}(E_{FB},n) = [\sqrt{A_{48}(E_{FB},n)}]$ and $Z_{44}(E_{FB},n) = \sum_{r=1}^{s_0} Z_B(r)[Y_{44}(E_{FB},n)]$.

3.2.7 *The Field Emission from Tellurium Under Magnetic Quantization*

The dispersion under magnetic quantization can be written following (1.81) as

$$E = \Psi_1 k_z^2 + \Psi_2 \frac{2eB}{\hbar}\left(n+\frac{1}{2}\right) \pm \left[\Psi_3^2 k_z^2 + \Psi_4^2 \frac{2eB}{\hbar}\left(n+\frac{1}{2}\right)\right]^{1/2}. \tag{3.75}$$

Therefore,

$$k_z^2 = A_{50,\pm}(E,n), \tag{3.76}$$

where $A_{50,\pm}(E,n) = (2\Psi_1^2)^{-1}\left[\Psi_5(E,n) \pm [\Psi_5^2(E,n) - 4\Psi_1^2\Psi_6(E,n)]^{\frac{1}{2}}\right]$,

$$\Psi_5(E,n) = \left[2\Psi_1\left[E - \Psi_2\frac{2eB}{\hbar}\left(n+\frac{1}{2}\right)\right] + \Psi_3^2\right], \quad \text{and}$$

$$\Psi_6(E,n) = \left[\left[E - \Psi_2\frac{2eB}{\hbar}\left(n+\frac{1}{2}\right)\right]^2 - \Psi_4^2\frac{2eB}{\hbar}\left(n+\frac{1}{2}\right)\right]$$

The field-emitted current density is given by

$$J = \frac{e^2 B g_v k_B T}{4\pi^2\hbar^2}\sum_{n=0}^{n_{\max}}[F_0(\eta_{44,+})\exp[-\beta_{44,+}] + F_0(\eta_{44,-})\exp[-\beta_{44,-}]], \tag{3.77}$$

where $\eta_{44,\pm} = (k_B T)^{-1}[E_{FB} - E_{44,\pm}]m$,

$$E_{44,\pm} = \left[\Psi_2\frac{2eB}{\hbar}\left(n+\frac{1}{2}\right) \pm \Psi_4\left[\frac{2eB}{\hbar}\left(n+\frac{1}{2}\right)\right]^{1/2}\right],$$

$$\beta_{44,\pm} = \frac{4[A_{50,\pm}(V_0,n)]^{3/2}}{3eF_{sz}[A_{51,\pm}(V_0,n)]}, \quad A_{50,\pm}(V_0,n)$$

$$= (2\Psi_1^2)^{-1}\left[\Psi_5(V_0,n) \pm [\Psi_5^2(V_0,n) - 4\Psi_1^2\Psi_6(V_0,n)]^{\frac{1}{2}}\right],$$

$$\Psi_5(V_0,n)=\left[2\Psi_1\left[V_0-\Psi_2\frac{2eB}{\hbar}\left(n+\frac{1}{2}\right)\right]+\Psi_3^2\right],$$

$$\Psi_6(V_0,n)=\left[\left[V_0-\Psi_2\frac{2eB}{\hbar}\left(n+\frac{1}{2}\right)\right]^2-\Psi_4^2\frac{2eB}{\hbar}\left(n+\frac{1}{2}\right)\right],$$

$$A_{51,\pm}(V_0,n)=(2\Psi_1^2)^{-1}\left[\Psi_7\pm\left[\Psi_5^2(V_0,n)-4\Psi_1^2\Psi_6(V_0,n)\right]^{-1}\right.$$
$$\left.\times\left[\Psi_5(V_0,n)\Psi_7-2\Psi_1^2\Psi_8(V_0,n)\right]\right],\quad\text{and}$$

$$\Psi_7=2\Psi_1,\quad\Psi_8(V_0,n)=2\left[V_0-\Psi_2\frac{2eB}{\hbar}\left(n+\frac{1}{2}\right)\right].$$

The electron concentration can be expressed as

$$n_0=\left(\frac{eBg_v}{2\pi^2\hbar}\right)\sum_{n=0}^{n_{\max}}\left[Y_{45}(E_{FB},n)+Z_{45}(E_{FB},n)\right],\tag{3.78}$$

where $Y_{45}(E_{FB},n)=[\sqrt{A_{50,+}(E_{FB},n)}+\sqrt{A_{50,-}(E_{FB},n)}]$ and $Z_{45}(E_{FB},n)=\sum_{r=1}^{s_0}Z_B(r)[Y_{45}(E_{FB},n)]$.

3.2.8 *The Field Emission from n-Gallium Phosphide Under Magnetic Quantization*

The magneto-electron energy spectrum can be written following (1.86) as

$$E=a_0\frac{2eB}{\hbar}\left(n+\frac{1}{2}\right)+b_0k_z^2-\left[\left[C\frac{2eB}{\hbar}\left(n+\frac{1}{2}\right)+|V_G|^2Ck_z^2\right]^{\frac{1}{2}}\right]+|V_G|,\tag{3.79}$$

where $a_0=\dfrac{\hbar^2}{2m_\perp^*}+\dfrac{A\hbar^2}{2m_\parallel^*}$, $b_0=\dfrac{\hbar^2}{2m_\parallel^*}$, and $C=\dfrac{\hbar^4k_0^2}{(m_\parallel^*)^2}$.

Therefore

$$k_z^2=A_{52,\pm}(E,n),\tag{3.80}$$

where $A_{52,\pm}(E,n)=(2b_0^2)^{-1}\left[\Psi_{11}(E,n)\pm\left[\Psi_{11}^2(E,n)-4b_0^2\Psi_{12}(E,n)\right]^{\frac{1}{2}}\right]$,

$$\Psi_{11}(E,n) = [2b_0\,[E - \Psi_9(n)] + C], \quad \Psi_{12}(E,n)$$
$$= \left[[E - \Psi_9(n)]^2 - \Psi_{10}(n)\right], \ \Psi_9(n) = |V_G| + a_0\frac{2eB}{\hbar}\left(n + \frac{1}{2}\right), \text{ and}$$
$$\Psi_{10}(n) = C\frac{2eB}{\hbar}\left(n + \frac{1}{2}\right) + |V_G|^2.$$

The field-emitted current density is given by

$$J = \frac{e^2 Bg_v k_B T}{4\pi^2\hbar^2}\sum_{n=0}^{n_{max}}[F_0(\eta_{45,+})\exp[-\beta_{45,+}] + F_0(\eta_{45,-})\exp[-\beta_{45,-}]], \quad (3.81)$$

where $\eta_{45,\pm} = (k_BT)^{-1}[E_{FB} - E_{45,\pm}]$,

$$E_{45,\pm} = \left[a_0\left(\frac{2eB}{\hbar}\left(n+\frac{1}{2}\right)\right) - \left[C\frac{2eB}{\hbar}\left(n+\frac{1}{2}\right) + |V_G|^2\right]^{1/2} + |V_G|\right], \quad (3.82)$$

$$\beta_{45,\pm} = \frac{4[A_{52,\pm}(V_0,n)]^{3/2}}{3eF_{sz}[A_{54,\pm}(V_0,n)]}, \quad A_{52,\pm}(V_0,n)$$
$$= (2b_0^2)^{-1}[\Psi_{11}(V_0,n) \pm [\Psi_{11}^2(V_0,n) - 4b_0^2\Psi_{12}(V_0,n)]^{\frac{1}{2}}],$$
$$\Psi_{11}(V_0,n) = [2b_0\,[V_0 - \Psi_9(n)] + C], \ \Psi_{12}(V_0,n) = \left[[V_0 - \Psi_9(n)]^2 - \Psi_{10}(n)\right],$$
$$A_{54,\pm}(V_0,n) = (2b_0^2)^{-1}[\Psi_{13} \pm [\Psi_{11}^2(V_0,n) - 4b_0^2\Psi_{12}(V_0,n)]^{-\frac{1}{2}}$$
$$\cdot [\Psi_{11}(V_0,n)\Psi_{13} - 2b_0^2\Psi_{14}(V_0,n)]],$$

and $\Psi_{13} = 2b_0$, $\Psi_{14}(V_0,n) = 2[V_0 - \Psi_9(n)]$.
The electron concentration can be expressed as

$$n_0 = \left(\frac{eBg_v}{2\pi^2\hbar}\right)\sum_{n=0}^{n_{max}}[Y_{46}(E_{FB},n) + Z_{46}(E_{FB},n)], \quad (3.83)$$

where $Y_{46}(E_{FB},n) = [\sqrt{A_{52,+}(E_{FB},n)} + \sqrt{A_{52,-}(E_{FB},n)}]$ and $Z_{46}(E_{FB},n) = \sum_{r=1}^{s_0} Z_B(r)[Y_{46}(E_{FB},n)]$.

3.2.9 *The Field Emission from Platinum Antimonide Under Magnetic Quantization*

The magneto-dispersion relation can be written following (1.91) as

$$\left[E + \frac{\bar{\lambda}_0(\bar{a})^2 eB}{2\hbar}\left(n + \frac{1}{2}\right) + \frac{\bar{\lambda}_0(\bar{a})^2}{4}k_z^2 - \frac{\bar{l}(\bar{a})^2 eB}{2\hbar}\left(n + \frac{1}{2}\right)\right]$$
$$\left[E + \bar{\delta}_0 - \frac{\bar{\upsilon}(\bar{a})^2 eB}{2\hbar}\left(n + \frac{1}{2}\right) - \frac{\bar{\upsilon}(\bar{a})^2}{4}k_z^2 - \frac{\bar{n}(\bar{a})^2 eB}{2\hbar}\left(n + \frac{1}{2}\right)\right]$$
$$= \frac{I(\bar{a})^4}{16}\left[k_z^2 + \frac{2eB}{\hbar}\left(n + \frac{1}{2}\right)\right]^2. \quad (3.84)$$

Therefore,

$$k_z^2 = A_{55,\pm}(E, n), \quad (3.85)$$

where $A_{55,\pm}(E,n) = (2\Psi_{17})^{-1}\left[-\Psi_{18}(E,n) \pm \left[\Psi_{18}^2(E,n) - 4\Psi_{17}\Psi_{19}(E,n)\right]^{\frac{1}{2}}\right]$, $\Psi_{17} = \left[\frac{I(\bar{a})^4}{16} + \frac{\bar{\lambda}_0\bar{\upsilon}(\bar{a})^4}{16}\right]$,

$$\Psi_{18}(E,n) = \left[\frac{I(\bar{a})^4 eB}{4\hbar}\left(n + \frac{1}{2}\right) + \Psi_{15}(E,n)\frac{\bar{\upsilon}(\bar{a})^2}{4} - \Psi_{16}(E,n)\frac{\bar{\lambda}_0(\bar{a})^2}{4}\right],$$

$$\Psi_{20} = \frac{\bar{\upsilon}(\bar{a})^2}{4},$$

$$\Psi_{21}(E,n) = [\Psi_{16}(E,n) + \Psi_{15}(E,n)],$$

$$\Psi_{19}(E,n) = \left(\Psi_{15}(E,n)\Psi_{16}(E,n) - \frac{I(\bar{a})^4 e^2 B^2\left(n + \frac{1}{2}\right)^2}{4\hbar^2}\right),$$

$$\Psi_{15}(E,n) = \left[E + \frac{\bar{\lambda}_0(\bar{a})^2 eB}{2\hbar}\left(n + \frac{1}{2}\right) - \frac{\bar{l}(\bar{a})^2 eB}{2\hbar}\left(n + \frac{1}{2}\right)\right],$$

and

$$\Psi_{16}(E,n) = \left[E + \bar{\delta}_0 - \frac{\bar{\upsilon}(\bar{a})^2 eB}{2\hbar}\left(n + \frac{1}{2}\right) - \frac{\bar{n}(\bar{a})^2 eB}{2\hbar}\left(n + \frac{1}{2}\right)\right].$$

The field-emitted current density is given by

$$J = \frac{e^2 B g_v k_B T}{4\pi^2\hbar^2}\sum_{n=0}^{n_{\max}}[F_0(\eta_{46,+})\exp[-\beta_{46,+}] + F_0(\eta_{46,-})\exp[-\beta_{46,-}]], \quad (3.86)$$

where $\eta_{46,\pm} = (k_B T)^{-1}[E_{FB} - E_{46,\pm}]$, and $E_{46,\pm}$ is the root of the equation.

$$E_{46,\pm} = \frac{\left[-\Psi_{22}(n) \pm \sqrt{\Psi_{22}^2(n) + 4\Psi_{23}(n)}\right]}{2}, \tag{3.87}$$

where $\Psi_{22}(n) = \left[\frac{\bar{\lambda}_0(\bar{a})^2 eB}{2\hbar}\left(n+\frac{1}{2}\right) - \frac{\bar{l}(\bar{a})^2 eB}{2\hbar}\left(n+\frac{1}{2}\right) + \bar{\delta}_0 - \frac{\bar{\upsilon}(\bar{a})^2 eB}{2\hbar}\left(n+\frac{1}{2}\right) - \frac{\bar{n}(\bar{a})^2 eB}{2\hbar}\left(n+\frac{1}{2}\right)\right]$,

$$\Psi_{23}(n) = \left[\frac{I(\bar{a})^4 e^2 B^2 \left(n+\frac{1}{2}\right)^2}{4\hbar^2} + \left[\frac{\bar{\lambda}_0(\bar{a})^2 eB}{2\hbar}\left(n+\frac{1}{2}\right) - \frac{\bar{l}(\bar{a})^2 eB}{2\hbar}\left(n+\frac{1}{2}\right)\right]\right.$$
$$\left.\times \left[-\bar{\delta}_0 + \frac{\bar{\upsilon}(\bar{a})^2 eB}{2\hbar}\left(n+\frac{1}{2}\right) + \frac{\bar{n}(\bar{a})^2 eB}{2\hbar}\left(n+\frac{1}{2}\right)\right]\right],$$

$$\beta_{46,\pm} = \frac{4[A_{55,\pm}(V_0,n)]^{3/2}}{3eF_{sz}[A_{56,\pm}(V_0,n)]},$$

$$A_{55,\pm}(V_0,n) = (2\Psi_{17})^{-1}\left[-\Psi_{18}(V_0,n) \pm \left[\Psi_{18}^2(V_0,n) - 4\Psi_{17}\Psi_{19}(V_0,n)\right]^{\frac{1}{2}}\right],$$

$$\Psi_{18}(V_0,n) = \left[\frac{I(\bar{a})^4 eB}{4\hbar}\left(n+\frac{1}{2}\right) + \Psi_{15}(V_0,n)\frac{\bar{\upsilon}(\bar{a})^2}{4} - \Psi_{16}(V_0,n)\frac{\bar{\lambda}_0(\bar{a})^2}{4}\right],$$

$$\Psi_{15}(V_0,n) = \left[V_0 + \frac{\bar{\lambda}_0(\bar{a})^2 eB}{2\hbar}\left(n+\frac{1}{2}\right) - \frac{\bar{l}(\bar{a})^2 eB}{2\hbar}\left(n+\frac{1}{2}\right)\right],$$

$$\Psi_{16}(V_0,n) = \left[V_0 + \bar{\delta}_0 - \frac{\bar{\upsilon}(\bar{a})^2 eB}{2\hbar}\left(n+\frac{1}{2}\right) - \frac{\bar{n}(\bar{a})^2 eB}{2\hbar}\left(n+\frac{1}{2}\right)\right],$$

$$\Psi_{19}(V_0,n) = \left(\Psi_{15}(V_0,n)\Psi_{16}(V_0,n) - \frac{I(\bar{a})^4 e^2 B^2\left(n+\frac{1}{2}\right)^2}{4\hbar^2}\right)$$

$$A_{56,\pm}(V_0,n) = (2\Psi_{17})^{-1}\left[-\Psi_{20} \pm \left[\Psi_{18}^2(V_0,n) - 4\Psi_{17}\Psi_{19}(V_0,n)\right]^{1/2}\right.$$
$$\left.\times \left[\Psi_{18}(V_0,n)\Psi_{20} + 2\Psi_{17}\Psi_{21}(V_0,n)\right]\right], \quad \text{and}$$

$$\Psi_{20} = \frac{\bar{\upsilon}(\bar{a})^2}{4}, \Psi_{21}(V_0,n) = [\Psi_{16}(V_0,n) + \Psi_{15}(V_0,n)].$$

The electron concentration can be expressed as

$$n_0 = \left(\frac{eBg_v}{2\pi^2\hbar}\right)\sum_{n=0}^{n_{\max}}[Y_{47}(E_{FB},n) + Z_{47}(E_{FB},n)], \tag{3.88}$$

where $Y_{47}(E_{FB}, n) = [\sqrt{A_{55,+}(E_{FB}, n)} + \sqrt{A_{55,-}(E_{FB}, n)}]$ and $Z_{47}(E_{FB}, n) = \sum_{r=1}^{s_0} Z_B(r)[Y_{47}(E_{FB}, n)]$.

3.2.10 The Field Emission from Bismuth Telluride Under Magnetic Quantization

In the presence of a quantizing magnetic field $\vec{B}$ along the k_x direction, the magneto-dispersion relation of the carriers in Bi_2Te_3 can be written following (1.97) as

$$E(1 + \alpha E) = \bar{\omega}_1 k_x^2 + \hbar\omega_{31}\left(n + \frac{1}{2}\right), \tag{3.89}$$

where $\omega_{31} = \frac{eB}{M_{31}}$ and $M_{31} = \dfrac{m_0}{\left[\bar{\alpha}_{22}\bar{\alpha}_{23} - \frac{(\bar{\alpha}_{23})^2}{4}\right]^{1/2}}$.

Therefore

$$k_x^2 = \frac{E(1 + \alpha E) - \hbar\omega_{31}\left(n + \frac{1}{2}\right)}{\bar{\omega}_1}.$$

The field-emitted current density is given by

$$J = \frac{e^2 B g_v k_B T}{2\pi^2\hbar^2} \sum_{n=0}^{n_{max}} [F_0(\eta_{47}) \exp[-\beta_{47}]], \tag{3.90}$$

where $\eta_{47} = (k_B T)^{-1}[E_{FB} - E_{47}]$, and E_{47} is the root of the equation.

$$E_{47} = (2\alpha)^{-1}\left[-1 + \sqrt{1 + 4\alpha\left(n + \frac{1}{2}\right)\hbar\omega_{31}}\right], \tag{3.91}$$

$$\beta_{47} = \frac{4[A_{57}(V_0, n)]^{3/2}}{3eF_{sz}[A_{58}(V_0)]}, \quad A_{57}(V_0, n) = \frac{V_0(1 + \alpha V_0) - \hbar\omega_{31}\left(n + \frac{1}{2}\right)}{\bar{\omega}_1}, \quad \text{and}$$

$A_{58}(V_0) = [(1 + 2\alpha V_0)/\bar{\omega}_1]$.

The electron concentration can be expressed as

$$n_0 = \left(\frac{eBg_v}{\pi^2\hbar}\right) \sum_{n=0}^{n_{max}} [Y_{48}(E_{FB}, n) + Z_{48}(E_{FB}, n)], \tag{3.92}$$

where $Y_{48}(E_{FB},n) = \left[E_{FB}(1+\alpha E_{FB}) - (n+\frac{1}{2})\hbar\omega_{31}/\bar{\omega}_1\right]$ and $Z_{48}(E_{FB},n) = \sum_{r=1}^{s_0} Z_B(r)[Y_{48}(E_{FB},n)]$.

3.2.11 The Field Emission from Germanium Under Magnetic Quantization

3.2.11.1 The Model of Cardona et al.

The dispersion relation of the conduction electrons in $n-Ge$ in accordance with the model of Cardona et al. in the presence of a quantizing magnetic field $\vec{B}$ the along z-direction can be written following (1.102) as

$$E(1+\alpha E) = \hbar\omega_\perp\left(n+\frac{1}{2}\right) + \frac{\hbar^2 k_z^2}{2m_\parallel^*} + 2\alpha E\left(\frac{\hbar^2 k_z^2}{2m_\parallel^*}\right) - \alpha\left(\frac{\hbar^2 k_z^2}{2m_\parallel^*}\right)^2, \quad (3.93)$$

where $\omega_\perp = \frac{eB}{m_\perp^*}$.

Equation (3.93) can be written as

$$k_z^2 = \frac{2m_\parallel^*}{\hbar^2} A_{69}(E,n), \quad (3.94)$$

where $A_{69}(E,n) = (2\alpha)^{-1}\left[1+2\alpha E - \left[1+4\alpha\left(n+\frac{1}{2}\right)\hbar\omega_\perp\right]^{1/2}\right]$.

The field-emitted current density is given by

$$J = \frac{e^2 B g_v k_B T}{2\pi^2\hbar^2}\sum_{n=0}^{n_{max}}[F_0(\eta_{48})\exp[-\beta_{48}]], \quad (3.95)$$

where $\eta_{48} = (k_B T)^{-1}[E_{FB} - E_{48}]$, and E_{48} is the root of the equation.

$$E_{48} = (2\alpha)^{-1}\left[-1+\sqrt{1+4\alpha\left(n+\frac{1}{2}\right)\hbar\omega_\perp}\right], \quad (3.96)$$

$$\beta_{48} = \frac{4\sqrt{2m_\parallel^*}[A_{69}(V_0,n)]^{3/2}}{3eF_{sz}\hbar}, \quad \text{and}$$

$$A_{69}(V_0,n) = (2\alpha)^{-1}\left[1+2\alpha V_0 - \left[1+4\alpha\left(n+\frac{1}{2}\right)\hbar\omega_\perp\right]^{1/2}\right].$$

The electron concentration can be expressed as

$$n_0 = \left(\frac{eBg_v}{\pi^2\hbar}\right)\sum_{n=0}^{n_{max}}[Y_{49}(E_{FB},n) + Z_{49}(E_{FB},n)], \tag{3.97}$$

where $Y_{49}(E_{FB},n) = \frac{\sqrt{2m_{\parallel}^*}}{\hbar}[\sqrt{A_{69}(E_{FB},n)}]$ and $Z_{49}(E_{FB},n) = \sum_{r=1}^{s_0} Z_B(r)[Y_{49}(E_{FB},n)]$.

3.2.11.2 The Model of Wang and Ressler

The magneto-dispersion law in $n - Ge$ in accordance with the model of Wang and Ressler can be written following (1.107) as

$$k_z^2 = \frac{2m_{\parallel}^*}{\hbar^2}[A_{71}(E,n)], \tag{3.98}$$

where $A_{71}(E,n) = \left[\Psi_{24}(n) - \frac{1}{2\bar{e}_1}\left[\Psi_{25}(n) - 4\bar{e}_1 E\right]^{1/2}\right]$, $\Psi_{24}(n) = (2\bar{e}_1)^{-1}\left[1 - \bar{d}_1\left(n+\frac{1}{2}\right)\hbar\omega_\perp\right]$, and $\Psi_{25}(n) = \left[\left\{1 - \bar{d}_1\left(n+\frac{1}{2}\right)\hbar\omega_\perp\right\}^2 + \bar{e}_1\left\{\left(n+\frac{1}{2}\right)\hbar\omega_\perp - \bar{c}_1\left\{\left(n+\frac{1}{2}\right)\hbar\omega_\perp\right\}^2\right\}\right]$. The field-emitted current density is given by

$$J = \frac{e^2 Bg_v k_B T}{2\pi^2\hbar^2}\sum_{n=0}^{n_{max}}[F_0(\eta_{49})\exp[-\beta_{49}]], \tag{3.99}$$

where $\eta_{49} = (k_B T)^{-1}[E_{FB} - E_{49}]$, and E_{49} is the root of the equation.

$$E_{49} = \left(n+\frac{1}{2}\right)\hbar\omega_\perp - \bar{c}_1\left\{\left(n+\frac{1}{2}\right)\hbar\omega_\perp\right\}^2, \tag{3.100}$$

$$\beta_{49} = \frac{4\sqrt{2m_{\parallel}^*}[A_{71}(V_0,n)]^{3/2}}{3eF_{sz}\hbar[A_{72}(V_0,n)]}, \quad A_{71}(V_0,n) = \left[\Psi_{24}(n) - \frac{1}{2\bar{e}_1}[\Psi_{25}(n) - 4\bar{e}_1 V_0]^{1/2}\right],$$

and $A_{72}(V_0,n) = [\Psi_{25}(n) - 4\bar{e}_1 V_0]^{-1/2}$

The electron concentration can be expressed as

$$n_0 = \left(\frac{eBg_v}{\pi^2\hbar}\right)\sum_{n=0}^{n_{max}}[Y_{50}(E_{FB},n) + Z_{50}(E_{FB},n)], \tag{3.101}$$

where $Y_{50}(E_{FB},n) = \frac{\sqrt{2m_{\parallel}^*}}{\hbar}[\sqrt{A_{71}(E_{FB},n)}]$ and $Z_{50}(E_{FB},n) = \sum_{r=1}^{s_0} Z_B(r)[Y_{50}(E_{FB},n)]$.

3.2.12 The Field Emission from Gallium Antimonide Under Magnetic Quantization

The magneto-dispersion relation in this case can be written as

$$k_z^2 = \frac{2m_c}{\hbar^2}\left[I_{16}(E) - \left(n+\frac{1}{2}\right)\hbar\omega_c\right], \tag{3.102}$$

where $I_{16}(E)$ has been defined in (1.113).

The magneto-field-emitted current density is given by

$$J = \frac{e^2 B g_v k_B T}{2\pi^2\hbar^2}\sum_{n=0}^{n_{\max}}[F_0(\eta_{50})\exp[-\beta_{50}]], \tag{3.103}$$

where $\eta_{50} = (k_B T)^{-1}[E_{FB} - E_{50}(n)]$,

$$E_{50}(n) = \left[\left(n+\frac{1}{2}\right)\frac{\hbar eB}{m_0} - \frac{E'_{g0}}{2} + \frac{E'_{g0}}{2}\left\{1+\frac{4\hbar eB}{E'_{g0}}\left(n+\frac{1}{2}\right)\left(\frac{1}{m_c}-\frac{1}{m_0}\right)\right\}^{1/2}\right],$$

$$\beta_{50} = \frac{4\sqrt{2m_c}[A_{73}(V_0,n)]^{3/2}}{3eF_{sz}\hbar[A_{74}(V_0,n)]}, \quad A_{73}(V_0,n) = \left[I_{16}(V_0) - \left(n+\frac{1}{2}\right)\hbar\omega_c\right], \tag{3.104}$$

and

$$A_{74}(V_0,n) = \left[1 - \frac{1}{2}E'_{g0}\left(1-\frac{m_c}{m_0}\right)\left[\left(\frac{E'_{g0}}{2}\right)^2 + \left[\frac{(E'_{g0})^2}{2}\left(1-\frac{m_c}{m_0}\right)\right]\right.\right.$$
$$\left.\left. + \left[\frac{E'_{g0}}{2}\left(1-\frac{m_c}{m_0}\right)\right]^2 + V_0 E'_{g0}\left(1-\frac{m_c}{m_0}\right)\right]^{-1/2}\right].$$

The electron concentration can be expressed as

$$n_0 = \left(\frac{eBg_v}{\pi^2\hbar}\right)\sum_{n=0}^{n_{\max}}[\overline{Y}_{50}(E_{FB},n) + \overline{Z}_{50}(E_{FB},n)], \tag{3.105}$$

where $\overline{Y}_{50}(E_{FB},n) = \frac{\sqrt{2m_c}}{\hbar}[\sqrt{A_{73}(E_{FB},n)}]$ and $\overline{Z}_{50}(E_{FB},n) = \sum_{r=1}^{s_0} Z_B(r)[\overline{Y}_{50}(E_{FB},n)]$.

3.2.13 The Field Emission from II–V Semiconductors Under Magnetic Quantization

The magneto-dispersion law in II–V semiconductors in the presence of a magnetic field $\vec{B}$ along the k_y direction can be written as

$$k_y^2 = A_{75,\pm}(E,n), \tag{3.106}$$

where $A_{75,\pm}(E,n) = [I_{35}E + I_{36,\pm}(n) \pm \sqrt{E^2 + EI_{38,\pm}(n) + I_{39,\pm}(n)}]$, $I_{35} = \dfrac{\theta_2}{(\theta_2^2 - \theta_5^2)}$, $I_{36,\pm}(n) = \dfrac{I_{33,\pm}(n)}{2(\theta_2^2 - \theta_5^2)}$,

$$I_{38,\pm}(n) = (4\theta_5^2)^{-1}\left[4\theta_2 I_{33,\pm}(n) + 8\theta_2^2 I_{31,\pm}(n) - \theta_5^2 I_{31,\pm}(n)\right],$$

$$I_{39,\pm}(n) = (4\theta_5^2)^{-1}\left[I_{33,\pm}^2(n) + 4\theta_2^2 I_{34,\pm}(n) - 4\theta_5^2 I_{34,\pm}(n)\right],$$

$$I_{33,\pm}(n) = \left[G_3^2 + 2\theta_5 I_{32}(n) - 2\theta_2 I_{31,\pm}(n)\right],$$

$$I_{34,\pm}(n) = \left[I_{32}^2(n) + \Delta_3^2 - I_{31,\pm}(n)\right], \quad I_{31,\pm}(n) = \left[\left(n+\frac{1}{2}\right)\hbar\omega_{31} - \frac{\delta_4^2}{4\theta_1} \pm \Delta_3\right],$$

$$I_{32}(n) = \left[\left(n+\frac{1}{2}\right)\hbar\omega_{32} - \frac{\delta_5^2}{4\theta_5}\right],$$

$$\omega_{31} = \frac{eB}{\sqrt{M_{31}M_{32}}}, \quad \omega_{32} = \frac{eB}{\sqrt{M_{33}M_{34}}}, \quad M_{31} = \frac{\hbar^2}{2\theta_1}, \quad M_{32} = \frac{\hbar^2}{2\theta_3},$$

$$M_{33} = \frac{\hbar^2}{2\theta_5}, \quad \text{and} \quad M_{34} = \frac{\hbar^2}{2\theta_7}$$

The magneto-field-emitted current density is given by

$$J = \frac{e^2 B g_v k_B T}{2\pi^2\hbar^2}\sum_{n=0}^{n_{max}}\left[F_0(\eta_{51,+})\exp[-\beta_{51,+}] + F_0(\eta_{51,-})\exp[-\beta_{51,-}]\right] \tag{3.107}$$

where $\eta_{51,\pm} = (k_B T)^{-1}[E_{FB} - E_{51,\pm}]$, and $E_{51,\pm}$ is the root of the equation.

$$E_{51,\pm} = I_{31,\pm}(n) + \left[I_{32}^2(n) + \Delta_3^2\right], \tag{3.108}$$

$$\beta_{51,\pm} = \frac{4[A_{75,\pm}(V_0,n)]^{3/2}}{3eF_{sz}\hbar[A_{76,\pm}(V_0,n)]}, \quad A_{75,\pm}(V_0,n) = \Big[I_{35}V_0 + I_{36,\pm}(n) \pm \sqrt{V_0^2 + V_0 I_{38,\pm}(n) + I_{39,\pm}(n)}\Big],$$

and

$$A_{76,\pm}(V_0,n) = \left[I_{35} \pm \frac{V_0 + \frac{1}{2}I_{38,\pm}(n)}{\sqrt{V_0^2 + V_0 I_{38,\pm}(n) + I_{39,\pm}(n)}} \right].$$

The electron concentration can be expressed as

$$n_0 = \left(\frac{eBg_v}{2\pi^2\hbar}\right) \sum_{n=0}^{n_{\max}} [Y_{51}(E_{FB},n) + Z_{51}(E_{FB},n)], \quad (3.109)$$

where $Y_{51}(E_{FB},n) = [\sqrt{A_{75,+}(E_{FB},n)} + \sqrt{A_{75,-}(E_{FB},n)}]$ and $Z_{51}(E_{FB},n) = \sum_{r=1}^{s_0} Z_B(r)[Y_{51}(E_{FB},n)]$.

3.3 Result and Discussions

Using the appropriate equations and taking the values of the energy band constants for n-$CdGeAs_2$, we have plotted J as a function of inverse quantizing magnetic field as shown in the plot of Fig. 3.1 in accordance with the generalized band model and two-band model of Kane. Figure 3.2 explores the current density as a function of $1/B$ for n-InSb for the models of Stillman et al., Newson et al., Palik et al., and Kane (both three and two bands), respectively. Figures 3.3 and 3.4 cover all the cases of Fig. 3.2 as functions of carrier degeneracy and electric field, respectively. Figures 3.5 and 3.6 exhibit the variation of J as functions of $1/B$ and alloy composition for $Hg_{1-x}Cd_xTe$ in accordance with three- and two-band models of Kane. Figures 3.7–3.8 explore the dependence of the current density on $1/B$, carrier concentration, and

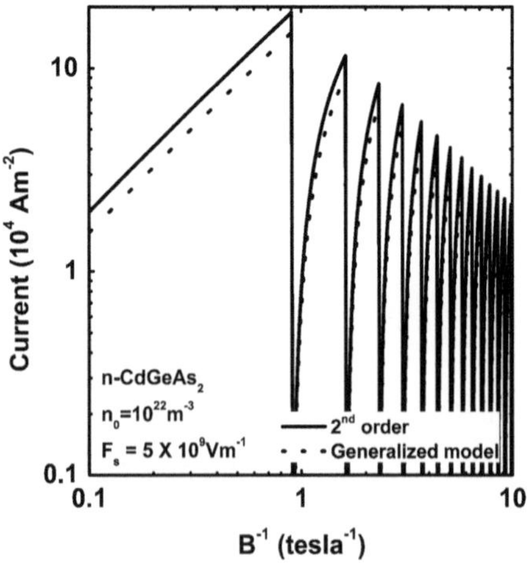

Fig. 3.1 Plot of the field-emitted current density as a function of inverse magnetic field for n-$CdGeAs_2$ in accordance with the generalized and two-band model of Kane

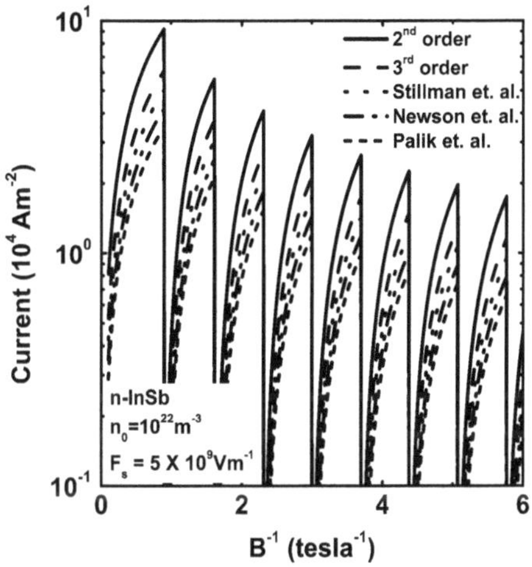

Fig. 3.2 Plot of the field-emitted current density as a function of inverse magnetic field for n-InSb in accordance with the models of Stillman et al., Newson et al., and Palik et al. together with the three- and two-band models of Kane

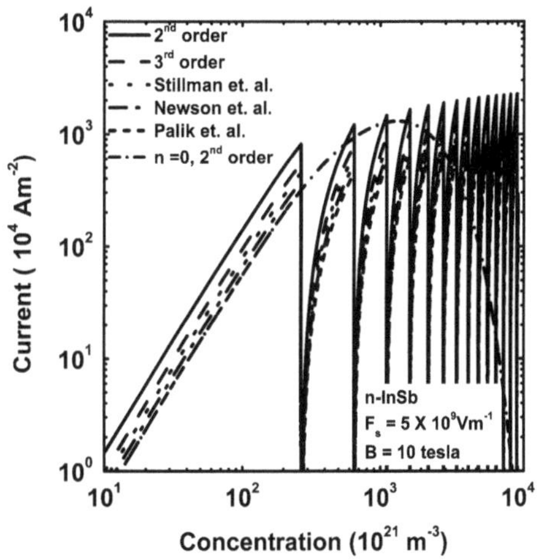

Fig. 3.3 Plot of the field-emitted current density as a function of carrier concentration for n-InSb in accordance with the models of Stillman et al., Newson et al., and Palik et al. together with the three- and two-band models of Kane. The magnetic quantum limit case has further been shown

electric field, respectively, for CdS. Figure 3.9 shows the plot of the term β as functions of $1/B$ and carrier concentration for Bismuth in accordance with the Cohen model. Figures 3.10–3.11 exhibit the field-emitted current density from stressed InSb as functions of $1/B$, carrier degeneracy, and electric field, respectively.

Figures 3.12–3.13 show the influence of $1/B$, carrier degeneracy, and electric field on J from Ge (in accordance with the models of Wang et al. and Cardona et al.), GaSb, Bi_2Te_3, GaP, and Te, respectively.

From Fig. 3.1, we observe that the current density is an oscillatory function of inverse quantizing magnetic field. The oscillatory dependence is due to the crossing of the Fermi level by the Landau subbands in steps, resulting in successive reduction of the number of occupied Landau levels as the magnetic field is increased. For each coincidence of the Landau level with the Fermi level, there would be a discontinuity in the density-of-states function, resulting in a peak of oscillation. These peaks

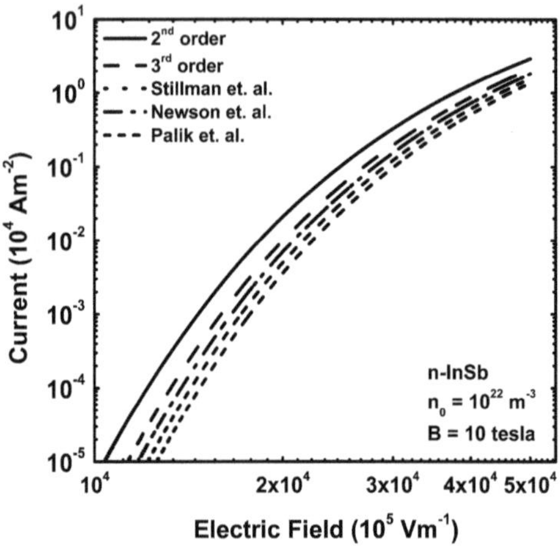

Fig. 3.4 Plot of the field-emitted current density as a function of electric field for n-InSb in accordance with the models of Stillman et al., Newson et al., and Palik et al. together with the three- and two-band models of Kane

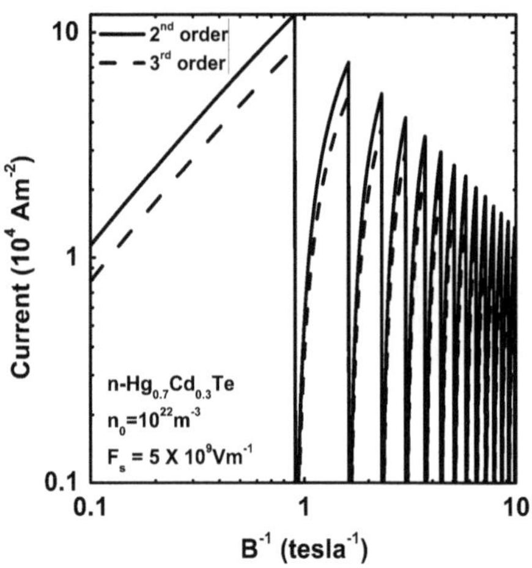

Fig. 3.5 Plot of the field-emitted current density as a function of inverse magnetic field for n-$Hg_{0.7}Cd_{0.3}$Te in accordance with the three- and two-band models of Kane

should occur whenever the Fermi energy is a multiple of the energy separation between the two consecutive Landau levels. Thus, we observe that the origin of the oscillations in the field-emitted current density is the same as the Shubnikov–de Haas oscillations. With the increase in magnetic field, the amplitude of oscillations will increase and ultimately at very large value of the magnetic field, the conditions for magnetic quantum limit will be reached.

From Fig. 3.2, we observe that J exhibits oscillatory dependence on $1/B$ and the influence of energy band models is to change the magnitude of J, although the periodicity remains same for all of the respective curves. We also note that with the application of a magnetic field, the current density reduces to a large extent about $10^5\,\mathrm{Am}^{-2}$. Fig. 3.3 expresses the fact that the J oscillates with the carrier concentration for n-InSb for all types of band models as considered in Fig. 3.2. Under the condition of magnetic quantum limit, the current density initially increases, reaches a peak, and finally decreases with increasing degeneracy.

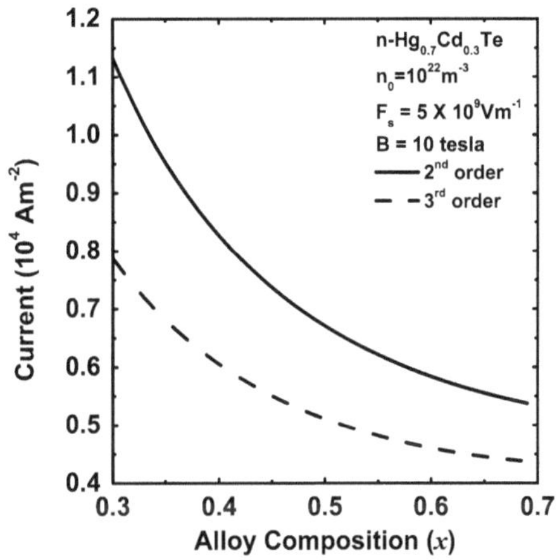

Fig. 3.6 Plot of the field-emitted current density as a function of alloy composition for $n\text{-}Hg_{01-x}Cd_xTe$ in accordance with the three and two-band models of Kane

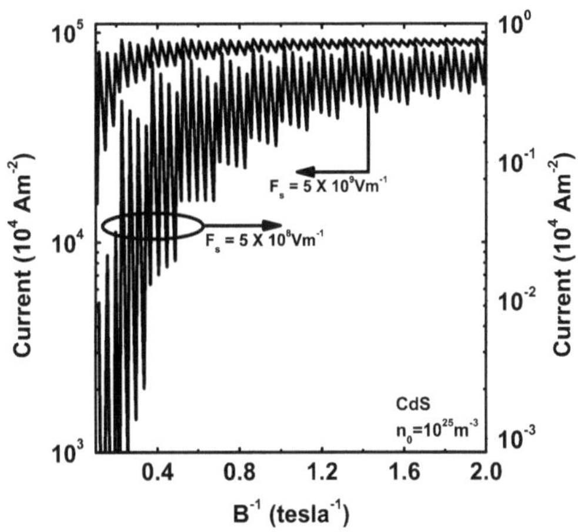

Fig. 3.7 Plot of the field-emitted current density as a function of $1/B$ for CdS

The reason for this has been discussed in "Result and Discussions" in Chap. 1. Incidentally, with more and more subband generation, the periodicity dominates over the falling nature of the J. With the application of the magnetic field, the cut-in values in n-InSb reach about $10^9\,\text{Vm}^{-1}$, a value comparable to that of the metals, as appears from Fig. 3.4. Similar nature has been observed for the material $n - Hg_{0.7}Cd_{0.3}Te$ with respect to the periodic variation of current density with $1/B$ in Fig. 3.5. As the alloy composition increases, J decreases which we note from Fig. 3.6. The influence of the spin–orbit splitting constant reduces J. At $x = 0.7$, we observe that J is being reduced to almost half of its value at $x = 0.3$.

Influence of spin on the field-emitted current density from Cds can be accessed from Fig. 3.7. We observe that the electron spin splits the peak like an "up-down" fashion. For the purpose of condensed presentation, we have not included the effect of spin on J in the previous figures. The influence of electron concentration on J from CdS at the magnetic quantum limit has been exhibited in Fig. 3.14. We note that with the increase in magnetic field, magnitude of J increases with a shift in the

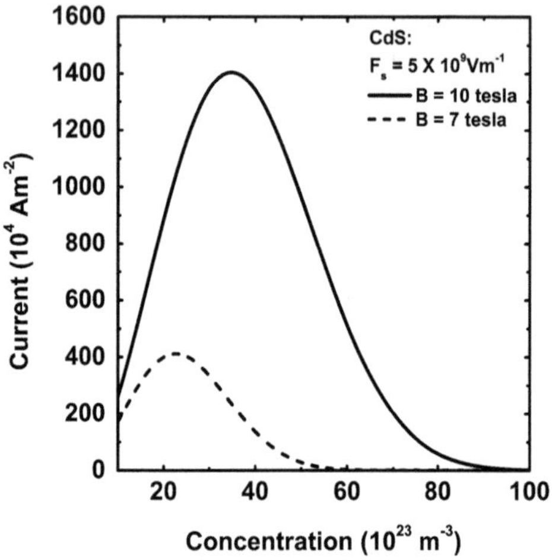

Fig. 3.8 Plot of the field-emitted current density as a function of electric field for CdS

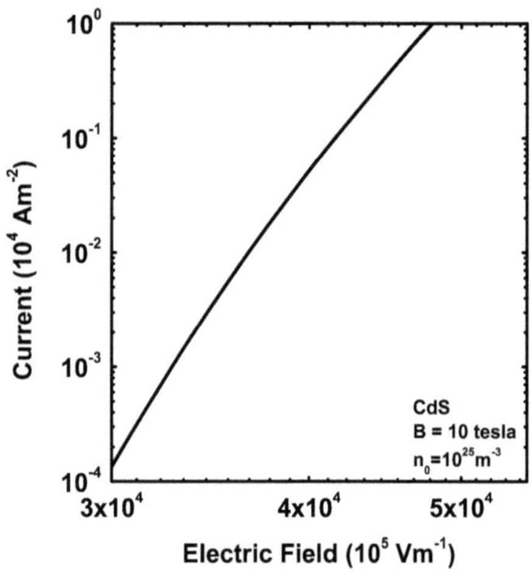

Fig. 3.9 Plot of the term β as a functions of $1/B$ and concentration for Bi in accordance with Cohen model

peak. The value reduces almost zero beyond $10^{25}\,m^{-3}$ at $B = 10$ Tesla, whereas for $B = 7$ Tesla, the same phenomenon happens beyond $6 \times 10^{24}\,m^{-3}$. Besides, the magnitude of the peak reduces almost one third when B changes by 3 Tesla. The influence of electric field on J from CdS has been shown in Fig. 3.8 where for high carrier degeneracy, very small current is obtained even at field strength nearly $3 \times 10^9\,Vm^{-1}$. Since the magnetic field is very prominent in reducing J, in Fig. 3.9, we have shown the effect of magnetic field and concentration on the function β for bismuth. We note that β increases nonlinearly with both the $1/B$ and concentration and is extremely large in both the cases, making the J very low.

The effect of magnetic field on J from stressed n-InSb has been shown in Fig. 3.10 where J exhibits the SdH oscillations as also previously discussed. It is important to note that stress enhances J and the application of stress of 4 Kbar results in nearly $10^9\,Am^{-2}$ field-emitted current density at $5 \times 10^{10}\,Vm^{-1}$ electric

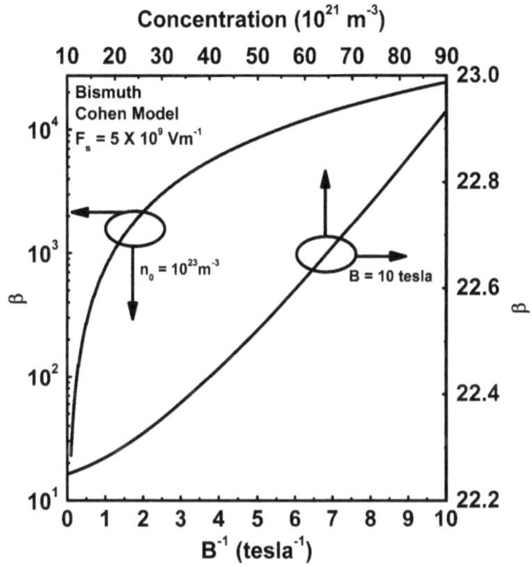

Fig. 3.10 Plot of the field-emitted current density as a function of $1/B$ for stressed n-InSb

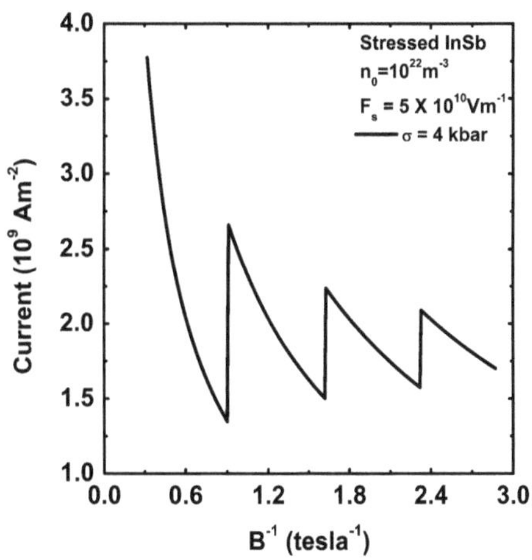

Fig. 3.11 Plot of the field-emitted current density as a function of electric field for stressed n-InSb

field. The effect of application on stress on the field-emitted current density can also be observed from Fig. 3.15. We note that J oscillates with the carrier degeneracy. Very high cut-in electric field is required to obtain the field-emitted current density from stressed n-InSb and about $10^9\,\text{Am}^{-2}$ current density can be achieved at $5 \times 10^{10}\,\text{Vm}^{-1}$ field strength, which appears from Fig. 3.11. From Fig. 3.12, we note that J oscillates with $1/B$ for Te, GaP, Bi_2Te_3, GaSb, and Ge (for both types of band models). The relative magnitudes are the signatures of the different dispersion relations of the different compounds. Same oscillatory dependence of J with degeneracy has been plotted for the aforementioned materials in Fig. 3.16 realizing different magnitudes in J. Relatively smaller cut-in field is needed for GaSb as can be observed from Fig. 3.13, whereas a higher cut-in field is required for Bi_2Te_3 in the present case.

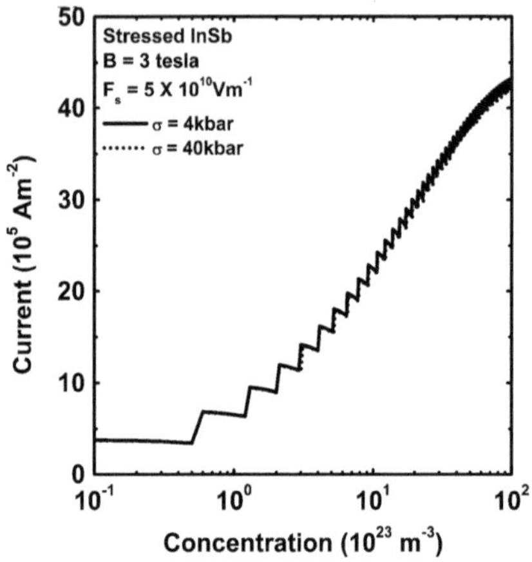

Fig. 3.12 Plot of the field-emitted current density as a function of 1/B for Te, GaP, Bi_2Te_3, GaSb, and Ge (in accordance with the models of Wang et al. and Cardona et al.)

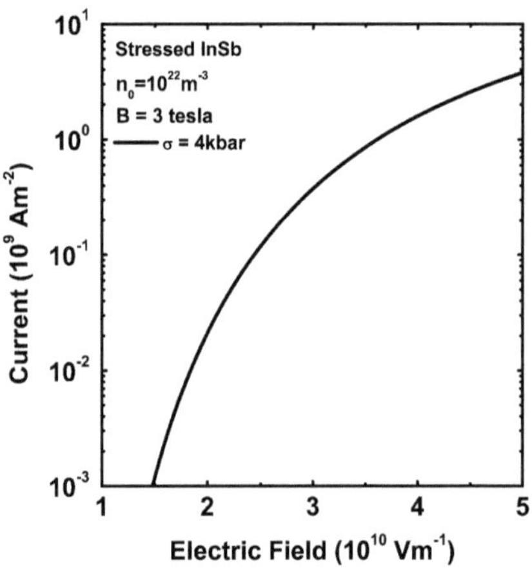

Fig. 3.13 Plot of the field-emitted current density as a function of electric field for Te, GaP, Bi_2Te_3, GaSb, and Ge (in accordance with the models of Wang et al. and Cardona et al.)

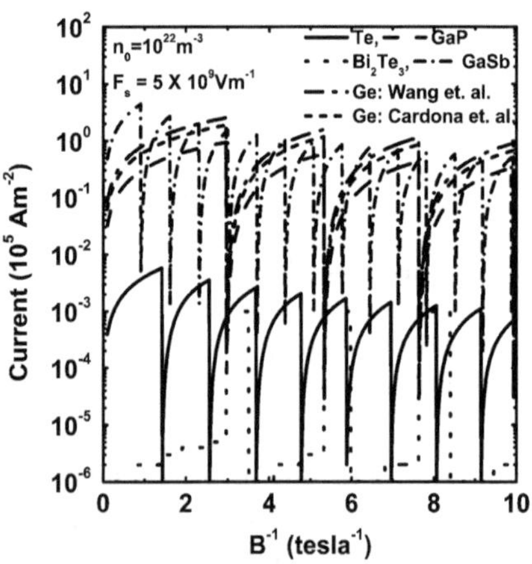

Fig. 3.14 Plot of the field-emitted current density as a function of carrier concentration for CdS

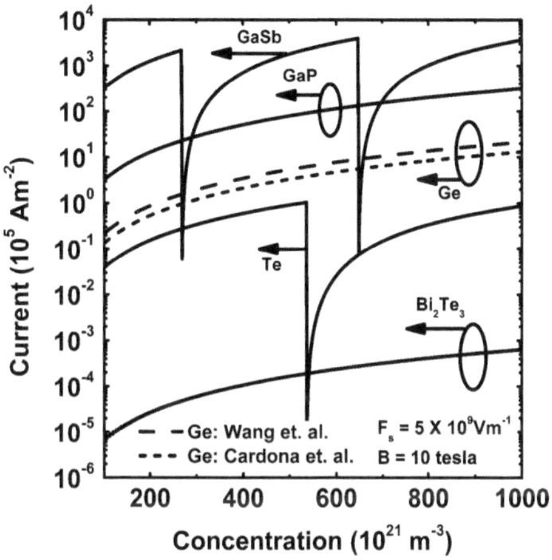

Fig. 3.15 Plot of the field-emitted current density as a function of carrier concentration for stressed n-InSb

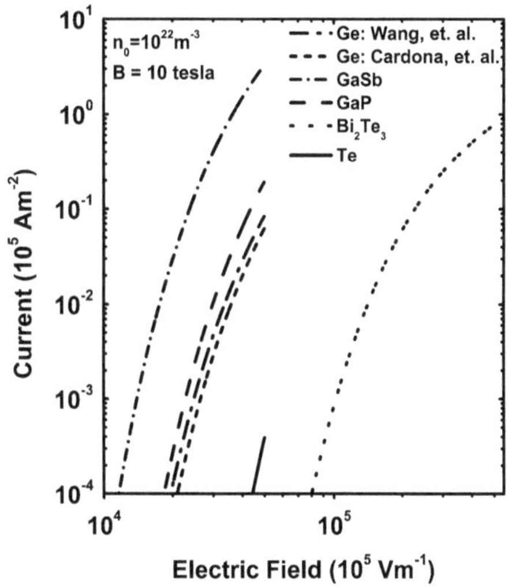

Fig. 3.16 Plot of the field-emitted current density as a function of carrier concentration for Te, GaP, Bi_2Te_3, GaSb, and Ge (in accordance with the models of Wang et al. and Cardona et al.)

Although the effects of collision are usually small at low temperatures at which the quantum effects are prominent, the sharpness of the amplitude of the oscillatory plots would be reduced by collision broadening. Nevertheless, the present analyses remain valid since the effect of collision broadening can be taken into account by an effective increase in temperature. Besides, in a more rigorous statement, the many body effects should be considered along with a self-consistent procedure. This simplified analysis exhibits the basic qualitative features of J in degenerate semiconductors under magnetic quantization with reasonable accuracy. For the purpose of condensed presentation, the specific carrier statistics for a specific semiconductor having a particular electron energy spectrum and the corresponding field-emitted current density under magnetic quantization have been presented in Table 3.1.

Table 3.1 The carrier statistics and the Fowler–Nordheim field emission under magnetic quantization from nonlinear optical, III–V, II–VI, Bismuth, IV–VI, stressed materials, Te, n-GaP, $PtSb_2$, Bi_2Te_3, n-Ge, GaSb, and II–V materials

Type of materials	The carrier statistics	The Fowler–Nordheim field emission
1. Nonlinear optical materials	In accordance with the generalized electron dispersion relation as given by (3.1), $n_0 = \frac{eBg_v}{2\pi^2\hbar}\sum_{n=0}^{n_{max}}[Y_{31}(E_{FB},n) + Z_{31}(E_{FB},n)]$ (3.7)	$J = \frac{e^2 g_v B k_B T}{4\pi^2\hbar^2}\sum_{n=0}^{n_{max}}[F_0(\eta_{31,+})\exp[-\beta_{31,+}] + F_0(\eta_{31,-})\exp[-\beta_{31,-}]]$ (3.6)
2. III–V materials, the conduction electrons of which can be defined by five types of energy wave vector dispersion relations as described in the column beside	(a) Three-band model of Kane: In accordance with the three-band model of Kane (3.8), which is the special case of (3.1) $n_0 = \frac{\sqrt{2m_c}eBg_v}{2\pi^2\hbar^2}\sum_{n=0}^{n_{max}}[\sqrt{A_{32,+}(E_{FB},n)} + \sqrt{A_{32,-}(E_{FB},n)}] + Z_{32}(E_{FB},n)$ (3.13)	$\therefore J = \frac{e^2 g_v B k_B T}{4\pi^2\hbar^2}\sum_{n=0}^{n_{max}}[F_0(\eta_{32,+})\exp[-\beta_{32,+}] + F_0(\eta_{32,-})\exp[-\beta_{32,-}]$ (3.11)
	(b) Two-band model of Kane: In accordance with the two-band model of Kane (3.14), $n_0 = \frac{\sqrt{2m_c}eBg_v}{2\pi^2\hbar^2}\sum_{n=0}^{n_{max}} Y_{32}(E_{FB},n) + Z_{32}(E_{FB},n)$ (3.19)	$J = \frac{e^2 g_v B k_B T}{4\pi^2\hbar^2}\sum_{n=0}^{n_{max}}[F_0(\eta_{33,+})\exp[-\beta_{33,+}] + F_0(\eta_{33,-})\exp[-\beta_{33,-}]]$ (3.17)
	(c) Parabolic energy bands $n_0 = (N_c\theta/2)\sum_{n=0}^{n_{max}} F_{-1/2}(\eta'_{B,+})$ (3.20)	$J = \frac{e^2 g_v B k_B T}{h^2}\sum_{n=0}^{n_{max}}[F_0(\eta'_{B,+})\exp[-\beta_{34,+}(V_0,n)] + F_0(\eta'_{B,-})\exp[-\beta_{34,-}(V_0,n)]]$ (3.21)
	(d) The model of Stillman et al.: In accordance with the model of Stillman et al (3.25), $n_0 = \frac{\sqrt{2m_c}eBg_v}{\pi^2\hbar^2}\sum_{n=0}^{n_{max}}[Y_{33}(E_{FB},n) + Z_{33}(E_{FB},n)]$ (3.29)	$J = \frac{e^2 g_v B k_B T}{2\pi^2\hbar^2}\sum_{n=0}^{n_{max}}[F_0(\eta_{34})\exp[-\beta_{34}]]$ (3.27)

	(e) The model of Palik et al.: In accordance with the model of Palik et al. (3.30), $n_0 = \frac{eBg_v}{2\pi^2 h}\sum_{n=0}^{n_{max}}[Y_{34}(E_{FB},n) + Z_{34}(E_{FB},n)]$ (3.35)	$J = \frac{e^2 g_v B k_B T}{h^2}\sum_{n=0}^{n_{max}}[F_0(\eta_{35,+})\exp[-\beta_{35,+}] + F_0(\eta_{35,-})\exp[-\beta_{35,-}]]$ (3.33)
3. II–VI materials as described by the Hopfield model	In accordance with (3.36a), $n_0 = \frac{eBg_v\sqrt{2m^*_{\parallel c}}}{2\pi^2\hbar^2}\sum_{n=0}^{n_{max}}[Y_{35}(E_{FB},n) + Z_{35}(E_{FB},n) + Z_{35}(E_{FB},n)]$ (3.39)	$J = \frac{e^2 g_v B k_B T}{h^2}\sum_{n=0}^{n_{max}}[F_0(\eta_{36,+})\exp[-\beta_{36,+}] + F_0(\eta_{36,-})\exp[-\beta_{36,-}]$ (3.37)
4. Bismuth, the carriers of which can be defined by five types of energy band models as described in the column of carrier statistics	(a) The McClure and Choi model: In accordance with (3.39), $n_0 = \frac{eBg_v\sqrt{2m_3}}{2\pi^2\hbar^2}\sum_{n=0}^{n_{max}}[Y_{36}(E_{FB},n) + Z_{36}(E_{FB},n)]$ (3.42b)	$J = \frac{e^2 g_v B k_B T}{h^2}\sum_{n=0}^{n_{max}}[F_0(\eta_{37,+})\exp[-\beta_{37,+}] + F_0(\eta_{37,-})\exp[-\beta_{37,-}]]$ (3.41)
	(b) The Cohen model: In accordance with (3.43), $n_0 = \frac{eBg_v\sqrt{2m_3}}{2\pi^2\hbar^2}\sum_{n=0}^{n_{max}}[Y_{37}(E_{FB},n) + Z_{37}(E_{FB},n)]$ (3.47)	$J = \frac{e^2 g_v B k_B T}{h^2}\sum_{n=0}^{n_{max}}[F_0(\eta_{38,+})\exp[-\beta_{38,+}] + F_0(\eta_{38,-})\exp[-\beta_{38,-}]]$ (3.46)
	(c) The Lax model: In accordance with (3.48), $n_0 = \frac{eBg_v\sqrt{2m_3}}{2\pi^2\hbar^2}\sum_{n=0}^{n_{max}}[Y_{40}(E_{FB},n) + Z_{40}(E_{FB},n)]$ (3.52)	$J = \frac{e^2 g_v B k_B T}{h^2}\sum_{n=0}^{n_{max}}[F_0(\eta_{39,+})\exp[-\beta_{39,+}] + F_0(\eta_{39,-})\exp[-\beta_{39,-}]]$ (3.50)
5. IV–VI materials, the carriers of which can be defined by the model of Dimmock, Bangert and Kastner, and Foley and Landenberg	(a) The Dimmock model: In accordance with (3.54), $n_0 = \left(\frac{eBg_v}{\pi^2\hbar}\right)\sum_{n=0}^{n_{max}}[Y_{41}(E_{FB},n) + Z_{41}(E_{FB},n)]$ (3.58)	$J = \frac{e^2 g_v B k_B T}{2\pi^2\hbar^2}\sum_{n=0}^{n_{max}}[F_0(\eta_{40})\exp[-\beta_{40}]]$ (3.56)

(continued)

Table 3.1 (continued)

Type of materials	The carrier statistics	The Fowler–Nordheim field emission
	(b) The Bangert and Kastner model In accordance with (3.61) $n_0 = \left(\frac{eBg_v}{\pi^2\hbar}\right)\sum_{n=0}^{n_{max}}[Y_{42}(E_{FB},n) + Z_{42}(E_{FB},n)]$ (3.64)	$J = \frac{e^2 g_v B k_B T}{2\pi^2\hbar^2}\sum_{n=0}^{n_{max}}[F_0(\eta_{41})\exp[-\beta_{41}]]$ (3.62)
	(c) The Foley and Landenberg model In accordance with (3.66) $n_0 = \left(\frac{eBg_v}{\pi^2\hbar}\right)\sum_{n=0}^{n_{max}}[Y_{43}(E_{FB},n) + Z_{43}(E_{FB},n)]$ (3.69)	$J = \frac{e^2 g_v B k_B T}{2\pi^2\hbar^2}\sum_{n=0}^{n_{max}}[F_0(\eta_{42})\exp[-\beta_{42}]]$ (3.67)
6. Stressed materials, as defined by the model of Seiler et al.	In accordance with (3.70), $n_0 = \left(\frac{eBg_v}{\pi^2\hbar}\right)\sum_{n=0}^{n_{max}}[Y_{44}(E_{FB},n) + Z_{44}(E_{FB},n)]$ (3.74)	$J = \frac{e^2 B g_v k_B T}{2\pi^2\hbar^2}\sum_{n=0}^{n_{max}}[F_0(\eta_{43})\exp[-\beta_{43}]]$ (3.72)
7. Tellurium the conduction electrons of which can be defined by the model of Bouat et al.	In accordance with (3.76), $n_0 = \left(\frac{eBg_v}{2\pi^2\hbar}\right)\sum_{n=0}^{n_{max}}[Y_{45}(E_{FB},n) + Z_{45}(E_{FB},n)]$ (3.78)	$J = \frac{e^2 B g_v k_B T}{4\pi^2\hbar^2}\sum_{n=0}^{n_{max}}[F_0(\eta_{44,+})\exp[-\beta_{44,+}] + F_0(\eta_{44,-})\exp[-\beta_{44,-}]]$ (3.77)
8. n-GaP as described by the Rees model	In accordance with (3.80), $n_0 = \left(\frac{eBg_v}{2\pi^2\hbar}\right)\sum_{n=0}^{n_{max}}[Y_{46}(E_{FB},n) + Z_{46}(E_{FB},n)]$ (3.83)	$J = \frac{e^2 B g_v k_B T}{4\pi^2\hbar^2}\sum_{n=0}^{n_{max}}[F_0(\eta_{45,+})\exp[-\beta_{45,+}] + F_0(\eta_{45,-})\exp[-\beta_{45,-}]]$ (3.81)
9. $PtSb_2$, as defined by the Emtage model	In accordance with (3.85), $n_0 = \left(\frac{eBg_v}{2\pi^2\hbar}\right)\sum_{n=0}^{n_{max}}[Y_{47}(E_{FB},n) + Z_{47}(E_{FB},n)]$ (3.88)	$J = \frac{e^2 B g_v k_B T}{4\pi^2\hbar^2}\sum_{n=0}^{n_{max}}[F_0(\eta_{46,+})\exp[-\beta_{46,+}] + F_0(\eta_{46,-})\exp[-\beta_{46,-}]]$ (3.86)

10. Bi_2Te_3, which follows the model of Stordeur et al.	In accordance with (3.89), $n_0 = \left(\frac{eBg_v}{\pi^2\hbar}\right)\sum_{n=0}^{n_{max}}[Y_{48}(E_{FB},n) + Z_{48}(E_{FB},n)]$ (3.92)	$J = \frac{e^2Bg_vk_BT}{2\pi^2\hbar^2}\sum_{n=0}^{n_{max}}[F_0(\eta_{47})\exp[-\beta_{47}]]$ (3.90)
11. n-Ge, the conduction electrons of which can be defined by two types of energy band models as described in the column of carrier statistics	(a) In accordance with the model of Cardona et al. (3.94), $n_0 = \left(\frac{eBg_v}{\pi^2\hbar}\right)\sum_{n=0}^{n_{max}}[Y_{49}(E_{FB},n) + Z_{49}(E_{FB},n)]$ (3.97)	$J = \frac{e^2Bg_vk_BT}{2\pi^2\hbar^2}\sum_{n=0}^{n_{max}}[F_0(\eta_{48})\exp[-\beta_{48}]]$ (3.95)
	(b) In accordance with the model of Wang and Ressler (3.98), $n_0 = \left(\frac{eBg_v}{\pi^2\hbar}\right)\sum_{n=0}^{n_{max}}[Y_{50}(E_{FB},n) + Z_{50}(E_{FB},n)]$ (3.101)	$J = \frac{e^2Bg_vk_BT}{2\pi^2\hbar^2}\sum_{n=0}^{n_{max}}[F_0(\eta_{49})\exp[-\beta_{49}]]$ (3.99)
12. Gallium Antimonide, the carriers of which can be defined by the model of Mathur et al.	In accordance with (3.102), $n_0 = \left(\frac{eBg_v}{\pi^2\hbar}\right)\sum_{n=0}^{n_{max}}[Y_{50}(E_{FB},n) + Z_{50}(E_{FB},n)]$ (3.104)	$J = \frac{e^2Bg_vk_BT}{2\pi^2\hbar^2}\sum_{n=0}^{n_{max}}[F_0(\eta_{50})\exp[-\beta_{50}]]$ (3.103)
13. II–V materials, as defined by the model of Yamada	In accordance with (3.106), $n_0 = \left(\frac{eBg_v}{2\pi^2\hbar}\right)\sum_{n=0}^{n_{max}}[Y_{51}(E_{FB},n) + Z_{51}(E_{FB},n)]$ (3.109)	$J = \frac{e^2Bg_vk_BT}{2\pi^2\hbar^2}\sum_{n=0}^{n_{max}}[F_0(\eta_{51,+})\exp[-\beta_{51,+}] + F_0(\eta_{51,-})\exp[-\beta_{51,-}]]$ (3.107)

3.4 Open Research Problems

(R.3.1) (a) Investigate the FNFE from all the bulk semiconductors as considered in this chapter in the absence of any field by converting the summations over the quantum numbers to the corresponding integrations by including the uniqueness conditions in the appropriate cases and considering the effect of image force in the subsequent study in each case.

(b) Investigate the FNFE both in the presence of an arbitrarily oriented quantizing magnetic field including broadening and the electron spin (applicable under magnetic quantization) for all the bulk semiconductors whose unperturbed carrier energy spectra are defined in Chap. 1.

(R.3.2) Investigate the FNFE in the presence of quantizing magnetic field under an arbitrarily oriented (a) nonuniform electric field and (b) alternating electric field respectively for all the semiconductors whose unperturbed carrier energy spectra are defined in Chap. 1 by including spin and broadening respectively.

(R.3.3) Investigate the FNFE under an arbitrarily oriented alternating quantizing magnetic field by including broadening and the electron spin for all the semiconductors whose unperturbed carrier energy spectra are defined in Chap. 1.

(R.3.4) Investigate the FNFE under an arbitrarily oriented alternating quantizing magnetic field and crossed alternating electric field by including broadening and the electron spin for all the semiconductors whose unperturbed carrier energy spectra are defined in Chap. 1.

(R.3.5) Investigate the FNFE under an arbitrarily oriented alternating quantizing magnetic field and crossed alternating nonuniform electric field by including broadening and the electron spin whose for all the semiconductors unperturbed carrier energy spectra are defined in Chap. 1.

(R.3.6) Investigate the FNFE in the presence and absence of an arbitrarily oriented quantizing magnetic field under exponential, Kane, Halperin, Lax, and Bonch-Bruevich band tails [71] for all the semiconductors whose unperturbed carrier energy spectra are defined in Chap. 1 by including spin and broadening (applicable under magnetic quantization).

(R.3.7) Investigate the FNFE in the presence of an arbitrarily oriented quantizing magnetic field for all the semiconductors as defined in (6) under an arbitrarily oriented (a) nonuniform electric field and (b) alternating electric field respectively whose unperturbed carrier energy spectra are defined in Chap. 1.

(R.3.8) Investigate the FNFE for all the semiconductors as described in (6) under an arbitrarily oriented alternating quantizing magnetic field by including broadening and the electron spin whose unperturbed carrier energy spectra as defined in Chapter 1.

(R.3.9) Investigate the FNFE as discussed in (6) under an arbitrarily oriented alternating quantizing magnetic field and crossed alternating electric field by including broadening and the electron spin for all the semiconductors whose unperturbed carrier energy spectra as defined in Chap. 1.

(R.3.10) Investigate all the appropriate problems of this chapter after proper modifications introducing new t

(R.3.11) heoretical formalisms for functional, negative refractive index, macro molecular, and organic and magnetic materials.

(R.3.12) Investigate all the appropriate problems of this chapter for p-InSb, p-CuCl, and stressed semiconductors having diamond structure valence bands whose dispersion relations of the carriers in bulk semiconductors are given by Cunningham [74], Yekimov et al. [75], and Roman and Ewald [76], respectively.

(R.3.13) (a) Formulate the minimum tunneling, Dwell and phase tunneling, Buttiker and Landauer and intrinsic times for all types of systems as discussed in this chapter.

(b) Investigate all the appropriate problems of this chapter for the Dirac electron.

(c) Investigate all the problems of this chapter by removing all the mathematical approximations and establishing the respective appropriate uniqueness conditions.

References

1. N. Miura, *Physics of Semiconductors in High Magnetic Fields*, Series on Semiconductor Science and Technology (Oxford University Press, USA, 2007)
2. K.H.J Buschow, F.R. de Boer, *Physics of Magnetism and Magnetic Materials* (Springer, New York, 2003)
3. D. Sellmyer, R. Skomski (Ed.) *Advanced Magnetic Nanostructures* (Springer, New York, 2005)
4. J.A.C. Bland, B. Heinrich (Ed.) *Ultrathin Magnetic Structures III: Fundamentals of Nanomagnetism (Pt. 3)* (Springer-Verlag, Germany, 2005)
5. B.K. Ridley, *Quantum Processes in Semiconductors*, Fourth Edition (Oxford publications, Oxford, 1999)
6. J.H. Davies, *Physics of Low Dimensional Semiconductors* (Cambridge University Press, Cambridge, 1998)
7. S. Blundell, *Magnetism in Condensed Matter, Oxford Master Series in Condensed Matter Physics* (Oxford University Press, Oxford, 2001)
8. C. Weisbuch, B. Vinter, *Quantum Semiconductor Structures: Fundamentals and Applications* (Academic Publishers, New York, 1991)
9. D. Ferry, *Semiconductor Transport* (CRC, New York, 2000)
10. M. Reed (Ed.) *Semiconductors and Semimetals: Nanostructured Systems* (Academic Press, New York, 1992)
11. T. Dittrich, *Quantum Transport and Dissipation* (Wiley-VCH Verlag GmbH, Berlin, 1998)
12. A.Y. Shik, *Quantum Wells: Physics & Electronics of Two Dimensional Systems* (World Scientific, New York, 1997)
13. K.P. Ghatak, M. Mondal, Z. *Naturforschung A* **41a**, 881 (1986)

14. K.P. Ghatak, M. Mondal, *J. Appl. Phys.* **62**, 922 (1987)
15. K.P. Ghatak, S.N. Biswas, *Phys. Stat. Sol. (b)* **140**, K107 (1987)
16. K.P. Ghatak, M. Mondal, *J. Mag. Mag. Mat.* **74**, 203 (1988)
17. K.P. Ghatak, M. Mondal, *Phys. Stat. Sol. (b)* **139**, 195 (1987)
18. K.P. Ghatak, M. Mondal, *Phys. Stat. Sol. (b)* **148**, 645 (1988)
19. K.P. Ghatak, B. Mitra, A. Ghoshal, *Phys. Stat. Sol. (b)* **154**, K121 (1989)
20. K.P. Ghatak, S.N. Biswas, *J. Low Temp. Phys.* **78**, 219 (1990)
21. K.P. Ghatak, M. Mondal, *Phys. Stat. Sol. (b)* **160**, 673 (1990)
22. K.P. Ghatak, B. Mitra, *Phys. Lett. A* **156**, 233 (1991)
23. K.P. Ghatak, A. Ghoshal, B. Mitra, *Nouvo Cimento D* **13D**, 867 (1991)
24. K.P. Ghatak, M. Mondal, *Phys. Stat. Sol. (b)* **148**, 645 (1989)
25. K.P. Ghatak, B. Mitra, *Int. J. Elect.* **70**, 345 (1991)
26. K.P. Ghatak, S.N. Biswas, *J. Appl. Phys.* **70**, 299 (1991)
27. K.P. Ghatak, A. Ghoshal, *Phys. Stat. Sol. (b)* **170**, K27 (1992)
28. K.P. Ghatak, *Nouvo Cimento D* **13D**, 1321 (1992)
29. K.P. Ghatak, B. Mitra, *Intt. J. Elect.* **72**, 541 (1992)
30. K.P. Ghatak, S.N. Biswas, *Nonlinear Optics* **4**, 347 (1993)
31. K.P. Ghatak, M. Mondal, *Phys. Stat. Sol. (b)* **175**, 113 (1993)
32. K.P. Ghatak, S.N. Biswas, *Nonlinear Optics* **4**, 39 (1993)
33. K.P. Ghatak, B. Mitra, *Nouvo Cimento* **15D**, 97 (1993)
34. K.P. Ghatak, S.N. Biswas, *Nanostruct. Mater.* **2**, 91 (1993)
35. K.P. Ghatak, M. Mondal, *Phys. Stat. Sol. (b)* **185**, K5 (1994)
36. K.P. Ghatak, B. Goswami, M. Mitra, B. Nag, *Nonlinear Optics* **16**, 9 (1996)
37. K.P. Ghatak, M. Mitra, B. Goswami, B. Nag, *Nonlinear Optics* **16**, 167 (1996)
38. K.P. Ghatak, D.K. Basu, B. Nag, *J. Phys. Chem. Sol.* **58**, 133 (1997)
39. K.P. Ghatak, B. Nag, *Nanostruct. Mater.* **10**, 923 (1998)
40. D. Roy Choudhury, A.K. Choudhury, K.P. Ghatak, A.N. Chakravarti, *Phys. Stat. Sol. (b)* **98**, K141 (1980)
41. A.N. Chakravarti, K.P. Ghatak, A. Dhar, S. Ghosh, *Phys. Stat. Sol. (b)* **105**, K55 (1981)
42. A.N. Chakravarti, A.K. Choudhury, K.P. Ghatak, *Phys. Stat. Sol. (a)* **63**, K97 (1981)
43. A.N. Chakravarti, A.K. Choudhury, K.P. Ghatak, S. Ghosh, A. Dhar, *Appl. Phys.* **25**, 105 (1981)
44. A.N. Chakravarti, K.P. Ghatak, G.B. Rao, K.K. Ghosh, *Phys. Stat. Sol. (b)* **112**, 75 (1982)
45. A.N. Chakravarti, K.P. Ghatak, K.K. Ghosh, H.M. Mukherjee, *Phys. Stat. Sol. (b)* **116**, 17 (1983)
46. M. Mondal, K.P. Ghatak, *Phys. Stat. Sol. (b)* **133**, K143 (1984)
47. M. Mondal, K.P. Ghatak, *Phys. Stat. Sol. (b)* **126**, K47 (1984)
48. M. Mondal, K.P. Ghatak, *Phys. Stat. Sol. (b)* **126**, K41 (1984)
49. M. Mondal, K.P. Ghatak, *Phys. Stat. Sol. (b)* **129**, K745 (1985)
50. M. Mondal, K.P. Ghatak, *Phys. Scr.* **31**, 615 (1985)
51. M. Mondal, K.P. Ghatak, *Phys. Stat. Sol. (b)* **135**, 239 (1986)
52. M. Mondal, K.P. Ghatak, *Phys. Stat. Sol. (b)* **93**, 377 (1986)
53. M. Mondal, K.P. Ghatak, *Phys. Stat. Sol. (b)* **135**, K21 (1986)
54. M. Mondal, S. Bhattacharyya, K.P. Ghatak, *Appl. Phys. A* **42A**, 331 (1987)
55. S.N. Biswas, N. Chattopadhyay, K.P. Ghatak,*Phys. Stat. Sol. (b)* **141**, K47 (1987)
56. B. Mitra, K.P. Ghatak, *Phys. Stat. Sol. (b)* **149**, K117 (1988)
57. B. Mitra, A. Ghoshal, K.P. Ghatak, *Phys. Stat. Sol. (b)* **150**, K67 (1988)
58. M. Mondal, K.P. Ghatak, *Phys. Stat. Sol. (b)* **147**, K179 (1988)
59. M. Mondal, K.P. Ghatak, *Phys. Stat. Sol. (b)* **146**, K97 (1988)
60. B. Mitra, A. Ghoshal, K.P. Ghatak, *Phys. Stat. Sol. (b)* **153**, K209 (1989)
61. B. Mitra, K.P. Ghatak, *Phys. Letts.* **142A**, 401 (1989)
62. B. Mitra, A. Ghoshal, K.P. Ghatak, *Phys. Stat. Sol. (b)* **154**, K147 (1989)
63. B. Mitra, K.P. Ghatak, *Sol. State Elect.* **32**, 515 (1989)
64. B. Mitra, A. Ghoshal, K.P. Ghatak, *Phys. Stat. Sol. (b)* **155**, K23 (1989)
65. B. Mitra, K.P. Ghatak, *Phys. Letts.* **135A**, 397 (1989)

66. B. Mitra, K.P. Ghatak, *Phys. Letts. A* **146A**, 357 (1990)
67. B. Mitra, K.P. Ghatak, *Phys. Stat. Sol. (b)* **164**, K13 (1991)
68. S.N. Biswas, K.P. Ghatak, *Int. J. Elect.* **70**, 125 (1991)
69. P.R. Wallace, *Phys. Stat. Sol. (b)*,**92**, 49 (1979)
70. K.P. Ghatak, S. Bhattacharya, D. De, *Einstein Relation in Compound Semiconductors and Their Nanostructures*, Springer Series in Materials Science, **Vol. 116** (Springer-Verlag, Germany, 2009)
71. B.R. Nag, *Electron Transport in Compound Semiconductors*, Springer Series in Solid-State Sciences, **Vol. 11** (Springer-Verlag, Germany, 1980).
72. C.C. Wu, C.J. Lin, *J. Low Temp. Phys.* **57**, 469 (1984)
73. M.H. Chen, C.C. Wu, C.J. Lin, *J. Low Temp. Phys.* **55**, 127 (1984)
74. R.W. Cunningham, *Phys. Rev.* **167**, 761 (1968)
75. A.I. Yekimov, A.A. Onushchenko, A.G. Plyukhin, Al.L. Efros, *J. Expt. Theor. Phys.* **88**, 1490 (1985).
76. B.J. Roman, A.W. Ewald, *Phys. Rev. B* **5**, 3914 (1972)

Chapter 4
Field Emission from Superlattices of Nonparabolic Semiconductors Under Magnetic Quantization

4.1 Introduction

The importance of magnetic quantization on the electronic properties of different semiconductors having various band structures has already been described in [1–12] and Chap. 3, respectively. In this chapter, we shall study the FNFE from III–V, II–VI, IV–VI, and HgTe/CdTe SLs with graded interfaces in Sects. 4.2.1–4.2.4 under magnetic quantization. From Sects. 4.2.5 to 4.2.8, we shall investigate the same from III–V, II–VI, IV–VI, and HgTe/CdTe effective mass SLs under magnetic quantization, respectively. In Sect. 4.3, the dependences of the FNFE with respect to various variables have been studied by taking GaAs/$Ga_{1-x}Al_xAs$, CdS/CdTe, PbTe/PbSnTe, and HgTe/CdTe SLs and the corresponding effective mass SLs as examples. Section 4.4 contains open research problems.

4.2 Theoretical Background

4.2.1 *The Field Emission from III–V Superlattices with Graded Interfaces Under Magnetic Quantization*

The dispersion law of the electrons of III–V SLs with graded interfaces can be written, following (2.2), under magnetic quantization as

$$k_z^2 = \left[\frac{1}{L_0^2}\left\{\cos^{-1}\left[\frac{1}{2}f_{21B}(n,E)\right]\right\}^2 - \phi_{1B}(n)\right], \tag{4.1}$$

S. Bhattacharya and K.P. Ghatak, *Fowler–Nordheim Field Emission*, Springer Series in Solid-State Sciences 170, DOI 10.1007/978-3-642-20493-7_4,

where

$$f_{21B}(n,E) = \Big[2\cosh\{g_{21B}(n,E)\}\cos\{g_{22}(n,E)\}+Q_{21B}(n,E)\sinh\{g_{21B}(n,E)\}$$

$$\cdot\sin\{g_{22B}(n,E)\} + \Delta_{21}\Big[\Big(\frac{\{J_{21B}(n,E)\}^2}{J_{22B}(n,E)} - 3J_{22B}(n,E)\Big)$$

$$\times\cosh\{g_{21B}(n,E)\}\cdot\sin\{g_{22B}(n,E)\} + \Big(3J_{21B}(n,E)$$

$$-\frac{\{J_{22B}(n,E)\}^2}{J_{21B}(n,E)}\Big)(\sinh\{g_{21B}(n,E)\}\cos\{g_{22B}(n,E)\}\Big]$$

$$+\Delta_{21}\Big[2(\{J_{21B}(n,E)\}^2 - \{J_{22B}(n,E)\}^2)\cosh\{g_{21B}(n,E)\}$$

$$\times\cos\{g_{22B}(n,E)\} + \frac{1}{12}\Big(\frac{5\{J_{22B}(n,E)\}^3}{J_{21B}(n,E)} + \frac{5\{J_{21B}(n,E)\}^3}{J_{22B}(n,E)}$$

$$-\{34J_{21B}(n,E)J_{22B}(n,E)\}\Big)\sinh\{g_{21B}(n,E)\}\sin\{g_{22B}(n,E)\}\Big]\Big],$$

$$g_{21B}(n,E) \equiv J_{21B}(n,E)[a_0 - \Delta_{21}],$$

$$J_{21B}(n,E) \equiv \left[\frac{2m_{c2}E'}{\hbar^2}G(E-\overline{V}_0,\alpha_2,\Delta_2) + \phi_{1B}(n)\right]^{1/2},$$

$$\phi_{1B}(n) = \frac{2eB}{\hbar}\left(n+\frac{1}{2}\right), \quad g_{22B}(n,E) = J_{22B}(n,E)[b_0 - \Delta_{21}],$$

$$J_{22B}(n,E) \equiv \left[\frac{2m_{c1}E}{\hbar^2}G(E,\alpha_1,\Delta_1) - \phi_{1B}(n)\right]^{1/2}, \quad \text{and}$$

$$Q_{21B}(n,E) \equiv \left[\frac{J_{21B}(n,E)}{J_{22B}(n,E)} - \frac{J_{22B}(n,E)}{J_{21B}(n,E)}\right].$$

Therefore, (4.1) can be expressed as

$$k_z^2 = A_{21B}(E,n), \tag{4.2}$$

where $A_{21B}(E,n) = \left[\frac{1}{L_0^2}\left\{\cos^{-1}\left[\frac{1}{2}f_{21B}(n,E)\right]\right\}^2 - \phi_{1B}(n)\right].$

The electron concentration is given by

$$n_0 = \frac{eBg_v}{\pi^2\hbar}\sum_{n=0}^{n_{\max}}[D_{21B}(E_{FB},n) + D_{22B}(E_{FB},n)], \tag{4.3}$$

where $D_{21B}(E_{FB}, n) = \sqrt{A_{21B}(E_{FB}, n)}$, E_{FB} is the Fermi energy in this case, and

$$D_{22B}(E_{FB}, n) = \sum_{r=1}^{s_o} Z_B(r)[D_{21B}(E_{FB}, n)].$$

The field emitted current density assumes the form

$$J = \frac{g_v B e^2 k_B T}{2\pi^2\hbar^2} \sum_{n=0}^{n_{\max}} F_0(\eta_{21B}) \exp(-\beta_{21B})], \tag{4.4}$$

where $\eta_{21B} = (E_{FB} - E_{21B})/k_B T$, and E_{21B} is the lowest positive root of the equation.

$$f_{21B}\left(E_{21B}, n\right) = 2\cos\left[L_0(\phi_{1B}(n))^{\frac{1}{2}}\right], \tag{4.5}$$

$$\beta_{21B} = \frac{4}{3}\left[A_{21B}\left(V_0, n\right)\right]^{3/2} \cdot \left[eF_{sz}\overline{A}_{21B}(V_0, n)\right]^{-1},$$

$$\overline{A}_{21B}(V_0, n) = \left(\frac{1}{L_0^2}\left\{\cos^{-1}\left[\frac{1}{2} f_{21B}(n, V_0)\right]\right\}\left\{1 - \frac{1}{4} f_{21B}^2\left(n, V_0\right)\right\}^{-1/2} \cdot \bar{f}_{21B}\left(n, V_0\right)\right),$$

$$\begin{aligned}
\bar{f}_{21B}(n, V_0) = \Bigg[& 2g'_{21B}(n, V_0) \sinh\{g_{21B}(n, V_0)\} \cos\{g_{22B}(n, V_0)\} - 2g'_{22B}(n, V_0) \\
& \times \sin\{g_{22B}(n, V_0)\} \cdot \cosh\{g_{21B}(n, V_0)\} + Q'_{21B}(n, V_0) \\
& \times \sinh\{g_{21B}(n, V_0)\} \sin\{g_{22B}(n, V_0)\} + Q_{21B}(n, V_0) \\
& \times \left[g'_{21B}(n, V_0) \cdot \cosh\{g_{21B}(n, V_0)\} \sin\{g_{22B}(n, V_0)\}\right. \\
& \left. + g'_{22B}(n, V_0) \cos\{g_{22B}(n, V_0)\} \sinh\{g_{21B}(n, V_0)\}\right] + \Delta_{21} \\
& \cdot \Bigg[\Bigg\{ \frac{2J_{21B}(n, V_0) J'_{21B}(n, V_0)}{J_{22B}(n, V_0)} - \frac{J_{21B}^2(n, V_0) J'_{22B}(n, V_0)}{J_{22B}^2(n, V_0)} \\
& - 3J'_{22B}(n, V_0)\Bigg\} \cosh\{g_{21B}(n, V_0)\} \sin\{g_{22B}(n, V_0)\} + \Bigg\{ \frac{J_{21B}^2(n, V_0)}{J_{22B}(n, V_0)} \\
& - 3J_{22B}(n, V_0)\Bigg\} \{g'_{22B}(n, V_0) \cosh\{g_{21B}(n, V_0)\} \cos\{g_{22B}(n, V_0)\} \\
& + g'_{21B}(n, V_0) \cdot \sin\{g_{22B}(n, V_0)\} \sinh\{g_{21B}(n, V_0)\}\} \\
& + \left\{3J_{22B}(n, V_0) - \frac{J_{22B}^2(n, V_0)}{J_{21B}(n, V_0)}\right\} \Big\{\{g'_{21B}(n, V_0) \cosh\{g_{21B}(n, V_0)\}
\end{aligned}$$

$$\times\cos\{g_{22B}(n,V_0)\} - g'_{22B}(n,V_0)\cdot\sin\{g_{22B}(n,V_0)\}$$

$$\times\sinh\{g_{21B}(n,V_0)\}\} + \left\{3J'_{21B}(n,V_0) - \frac{2J_{22B}(n,V_0)J'_{22B}(n,V_0)}{J_{21B}(n,V_0)}\right.$$

$$\left. + \frac{J'_{21B}(n,V_0)J^2_{22B}(n,V_0)}{J^2_{21B}(n,V_0)}\right\} \sinh\{g_{21B}(n,V_0)\}\cos\{g_{22B}(n,V_0)\}\Big\}\Big]$$

$$+ \Delta_{21}\left[\left[2J^2_{21B}(n,V_0) - 2J^2_{22B}(n,V_0)\right]\{g'_{21B}(n,V_0)\right.$$

$$\times\sinh\{g_{21B}(n,V_0)\}\cdot\cos\{g_{22B}(n,V_0)\} - g'_{22B}(n,V_0)$$

$$\times\sin\{g_{22B}(n,V_0)\}\cosh\{g_{21B}(n,V_0)\} + \{4J_{21B}(n,V_0)J'_{21B}(n,V_0)$$

$$- 4J_{22B}(n,V_0)J'_{22B}(n,V_0)\}\{\cosh\{g_{21B}(n,V_0)\}\cos\{g_{22B}(n,V_0)\}\}$$

$$+ \frac{1}{12}\left[\frac{15J^2_{22B}(n,V_0)J'_{22}(n,V_0)}{J_{21B}(n,V_0)} - \frac{5J^3_{22B}(n,V_0)J'_{21B}(n,V_0)}{J^2_{21B}(n,V_0)}\right.$$

$$- \frac{5J^3_{21B}(n,V_0)J'_{22}(n_x,n_y,V_0)}{J^2_{22B}(n,V_0)} + \frac{15J^2_{21B}(n,V_0)J'_{21B}(n,V_0)}{J_{22B}(n,V_0)}$$

$$\left. - 34J'_{21B}(n,V_0)\cdot J_{22B}(n,V_0) - 34J_{21B}(n,V_0)J'_{22B}(n,V_0)\right]$$

$$\times\{\sinh\{g_{21B}(n,V_0)\}\sin\{g_{22B}(n,V_0)\}\}$$

$$+ \left\{\frac{5J^3_{22B}(n,V_0)}{J_{21B}(n,V_0)} + \frac{5J^3_{21B}(n,V_0)}{J_{22B}(n,V_0)} - 34J_{21B}(n,V_0)J_{22B}(n,V_0)\right\}$$

$$\times\{g'_{21B}(n,V_0)\cosh\{g_{21B}(n,V_0)\}\sin\{g_{22B}(n,V_0)\} + g'_{22B}(n,V_0)$$

$$\left.\left.\times\cos\{g_{22B}(n,V_0)\}\sinh\{g_{21B}(n,V_0)\}\}\right]\right],$$

$$g'_{21B}(n,V_0) = J'_{21B}(n,V_0)\left[a_0 - \Delta_{21}\right],$$

$$J'_{21B}(n,V_0) = [J_{21B}(n,V_0)]^{-1}\left[-\frac{m_{c2}}{\hbar^2}G(V_0-\overline{V}_0,\alpha_2,\Delta_2) + \frac{m_{c2}}{\hbar^2}(\overline{V}_0 - V_0)\right.$$

$$\left.\times G'(V_0-\overline{V}_0,\alpha_2,\Delta_2)\right],$$

$$G'(V_0-\overline{V}_0,\alpha_2,\Delta_2) = G(V_0-\overline{V}_0,\alpha_2,\Delta_2)\cdot\left[\frac{1}{(V_0-\overline{V}_0+E_{g2}+\Delta_2)}\right.$$

$$\left. + \frac{1}{(V_0-\overline{V}_0+E_{g2})} - \frac{1}{(V_0-\overline{V}_0+E_{g2}+\frac{2}{3}\Delta_2)}\right],$$

$$g'_{22B}(n, V_0) = J'_{22B}(n, V_0)\,[b_0 - \Delta_{21}], \quad J'_{22B}(n, V_0) = [J_{22B}(n, V_0)]^{-1} \times \left[\frac{m_{c1}}{\hbar^2} G(V_0, \alpha_1, \Delta_1) + \frac{m_{c1}}{\hbar^2} V_0 G'(V_0, \alpha_1, \Delta_1)\right],$$

$$G'(V_0, \alpha_1, \Delta_1) = G(V_0, \alpha_1, \Delta_1) \cdot \left[\frac{1}{(V_0 + E_{g1})} + \frac{1}{(V_0 + E_{g1} + \Delta_1)} - \frac{1}{\left(V_0 + E_{g1} + \frac{2}{3}\Delta_1\right)}\right],$$

and

$$Q'_{21B}(n, V_0) \equiv \left[\frac{J'_{21B}(n, V_0)}{J_{22B}(V_0, n)} - \frac{J'_{22B}(V_0, n)}{J_{21B}(n, V_0)} - \frac{J_{21B}(n, V_0) \cdot J'_{22B}(V_0, n)}{J^2_{22B}(V_0, n)} + \frac{J'_{21B}(n, V_0) \cdot J_{22B}(V_0, n)}{J^2_{21B}(V_0, n)}\right].$$

4.2.2 *The Field Emission from II–VI Superlattices with Graded Interfaces Under Magnetic Quantization*

The electron dispersion law in II–VI superlattices can be written following (2.9) under magnetic quantization as

$$k_z^2 = \left[\frac{1}{L_0^2}\left\{\cos^{-1}\left[\frac{1}{2} f_{212B}(n, E)\right]\right\}^2 - \phi_{1B}(n)\right], \tag{4.6}$$

where

$$\begin{aligned} f_{212B}(n, E) = &\Big[2\cosh\{g_{212B}(n, E)\}\cos\{g_{222B}(n, E)\} + Q_{212B}(n, E) \\ &\times \sinh\{g_{212B}(n, E)\} \cdot \sin\{g_{222B}(n, E)\} \\ &+ \Delta_{21}\left[\left(\frac{\{J_{212B}(n, E)\}^2}{J_{222B}(n, E)} - 3J_{222B}(n, E)\right)\cosh\{g_{212B}(n, E)\}\right. \\ &\cdot \sin\{g_{222B}(n, E)\} + \left(3J_{212B}(n, E) - \frac{\{J_{222B}(n, E)\}^2}{J_{212B}(n, E)}\right) \\ &\left.\times \sinh\{g_{212B}(n, E)\}\cos\{g_{222B}(n, E)\}\right] + \Delta_{21}\Big[2\left(\{J_{212B}(n, E)\}^2\right. \end{aligned}$$

$$- \left.\{J_{222B}(n,E)\}^2\right) \cosh\{g_{212B}(n,E)\} \cos\{g_{222B}(n,E)\}$$

$$+ \frac{1}{12}\left(\frac{5\{J_{222B}(n,E)\}^3}{J_{212B}(n,E)} + \frac{5\{J_{212B}(n,E)\}^3}{J_{222B}(n,E)} - \{34 J_{212B}(n,E)\right.$$

$$\left.\left.\left. \times J_{222B}(n,E)\}\right) \sinh\{g_{212B}(n,E)\} \sin\{g_{222B}(n,E)\}\right]\right],$$

$$Q_{212B}(E,n) = \left[\frac{J_{212B}(E,n)}{J_{222B}(E,n)} - \frac{J_{222B}(E,n)}{J_{212B}(E,n)}\right],$$

$$g_{212B}(n,E) \equiv J_{212B}(n,E)\left[a_0 - \Delta_{21}\right],$$

$$J_{212B}(n,E) \equiv \left[\frac{2m_{c2}E'}{\hbar^2} G(E - \overline{V}_0, \alpha_2, \Delta_2) + \phi_{1B}(n)\right]^{1/2},$$

$$g_{222B}(n,E) = J_{222B}(n,E)\left[b_0 - \Delta_{21}\right],$$

$$J_{222B}(n,E) \equiv \left[\frac{2m_{\parallel,1}}{\hbar^2}\left[E - \frac{\hbar^2}{2m_{\perp,1}}\phi_{1B}(n) \mp C_0\sqrt{\phi_{1B}(n)}\right]\right]^{1/2}.$$

Therefore, (4.6) can be expressed as

$$k_z^2 = A_{212B}(E,n), \tag{4.7}$$

where $A_{212B}(E,n) = \left[\frac{1}{L_0^2}\left\{\cos^{-1}\left[\frac{1}{2} f_{212B}(n,E)\right]\right\}^2 - \phi_{1B}(n)\right]$.

The electron concentration is given by

$$n_0 = \frac{eBg_v}{2\hbar\pi^2}\sum_{n=0}^{n_{max}}\left[D_{212B}(E_{FB},n) + D_{222B}(E_{FB},n)\right], \tag{4.8}$$

where $D_{212B}(E_{FB},n) = \sqrt{A_{212B}(E_{FB},n)}$ and

$$D_{222B}(E_{FB},n) = \sum_{r=1}^{s_o} Z_B(r)\left[D_{212B}(E_{FB},n)\right].$$

The field emitted current density can be written as

$$J = \frac{g_v B e^2 k_B T}{h^2}\sum_{n=0}^{n_{max}} F_0(\eta_{22B})\exp(-\beta_{22B})] \tag{4.9}$$

where $\eta_{22B} = (E_{F1D} - E_{22B})/k_B T$, and E_{22B} is the root of the equation.

$$\left\{\cos^{-1}\left[\frac{1}{2}f_{212B}(E_{22B},n)\right]\right\}^2 = L_0^2\phi_{1B}(n) \tag{4.10}$$

$$\beta_{22B} = \frac{4}{3}\left[A_{212B}(V_0,n)\right]^{3/2}\cdot\left[3eF_{sz}\overline{A}_{212B}(V_0,n)\right]^{-1},$$

$$\overline{A}_{212B}(V_0,n) = \left(\frac{1}{L_0^2}\left\{\cos^{-1}\left[\frac{1}{2}f_{212B}(n,V_0)\right]\right\}\left\{1-\frac{1}{4}f_{212B}^2(n,V_0)\right\}^{-1/2}\right.$$
$$\left.\cdot\,\bar{f}_{212B}(n,V_0)\right)$$

$$\bar{f}_{212B}(n,V_0)$$
$$= \Big[2g'_{212B}(n,V_0)\sinh\{g_{212B}(n,V_0)\}\cos\{g_{222B}(n,V_0)\}$$
$$-2g'_{222B}(n,V_0)\sin\{g_{222B}(n,V_0)\}\cdot\cosh\{g_{212B}(n,V_0)\}+Q'_{212B}(n,V_0)$$
$$\times\sinh\{g_{212B}(n,V_0)\}\sin\{g_{222B}(n,V_0)\}+Q_{212B}(n,V_0)\left[g'_{212B}(n,V_0)\right.$$
$$\cdot\cosh\{g_{212B}(n,V_0)\}\sin\{g_{222B}(n,V_0)\}+g'_{222B}(n,V_0)\cdot\cos\{g_{222B}(n,V_0)\}$$
$$\times\sinh\{g_{212B}(n,V_0)\}]+\Delta_{21}\left[\left\{\frac{2J_{212B}(n,V_0)J'_{212B}(n,V_0)}{J_{222B}(n,V_0)}\right.\right.$$
$$\left.-\frac{J_{212B}^2(n,V_0)J'_{222B}(n,V_0)}{J_{222B}^2(n,V_0)}-3J'_{222B}(n,V_0)\right\}\cosh\{g_{212B}(n,V_0)\}$$
$$\times\sin\{g_{222B}(n,V_0)\}+\left\{\frac{J_{212B}^2(n,V_0)}{J_{222B}(n,V_0)}-3J_{222B}(n,V_0)\right\}\{g'_{212B}(n,V_0)$$
$$\cdot\sinh\{g_{212B}(n,V_0)\}\sin\{g_{222B}(n,V_0)\}+g'_{222B}(n,V_0)\cdot\cos\{g_{222B}(n,V_0)\}$$
$$\times\cosh\{g_{212B}(n,V_0)\}\}+\left\{3J'_{212B}(n,V_0)+\frac{J_{222B}^2(n,V_0)J'_{212B}(n,V_0)}{J_{212B}^2(n,V_0)}\right.$$
$$-\{\{J_{212B}(n,V_0)\}^{-1}\cdot\{2J'_{222B}(n,V_0)\cdot J_{222B}(n,V_0)\}\}\{\sinh[g_{212B}(n,V_0)]$$
$$\cdot\cos\{g_{222B}(n,V_0)\}]\}+\{g'_{212B}(n,V_0)\cdot\cosh\{g_{212B}(n,V_0)\}\cos\{g_{222B}(n,V_0)\}$$
$$-g'_{222B}(n,V_0)\cdot\sin\{g_{222B}(n,,V_0)\}\cdot\sinh[g_{212B}(n,V_0)]\}\left\{3J_{212B}(n,V_0)\right.$$
$$\left.\left.-\frac{J_{222B}^2(n,V_0)}{J_{212B}(n,V_0)}\right\}\right]+\Delta_{21}\Big[\Big[4J_{212B}(n,V_0)\cdot J'_{212B}(n,V_0)-4J_{222B}(n,V_0)$$
$$\times J'_{222B}(n,V_0)\}\{\cosh\{g_{212B}(n,V_0)\}\cos\{g_{222B}(n,V_0)\}\}+\{2J_{212B}^2(n,V_0)$$
$$-2J_{222B}^2(n,V_0)\}\{g'_{212B}(n,V_0)\sinh[g_{212B}(n,V_0)]\cos\{g_{222B}(n,V_0)\}$$

$$- g'_{222B}(n, V_0)\sin\{g_{222B}(n, V_0)\}\sinh[g_{212B}(n, V_0)]$$

$$+ \frac{1}{12}\left[\frac{15J^2_{222B}(n, V_0)J'_{222B}(n, V_0)}{J_{212B}(n, V_0)} - \frac{5J^3_{222B}(n, V_0)J'_{212B}(n, V_0)}{J^2_{212B}(n, V_0)}\right.$$

$$- \frac{5J^3_{212B}(n,V_0)J'_{222B}(n,V_0)}{J^2_{222B}(n, V_0)} + \frac{15J^2_{212B}(n,V_0)J'_{212B}(n,V_0)}{J_{222B}(n, V_0)} - 34J'_{212B}(n,V_0)$$

$$\left.\cdot J_{222B}(n, V_0) - 34J_{212B}(n, V_0)J'_{222B}(n, V_0)\right]\{\sinh\{g_{212B}(n, V_0)\}$$

$$\times \sin\{g_{222B}(n, V_0)\}\} + \frac{1}{12}\left\{\frac{5J^3_{222B}(n, V_0)}{J_{212B}(n, V_0)} + \frac{5J^3_{212B}(n, V_0)}{J_{222B}(n, V_0)}\right.$$

$$- 34J_{222B}(n, V_0)J_{212B}(n, V_0)\} + \{g'_{212B}(n, V_0)\cosh\{g_{212B}(n, V_0)\}$$

$$\left.\left.\left.\times \sin\{g_{222B}(n, V_0)\}\cdot g'_{222B}(n, V_0)\cdot\cos\{g_{222B}(n, V_0)\}\sinh\{g_{212B}(n, V_0)\}\right\}\right]\right]$$

$$g'_{212B}(n, V_0) = J'_{212B}(n, V_0)\,[a_0 - \Delta_{21}],$$

$$J_{212B}(V_0, n) = \left[\left(\frac{2m_{c2}}{\hbar^2}\right)(\overline{V}_0 - V_0)G(V_0 - \overline{V}_0, \alpha_2, \Delta_2) + \phi_{1B}(n)\right]^{1/2},$$

$$J'_{212B}(n, V_0) = [J_{212B}(n, V_0)]^{-1}\left[-\frac{m_{c2}}{\hbar^2}G(V_0 - \overline{V}_0, \alpha_2, \Delta_2)\right.$$
$$\left. + \frac{m_{c2}}{\hbar^2}(\overline{V}_0 - V_0)G'(V_0 - \overline{V}_0, \alpha_2, \Delta_2)\right]$$

$$G'(V_0 - \overline{V}_0, \alpha_2, \Delta_2) = G(V_0 - \overline{V}_0, \alpha_2, \Delta_2)\cdot\left[\frac{1}{(V_0 - \overline{V}_0 + E_{g2} + \Delta_2)}\right.$$
$$\left. + \frac{1}{(V_0 - \overline{V}_0 + E_{g2})} - \frac{1}{\left(V_0 - \overline{V}_0 + E_{g2} + \frac{2}{3}\Delta_2\right)}\right]$$

$$g'_{222B}(n, V_0) = J'_{222}(n, V_0)\,[b_0 - \Delta_{21}], \quad J'_{222B}(n, V_0) = [J_{222B}(n, V_0)]^{-1}\left[\frac{m_{\parallel,1}}{\hbar^2}\right],$$

$$J_{222B}(n, V_0) = \left[\frac{2m_{\parallel,1}}{\hbar^2}\left[V_0 - \frac{\hbar^2}{2m_{\perp,1}}\phi_{1B}(n) \mp C_0\,[\phi_{1B}(n)]^{1/2}\right]\right]^{1/2}$$

$$Q'_{212B}(n, V_0) \equiv \left[\frac{J'_{212B}(n, V_0)}{J_{222B}(n, V_0)} - \frac{J'_{222B}(n, V_0)}{J_{212B}(n, V_0)} - \frac{J_{212B}(n, V_0)\cdot J'_{222B}(n, V_0)}{J^2_{222B}(n, V_0)}\right.$$
$$\left. + \frac{J_{222B}(n, V_0)\cdot J'_{212B}(n, V_0)}{J^2_{212}(n, V_0)}\right].$$

4.2.3 The Field Emission from IV–VI Superlattices with Graded Interfaces Under Magnetic Quantization

The electron dispersion law under magnetic quantization in IV–VI SLs with graded interfaces can be expressed following (2.16) as

$$k_z^2 = \left[\frac{1}{L_0^2}\left\{\cos^{-1}\left[\frac{1}{2}I_{23B}(n,E)\right]\right\}^2 - \phi_{1B}(n)\right], \tag{4.11}$$

where

$$\begin{aligned}
&I_{23B}(n,E)|_{E=V_0}\\
&= \Bigg[2\cosh\left\{\bar{\beta}_{23B}(n,V_0)\right\}\cos\left\{\bar{\gamma}_{23B}(n,V_0)\right\} + \bar{\varepsilon}_{23B}(n,V_0)\\
&\quad \times \sinh\left\{\bar{\beta}_{23B}(n,V_0)\right\}\cdot\sin\left\{\bar{\gamma}_{23B}(n,V_0)\right\} + \Delta_{21}\Bigg[\Bigg(\frac{\left\{\overline{K}_{53B}(n,V_0)\right\}^2}{\overline{K}_{63}(n,V_0)}\\
&\quad - 3\overline{K}_{63B}(n,V_0)\Bigg)\cosh\left\{\bar{\beta}_{23B}(n,V_0)\right\}\cdot\sin\left\{\bar{\gamma}_{23B}(n,V_0)\right\} + \Bigg(3\overline{K}_{53B}(n,V_0)\\
&\quad - \frac{\left\{\overline{K}_{63B}(n,V_0)\right\}^2}{\overline{K}_{53B}(n,V_0)}\Bigg)\sinh\left\{\bar{\beta}_{23B}(n,V_0)\right\}\cos\left\{\bar{\gamma}_{23B}(n,V_0)\right\}\Bigg]\\
&\quad + \Delta_{21}\Bigg[2\left(\left\{\overline{K}_{53B}(n,V_0)\right\}^2 - \left\{\overline{K}_{63B}(n,V_0)\right\}^2\right)\cosh\left\{\bar{\beta}_{23B}(n,V_0)\right\}\\
&\quad \times \cos\left\{\bar{\gamma}_{23B}(n,V_0)\right\} + \frac{1}{12}\Bigg(\frac{5\left\{\overline{K}_{53B}(n,V_0)\right\}^3}{\overline{K}_{63B}(n,V_0)} + \frac{5\left\{\overline{K}_{63B}(n,V_0)\right\}^3}{\overline{K}_{53B}(n,V_0)}\\
&\quad - \left\{34\overline{K}_{53B}(n,V_0)\,\overline{K}_{63B}(n,V_0)\right\}\Bigg)\sinh\left\{\bar{\beta}_{23B}(n,V_0)\right\}\sin\left\{\bar{\gamma}_{23B}(n,V_0)\right\}\Bigg]\Bigg]
\end{aligned}$$

$$\bar{\beta}_{23B}(n,V_0) \equiv \overline{K}_{53B}(n,V_0)\left[a_0 - \Delta_{21}\right],$$

$$\begin{aligned}
\overline{K}_{53B}(n,E) \equiv \Big[&-\left[(E-\overline{V}_0)H_{12} + H_{22}(n)\right] + \left[(E-\overline{V}_0)^2 H_{32}\right.\\
&\left.+\ (E-\overline{V}_0)H_{42}(n) + H_{52}(n)\right]^{1/2}\Big]^{1/2},
\end{aligned}$$

$$H_{3i} = \frac{f_i^2}{\left(b_i^2 - f_i^2\right)^2},\ i = 1,2, \quad f_i = \frac{\hbar^2}{2m_{\parallel,i}^+}, \quad b_i = \frac{\hbar^2}{2m_{\parallel,i}^-},$$

$$H_{4i}(n) = \left[4(b_i^2 - f_i^2)^2\right]^{-1}\left[4b_i^2 E_{g_i} + 4b_i d_i + 4b_i f_i E_{g_i} + 4f_i^2 E_{g_i}\right.$$

$$+\,\phi_{1B}(n)\left[8b_i^2a_i+8C_if_ib_i-8a_ib_i\right]\Big],$$

$$d_i=P_{\|,i}^2,\quad a_i=\frac{\hbar^2}{2m_{\perp,i}^-},\quad C_i=P_{\perp,i}^2,\quad a_i=\frac{\hbar^2}{2m_{\perp,i}^+},$$

$$H_{5i}(n)\equiv\left[4(b_i^2-f_i^2)^2\right]^{-1}\left[\phi_{1B}^2(n)\left[-8a_ib_iC_if_i+4b_i^2C_i^2+4f_i^2a_i^2-4f_i^2C_i^2\right]\right.$$
$$+\,\phi_{1B}(n)\left[8d_iC_if_i-4a_ib_id_i-4a_ib_if_iE_{g_i}+4b_i^2C_i+4b_i^2e_iE_{g_i}\right.$$
$$\left.\left.-\,4a_if_i^2E_{g_i}4f_i^2e_iE_{g_i}\right]\right]+\left[E_{g_i}^2b_i^2+d_i^2+f_i^2E_{g_i}^2+2E_{g_i}f_id_i\right]\Big],$$

$$H_{1i}=\frac{b_i}{(b_i^2-f_i^2)},\;H_{2i}(n)=\left[2(b_i^2-f_i^2)\right]^{-1}\left[E_{g_i}b_i+d_i+f_iE_{g_i}\right.$$
$$\left.+\,2\left(C_if_i-a_ib_i\right)\phi_{1B}(n)\right],$$

$$\bar{\gamma}_{23B}(n,E)\equiv\overline{K}_{63B}(n,E)\left[b_0-\Delta_{21}\right],\overline{K}_{63B}(n,E)\equiv\left[-\left[E^2H_{31}+EH_{41}(n)\right.\right.$$
$$\left.+\,H_{51}(n)\right]^{1/2}+\left[EH_{11}+H_{21}(n)\right]\Big]^{1/2},$$

$$\bar{\varepsilon}_{23B}(E,n)=\left[\frac{\overline{K}_{53B}(E,n)}{\overline{K}_{63B}(E,n)}-\frac{\overline{K}_{63B}(E,n)}{\overline{K}_{53B}(E,n)}\right],$$

$$\therefore k_z^2=\overline{A}_{3B}(E,n),\tag{4.12}$$

where $\overline{A}_{3B}(E,n)=\left[\frac{1}{L_0^2}\left\{\cos^{-1}\left[\frac{1}{2}I_{23}(n,E)\right]\right\}^2-\phi_{1B}(n)\right]$.

The electron concentration is given by

$$n_0=\frac{eBg_\text{v}}{\pi^2\hbar}\sum_{n=0}^{n_\text{max}}\left[\overline{D}_{3B}(E_{\text{F}B},n)+\overline{D}_{4B}(E_{\text{F}B},n)\right],\tag{4.13}$$

where $\overline{D}_{3B}(E_{\text{F}B},n)=\sqrt{\overline{A}_{3B}(E_{\text{F}B},n)}$ and $\overline{D}_{4B}(E_{\text{F}B},n)=\sum_{r=1}^{s_0}Z_B(r)[\overline{D}_{3B}(E_{\text{F}B},n)]$.

The field-emitted current density can be written as

$$J=\frac{Bg_\text{v}e^2k_\text{B}T}{2\pi^2\hbar^2}\sum_{n=0}^{n_\text{max}}F_0(\eta_{23B})\exp(-\beta_{23B}),\tag{4.14}$$

where $\eta_{23B}=(E_{\text{F}B}-E_{23B})/k_\text{B}T$, and E_{23B} is the root of the equation.

$$I_{23B}(E_{23B},n)=2\cos\left[L_0(\phi_{1B}(n))^{\frac{1}{2}}\right],\tag{4.15}$$

$$\beta_{23B}=\frac{4}{3}\left[\overline{A}_{3B}(V_0,n)\right]^{3/2}\cdot\left[3eF_{\text{s}z}\overline{A}_{4B}(V_0,n)\right]^{-1},$$

$$\overline{A}_{4B}(V_0,n) = \left(\frac{1}{L_0^2}\left\{\cos^{-1}\left[\frac{1}{2}I_{23B}(n,V_0)\right]\right\}\left\{1-\frac{1}{4}I_{23B}^2(n,V_0)\right\}^{-1/2} \cdot \overline{I}_{23B}(n,V_0)\right),$$

$$\begin{aligned}
&\overline{I}_{23B}(n,V_0)\\
&= \Bigg[2\bar{\beta}_{023B}(n,V_0)\sinh\left\{\bar{\beta}_{23B}(n,V_0)\right\}\cos\left\{\bar{\gamma}_{23B}(n,V_0)\right\}\\
&\quad - 2\bar{\gamma}_{023B}(n,V_0)\sin\left\{\bar{\gamma}_{23B}(n,V_0)\right\}\cdot\cosh\left\{\bar{\beta}_{23B}(n,V_0)\right\} + \bar{\varepsilon}_{023B}(n,V_0)\\
&\quad \times\sinh\left\{\bar{\beta}_{23B}(n,V_0)\right\}\sin\left\{\bar{\gamma}_{23B}(n,V_0)\right\} + \bar{\varepsilon}_{23B}(n,V_0)\left[\bar{\beta}_{023B}(n,V_0)\right.\\
&\quad \cdot\cosh\left\{\bar{\beta}_{23B}(n,V_0)\right\}\sin\left\{\bar{\gamma}_{23B}(n,V_0)\right\} + \bar{\gamma}_{023B}(n,V_0)\cdot\cos\left\{\bar{\gamma}_{23B}(n,V_0)\right\}\\
&\quad \times\left.\sinh\left\{\bar{\beta}_{23B}(n,V_0)\right\}\right] + \Delta_{21}\cdot\Bigg[\Bigg\{\frac{\overline{K}_{53B}^2(n,V_0)\overline{K}_{063B}(n,V_0)}{\overline{K}_{63B}^2(n,V_0)}\\
&\quad + \frac{2\overline{K}_{53B}(n,V_0)\overline{K}_{053B}(n,V_0)}{\overline{K}_{63}(n,V_0)} - 3\overline{K}_{063B}(n,V_0)\Bigg\}\\
&\quad \times\cosh\left\{\bar{\beta}_{23B}(n,V_0)\right\}\sin\left\{\bar{\gamma}_{23B}(n,V_0)\right\} + \left\{\frac{\overline{K}_{53B}^2(n,V_0)}{\overline{K}_{63B}(n,V_0)} - 3\overline{K}_{63B}(n,V_0)\right\}\\
&\quad \times\left\{\bar{\beta}_{023B}(n,V_0)\sinh\left\{\bar{\beta}_{23B}(n,V_0)\right\}\sin\left\{\bar{\gamma}_{23B}(n,V_0)\right\} + \bar{\gamma}_{023B}(n,V_0)\right.\\
&\quad \cdot\left.\cos\left\{\bar{\gamma}_{23B}(n,V_0)\right\}\cosh\left\{\bar{\beta}_{23B}(n,V_0)\right\}\right\} + \Bigg\{3\overline{K}_{053B}(n,V_0) + \frac{\overline{K}_{63B}(n,V_0)}{\overline{K}_{53B}^2(n,V_0)}\\
&\quad \cdot\overline{K}_{053B}(n,V_0) - \left\{\left\{\overline{K}_{53B}(n,V_0)\right\}^{-1}\cdot\left\{2\overline{K}_{63B}(n,V_0)\cdot\overline{K}_{063B}(n,V_0)\right\}\right\}\\
&\quad \cdot\left\{\sinh[\bar{\beta}_{23B}(n,V_0)]\cdot\cos\{\bar{\gamma}_{23B}(n,V_0)\}\right]\Bigg\} + \left\{\bar{\beta}_{023B}(n,V_0)\cdot\cosh\left\{\bar{\beta}_{23B}(n,V_0)\right\}\right.\\
&\quad \times\cos\left\{\bar{\gamma}_{23B}(n,V_0)\right\} - \bar{\gamma}_{023B}(n,V_0)\sin\left\{\bar{\gamma}_{23B}(n,V_0)\right\}\cdot\left.\sinh\left[\bar{\beta}_{23B}(n,V_0)\right]\right\}\\
&\quad \times\left\{3\overline{K}_{53B}(n,V_0) - \frac{\overline{K}_{63B}^2(n,V_0)}{\overline{K}_{53B}(n,V_0)}\right\}\Bigg] + \Delta_{21}\Bigg[\Bigg[4\overline{K}_{53B}(n,V_0)\cdot\overline{K}_{053B}(n,V_0)\\
&\quad - 4\overline{K}_{63B}(n,V_0)\overline{K}_{063B}(n,V_0)\}\left\{\cosh\left\{\bar{\beta}_{23B}(n,V_0)\right\}\cos\left\{\bar{\gamma}_{23B}(n,V_0)\right\}\right\}\\
&\quad \times\left\{2\overline{K}_{53B}^2(n,V_0) - 2\overline{K}_{63B}^2(n,V_0)\right\}\left\{\bar{\beta}_{023B}(n,V_0)\sinh[\bar{\beta}_{23B}(n,V_0)]\right.\\
&\quad \times\cos\left\{\bar{\gamma}_{23B}(n,V_0)\right\} - \bar{\gamma}_{023B}(n,V_0)\sin\left\{\bar{\gamma}_{23B}(n,V_0)\right\}\cdot\cosh\left[\bar{\beta}_{23B}(n,V_0)\right]
\end{aligned}$$

$$+\frac{1}{12}\left[\frac{15\overline{K}^2_{53B}(n,V_0)\overline{K}_{063B}(n,V_0)}{\overline{K}_{63B}(n,V_0)}-\frac{5\overline{K}^3_{53B}(n,V_0)\overline{K}_{063B}(n,V_0)}{\overline{K}^2_{63B}(n,V_0)}\right.$$

$$-\frac{5\overline{K}^2_{63B}(n,V_0)\overline{K}_{053B}(n,V_0)}{\overline{K}^2_{53B}(n,V_0)}+\frac{15\overline{K}^2_{53B}(n,V_0)\overline{K}_{063B}(n,V_0)}{\overline{K}_{53B}(n,V_0)}-34\overline{K}_{053B}(n,V_0)$$

$$\left.\cdot\overline{K}_{63B}(n,V_0)-34\overline{K}_{063B}(n,V_0)\overline{K}_{53B}(n,V_0)\right]\{\sinh\{\bar{\beta}_{23B}(n,V_0)\}$$

$$\times\sin\{\bar{\gamma}_{23B}(n,V_0)\}\}+\frac{1}{12}\left\{\frac{5\overline{K}^3_{63B}(n,V_0)}{\overline{K}_{53B}(n,V_0)}+\frac{5\overline{K}^3_{53B}(n,V_0)}{\overline{K}_{63B}(n,V_0)}-34\overline{K}_{53B}(n,V_0)\right.$$

$$\left.\times\overline{K}_{63B}(n,V_0)\right\}\cdot\{\bar{\beta}_{023B}(n,V_0)\cosh\{\bar{\beta}_{23B}(n,V_0)\}\cdot\sin\{\bar{\gamma}_{23B}(n,V_0)\}$$

$$\left.\left.+\bar{\gamma}_{023B}(n,V_0)\cos\{\bar{\gamma}_{23B}(n,V_0)\}\sinh\{\bar{\beta}_{23B}(n,V_0)\}\}\right]\right],$$

$$\bar{\beta}_{023B}(n,V_0)=\overline{K}_{063B}(n,V_0)\left[a_0-\Delta_{21}\right],$$

$$\overline{K}_{063B}(n,V_0)=\left[2\overline{K}_{53B}(n,V_0)\right]^{-1}\left[\frac{1}{2}\left[\left(V_0-\overline{V}_0\right)^2H_{32}+\left(V_0-\overline{V}_0\right)H_{42}(V_0,n)\right.\right.$$

$$\left.\left.+H_{52}(V_0,n)\right]^{-1/2}\left[2\left(V_0-\overline{V}_0\right)H_{32}+H_{42}(V_0,n)-H_{12}\right]\right],$$

$$\bar{\gamma}_{023B}(n,V_0)=\overline{K}_{063B}(n,V_0)\left[b_0-\Delta_{21}\right],$$

$$\overline{K}_{063B}(V_0,n)=\left[2\overline{K}_{63B}(V_0,n)\right]^{-1}\left[H_{11}-\frac{\left[2V_0H_{31}+H_{41}(n)\right]}{\left[V_0^2H_{31}+V_0H_{41}(n)+H_{51}(n)\right]^{1/2}}\right],$$

$$\bar{\varepsilon}_{023B}(n,V_0)\equiv\left[\frac{\overline{K}_{053B}(n,V_0)}{\overline{K}_{063B}(n,V_0)}-\frac{\overline{K}_{063B}(n,V_0)}{\overline{K}_{53B}(n,V_0)}-\frac{\overline{K}_{53B}(n,V_0)\cdot\overline{K}_{063B}(n,V_0)}{\overline{K}^2_{63B}(n,V_0)}\right.$$

$$\left.-\frac{\overline{K}_{63B}(n,V_0)\cdot\overline{K}_{053B}(n,V_0)}{\overline{K}^2_{53B}(n,V_0)}\right].$$

4.2.4 The Field Emission from HgTe/CdTe Superlattices with Graded Interfaces Under Magnetic Quantization

The electron dispersion law in HgTe/CdTe SLs under magnetic quantization can be expressed following (2.23) as

$$k_z = \left[\frac{1}{L_0^2} \{\rho_{11B}(n, E)\} - \{\phi_{1B}(n)\} \right]^{1/2}, \tag{4.16}$$

where $\rho_{11B}(n, E) = \left[\cos^{-1} \left\{ \frac{1}{2} \psi_{11B}(n, E) \right\} \right]^2$,

$$\psi_{11B}(n, E)$$

$$= \Bigg[2\cosh\{\beta_{11B}(n, E)\} \cos\{\gamma_{11B}(n, E)\} + \varepsilon_{11B}(n, E)$$

$$\times \sinh\{\beta_{11B}(n, E)\} \sin\{\gamma_{11B}(n, E)\} + \Delta_{21} \left[\left(\frac{\{K_{13B}(n, E)\}^2}{K_{14B}(n, E)} - 3K_{14B}(n, E) \right) \right.$$

$$\times \cosh\{\beta_{11B}(n, E)\} \sin\{\gamma_{11B}(n, E)\} + \left(3K_{13B}(n, E) - \frac{\{K_{14B}(n, E)\}^2}{K_{13B}(n, E)} \right)$$

$$\times \sinh\{\beta_{11B}(n, E)\} \cos\{\gamma_{11B}(n, E)\} + \Delta_{21} \Big[2 \Big(\{K_{13B}(n, E)\}^2$$

$$- \{K_{14B}(n, E)\}^2 \Big) \cosh\{\beta_{11B}(n, E)\} \cos\{\gamma_{11B}(n, E)\}$$

$$+ \frac{1}{12} \left(\frac{5\{K_{13B}(n, E)\}^3}{K_{14B}(n, E)} + \frac{5\{K_{14B}(n, E)\}^3}{K_{13B}(n, E)} - \{34K_{14B}(n, E) K_{13B}(n, E)\} \right)$$

$$\left. \times \sinh\{\beta_{11B}(n, E)\} \sin\{\gamma_{11B}(n, E)\} \Big] \right] \Bigg],$$

$$\varepsilon_{11B}(n, E) \equiv \left[\frac{K_{13B}(n, E)}{K_{14B}(n, E)} - \frac{K_{14B}(n, E)}{K_{13B}(n, E)} \right],$$

$$\beta_{11B}(n, E) \equiv K_{13B}(n, E) [a_0 - \Delta_{21}],$$

$$\gamma_{11B}(n, E) = K_{14B}(n, E) [b_0 - \Delta_{21}],$$

$$K_{14B}(n, E) \equiv \left[\frac{B_0^2 + 2A_0 E - B_0 \sqrt{B_0^2 + 4A_0 E}}{2A_0^2} - \{\phi_{1B}(n)\} \right]^{1/2},$$

$$K_{13B}(n, E) \equiv \left[\left(\frac{2m_{c2} E'}{\hbar^2} \right) G\left(E - \overline{V}_0, E_{g2}, \Delta_2\right) + \{\phi_{1B}(n)\} \right]^{1/2},$$

$$E' = \overline{V}_0 - E.$$

Therefore, (4.16) can be written as

$$k_z^2 = \overline{A}_{5B}(E,n), \tag{4.17}$$

where $\overline{A}_{5B}(E,n) = \left[\frac{1}{L_0^2}\left\{\cos^{-1}\left[\frac{1}{2}\psi_{11B}(n,E)\right]\right\}^2 - \phi_{1B}(n)\right]$.

The electron concentration is given by

$$n_0 = \frac{eBg_v}{\pi^2\hbar}\sum_{n=0}^{n_{\max}}\left[\overline{D}_{5B}(E_{FB},n) + \overline{D}_{6B}(E_{FB},n)\right], \tag{4.18}$$

where $\overline{D}_{5B}(E_{FB},n) = \sqrt{\overline{A}_{5B}(E_{FB},n)}$ and $\overline{D}_{6B}(E_{FB},n) = \sum_{r=1}^{s_o} Z_{1D}(r)[\overline{D}_{5B}(E_{FB},n)]$.

The field-emitted current density assumes the form

$$J = \frac{g_v Be^2 k_B T}{2\pi^2\hbar^2}\sum_{n=0}^{n_{\max}} F_0(\eta_{24B})\exp(-\beta_{24B}), \tag{4.19}$$

where $\eta_{24B} = (E_{FB} - E_{24B})/k_B T$, and E_{24B} is the root of the equation.

$$\overline{A}_{5B}(E_{24B},n) = 0, \tag{4.20}$$

$$\beta_{24B} = \frac{4}{3}\left[\overline{A}_{5B}(V_0,n)\right]^{3/2}\cdot\left[3eF_{sz}\overline{A}_{6B}(V_0,n)\right]^{-1},$$

$$\overline{A}_{6B}(V_0,n) = \left(\frac{1}{L_0^2}\left\{\cos^{-1}\left[\frac{1}{2}\psi_{11B}(n,V_0)\right]\right\}\bar{\psi}_{11B}(V_0,n)\cdot\left\{1-\frac{1}{4}\psi_{11B}^2(n,V_0)\right\}^{-1/2}\right),$$

$$\begin{aligned}
&\bar{\psi}_{11B}(n,V_0)\\
&= \Big[2\beta_{011B}(n,V_0)\sinh\{\beta_{11B}(n,V_0)\}\cos\{\gamma_{11B}(n,V_0)\} - 2\gamma_{011B}(n,V_0)\\
&\quad\times\sin\{\gamma_{11B}(n,V_0)\}\cdot\cosh\{\beta_{11B}(n,V_0)\} + \varepsilon'_{11B}(n,V_0)\sinh\{\beta_{11B}(n,V_0)\}\\
&\quad\times\sin\{\gamma_{11B}(n,V_0)\} + \varepsilon_{11B}(n,V_0)\left[\beta_{011B}(n,V_0)\cdot\cosh\{\beta_{11B}(n,V_0)\}\right.\\
&\quad\times\left.\sin\{\gamma_{11B}(n,V_0)\} + \gamma_{011B}(n,V_0)\cdot\cos\{\gamma_{11B}(n,V_0)\}\sinh\{\beta_{11B}(n,V_0)\}\right]\\
&\quad+\Delta_{21}\cdot\left[\left\{\frac{2K_{13B}(n,V_0)K_{013B}(n,V_0)}{K_{14B}(n,V_0)} - 3K_{014B}(n,V_0)\right.\right.\\
&\quad\left.-\frac{K_{13B}^2(n,V_0)K_{014B}(n,V_0)}{K_{14B}(n,V_0)}\right\}\cosh\{\beta_{11B}(n,V_0)\}\sin\{\gamma_{11B}(n,V_0)\}
\end{aligned}$$

$$+\left\{\frac{K_{13B}^2(n,V_0)}{K_{14B}(n,V_0)}-3K_{14B}(n,V_0)\right\}\cdot\{\beta_{011B}(n,V_0)\sinh\{\beta_{11B}(n,V_0)\}$$

$$\times\sin\{\gamma_{11B}(n,V_0)\}+\gamma_{011B}(n,V_0)\cdot\cos\{\gamma_{11B}(n,V_0)\}\cosh\{\beta_{11B}(n,V_0)\}\}$$

$$+\left\{3K_{013B}(n,V_0)+\frac{K_{14B}^2(n,V_0)}{K_{13B}^2(n,V_0)}\cdot K_{013B}(n,V_0)-\{\{K_{13B}(n,V_0)\}^{-1}\right.$$

$$\cdot\{2K_{14B}(n,V_0).K_{014B}(n,V_0)\}\}\cdot\{\sinh[\beta_{11B}(n,V_0)]\cdot\cos\{\gamma_{11B}(n,V_0)\}\}$$

$$+\left\{3K_{13B}(n,V_0)-\frac{K_{14B}^2(n,V_0)}{K_{13B}(n,V_0)}\right\}\cdot[\beta_{011B}(n,V_0)\cos\{\gamma_{11B}(n,V_0)\}$$

$$\cdot\cosh\{\beta_{11B}(n,V_0)\}-\gamma_{011B}(n,V_0)\sin\{\gamma_{11B}(n,V_0)\}\cdot\sinh[\beta_{11B}(n,V_0)]$$

$$+\Delta_{21}\left[\left[4K_{13B}(n,V_0)\cdot K_{013B}(n,V_0)-4K_{14B}(n,V_0)K_{014B}(n,V_0)\}\right.\right.$$

$$\times\{\cosh\{\beta_{11B}(n,V_0)\}\cos\{\gamma_{11B}(n,V_0)\}\}+\left\{2K_{13B}^2(n,V_0)-2K_{14B}^2(n,V_0)\right\}$$

$$\times\{\beta_{011B}(n,V_0)\sinh[\beta_{11B}(n,V_0)]\cos\{\gamma_{11B}(n,V_0)\}-\gamma_{011B}(n,V_0)$$

$$\times\sin\{\gamma_{11B}(n,V_0)\}\cdot\cosh[\beta_{11B}(n,V_0)]+\frac{1}{12}\left[\frac{15K_{13B}^2(n,V_0)K_{013B}(n,V_0)}{K_{14B}(n,V_0)}\right.$$

$$-\frac{5K_{13B}^3(n,V_0)K_{014B}(n,V_0)}{K_{14B}^2(n,V_0)}-\frac{5K_{14B}^3(n,V_0)K_{013B}(n,V_0)}{K_{13B}^2(n,V_0)}$$

$$+\frac{15K_{14B}^2(n,V_0)K_{014B}(n,V_0)}{K_{13B}(n,V_0)}-34K_{013B}(n,V_0)K_{14B}(n,V_0)$$

$$\left.-34K_{13B}(n,V_0)K_{014B}(n,V_0)\right]\{\sinh\{\beta_{11B}(n,V_0)\}\cdot\sin\{\gamma_{11B}(n,V_0)\}\}$$

$$+\{\beta_{011B}(n,V_0)\cosh\{\beta_{11B}(n,V_0)\}\sin\{\gamma_{11B}(n,V_0)\}+\{\gamma_{011B}(n,V_0)$$

$$\times\cos\{\gamma_{11B}(n,V_0)\}\cdot\sinh\{\beta_{11B}(n,V_0)\}\cdot\frac{1}{12}\left\{\frac{5K_{13B}^3(n,V_0)}{K_{14B}(n,V_0)}+\frac{5K_{14B}^3(n,V_0)}{K_{13B}(n,V_0)}\right.$$

$$\left.\left.\left.-34K_{13}(n,V_0)K_{14}(n,V_0)\right\}\right]\right],$$

$$\beta_{011B}(n,V_0)=K_{013B}(n,V_0)[a_0-\Delta_{21}],$$

$$K_{013B}(n,V_0)=[K_{13B}(n,V_0)]^{-1}\left[-\frac{m_{c2}}{\hbar^2}G\left(V_0-\overline{V}_0,\alpha_2,\Delta_2\right)\right.$$

$$+\frac{m_{c2}}{\hbar^2}(\overline{V}_0-V_0)G\left(V_0-\overline{V}_0,\alpha_2,\Delta_2\right)\cdot\left[\frac{1}{(V_0-\overline{V}_0+E_{g2}+\Delta_2)}\right.$$

$$+\frac{1}{(V_0-\overline{V}_0+E_{g2})}-\frac{1}{(V_0-\overline{V}_0+E_{g2}+\frac{2}{3}\Delta_2)}\Bigg]\Bigg],$$

$$\gamma_{011B}(n,V_0)=K_{014B}(n,V_0)\left[b_0-\Delta_{21}\right],$$

$$K_{014B}(n,V_0)=\left[K_{14B}(n,V_0)\right]^{-1}\left[\frac{V_0}{A_0}-\frac{B_0}{A_0}\{B_0^2+4A_0V_0\}^{-1/2}\right],$$

and

$$\varepsilon'_{11B}(n,V_0)\equiv\left[\frac{K_{013B}(n,V_0)}{K_{14B}(n,V_0)}-\frac{K_{014B}(n,V_0)}{K_{13B}(n,V_0)}-\frac{K_{13B}(n,V_0)\cdot K_{014B}(n,V_0)}{K_{14B}^2(n,V_0)}\right.$$
$$\left.-\frac{K_{14B}(n,V_0)\cdot K_{013B}(n,V_0)}{K_{13B}^2(n,V_0)}\right].$$

4.2.5 The Field Emission from III–V Effective Mass Superlattices Under Magnetic Quantization

The electron dispersion law in III–V effective mass superlattices (EMSLs) can be written following (2.29) under magnetic quantization as

$$k_x^2=[\rho_{21B}(n,E)], \tag{4.21}$$

where $\rho_{21B}(n,E)=\frac{1}{L_0^2}\left[\cos^{-1}(f_{22B}(n,E))\right]^2-\{\phi_{1B}(n)\}$,

$$f_{22B}(n,E)=a_1\cos[a_0C_{22B}(n,E)+b_0D_{22B}(n,E)]-a_2\cos[a_0C_{22B}(n,E)$$
$$-b_0D_{22B}(n,E)],$$

$$C_{22B}(n,E)\equiv\left[\left(\frac{2m_{c1}E}{\hbar^2}\right)G\left(E,E_{g_1},\Delta_1\right)-\{\phi_{1B}(n)\}\right]^{1/2},\quad\text{and}$$

$$D_{22B}(n,E)\equiv\left[\left(\frac{2m_{c2}E}{\hbar^2}\right)G\left(E,E_{g_1},\Delta_2\right)-\{\phi_{1B}(n)\}\right]^{1/2}.$$

The electron concentration is given by

$$n_0=\frac{eBg_v}{\pi^2\hbar}\sum_{n=0}^{n_{\max}}[\overline{D}_{7B}(E_{FB},n)+\overline{D}_{8B}(E_{FB},n)], \tag{4.22}$$

where $\overline{D}_{7B}(E_{FB},n)=\sqrt{\rho_{21B}(E_{FB},n)}$ and $\overline{D}_{8B}(E_{FB},n)=\sum_{r=1}^{s_0}Z_B(r)[\overline{D}_{7B}(E_{FB},n)]$.

The field-emitted current density assumes the form

$$J = \frac{Bg_v e^2 k_B T}{2\pi^2\hbar^2} \sum_{n_z=1}^{n_{z\max}} F_0(\eta_{26B}) \exp(-\beta_{26B})], \tag{4.23}$$

where $\eta_{26B} = (E_{FB} - E_{26B})/k_BT$, and E_{26B} is the root of the equation.

$$\rho_{21B}(E_{26B}, n) = 0, \tag{4.24}$$

$$\beta_{26B} = \frac{4}{3}\left[\rho_{21B}(V_0, n)\right]^{3/2} \cdot \left[3eF_{sz}\bar{\rho}_{21B}(V_0, n)\right]^{-1},$$

$$\bar{\rho}_{21B}(V_0, n) = \left(\frac{2}{L_0^2}\left\{\cos^{-1}\left[f_{22B}(n, V_0)\right]\right\} \bar{f}_{22B}(V_0, n). \left\{1 - f_{22B}^2(n, V_0)\right\}^{1/2}\right),$$

$$\begin{aligned}\bar{f}_{22B}(V_0, n) = &\left[a_1 \sin\left\{a_0 C_{22B}(V_0, n) + b_0 D_{22B}(V_0, n)\right\}\left\{a_0 \overline{C}_{22B}(V_0, n)\right.\right.\\ &\left.+ b_0 \overline{D}_{22B}(V_0, n)\right\} - a_2 \sin\left\{a_0 C_{22B}(V_0, n) - b_0 D_{22B}(V_0, n)\right\}\\ &\left.\times \left\{a_0 \overline{C}_{22B}(V_0, n) - b_0 \overline{D}_{22B}(V_0, n)\right\}\right],\end{aligned}$$

$$\begin{aligned}&\overline{C}_{22B}(V_0, n)\\ &= \left(\frac{m_{c1}}{\hbar^2}\right)\left[C_{22B}(V_0, n)\right]^{-1}\left[G(V_0, E_{g1}, \Delta_1)\right.\\ &\left.\cdot\left[1 + V_0\left[\frac{1}{(V_0 + E_{g1} + \Delta_1)} + \frac{1}{(V_0 + E_{g1})} - \frac{1}{(V_0 + E_{g1} + \frac{2}{3}\Delta_1)}\right]\right]\right],\end{aligned}$$

$$\begin{aligned}&\overline{D}_{22B}(V_0, n)\\ &= \left(\frac{m_{c2}}{\hbar^2}\right)\left[D_{22B}(V_0, n)\right]^{-1}\left[G(V_0, E_{g2}, \Delta_2)\right.\\ &\left.\cdot\left[1 + V_0\left[\frac{1}{(V_0 + E_{g2} + \Delta_2)} + \frac{1}{(V_0 + E_{g2})} - \frac{1}{(V_0 + E_{g2} + \frac{2}{3}\Delta_2)}\right]\right]\right].\end{aligned}$$

4.2.6 The Field Emission from II–VI Effective Mass Superlattices Under Magnetic Quantization

The electron dispersion law in II–VI EMSLs can be written following (2.34) under magnetic quantization as

$$k_z^2 = \frac{1}{L_0^2}\left[\cos^{-1}\left(f_{24B}(n, E)\right)\right]^2 - \left\{\phi_{1B}(n)\right\}, \tag{4.25}$$

where

$$f_{24B}(n,E) = a_3 \cos\left[a_0 C_{24B}(n,E) + b_0 D_{24B}(n,E)\right] - a_4 \cos\left[a_0 C_{24B}(n,E) + b_0 D_{24B}(n,E)\right],$$

$$C_{24B}(n,E) \equiv \left(\frac{2m^*_{\parallel,1}}{\hbar^2}\right)^{1/2} \left[E - \left\{\frac{\hbar^2}{2m^*_{\perp,1}}\phi_{1B}(n)\right\} \mp C_0 \left\{\phi_{1B}(n)\right\}^{1/2}\right]^{1/2},$$

and

$$D_{24B}(n,E) \equiv \left[\left(\frac{2m_{c2}}{\hbar^2}\right) EG(E, E_{g_2}, \Delta_2) - \phi_{1B}(n)\right]^{1/2}.$$

Equation (4.25) can be expressed as

$$k_z^2 = \left[\rho_{22B}(n,E)\right], \tag{4.26}$$

where $\rho_{22B}(n,E) = \frac{1}{L_0^2}[\cos^{-1}(f_{24B}(n,E))]^2 - \{\phi_{1B}(n)\}$.

The electron concentration is given by

$$n_0 = \frac{eBg_v}{2\pi^2\hbar} \sum_{n=0}^{n_{max}} \left[\overline{D}_{9B}(E_{FB},n) + \overline{D}_{10B}(E_{FB},n)\right], \tag{4.27}$$

where $\overline{D}_{9B}(E_{FB},n) = \sqrt{\rho_{22B}(E_{FB},n)}$ and $\overline{D}_{10B}(E_{FB},n) = \sum_{r=1}^{s_o} Z_B(r)[\overline{D}_{9B}(E_{FB},n)]$.

The field-emitted current density can be written as

$$J = \frac{g_v e^2 B k_B T}{h^2} \sum_{n=0}^{n_{max}} F_0(\eta_{27B}) \exp(-\beta_{27B})], \tag{4.28}$$

where $\eta_{27B} = (E_{FB} - E_{27B})/k_B T$, and E_{27B} is the root of the equation.

$$\rho_{22B}(E_{27B},n) = 0, \tag{4.29}$$

$$\beta_{27B} = \frac{4}{3}\left[\rho_{22B}(V_0,n)\right]^{3/2} \cdot \left[3eF_{sz}\bar{\rho}_{22B}(V_0,n)\right]^{-1},$$

$$\bar{\rho}_{22B}(V_0,n) = \left(\frac{2}{L_0^2}\left\{\cos^{-1}[f_{24B}(n,V_0)]\right\} \bar{f}_{24B}(V_0,n)\{1 - f_{24B}^2(n,V_0)\}^{1/2}\right),$$

$$\bar{f}_{24B}(V_0,n) = \left[a_3 \sin\{a_0 C_{24B}(V_0,n) + b_0 D_{24B}(V_0,n)\}\{a_0 \overline{C}_{24B}(V_0,n) + b_0 \overline{D}_{24B}(V_0,n)\} - a_4 \sin\{a_0 C_{24B}(V_0,n) - b_0 D_{24B}(V_0,n)\}\right.$$

$$\times \{a_0 \overline{C}_{24B}(V_0, n) - b_0 \overline{D}_{24B}(V_0, n)\}],$$

$$\overline{C}_{24B}(V_0, n) = \left(\frac{m^*_{\|,1}}{\hbar^2}\right) [C_{24B}(V_0, n)]^{-1},$$

$$\overline{D}_{24B}(V_0, n)$$

$$= \left(\frac{m_{c2}}{\hbar^2}\right) [D_{24B}(V_0, n)]^{-1} \Bigg[G(V_0, E_{g2}, \Delta_2)$$

$$\times \left[1 + V_0 \left[\frac{1}{(V_0 + E_{g2} + \Delta_2)} + \frac{1}{(V_0 + E_{g2})} - \frac{1}{\left(V_0 + E_{g2} + \frac{2}{3}\Delta_2\right)}\right]\right]\Bigg].$$

4.2.7 The Field Emission from IV–VI Effective Mass Superlattices Under Magnetic Quantization

The electron dispersion law in IV–VI, EMSLs can be written under magnetic quantization following (2.40) as

$$k_z^2 = \frac{1}{L_0^2}[\cos^{-1}(f_{26B}(n, E))]^2 - \{\phi_{1B}(n)\}, \tag{4.30}$$

where $f_{26B}(n, E) = a_5 \cos[a_0 C_{42B}(n, E) + b_0 D_{42B}(n, E)] - a_6 \cos[a_0 C_{42B}(n, E) -b_0 D_{42B}(n, E)]$,

$$a_5 = \left[\sqrt{\frac{m_2^*}{m_1^*}} + 1\right]^2 \left[4\left(\frac{m_2^*}{m_1^*}\right)^{1/2}\right]^{-1}, \quad a_6 = \left[-1 + \sqrt{\frac{m_2^*}{m_1^*}}\right]^2 \left[4\left(\frac{m_2^*}{m_1^*}\right)^{1/2}\right]^{-1},$$

$$m_i^* = \left[\frac{\hbar^2}{2\{b_i^2 - f_i^2\}}\right] \Bigg[b_i - \left[b_i d_i + b_i f_i E_{gi} + b_i^2 E_{gi} - E_{gi} b_i^2 + E_{gi} f_i^2\right]$$

$$\times \left[b_i^2 E_{gi}^2 + d_i^2 + f_i^2 E_{gi}^2 + 2 b_i d_i E_{g_i} + 2 E_{gi} d_i f_i + 2 b_i f_i E_{gi}^2\right]^{-1/2} \Bigg],$$

$$C_{42B}(E, n) \equiv \left[\left(U_1(E, n) - \left[U_1^2(E, n) - 4V_1(E, n)\right]^{1/2}\right)/2\right]^{1/2}, \quad i = 1, 2,$$

$$U_i(E, n) = \left[2b_i \left[E - a_i \frac{2eB}{\hbar}\left(n + \frac{1}{2}\right) + \frac{E_{gi}}{2}\right] + d_i + 2\left[\frac{2eB}{\hbar} e_i \left(n + \frac{1}{2}\right)\right.\right.$$

$$\left.\left. + \frac{E_{gi}}{2}\right] f_i\right] (b_i^2 - f_i^2)^{-1},$$

$$V_i(E,n) = \left[\left[E - a_i\frac{2eB}{\hbar}\left(n+\frac{1}{2}\right) + \frac{E_{gi}}{2}\right]^2 - C_i\frac{2eB}{\hbar}\left(n+\frac{1}{2}\right)\right.$$
$$\left. - \left[\frac{2eB}{\hbar}e_i(n+\frac{1}{2}) + \frac{E_{gi}}{2}\right]^2\right]\left(b_i^2 - f_i^2\right)^{-1},$$

$$D_{42B}(E,n) \equiv \left[\left(U_2(E,n) - \left[U_2^2(E,n) - 4V_2(E,n)\right]^{1/2}\right)/2\right]^{1/2}.$$

From (4.30), we get

$$k_z^2 = [\rho_{26B}(n,E)], \tag{4.31}$$

where $\rho_{26B}(n,E) = \frac{1}{L_0^2}\left[\cos^{-1}(f_{26B}(n,E))\right]^2 - \{\phi_{1B}(n)\}$.

The electron concentration is given by

$$n_0 = \frac{eBg_v}{\pi^2\hbar}\sum_{n=0}^{n_{\max}}[\overline{D}_{19B}(E_{FB},n) + \overline{D}_{20B}(E_{FB},n)], \tag{4.32}$$

where $\overline{D}_{19B}(E_{FB},n) = \sqrt{\rho_{26B}(E_{FB},n)}$ and $\overline{D}_{20B}(E_{FB},n) = \sum_{r=1}^{s_0} Z_B(r)[\overline{D}_{19B}(E_{FB},n)]$.

The field-emitted current density can be written as

$$J = \frac{2g_v e^2 B k_B T}{h^2}\sum_{n=0}^{n_{\max}} F_0(\eta_{28B})\exp(-\beta_{28B})], \tag{4.33}$$

where $\eta_{28B} = (E_{FB} - E_{28B})/k_B T$, and E_{28B} is the root of the equation.

$$\rho_{26B}(E_{28B},n) = 0, \tag{4.34}$$

$$\beta_{28B} = \frac{4}{3}[\rho_{26B}(V_0,n)]^{3/2}\cdot[3eF_{sz}\bar{\rho}_{26B}(V_0,n)]^{-1},$$

$$\bar{\rho}_{26B}(V_0,n) = \left(\frac{2}{L_0^2}\left\{\cos^{-1}\left[f_{26B}(n,V_0)\right]\right\}\bar{f}_{26B}(V_0,n).\left\{1 - f_{26B}^2(n,V_0)\right\}^{1/2}\right),$$

$$\bar{f}_{26B}(V_0,n) = \left[-a_6\sin\{a_0C_{42B}(V_0,n) - b_0D_{42B}(V_0,n)\}\{a_0\overline{C}_{42B}(V_0,n)\right.$$
$$- b_0\overline{D}_{42B}(V_0,n)\} + a_5\sin\{a_0C_{42B}(V_0,n) + b_0D_{42B}(V_0,n)\}$$
$$\left.\times\{a_0\overline{C}_{42B}(V_0,n) + b_0\overline{D}_{42B}(V_0,n)\}\right],$$

$$\overline{C}_{42B}(V_0,n) = [2\sqrt{2}C_{42B}(V_0,n)]^{-1}\cdot\left[\varsigma_{01} - \frac{[U_1(V_0,n)\varsigma_{01} - 2\overline{V}_1(V_0,n)]}{\sqrt{U_1^2(V_0,n) - 4V_1(V_0,n)}}\right],$$

$$\varsigma_{0i} = 2b_i\left(b_i^2 - f_i^2\right), \overline{V}_i(V_0, n) = 2\left(b_i^2 - f_i^2\right)^{-1}\left[V_0 - a_i\frac{2eB}{\hbar}\left(n + \frac{1}{2}\right) + \frac{1}{2}E_{gi}\right],$$

$$\overline{D}_{42B}(V_0, n) = [2\sqrt{2}D_{42B}(V_0, n)]^{-1} \cdot \left[\varsigma_{02} - \frac{\left[U_2(V_0, n)\varsigma_{02} - 2\overline{V}_2(V_0, n)\right]}{\sqrt{U_2^2(V_0, n) - 4V_2(V_0, n)}}\right].$$

4.2.8 The Field Emission from HgTe/CdTe effective mass superlattices under magnetic quantization

The electron energy spectrum under magnetic quantization in HgTe/CdTe, EMSLs in the presence of a quantizing magnetic field $\vec{B}$ along x-direction can be written from (2.45) as

$$k_x^2 = \left[\frac{1}{L_0^2}\left\{\cos^{-1}\left(f_{27B}(E, n)\right)\right\}^2 - \phi_{1B}(n)\right], \tag{4.35}$$

where $f_{27B}(n, E) = a_7\cos[a_0C_{43B}(n, E) + b_0D_{43}(n, E)] - a_8\cos[a_0C_{43B}(n, E) - b_0D_{43B}(n, E)]$,

$$a_7 = \left[\sqrt{\frac{m_{c2}}{m_{c1}}} + 1\right]^2\left[4\left(\frac{m_{c2}}{m_{c1}}\right)^{1/2}\right]^{-1}, \quad a_8 = \left[-1 + \sqrt{\frac{m_{c2}}{m_{c1}}}\right]^2\left[4\left(\frac{m_{c2}}{m_{c1}}\right)^{1/2}\right]^{-1},$$

$$C_{43B}(n, E) \equiv \left[\frac{B_0^2 + 2A_0E - B_0\sqrt{B_0^2 + 4A_0E}}{2A_0^2} - \frac{2eB}{\hbar}\left(n + \frac{1}{2}\right)\right]^{1/2},$$

$$D_{43B}(n, E) \equiv \left[\left(\frac{2Em_{c2}}{\hbar^2}\right)G(E, E_{g2}, \Delta_2) - \frac{2eB}{\hbar}\left(n + \frac{1}{2}\right)\right]^{1/2}.$$

$$G\left(E, E_{g_2}, \Delta_2\right) = \frac{\left(E + E_{g2}\right)\left(E + E_{g2} + \Delta_2\right)\left(E_{g2} + \frac{2}{3}\Delta_2\right)}{E_{g2}\left(E_{g2} + \Delta_2\right)\left(E + E_{g2} + \frac{2}{3}\Delta_2\right)}.$$

Equation (4.35) can be expressed as

$$k_x^2 = [\rho_{24B}(n, E)], \tag{4.36}$$

where $\rho_{24B}(n, E) = \left[\frac{1}{L_0^2}\left\{\cos^{-1}(f_{27B}(E, n))\right\}^2 - \phi_{1B}(n)\right]$.

The electron concentration can be written as

$$n_0 = \frac{g_v eB}{\pi^2 \hbar} \sum_{n=0}^{n_{\max}} \left[\overline{G}_{14B}\left(E_{FB}, n\right) + \overline{G}_{15B}(E_{FB}, n) \right], \tag{4.37}$$

where $\overline{G}_{14}(E_{FB}, n) = \sqrt{\rho_{24B}(E_{FB}, n)}$ and $\overline{G}_{15B}(E_{FB}, n) = \sum_{r=1}^{s_0} Z_B(r)[\overline{G}_{14B}(E_{FB}, n)]$.

The field emitted current density is given by

$$J = \frac{g_v e^2 B k_B T}{2\pi^2 \hbar^2} \sum_{n=0}^{n_{\max}} F_0(\eta_{29B}) \exp(-\beta_{29B})], \tag{4.38}$$

where $\eta_{29B} = (E_{FB} - E_{29B})/k_B T$, and E_{29B} is the root of the equation.

$$\rho_{24B}(E_{29B}, n) = 0, \tag{4.39}$$

$$\beta_{29B} = \frac{4}{3} \left[\rho_{24B}(V_0, n)\right]^{3/2} \cdot \left[3eF_{sz} \bar{\rho}_{24B}(V_0, n)\right]^{-1},$$

$$\bar{\rho}_{24}(V_0, n) = \left(\frac{2}{L_0^2} \left\{ \cos^{-1}\left[f_{27B}\left(n, V_0\right) \right] \right\} \bar{f}_{27B}(V_0, n) \left\{ 1 - f_{27B}^2\left(n, V_0\right) \right\}^{1/2} \right),$$

$$\begin{aligned}\bar{f}_{27B}(V_0, n) = \Big[&-a_8 \sin\left\{ a_0 C_{43B}(V_0, n) - b_0 D_{43B}(V_0, n) \right\} \left\{ a_0 \overline{C}_{43B}(V_0, n) \right. \\ &\left. - b_0 \overline{D}_{43B}(V_0, n) \right\} + a_7 \sin\left\{ a_0 C_{43B}(V_0, n) + b_0 D_{43B}(V_0, n) \right\} \\ &\times \left\{ a_0 \overline{C}_{43B}(V_0, n) + b_0 \overline{D}_{43B}(V_0, n) \right\} \Big],\end{aligned}$$

$$\overline{C}_{43B}(V_0, n) = \left[2C_{43B}(V_0, n)\right]^{-1} \cdot \left[\frac{1}{A_0} - \frac{B_0}{A_0} \left(B_0^2 + 4A_0 V_0\right)^{-1/2} \right],$$

$$\begin{aligned}\overline{D}_{43B}(V_0, n) = \left[2D_{43B}(V_0, n)\right]^{-1} \Bigg[&\frac{1}{(V_0 + E_{g2} + \Delta_2)} + \frac{1}{(V_0 + E_{g2})} \\ &+ \frac{1}{V_0} \frac{1}{\left(V_0 + E_{g2} + \frac{2}{3}\Delta_2\right)} \Bigg].\end{aligned}$$

4.3 Result and Discussions

Using Table 1.1 and $\Delta_{21} = 0.4$ nm together with (4.3), (4.4); (4.8), (4.9); (4.13), (4.14); and (4.18), (4.19), we have plotted the field-emitted current density as functions of $1/B$, electron concentration, and electric field in Figs. 4.1–4.3 for GaAs/AlGaAs, CdS/CdTe, and PbTe/PbSe superlattices with graded interfaces.

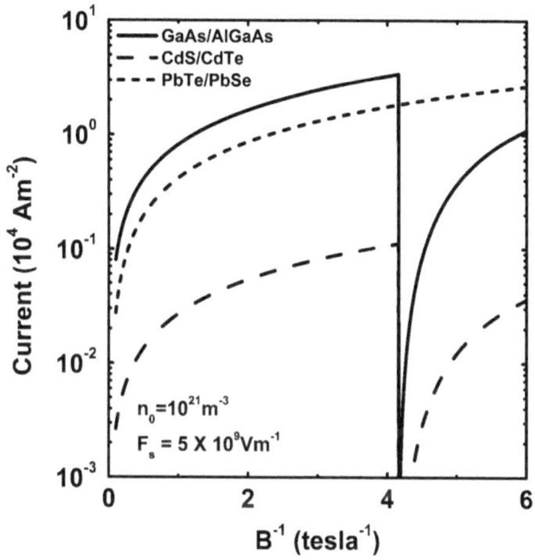

Fig. 4.1 Plot of the current density as a function of $1/B$ for GaAs/AlGaAs, CdS/CdTe, and PbTe/PbSe superlattices with graded interfaces in accordance with two-band energy band model of Kane

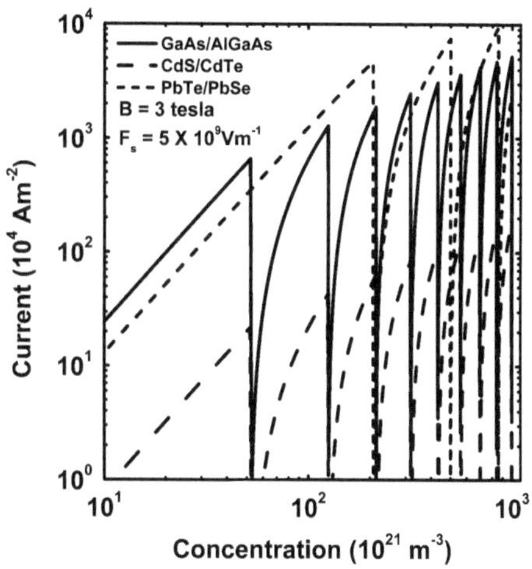

Fig. 4.2 Plot of the current density as a function of carrier concentration for GaAs/AlGaAs, CdS/CdTe, andPbTe/PbSe superlattices with graded interfaces in accordance with two-band energy band model of Kane

Using (4.22), (4.23), (4.27), (4.28), (4.32), (4.33); and (4.37), (4.38), we have plotted the field-emitted current density as functions of the said variables in Figs. 4.4–4.6 for GaAs/AlGaAs, CdS/CdTe and PbTe/PbSe effective mass superlattices respectively.

From Fig. 4.1, it appears that the current density oscillates with the inverse quantizing magnetic field due to SdH effect. We observe that in superlattices with graded interface under magnetic quantization, the frequency of oscillations is less as compared with the constituent materials of each of them. The magnitude of the current density is greatest in the case of GaAs/AlGaAs SLs and least for CdS/CdTe SLs, which is the signature of the band structure of the respective superlattices. Figure 4.2 shows that J oscillates with carrier degeneracy under the application of a strong magnetic field for superlattices with graded interfaces. The effect of subbands

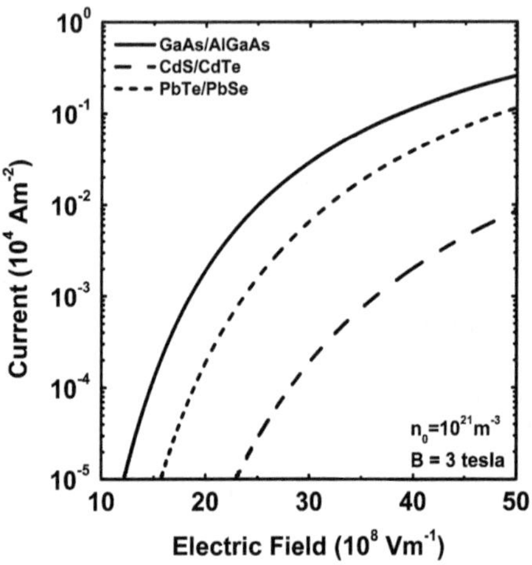

Fig. 4.3 Plot of the current density as a function of electric field for GaAs/AlGaAs, CdS/CdTe, and PbTe/PbSe superlattices with graded interfaces

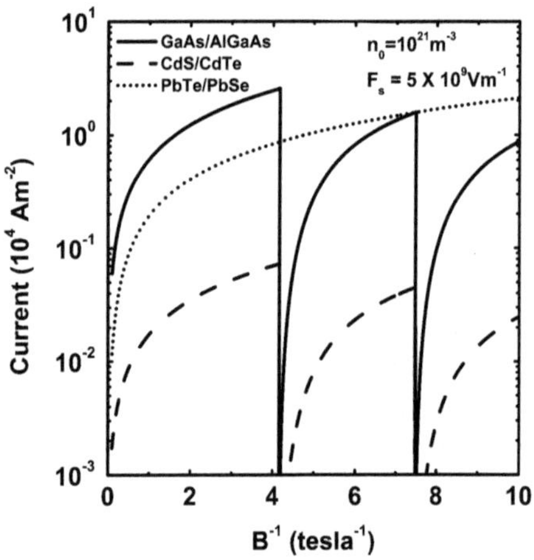

Fig. 4.4 Plot of the current density as a function of $1/B$ for GaAs/AlGaAs, CdS/CdTe, and PbTe/PbSe effective mass superlattices in accordance with two-band energy band model of Kane

enhances the field-emitted current density even at the degeneracy of $10^{24}\,\mathrm{m}^{-3}$. The effect of electric field on J at the lowest subband is shown in the Fig. 4.3. It appears that the cut-in field for GaAs/AlGaAs SLs is lowest and highest for CdS/CdTe SLs.

From Fig. 4.4, we observe that the field-emitted current density oscillates with $1/B$ in the case of GaAs/AlGaAs, CdS/CdTe, and PbTe/PbSe effective mass superlattices under magnetic quantization and can be compared with that of the corresponding superlattices with graded interfaces. This is particularly due to the difference in the analytical and fabrication techniques that mismatches the current densities in the two cases respectively as we observe from Fig. 4.5. We note that the current density is less in all types of effective mass superlattices as considered in this chapter. From Fig. 4.6, it appears that the current density in effective mass superlattices needs relatively higher electric cut-in field to generate a

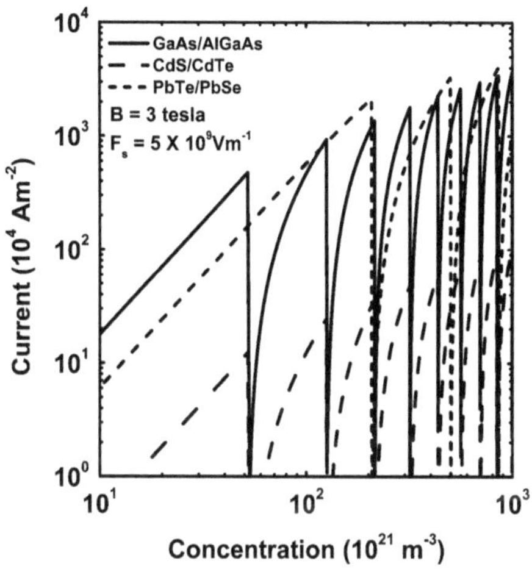

Fig. 4.5 Plot of the current density as a function of carrier concentration for GaAs/AlGaAs, CdS/CdTe, and PbTe/PbSe effective mass superlattices in accordance with two-band energy band model of Kane

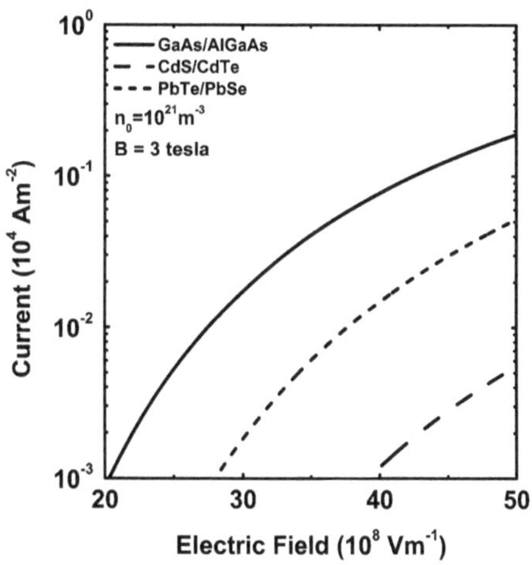

Fig. 4.6 Plot of the current density as a function of electric field for GaAs/AlGaAs, CdS/CdTe, and PbTe/PbSe effective mass superlattices in accordance with two-band energy band model of Kane

considerable amount of current density as compared with the corresponding cases of the superlattices with graded interfaces. For the purpose of condensed presentation, the specific carrier statistics for a specific superlattice having a particular electron energy spectrum and the corresponding field-emitted current density under magnetic quantization have been presented in Table 4.1.

4.4 Open Research Problems

(R.4.1) (a) Investigate the FNFE from all the superlattices as considered in this chapter in the absence of any field by converting the summations over

Table 4.1 The carrier statistics and the Fowler–Nordeim Field Emmission in superlattices under magnetic quantization

The Carrier Statistics		The Fowler–Nordeim Field Emmission	
III–V superlattices with graded interfaces			
$n_0 = \frac{eBg_v}{\pi^2\hbar}\sum_{n=0}^{n_{max}}[D_{21B}(E_{FB},n) + D_{22B}(E_{FB},n)]$	(4.3)	$J = \frac{g_v Be^2 k_B T}{2\pi^2\hbar^2}\sum_{n=0}^{n_{max}} F_0(\eta_{21B})\exp(-\beta_{21B})]$	(4.4)
II–VI superlattices with graded interfaces			
$n_0 = \frac{eBg_v}{2\hbar\pi^2}\sum_{n=0}^{n_{max}}[D_{212B}(E_{FB},n) + D_{222B}(E_{FB},n)]$	(4.8)	$J = \frac{g_v Be^2 k_B T}{h^2}\sum_{n=0}^{n_{max}} F_0(\eta_{22B})\exp(-\beta_{22B})]$	(4.9)
IV–VI superlattices with graded interfaces			
$n_0 = \frac{eBg_v}{\pi^2\hbar}\sum_{n=0}^{n_{max}}\left[\overline{D}_{3B}(E_{FB},n) + \overline{D}_{4B}(E_{FB},n)\right]$	(4.13)	$J = \frac{Bg_v e^2 k_B T}{2\pi^2\hbar^2}\sum_{n=0}^{n_{max}} F_0(\eta_{23B})\exp(-\beta_{23B})]$	(4.14)
HgTe/CdTe superlattices with graded interfaces			
$n_0 = \frac{eBg_v}{\pi^2\hbar}\sum_{n=0}^{n_{max}}\left[\overline{D}_{5B}(E_{FB},n) + \overline{D}_{6B}(E_{FB},n)\right]$	(4.18)	$J = \frac{g_v Be^2 k_B T}{2\pi^2\hbar^2}\sum_{n=0}^{n_{max}} F_0(\eta_{24B})\exp(-\beta_{24B})]$	(4.19)
III–V effective mass superlattices			
$n_0 = \frac{eBg_v}{\pi^2\hbar}\sum_{n=0}^{n_{max}}\left[\overline{D}_{7B}(E_{FB},n) + \overline{D}_{8B}(E_{FB},n)\right]$	(4.22)	$J = \frac{Bg_v e^2 k_B T}{2\pi^2\hbar^2}\sum_{n_z=1}^{n_{zmax}} F_0(\eta_{26B})\exp(-\beta_{26B})]$	(4.23)
II–VI effective mass superlattices			
$n_0 = \frac{eBg_v}{2\pi^2\hbar}\sum_{n=0}^{n_{max}}\left[\overline{D}_{9B}(E_{FB},n) + \overline{D}_{10B}(E_{FB},n)\right]$	(4.27)	$J = \frac{g_v e^2 Bk_B T}{h^2}\sum_{n=0}^{n_{max}} F_0(\eta_{27B})\exp(-\beta_{27B})]$	(4.28)
IV–VI effective mass superlattices			
$n_0 = \frac{eBg_v}{\pi^2\hbar}\sum_{n=0}^{n_{max}}\left[\overline{D}_{19B}(E_{FB},n) + \overline{D}_{20B}(E_{FB},n)\right]$	(4.32)	$J = \frac{2g_v e^2 Bk_B T}{h^2}\sum_{n=0}^{n_{max}} F_0(\eta_{28B})\exp(-\beta_{28B})]$	(4.33)
HgTe/CdTe effective mass superlattices			
$n_0 = \frac{g_v eB}{\pi^2\hbar}\sum_{n=0}^{n_{max}}\left[\overline{G}_{14B}(E_{FB},n) + \overline{G}_{15B}(E_{FB},n)\right]$	(4.37)	$J = \frac{g_v e^2 Bk_B T}{2\pi^2\hbar^2}\sum_{n=0}^{n_{max}} F_0(\eta_{29B})\exp(-\beta_{29B})]$	(4.38)

the quantum numbers to the corresponding integrations by including the uniqueness conditions in the appropriate cases and considering the effect of image force in the subsequent study in each case.

(b) Investigate the FNFE in the presence of an arbitrarily oriented non-quantizing magnetic field for all types of quantum wire superlattices as considered in this chapter by considering the electron spin.

(R.4.2) Investigate the FNFE in the presence of an additional arbitrarily oriented electric field for all types of quantum wire superlattices.

(R.4.3) Investigate the FNFE for all types of quantum wire superlattices as considered in this chapter under arbitrarily oriented crossed electric and magnetic fields.

(R.4.4) Investigate the FNFE in III–V, II–VI, IV–VI, and HgTe/CdTe quantum well superlattices with graded interfaces.

(R.4.5) Investigate the FNFE for all problems of R.2.1 to R.2.3 for III–V, II–VI, IV–VI, and HgTe/CdTe quantum well superlattices with graded interfaces.

(R.4.6) Investigate the FNFE for III–V, II–VI, IV–VI, and HgTe/CdTe quantum well effective mass superlattices.

(R.4.7) Investigate the FNFE for all problems of R.2.1 to R.2.3 for III–V, II–VI, IV–VI, and HgTe/CdTe quantum well effective mass superlattices.

(R.4.8) Investigate the FNFE for short period, strained layer, random, and Fibonacci quantum wire superlattices.

(R.4.9) Investigate the FNFE for short period, strained layer, random, and Fibonacci quantum well superlattices in the presence of an arbitrarily oriented alternating magnetic field by considering electron spin and broadening.

(R.4.10) Investigate the FNFE for strained layer, random, Fibonacci, polytype, and sawtooth superlattices in the presence of an arbitrarily oriented alternating electric field.

(R.4.11) Investigate the FNFE for strained layer, random, Fibonacci, polytype, and sawtooth superlattices in the presence of an arbitrarily oriented crossed electric and magnetic fields.

(R.4.12) Investigate the FNFE for strained layer, random, Fibonacci, polytype, and sawtooth quantum wells and quantum wires superlattices in the presence of an arbitrarily oriented electric field.

(R.4.13) Investigate the FNFE for strained layer, random, Fibonacci, polytype, and sawtooth quantum wells and quantum wires superlattices in the presence of arbitrarily oriented crossed electric and quantizing magnetic fields.

(R.4.14) (a) Formulate the minimum tunneling, Dwell and phase tunneling, Buttiker and Landauer, and intrinsic times for all types of superlattices as discussed in this chapter

(b) Investigate all the appropriate problems of this chapter for the Dirac electron.

(c) Investigate all the problems of this chapter by removing all the mathematical approximations and establishing the respective appropriate uniqueness conditions.

References

1. N. Miura, *Physics of Semiconductors in High Magnetic Fields,Series on Semiconductor Science and Technology* (Oxford University Press, Oxford, 2007)
2. K.H.J Buschow, F.R. de Boer, *Physics of Magnetism and Magnetic Materials* (Springer, New York, 2003)
3. D. Sellmyer, R. Skomski (eds.), *Advanced Magnetic Nanostructures* (Springer, New York, 2005)
4. J.A.C. Bland, B. Heinrich (eds.), *Ultrathin Magnetic Structures III: Fundamentals of Nanomagnetism (Pt. 3)* (Springer-Verlag, Berlin, 2005)
5. B.K. Ridley, *Quantum Processes in Semiconductors*, 4th edn. (Oxford, Oxford, 1999)
6. J.H. Davies, *Physics of Low Dimensional Semiconductors* (Cambridge University Press, UK, 1998)
7. S. Blundell, *Magnetism in Condensed Matter, Oxford Master Series in Condensed Matter Physics*(Oxford University Press, Oxford, 2001)
8. C. Weisbuch, B. Vinter, *Quantum Semiconductor Structures: Fundamentals and Applications* (Academic, New York, 1991)
9. D. Ferry, *Semiconductor Transport* (CRC, New York, 2000)
10. M. Reed (ed.), *Semiconductors and Semimetals: Nanostructured Systems* (Academic, New York, 1992)
11. T. Dittrich, *Quantum Transport and Dissipation* (Wiley-VCH, Berlin, 1998)
12. A.Y. Shik, *Quantum Wells: Physics and Electronics of Twodimensional Systems* (World Scientific, New York, 1997)

Part II
Fowler–Nordheim Field Emission from Quantum-Confined III–V Semiconductors in the Presence of Light Waves

Chapter 5
Field Emission from Quantum-Confined III–V Semiconductors in the Presence of Light Waves

5.1 Introduction

In the presence of strong light waves, the basic band structure of a semiconductor changes profoundly and consequently all the physical properties get radically modified. In this chapter, in Sect. 5.2.1, the field emission under magnetic quantization from the III–V compounds has been investigated in the presence of external photoexcitation whose unperturbed electron energy spectra are, respectively, defined by the three- and two-band models of Kane together with parabolic energy bands and the importance of III–V semiconductors have already been written in Chap. 1. In Sect. 5.2.2, the FNFE in the presence of light waves from quantum wires of III–V semiconductors has been studied. In Sect. 5.2.3, the FNFE from effective mass super lattices whose constituent materials are III–V semiconductors has been investigated in the presence of light waves under magnetic quantization. The FNFE from quantum wire effective mass super lattices of the said materials in the presence of light waves has been studied in Sect. 5.2.4. The FNFE from superlattices of III–V semiconductors with graded interfaces in the presence of light waves under magnetic quantization has been investigated in Sect. 5.2.5. The FNFE from quantum wire superlattices of the said materials with graded interfaces in the presence of light waves has been studied in Sect. 5.2.6. Section 5.3 contains result and discussions. Section 5.4 presents open research problems pertinent to this chapter.

5.2 Theoretical Background

5.2.1 Field Emission from III–V Semiconductors Under Magnetic Quantization in the Presence of Light Waves

The simplified electron energy spectra in III–V materials up to the second order in the presence of external photoexcitation whose unperturbed dispersion relations of

S. Bhattacharya and K.P. Ghatak, *Fowler–Nordheim Field Emission*, Springer Series in Solid-State Sciences 170, DOI 10.1007/978-3-642-20493-7_5,

the conduction electrons are defined by the three- and two-band models of Kane together with parabolic energy bands can, respectively, be expressed as [1]

$$\frac{\hbar^2 k^2}{2m_c} = \beta_{50}\left(E, \lambda, E_{g_0}, \Delta\right), \tag{5.1}$$

$$\frac{\hbar^2 k^2}{2m_c} = \omega_{50}\left(E, \lambda, E_{g_0}\right), \tag{5.2}$$

$$\frac{\hbar^2 k^2}{2m_c} = \rho_{50}\left(E, \lambda, E_{g_0}\right), \tag{5.3}$$

where

$$\beta_{50}\left(E, \lambda, E_{g_0}, \Delta\right) \equiv \left[I_{50}\left(E, E_{g_0}, \Delta\right) - \theta_{50}\left(E, \lambda, E_{g_0}, \Delta\right)\right],$$

$$I_{50}(E, E_{g_0}, \Delta) \equiv \frac{E\left(E + E_{g_0}\right)\left(E + E_{g_0} + \Delta\right)\left(E_{g_0} + \frac{2}{3}\Delta\right)}{E_{g_0}\left(E_{g_0} + \Delta\right)\left(E + E_{g_0} + \frac{2}{3}\Delta\right)},$$

$$\theta_{50}\left(E, \lambda, E_{g_0}, \Delta\right) = \frac{C_{50}(\lambda, E_{g0}, \Delta)}{\varphi_{50}(E)}\psi_{50}^2(E),$$

$$C_{50}(\lambda, E_{g_0}, \Delta) \equiv \frac{e^2}{96 m_r \pi c^3}\frac{I_0\lambda^2}{\sqrt{\varepsilon_{sc}\varepsilon_0}}\frac{E_{g_0}\left(E_{g_0} + \Delta\right)\overline{\beta}_{50}^{-2}}{\left(E_{g_0} + \frac{2}{3}\Delta\right)4}\left(t_{50} + \frac{\overline{\rho}_{50}}{\sqrt{2}}\right)^2,$$

m_r is the reduced mass and is given by $m_r^{-1} = (m_c)^{-1} + m_v^{-1}$, m_v is the effective mass of the heavy hole at the top of the valance band in the absence of any field, I_0 is the light intensity of wavelength λ, c is the velocity of light, ε_0 is the permittivity of vacuum, ε_{sc} is the permittivity of the material,

$$\overline{\beta}_{50} \equiv \left[\left(6(E_{g_0} + 2\Delta/3)(E_{g_0} + \Delta)\right)/\chi_{50}\right]^{1/2}, \quad \chi_{50} \equiv \left(6E_{g_0}^2 + 9E_{g_0}\Delta + 4\Delta^2\right),$$

$$t_{50} \equiv \left[6(E_{g_0} + 2\Delta/3)^2/\chi_{50}\right]^{1/2}, \quad \overline{\rho}_{50} \equiv \left(4\Delta^2/3\chi_{50}\right)^{1/2},$$

$$\phi_{50}(E) \equiv E_{g_0}\left(1 + 2\left(1 + \frac{m_c}{m_v}\right)\frac{I_{50}\left(E, E_{g_0}, \Delta\right)}{E_{g_0}}\right)^{1/2},$$

$$\psi_{50}(E) = \left[\left(1 + \frac{E_{g_0} - \delta'}{\phi_{50}(E) + \delta'}\right) + (E_{g_0} - \delta')\left[\frac{1}{\phi_{50}(E) + \delta'} - \frac{1}{E_{g_0} + \delta'}\right]^{\frac{1}{2}}\right.$$
$$\left.\times\left[\frac{1}{\phi_{50}(E) + \delta'} - \frac{E_{g_0} + \delta'}{(E_{g_0} - \delta')^2}\right]^{\frac{1}{2}}\right]$$

$$\delta' \equiv \left(E_{g_0}^2\Delta\right)(\chi_{50})^{-1}, \quad \omega_{50}\left(E, \lambda, E_{g_0}\right) \equiv E(1 + \alpha E) - B_{50}(E, \lambda),$$

$$B_{50}(E,\lambda) = \frac{C_{51}(\lambda, E_{g_0})\psi_{51}^2(E)}{\phi_{51}(E)}, \quad C_{51}(\lambda, E_{g_0}) \equiv \frac{e^2 I_0 \lambda^2 E_{g_0}}{384\pi c^3 m_r \sqrt{\varepsilon_{sc}\varepsilon_0}},$$

$$\phi_{51}(E) \equiv E_{g_0}\left\{1 + \frac{2m_c}{m_r}\frac{E(1+\alpha E)}{E_{g_0}}\right\}^{1/2}, \quad \psi_{51}(E) = \frac{2E_{g_0}}{\phi_{51}(E)},$$

$$\rho_{50}\left(E,\lambda,E_{g_0}\right) \equiv E - C_{52}\left[1 + \left(\frac{2m_c}{m_r}\right)\frac{E}{E_{g_0}}\right]^{-3/2} \quad \text{and}$$

$$C_{52} = \frac{e^2 I_0 \lambda^2}{96\pi c^3 m_r \sqrt{\varepsilon_{sc}\varepsilon_0}}.$$

Thus, under the limiting condition $\vec{k} \to 0$, from (5.1), (5.2), and (5.3), we observe that $E \neq 0$ and is positive. Therefore, in the presence of external light waves, the energy of the electron does not tend to zero when $\vec{k} \to 0$, whereas for the unperturbed three- and two-band models of Kane together with parabolic energy bands reflects the fact that for $\vec{k} \to 0$, $E \to 0$. As the conduction band is taken as the reference level of energy, the lowest positive value of E for $\vec{k} \to 0$ provides the increased band gap (ΔE_g) of the materials due to photon excitation. The values of the increased band gap can be obtained by computer iteration processes for various values of I_0 and λ.

The dispersion relation of the conduction electrons in III–V semiconductors under light waves can be written in the presence of quantizing magnetic field B along x-direction whose electron energy spectra are defined by (5.1)–(5.3) as

$$\left(n+\frac{1}{2}\right)\hbar\omega_c + \frac{\hbar^2 k_x^2}{2m_c} = \beta_{50}(E,\lambda,E_{g_0},\Delta), \tag{5.4}$$

$$\left(n+\frac{1}{2}\right)\hbar\omega_c + \frac{\hbar^2 k_x^2}{2m_c} = \omega_{50}(E,\lambda,E_{g_0}), \tag{5.5}$$

$$\left(n+\frac{1}{2}\right)\hbar\omega_c + \frac{\hbar^2 k_x^2}{2m_c} = \rho_{50}(E,\lambda,E_{g_0}). \tag{5.6}$$

The density-of-states function per valley assumes the forms

$$N_B(E) = \frac{eB\sqrt{2m_c}}{2\pi^2\hbar^2}\sum_{n=0}^{n_{max}} \frac{\beta'_{50}(E,\lambda,E_{g_0},\Delta)}{\sqrt{\beta_{50}(E,\lambda,E_{g_0},\Delta) - \left(n+\frac{1}{2}\right)\hbar\omega_c}}, \tag{5.7}$$

$$N_B(E) = \frac{eB\sqrt{2m_c}}{2\pi^2\hbar^2}\sum_{n=0}^{n_{max}} \frac{\omega'_{50}(E,\lambda,E_{g_0})}{\sqrt{\omega_{50}(E,\lambda,E_{g_0}) - \left(n+\frac{1}{2}\right)\hbar\omega_c}}, \tag{5.8}$$

and

$$N_B(E) = \frac{eB\sqrt{2m_c}}{2\pi^2\hbar^2}\sum_{n=0}^{n_{max}} \frac{\rho'_{50}(E,\lambda,E_{g_0})}{\sqrt{\rho_{50}(E,\lambda,E_{g_0}) - \left(n+\frac{1}{2}\right)\hbar\omega_c}}, \tag{5.9}$$

where

$$\beta'_{50}\left(E,\lambda,E_{g_0},\Delta\right) \equiv \left[I'_{50}\left(E,E_{g_0},\Delta\right) - \theta'_{50}\left(E,\lambda,E_{g_0},\Delta\right)\right],$$

$$I'_{50}(E,E_{g_0},\Delta) = I_{50}(E,E_{g_0},\Delta)\left[\frac{1}{E}+\frac{1}{E+E_{g_0}}+\frac{1}{E+E_{g_0}+\Delta}-\frac{1}{E+E_{g_0}+\frac{2}{3}\Delta}\right],$$

$$\theta'_{50}\left(E,\lambda,E_{g_0},\Delta\right) = \theta_{50}\left(E,\lambda,E_{g_0},\Delta\right)\left[-\frac{\phi'_{50}(E)}{\phi_{50}(E)}+\frac{2\phi'_{50}(E)}{\phi_{50}(E)}\right],$$

$$\phi'_{50}(E) = E_{g_0}\left(1+\frac{m_c}{m_v}\right)\left[\phi_{50}(E)\right]^{-1}\cdot I'_{50}(E,E_{g_0},\Delta),$$

$$\psi'_{50}(E) = \frac{-(E_{g_0}-\delta')\phi'_{50}(E)}{(\phi_{50}(E)+\delta')^2}\left[1+\frac{1}{2}\left[\frac{1}{\phi_{50}(E)+\delta'}-\frac{1}{E_{g_0}+\delta'}\right]^{-\frac{1}{2}}\right.$$

$$\cdot\left[\frac{1}{\phi_{50}(E)+\delta'}-\frac{(E_{g_0}+\delta')}{(E_{g_0}-\delta')^2}\right]^{\frac{1}{2}}$$

$$\left.+\frac{1}{2}\left[\frac{1}{\phi_{50}(E)+\delta'}-\frac{E_{g_0}+\delta'}{(E_{g_0}-\delta')^2}\right]^{-\frac{1}{2}}\cdot\left[\frac{1}{\phi_{50}(E)+\delta'}-\frac{1}{E_{g_0}+\delta'}\right]^{\frac{1}{2}}\right]$$

$$\omega'_{50}(E,\lambda,E_{g_0}) = \left[1+2\alpha E - B'_{50}(E,\lambda)\right],$$

$$B'_{50}(E,\lambda) = \left[\frac{-B_{50}(E,\lambda)\phi'_{51}(E)}{\phi_{51}(E)}+\frac{2B_{50}(E,\lambda)\psi'_{51}(E)}{\psi_{51}(E)}\right],$$

$$\psi'_{51}(E) = \left[\frac{-\psi_{51}(E)}{\phi_{51}(E)}\phi'_{51}(E)\right],\quad \phi'_{51}(E) = \frac{m_c}{m_r}\cdot\frac{E_{g_0}(1+2\alpha E)}{\phi_{51}(E)},$$

$$\rho'_{50}\left(E,\lambda,E_{g_0}\right) = \left[1+\frac{C_{52}\cdot 3m_c}{m_r}\left[1+\left(\frac{2m_c}{m_r}\right)\frac{E}{E_{g_0}}\right]^{-3/2}\right].$$

Combining (5.7), (5.8), and (5.9) with the Fermi–Dirac occupation probability factor, the electron concentration can, respectively, be expressed as

$$n_0 = \frac{g_v eB\sqrt{2m_c}}{\pi^2\hbar^2}\sum_{n=0}^{n_{max}}\left[\overline{M}_{51}(E_{FB},\lambda,n)+\overline{N}_{51}(E_{FB},\lambda,n)\right], \tag{5.10}$$

$$n_0 = \frac{g_v eB\sqrt{2m_c}}{\pi^2\hbar^2}\sum_{n=0}^{n_{max}}\left[\overline{M}_{52}(E_{FB},\lambda,n)+\overline{N}_{52}(E_{FB},\lambda,n)\right], \tag{5.11}$$

and

$$n_0 = \frac{g_v eB\sqrt{2m_c}}{\pi^2\hbar^2}\sum_{n=0}^{n_{max}}\left[\overline{M}_{53}(E_{FB},\lambda,n)+\overline{N}_{53}(E_{FB},\lambda,n)\right], \tag{5.12}$$

where

$$\overline{M}_{51}(E_{FB},\lambda,n) = \left[\beta_{50}(E_{FB},\lambda,E_{g0},\Delta) - \left(n+\frac{1}{2}\right)\hbar\omega_c\right]^{\frac{1}{2}},$$

$$\overline{N}_{51}(E_{FB},\lambda,n) = \sum_{r=1}^{s_0} Z_B(r)[\overline{M}_{51}(E_{FB},n)],$$

$$\overline{M}_{52}(E_{FB},\lambda,n) = \left[\omega_{50}(E_{FB},\lambda,E_{g0}) - \left(n+\frac{1}{2}\right)\hbar\omega_c\right]^{\frac{1}{2}},$$

$$\overline{N}_{52}(E_{FB},\lambda,n) = \sum_{r=1}^{s_0} Z_B(r)[\overline{M}_{52}(E_{FB},\lambda,n)],$$

$$\overline{M}_{53}(E_{FB},\lambda,n) = \left[\rho_{53}(E_{FB},\lambda,E_{g0}) - \left(n+\frac{1}{2}\right)\hbar\omega_c\right]^{\frac{1}{2}} \quad \text{and}$$

$$\overline{N}_{53}(E_{FB},\lambda,n) = \sum_{r=1}^{s_0} Z_B(r)[\overline{M}_{53}(E_{FB},\lambda,n)].$$

The field emitted current density in three cases can, respectively, be written as

$$J = \frac{e^2 B k_B T g_v}{2\pi^2\hbar^2}\sum_{n=0}^{n_{max}} F_0(\overline{\eta}_{51,B})\exp(-\overline{\beta}_{51,B}), \tag{5.13}$$

$$J = \frac{e^2 B k_B T g_v}{2\pi^2\hbar^2}\sum_{n=0}^{n_{max}} F_0(\overline{\eta}_{52,B})\exp(-\overline{\beta}_{52,B}), \tag{5.14}$$

and

$$J = \frac{e^2 B k_B T g_v}{2\pi^2\hbar^2}\sum_{n=0}^{n_{max}} F_0(\overline{\eta}_{53,B})\exp(-\overline{\beta}_{53,B}), \tag{5.15}$$

where $\overline{\eta}_{51,B} = \frac{E_{FB}-\overline{E}_{51,B}}{k_B T}$, where $\overline{E}_{51,B}$ is the lowest positive root of the equation

$$\beta_{50}(\overline{E}_{51,B},\lambda,E_{g0},\Delta) - \left(n+\frac{1}{2}\right)\hbar\omega_c = 0, \tag{5.16}$$

$\overline{\beta}_{51,B} = \frac{4\sqrt{2m_c}\left[\beta_{50}(V_0,\lambda,E_{g0},\Delta)-\left(n+\frac{1}{2}\right)\hbar\omega_c\right]^{\frac{3}{2}}}{3eF_{sx}\hbar\beta'_{50}(V_0,\lambda,E_{g0},\Delta)}$, $\overline{\eta}_{52,B} = \frac{E_{FB}-\overline{E}_{52,B}}{k_B T}$, F_{sx} is the electric field along x-direction, $\overline{E}_{52,B}$ is the lowest positive root of the equation

$$\omega_{50}(\overline{E}_{52,B},\lambda,E_{g0}) - \left(n+\frac{1}{2}\right)\hbar\omega_c = 0, \tag{5.17}$$

$\overline{\beta}_{52,B} = \frac{4\sqrt{2m_c}\left[\omega_{50}(V_0,\lambda,E_{g0})-\left(n+\frac{1}{2}\right)\hbar\omega_c\right]^{\frac{3}{2}}}{3eF_{sx}\hbar\omega'_{50}(V_0,\lambda,E_{g0})}$, $\overline{\eta}_{53,B} = \frac{E_{FB}-\overline{E}_{53,B}}{k_B T}$, $\overline{E}_{53,B}$ is the lowest positive root of the equation

$$\rho_{50}(\overline{E}_{53,B},\lambda,E_{g0}) - \left(n+\frac{1}{2}\right)\hbar\omega_c = 0 \tag{5.18}$$

$$\overline{\beta}_{53,B} = \frac{4\sqrt{2m_c}\left[\rho_{50}(V_0,\lambda,E_{g0}) - \left(n+\frac{1}{2}\right)\hbar\omega_c\right]^{\frac{3}{2}}}{3eF_{sx}\hbar\rho'_{50}(V_0,\lambda,E_{g0})},$$

$$\beta'_{50}(V_0,\lambda,E_{g0},\Delta) = \beta'_{50}(V_0,\lambda,E_{g0},\Delta)\big|_{E=V_0},$$

$$\omega'_{50}(V_0,\lambda,E_{g0}) = \omega'_{50}(V_0,\lambda,E_{g0})\big|_{E=V_0},\quad \rho'_{50}(V_0,\lambda,E_{g0}) = \rho'_{50}(V_0,\lambda,E_{g0})\big|_{E=V_0}.$$

5.2.2 *Field Emission from Quantum Wires of III–V Semiconductors*

From (5.1), (5.2), and (5.3), the dispersion relation of the conduction electrons for quantum wires of III–V materials in the presence of light waves can be expressed as

$$G_5(n_x,n_y) + \frac{\hbar^2 k_z^2}{2m_c} = \beta_{50}(E,\lambda,E_{g0},\Delta), \tag{5.19}$$

$$G_5(n_x,n_y) + \frac{\hbar^2 k_z^2}{2m_c} = \omega_{50}(E,\lambda,E_{g0}), \tag{5.20}$$

and

$$G_5(n_x,n_y) + \frac{\hbar^2 k_z^2}{2m_c} = \rho_{50}(E,\lambda,E_{g0}), \tag{5.21}$$

where $G_5(n_x,n_y) = \frac{\hbar^2\pi^2}{2m_c}[(n_y/d_y)^2 + (n_x/d_x)^2]$,

The electron concentration per unit length assumes the forms

$$n_0 = \frac{\sqrt{2m_c}2g_v}{\hbar\pi}\sum_{n_x=1}^{n_{x\max}}\sum_{n_y=1}^{n_{y\max}}\left[S_{51}(E_{F1D},\lambda,n_x,n_y) + T_{51}(E_{F1D},\lambda,n_x,n_y)\right] \tag{5.22}$$

where

$$S_{51}(E_{F1D},\lambda,n_x,n_y) = \left[\beta_{50}(E_{F1D},\lambda,E_{g0},\Delta) - G_5(n_x,n_y)\right]^{1/2}$$

$$T_{51}(E_{F1D},\lambda,n_x,n_y)\sum_{r=1}^{s_0} Z_{1D}(r)\left[S_{51}(E_{F1D},\lambda,n_x,n_y)\right],$$

$$Z_{1D}(r) = 2(k_B T)^{2r}(1-2^{1-2r})\xi(2r)\cdot\frac{\partial^{2r}}{\partial {E_{1D}}^{2r}},$$

$$n_0 = \frac{\sqrt{2m_c}2g_v}{\hbar\pi} \sum_{n_x=1}^{n_{x\max}} \sum_{n_y=1}^{n_{y\max}} [S_{52}(E_{F1D}, \lambda, n_x, n_y) + T_{52}(E_{F1D}, \lambda, n_x, n_y)] \quad (5.23)$$

where

$$S_{52}(E_{F1D}, \lambda, n_x, n_y) = \left[\omega_{50}(E_{F1D}, \lambda, E_{g_0}) - G_5(n_x, n_y)\right]^{1/2},$$

$$T_{52}(E_{F1D}, \lambda, n_x, n_y) \sum_{r=1}^{s_0} Z_{1D}(r) \left[S_{52}(E_{F1D}, \lambda, n_x, n_y)\right],$$

and

$$n_0 = \frac{\sqrt{2m_c}2g_v}{\hbar\pi} \sum_{n_x=1}^{n_{x\max}} \sum_{n_y=1}^{n_{y\max}} [S_{53}(E_{F1D}, \lambda, n_x, n_y) + T_{53}(E_{F1D}, \lambda, n_x, n_y)] \quad (5.24)$$

where

$$S_{53}(E_{F1D}, \lambda, n_x, n_y) = \left[\rho_{50}(E_{F1D}, \lambda, E_{g_0}) - G_5(n_x, n_y)\right]^{1/2},$$

$$T_{53}(E_{F1D}, \lambda, n_x, n_y) \sum_{r=1}^{s_0} Z_{1D}(r) \left[S_{53}(E_{F1D}, \lambda, n_x, n_y)\right],$$

The field emitted current in the three cases can respectively be written as

$$I = \frac{eg_v k_B T}{\pi\hbar} \sum_{n_x=1}^{n_{x\max}} \sum_{n_y=1}^{n_{y\max}} F_0(\overline{\eta}_{51,1D}) \exp(-\overline{\beta}_{51,1D}) \quad (5.25)$$

$$I = \frac{eg_v k_B T}{\pi\hbar} \sum_{n_x=1}^{n_{x\max}} \sum_{n_y=1}^{n_{y\max}} F_0(\overline{\eta}_{52,1D}) \exp(-\overline{\beta}_{52,1D}) \quad (5.26)$$

and

$$I = \frac{eg_v k_B T}{\pi\hbar} \sum_{n_x=1}^{n_{x\max}} \sum_{n_y=1}^{n_{y\max}} F_0(\overline{\eta}_{53,1D}) \exp(-\overline{\beta}_{53,1D}) \quad (5.27)$$

$\overline{\eta}_{51,1D} = \frac{E_{F1D} - E_{51,n_x,n_y}}{k_B T}$, E_{51,n_x,n_y} is the root of the equation

$$\beta_{50}(E_{51}, n_x, n_y, \lambda, E_{g_0}, \Delta) - G_5(n_x, n_y) = 0 \quad (5.28)$$

$$\overline{\beta}_{51,1D} = \frac{4\left[S_{51}(V_0, \lambda, n_x, n_y)\right]^{3/2} \sqrt{2m_c}}{3eF_{sz}\beta'_{50}(V_0, \lambda, E_{g_0}, \Delta)\hbar}$$

$\overline{\eta}_{52,1D} = \frac{E_{F1D}-E_{52,n_x,n_y}}{k_B T}$, E_{52,n_x,n_y} is the root of the equation

$$\omega_{50}(E_{52}, n_x, n_y, \lambda, E_{g0}) - G_5(n_x, n_y) = 0 \quad (5.29)$$

$$\overline{\beta}_{52,1D} = \frac{4\left[S_{52}(V_0, \lambda, n_x, n_y)\right]^{3/2} \sqrt{2m_c}}{3eF_{sz}\omega'_{50}(V_0, \lambda, E_{g0})\hbar}$$

$\overline{\eta}_{53,1D} = \frac{E_{F1D}-E_{53,n_x,n_y}}{k_B T}$, E_{53,n_x,n_y} is the root of the equation

$$\rho_{50}(E_{53}, n_x, n_y, \lambda, E_{g0}) - G_5(n_x, n_y) = 0 \quad (5.30)$$

$$\overline{\beta}_{53,1D} = \frac{4\left[S_{53}(V_0, \lambda, n_x, n_y)\right]^{3/2} \sqrt{2m_c}}{3eF_{sz}\rho'_{50}(V_0, \lambda, E_{g0})\hbar}$$

5.2.3 Field Emission from Effective Mass Superlattices of III–V Semiconductors in the Presence of Light Waves Under Magnetic Quantization

The dispersion relation of the conduction electrons in effective mass superlattices of III–V materials can be expressed following Sasaki [2] as

$$a_1 \cdot \cos[c_{15}(E, \lambda, E_{g_1}, \Delta_1)a_0 + c_{25}(E, \lambda, E_{g_2}, \Delta_2)b_0] - a_2 \cdot \cos[c_{15}(E, \lambda, E_{g_1}, \Delta_1)a_0 - c_{25}(E, \lambda, E_{g_2}, \Delta_2)b_0] = \cos(L_0 k) \quad (5.31)$$

where

$$a_1 = \left[\left[1 + \sqrt{\frac{m_{c_2}}{m_{c_1}}}\right]^2 \cdot \left[4\left(\sqrt{\frac{m_{c_2}}{m_{c_1}}}\right)^{1/2}\right]^{-1}\right],$$

$$a_2 = \left[\left[-1 + \sqrt{\frac{m_{c_2}}{m_{c_1}}}\right]^2 \cdot \left[4\left(\sqrt{\frac{m_{c_2}}{m_{c_1}}}\right)^{1/2}\right]^{-1}\right],$$

$$c_{i5}^2\left(E, \lambda, E_{g0i}, \Delta_i\right) = \frac{2m_{ci}}{\hbar^2}[\beta_{i50}\left(E, \lambda, E_{g0i}, \Delta_i\right) - k_\perp^2], \quad i=1,2, \quad k_\perp^2 = k_y^2 + k_z^2,$$

$$\beta_{i50}\left(E, \lambda, E_{g0i}, \Delta_i\right) = [I_{i50}\left(E, E_{g0i}, \Delta_i\right) - \theta_{i50}(E, \lambda, E_{g0i}, \Delta_i)],$$

$$I_{i50}\left(E, E_{g0i}, \Delta_i\right) = \frac{E\left(E + E_{g0i}\right)\left(E + E_{g0i} + \Delta_i\right)\left(E_{g0i} + \frac{2}{3}\Delta_i\right)}{E_{g0i}\left(E_{g0i} + \Delta_i\right)\left(E + E_{g0i} + \frac{2}{3}\Delta_i\right)},$$

$$\theta_{i50}\left(E,\lambda,E_{g0i},\Delta_i\right)=\frac{C_{i50}(\lambda,E_{g0i},\Delta_i)}{\varphi_{i50}(E)}\psi_{i50}^2(E),$$

$$C_{i50}\left(\lambda,E_{g0i},\Delta_i\right)=\frac{e^2}{96m_{ri}\pi c^3}\frac{I_0\lambda^2}{\sqrt{\varepsilon_{sci}\varepsilon_0}}\frac{E_{g0i}\left(E_{g0i}+\Delta_i\right)\overline{\beta}_{i50}^2}{\left(E_{g0i}+\frac{2}{3}\Delta_i\right)4}\left(t_{i50}+\frac{\overline{\rho}_{i50}}{\sqrt{2}}\right)^2,$$

m_{ri} is the reduced mass and is given by $m_{ri}^{-1}=(m_{ci})^{-1}+m_{vi}^{-1}$, m_{vi} is the effective mass of the heavy hole at the top of the valance band in the absence of any field,

$$\overline{\beta}_{i50}\equiv\left[\left(6(E_{g0i}+2\Delta_i/3)(E_{g0i}+\Delta_i)\right)/\chi_{i50}\right]^{1/2},$$

$$\chi_{i50}=\left(6E_{g0i}^2+9E_{g0i}\Delta_i+4\Delta_i^2\right),$$

$$t_{i50}=\left[6(E_{g0i}+2\Delta_i/3)^2/\chi_{i50}\right]^{1/2},\quad \overline{\rho}_{i50}=\left(4\Delta_i^2/3\chi_{i50}\right)^{1/2},$$

$$\phi_{i50}\left(E\right)=E_{g0i}\left(1+2\left(1+\frac{m_{ci}}{m_{vi}}\right)\frac{I_{i50}\left(E\right)}{E_{g0i}}\right)^{1/2},$$

$$\psi_{i50}(E)=\left[\left(1+\frac{E_{g0i}-\delta'_{i50}}{\phi_{i50}(E)+\delta'_{i50}}\right)+(E_{g0i}-\delta'_{i50})\left[\frac{1}{\phi_{i50}(E)+\delta'_{i50}}-\frac{1}{E_{g0i}+\delta'_{i50}}\right]^{1/2}\right.$$
$$\left.\times\left[\frac{1}{\phi_{i50}(E)+\delta'_{i50}}-\frac{E_{g0i}+\delta'_{i50}}{(E_{g0i}-\delta'_{i50})^2}\right]^{\frac{1}{2}}\right]$$

$$\delta'_{i50}=\left(E_{g0i}^2\Delta_i\right)(\chi_{i50})^{-1}$$

In the presence of a quantizing magnetic field B, along x-direction the magneto-energy spectrum assumes the form

$$k_x^2=\overline{\omega}_{15}(E,\lambda,n) \tag{5.32}$$

where

$$\overline{\omega}_{15}(E,\lambda,n)=\frac{1}{L_0^2}\left[\left[\cos^{-1}\{\overline{f}_{15}(E,\lambda,n)\}\right]^2-\frac{2eB}{\hbar}\left(n+\frac{1}{2}\right)L_0^2\right],$$

$$\overline{f}_{15}(E,\lambda,n)=[a_1\cos[\overline{c}_{15}(E,\lambda,E_{g01},\Delta_1,n)a_0+b_0\overline{c}_{25}(E,\lambda,E_{g02},\Delta_2,n)]$$
$$-a_2\cos[\overline{c}_{15}(E,\lambda,E_{g01},\Delta_1,n)a_0-b_0\overline{c}_{25}(E,\lambda,E_{g02},\Delta_2,n)]],$$

$$\overline{c}_{15}(E,\lambda,E_{g01},\Delta_1,n)=\left[\left(\frac{2m_{c1}}{\hbar^2}\right)[\beta_{150}(E,\lambda,E_{g01},\Delta_1)]-\frac{2eB}{\hbar}\left(n+\frac{1}{2}\right)\right]^{1/2},$$

$$\overline{c}_{25}(E,\lambda,E_{g02},\Delta_2,n)=\left[\left(\frac{2m_{c2}}{\hbar^2}\right)[\beta_{250}(E,\lambda,E_{g02},\Delta_2)]-\frac{2eB}{\hbar}\left(n+\frac{1}{2}\right)\right]^{1/2}.$$

The electron concentration assumes the form

$$n_0 = \frac{g_v eB}{\pi^2 \hbar L_0} \sum_{n=0}^{n_{\max}} [S_{54}(E_{FB}, \lambda, n) + T_{54}(E_{FB}, \lambda, n)] \tag{5.33}$$

where

$$S_{54}(E_{FB}, \lambda, n) = \left[\left[\cos^{-1} \overline{f}_{15}(E_{FB}, \lambda, n)\right]^2 - \frac{2eB}{\hbar}\left(n + \frac{1}{2}\right) L_0^2\right]^{1/2}$$

$$T_{54}(E_{FB}, \lambda, n) = \sum_{r=1}^{s_0} Z_B(r)\,[S_{54}(E_{FB}, \lambda, n)]$$

The transmission coefficient in this case can be written as

$$t = \exp(-\overline{\theta}_{15}) \tag{5.34}$$

where

$$\overline{\theta}_{15} = \frac{4[\overline{\omega}_{15}(V_0, \lambda, n)]^{3/2}}{3eF_{sx}\{\overline{\omega}_{15}(V_0, \lambda, n)\}'},$$

$$\{\overline{\omega}_{15}(V_0, \lambda, n)\}' = 2\{\overline{f}_{15}(V_0, \lambda, n)\}' L_0^{-2} \cdot \cos^{-1}[\overline{f}_{15}(V_0, \lambda, n)][1 - \overline{f}_{15}(V_0, \lambda, n)]^{-1/2}],$$

$$\begin{aligned}\{\overline{f}_{15}(V_0, \lambda, n)\}' = {} & [a_1 \sin[\overline{c}_{15}(V_0, \lambda, E_{g01}, \Delta_1, n)a_0 + b_0\overline{c}_{25}(V_0, \lambda, E_{g02}, \Delta_2, n)] \\ & \cdot [[\overline{c}_{15}(V_0, \lambda, E_{g01}, \Delta_1, n)]' a_0 + b_0[\overline{c}_{25}(V_0, \lambda, E_{g02}, \Delta_2, n)]'] \\ & - a_2 \sin[\overline{c}_{15}(V_0, \lambda, E_{g01}, \Delta_1, n)a_0 - b_0\overline{c}_{25}(V_0, \lambda, E_{g02}, \Delta_2, n)] \\ & \times [[\overline{c}_{15}(V_0, \lambda, E_{g01}, \Delta_1, n)]' a_0 - b_0[\overline{c}_{25}(V_0, \lambda, E_{g02}, \Delta_2, n)]']]\end{aligned}$$

$$\{\overline{c}_{15}(V_0, \lambda, E_{g01}, \Delta_1, n)\}' = \frac{m_{c1}}{\hbar^2}[\beta'_{150}(V_0, \lambda, E_{g01}, \Delta_1)/\{\overline{c}_{15}(V_0, \lambda, E_{g01}, \Delta_1, n)\}],$$

$$\{\overline{c}_{25}(V_0, \lambda, n)\}' = \frac{m_{c2}}{\hbar^2}[\beta'_{250}(V_0, \lambda, E_{g02}, \Delta_2)/\{\overline{c}_{25}(V_0, \lambda, E_{g02}, \Delta_2, n)\}],$$

$$\beta'_{i50}(V_0, \lambda, E_{g0i}, \Delta_i) = \left[I'_{i50}(V_0, E_{g0i}, \Delta_i) - \theta'_{i50}(V_0, \lambda, E_{g0i}, \Delta_i)\right],$$

$$I'_{i50}(V_0, E_{g0i}, \Delta_i) = I_{i50}(V_0, E_{g0i}, \Delta_i)\left[\frac{1}{V_0} + \frac{1}{V_0 + E_{g0i}} + \frac{1}{V_0 + E_{g0i} + \Delta_i} - \frac{1}{V_0 + E_{g0i} + \frac{2}{3}\Delta_i}\right], \quad i = 1, 2,$$

$$\theta'_{i50}(V_0,\lambda,E_{g_{0i}},\Delta_i)=\theta_{i50}(V_0,\lambda,E_{g_{0i}},\Delta_i)\left[\frac{-\theta'_{i50}(V_0,\lambda,E_{g_{0i}},\Delta_i)}{\theta_{i50}(V_0,\lambda,E_{g_{0i}},\Delta_i)}+\frac{2\psi'_{i50}(V_0)}{\psi_{i50}(V_0)}\right],$$

$$\phi'_{i50}(V_0)=E_{g_{0i}}\left(\left(1+\frac{m_{ci}}{m_{vi}}\right)\frac{I'_{i50}\left(V_0,E_{g_{0i}},\Delta_i\right)}{\phi_{i50}(V_0)}\right)^{1/2},$$

$$\psi'_{i50}(V_0)=\frac{-(E_{g_{0i}}-\delta'_{i50})\phi'_{i50}(V_0)}{(\phi_{i50}(V_0)+\delta'_{i50})^2}\left[1+\frac{1}{2}\left[\frac{1}{\phi_{i50}(V_0)+\delta'_{i50}}-\frac{1}{E_{g_{0i}}+\delta'_{i50}}\right]^{-\frac{1}{2}}\right.$$
$$\cdot\left[\frac{1}{\phi_{i50}(V_0)+\delta'_{i50}}-\frac{(E_{g_{0i}}+\delta'_{i50})}{(E_{g_{0i}}-\delta'_{i50})^2}\right]^{\frac{1}{2}}$$
$$+\frac{1}{2}\left[\frac{1}{\phi_{i50}(V_0)+\delta'_{i50}}-\frac{E_{g_{0i}}+\delta'_{i50}}{(E_{g_{0i}}-\delta'_{i50})^2}\right]^{-\frac{1}{2}}$$
$$\left.\cdot\left[\frac{1}{\phi_{i50}(V_0)+\delta'_{i50}}-\frac{1}{E_{g_{0i}}+\delta'_{i50}}\right]^{\frac{1}{2}}\right]$$

The field-emitted current density in this case is given by

$$J=\frac{e^2Bk_BTg_v}{2\pi^2\hbar^2}\sum_{n=0}^{n_{\max}}F_0(\overline{\eta}_{51BSL})\exp(-\overline{\beta}_{51BSL})\tag{5.35}$$

where $\overline{\eta}_{51BSL}=\frac{E_{FB}-E_{51BSL}}{k_BT}$, E_{51BSL} is the root of the equation

$$\overline{\omega}_{15}(E_{51BSL},\lambda,n)=0$$
$$\overline{\beta}_{51BSL}=\overline{\theta}_{15}$$

The electron concentration when the dispersion relations of the constituent materials are defined by the perturbed two-band model of Kane can be expressed as

$$n_0=\frac{g_veB}{\pi^2\hbar L_0}\sum_{n=0}^{n_{\max}}[S_{55}(E_{FB},\lambda,n)+T_{55}(E_{FB},\lambda,n)]\tag{5.36}$$

where

$$S_{55}(E_{FB},\lambda,n)=\left[\left[\cos^{-1}\overline{f}_{151}(E_{FB},\lambda,n)\right]^2-\frac{2eB}{\hbar}\left(n+\frac{1}{2}\right)L_0^2\right]^{1/2},$$

$$\overline{f}_{151}(E,\lambda,n)=[a_1\cos[\overline{c}_{151}(E,\lambda,E_{g_{01}},n)a_0+b_0\overline{c}_{251}(E,\lambda,E_{g_{02}},n)]$$
$$-a_2\cos[\overline{c}_{151}(E,\lambda,E_{g_{01}},n)a_0-b_0\overline{c}_{251}(E,\lambda,E_{g_{02}},n)]],$$

$$\overline{c}_{151}(E,\lambda,n) = \left[\left(\frac{2m_{c1}}{\hbar^2}\right)\left[\omega_{150}\left(E,\lambda,E_{g_{01}}\right)\right] - \frac{2eB}{\hbar}\left(n+\frac{1}{2}\right)\right]^{1/2},$$

$$\overline{c}_{251}(E,\lambda,n) = \left[\left(\frac{2m_{c2}}{\hbar^2}\right)\left[\omega_{250}\left(E,\lambda,E_{g_{02}}\right)\right] - \frac{2eB}{\hbar}\left(n+\frac{1}{2}\right)\right]^{1/2},$$

$$\omega_{i50}\left(E,\lambda,E_{g_{0i}}\right) \equiv E\,(1+\alpha_i E) - B_{i50}\,(E,\lambda),\quad B_{i50}(E,\lambda) = \frac{C_{i51}(\lambda,E_{g_{0i}})\psi_{i51}^2(E)}{\phi_{i51}(E)},$$

$$C_{i51}(\lambda,E_{g_{0i}}) \equiv \frac{e^2 I_0 \lambda^2 E_{g_{0i}}}{384\pi c^3 m_{ri}\sqrt{\varepsilon_{sci}\varepsilon_0}},$$

$$\phi_{i51}(E) \equiv E_{g_{0i}}\left\{1+\frac{2m_{ci}}{m_{ri}}\frac{E(1+\alpha_i E)}{E_{g_{0i}}}\right\}^{1/2},\quad \alpha_i = \frac{1}{E_{g_{0i}}},\quad \psi_{i51}(E) = \frac{2E_{g_{0i}}}{\phi_{i51}(E)},$$

$$T_{55}(E_{FB},\lambda,n) = \sum_{r=1}^{S_0} Z_B(r)\,[S_{55}(E_{FB},\lambda,n)]$$

The transmission coefficient in this case can be written as

$$t = \exp(-\overline{\theta}_{16}) \tag{5.37}$$

where $\overline{\theta}_{16} = \frac{4[\overline{\omega}_{16}(V_0,\lambda,n)]^{3/2}}{3eF_{sx}\{\overline{\omega}_{16}(V_0,\lambda,n)\}'}$, $\overline{\omega}_{16}(V_0,\lambda,n) = \frac{1}{L_0^2}\left[[\cos^{-1}\{\overline{f}_{151}(V_0,\lambda,n)\}]^2 - \frac{2eB}{\hbar}\right.$ $\left.\left(n+\frac{1}{2}L_0^2\right)\right]$,

$$\{\overline{\omega}_{16}(V_0,\lambda,n)\}' = 2\{\overline{f}_{151}(V_0,\lambda,n)\}' L_0^{-2}$$
$$\cdot\cos^{-1}[\overline{f}_{151}(V_0,\lambda,n)][1-\overline{f}_{151}(V_0,\lambda,n)]^{-1/2}],$$

$$\begin{aligned}[\overline{f}_{151}(V_0,\lambda,n)] = {} & [a_1\sin[\overline{c}_{151}(V_0,\lambda,E_{g_{01}},n)a_0 + b_0\overline{c}_{251}(V_0,\lambda,E_{g_{02}},n)] \\ & \cdot[[\overline{c}_{151}(V_0,\lambda,E_{g_{01}},n)]'a_0 + b_0[\overline{c}_{251}(V_0,\lambda,E_{g_{02}},n)]'] \\ & - a_2\sin[\overline{c}_{151}(V_0,\lambda,E_{g_{01}},n)a_0 - b_0\overline{c}_{251}(V_0,\lambda,E_{g_{02}},n)] \\ & \times[[\overline{c}_{151}(V_0,\lambda,E_{g_{01}},n)]'a_0 - b_0[\overline{c}_{251}(V_0,\lambda,E_{g_{02}},n)]']]\end{aligned}$$

$$\{\overline{c}_{151}(V_0,\lambda,E_{g_{01}},n)\}' = \frac{m_{c1}}{\hbar^2}[\omega_{150}(V_0,\lambda,E_{g_{01}})]'/\{\overline{c}_{151}(V_0,\lambda,E_{g_{01}},n)\},$$

$$\{\overline{c}_{251}(V_0,\lambda,E_{g_{02}},n)\}' = \frac{m_{c2}}{\hbar^2}[\omega_{250}(V_0,\lambda,E_{g_{01}})]'/\{\overline{c}_{251}(V_0,\lambda,E_{g_{02}},n)\},$$

$$\omega'_{i50}(V_0,\lambda,E_{g_{0i}}) = \left[1+2\alpha_i V_0 - B'_{i50}(V_0,\lambda)\right],$$

$$B'_{i50}(V_0,\lambda) = \left[\frac{-B_{i50}(V_0,\lambda)\phi'_{i51}(V_0)}{\phi_{i51}(V_0)} + \frac{2B_{i50}(V_0,\lambda)\psi'_{i51}(V_0)}{\psi_{i51}(V_0)}\right],$$

$$\psi'_{i51}(V_0) = \left[\frac{-\psi_{i51}(V_0)}{\phi_{i51}(V_0)}\phi'_{i51}(V_0)\right], \quad \phi'_{i51}(V_0) = \frac{m_{ci}}{m_{ri}} \cdot \frac{E_{g_{0i}}(1+2\alpha_i V_0)}{\phi_{i51}(V_0)}.$$

The field emitted current density in this case is given by

$$J = \frac{e^2 B k_B T g_v}{2\pi^2\hbar^2}\sum_{n=0}^{n_{\max}} F_0(\overline{\eta}_{52BSL})\exp(-\overline{\beta}_{52BSL}) \tag{5.38}$$

where $\overline{\eta}_{52BSL} = \frac{E_{FB}-E_{52BSL}}{k_B T}$, E_{52BSL} is the root of the equation

$$\overline{\omega}_{16}(E_{52BSL},\lambda,n) = 0 \tag{5.39}$$

$$\overline{\beta}_{52BSL} = \overline{\theta}_{16}$$

The electron concentration when the dispersion relations of the constituent materials are defined by the perturbed parabolic energy bands can be expressed as

$$n_0 = \frac{g_v eB}{\pi^2\hbar L_0}\sum_{n=0}^{n_{\max}}[S_{56}(E_{FB},\lambda,n) + T_{56}(E_{FB},\lambda,n)] \tag{5.40}$$

where

$$S_{56}(E_{FB},\lambda,n) = \left[\left[\cos^{-1}\overline{f}_{152}(E_{FB},\lambda,n)\right]^2 - \frac{2eB}{\hbar}\left(n+\frac{1}{2}\right)L_0^2\right]^{1/2}$$

$$\overline{f}_{152}(E,\lambda,n) = [a_1\cos[\overline{c}_{152}(E,\lambda,E_{g_1},n)a_0 + b_0\overline{c}_{252}(E,\lambda,E_{g_2},n)]$$
$$- a_2\cos[\overline{c}_{152}(E,\lambda,E_{g_1},n)a_0 - b_0\overline{c}_{252}(E,\lambda,E_{g_2},n)]]$$

$$\overline{c}_{152}(E,\lambda,E_{g_1},n) = \left[\left(\frac{2m_{c1}}{\hbar^2}\right)[\rho_{150}(E,\lambda,E_{g_{01}})] - \frac{2eB}{\hbar}\left(n+\frac{1}{2}\right)\right]^{\frac{1}{2}},$$

$$\overline{c}_{252}(E,\lambda,E_{g_2},n) = \left[\left(\frac{2m_{c2}}{\hbar^2}\right)[\rho_{250}(E,\lambda,E_{g_{02}})] - \frac{2eB}{\hbar}\left(n+\frac{1}{2}\right)\right]^{\frac{1}{2}},$$

$$\rho_{i50}\left(E,E_{g_{0i}},\lambda\right) = E - C_{i52}\left[1+\left(\frac{2m_{ci}}{m_{ri}}\right)\left(\frac{E}{E_{g_{0i}}}\right)\right]^{-3/2},$$

$$C_{i52} \equiv \frac{e^2 I_0\lambda^2}{96\pi c^3 m_{ri}\sqrt{\varepsilon_{sci}\varepsilon_0}}, \quad \alpha_i = \frac{1}{E_{g_{0i}}},$$

$$T_{56}(E_{FB},\lambda,n) = \sum_{r=1}^{s_0} Z_B(r)\,[S_{56}(E_{FB},\lambda,n)]$$

The transmission coefficient in this case can be written as

$$t = \exp(-\overline{\theta}_{17}) \tag{5.41}$$

where

$$\overline{\theta}_{17} = \frac{4[\overline{\omega}_{17}(V_0,\lambda,n)]^{3/2}}{3eF_{sx}\{\overline{\omega}_{17}(V_0,\lambda,n)\}'},$$

$$\overline{\omega}_{17}(V_0,\lambda,n) = \frac{1}{L_0^2}\left[[\cos^{-1}\{\overline{f}_{152}(V_0,\lambda,n)\}]^2 - \frac{2eB}{\hbar}\left(n+\frac{1}{2}\right)L_0^2\right],$$

$$\{\overline{\omega}_{17}(V_0,\lambda,n)\}' = 2\{\overline{f}_{152}(V_0,\lambda,n)\}'L_0^{-2} \cdot \cos^{-1}[\overline{f}_{152}(V_0,\lambda,n)][1-\overline{f}_{152}(V_0,\lambda,n)]^{-1/2}],$$

$$[\overline{f}_{152}(V_0,\lambda,n)] = [a_1 \sin[\overline{c}_{152}(V_0,\lambda,E_{g_1},n)a_0 + b_0\overline{c}_{252}(V_0,\lambda,E_{g_2},n)] \cdot [[\overline{c}_{152}(V_0,\lambda,E_{g_1},n)]'a_0 + b_0[\overline{c}_{252}(V_0,\lambda,E_{g_2},n)]'] - a_2\sin[\overline{c}_{152}(V_0,\lambda,E_{g_1},n)a_0 - b_0\overline{c}_{252}(V_0,\lambda,E_{g_2},n)] \times [[\overline{c}_{152}(V_0,\lambda,E_{g_1},n)]'a_0 - b_0[\overline{c}_{252}(V_0,\lambda,E_{g_2},n)]']]$$

$$\{\overline{c}_{152}(V_0,\lambda,E_{g_1},n)\}' = \frac{m_{c1}}{\hbar^2}[\rho_{150}(V_0,\lambda)]'/\{\overline{c}_{152}(V_0,\lambda,E_{g_1},n)\},$$

$$\{\overline{c}_{252}(V_0,\lambda,E_{g_2},n)\}' = \frac{m_{c2}}{\hbar^2}[\rho_{250}(V_0,\lambda)]'/\{\overline{c}_{252}(V_0,\lambda,E_{g2},n)\},$$

$$\rho'_{i50}\left(E,E_{g0i},\lambda\right) = 1 + C_{i52}\left(\frac{3m_{ci}}{m_{ri}}\right)\left[1+\left(\frac{2m_{ci}}{m_{ri}}\right)\left(\frac{E}{E_{g0i}}\right)\right]^{-5/2}$$

The field emitted current density in this case is given by

$$J = \frac{e^2Bk_BTg_v}{2\pi^2\hbar^2}\sum_{n=0}^{n_{max}} F_0(\overline{\eta}_{53BSL})\exp(-\overline{\beta}_{53B}) \tag{5.42}$$

where $\overline{\eta}_{53BSL} = \frac{E_{FB}-E_{53BSL}}{k_BT}$, E_{53BSL} is the root of the equation

$$\overline{\omega}_{17}(E_{53BSL},\lambda,n) = 0$$
$$\overline{\beta}_{53BSL} = \overline{\theta}_{17} \tag{5.43}$$

5.2.4 Field Emission from Quantum Wire Effective Mass Superlattices of III–V Semiconductors

The dispersion relation of the conduction electrons for quantum wire effective mass superlattices in accordance with the perturbed three-band model of Kane is given by

$$k_x^2 = \overline{\omega}_{19}(E,\lambda,n_y,n_z) \tag{5.44}$$

where $\overline{\omega}_{19}(E,\lambda,n_y,n_z) = \left[\frac{1}{L_0^2}\left[\cos^{-1}\{\overline{f}_3\left(E,\lambda,n_y,n_z\right)\}\right]^2 - H(n_y,n_z)\right]$.

$$\begin{aligned}
&\overline{f}_3(E,\lambda,n_y,n_z) \\
&\quad = [a_1\cos[\overline{e}_1(E,\lambda,E_{g_1},\Delta_1,n_y,n_z)a_0 + b_0\overline{e}_2(E,\lambda,E_{g_2},\Delta_2,n_y,n_z)] \\
&\quad\quad - a_2\cos[\overline{e}_1(E,\lambda,E_{g_1},\Delta_1,n_y,n_z)a_0 - b_0\overline{e}_2(E,\lambda,E_{g_2},\Delta_2,n_{y,}n_z)]]
\end{aligned} \tag{5.45}$$

$$\overline{e}_i^2(E,\lambda,E_{g0i},\Delta_i,n_y,n_z) = \left[\left(\frac{2m_{ci}}{\hbar^2}\right)[\beta_{i50}(E,\lambda,E_{g0i},\Delta_i)] - H\left(n_y,n_z\right)\right]$$

and

where $H\left(n_y,n_z\right) = \left[\left(\frac{n_y\pi}{d_y}\right)^2 + \left(\frac{n_z\pi}{d_z}\right)^2\right]$.

The expression of the electron concentration in this case can be written as

$$n_0 = \frac{2g_v}{\pi}\sum_{n_y=1}^{n_{y\max}}\sum_{n_z=1}^{n_{z\max}}\left[\overline{Q}_{23}\left(E_{\text{FIDEMSL}},\lambda,n_y,n_z\right) + \overline{Q}_{24}\left(E_{\text{FIDEMSL}},\lambda,n_y,n_z\right)\right] \tag{5.46}$$

where

$$\overline{Q}_{23}\left(E_{\text{FIDEMSL}},\lambda,n_y,n_z\right) = \sqrt{\overline{\omega}_{19}\left(E_{\text{FIDEMSL}},\lambda,n_y,n_z\right)},$$

$$\overline{Q}_{24}\left(E_{\text{FIDEMSL}},\lambda,n_y,n_z\right) = \sum_{R=1}^{R=R_0} Z\left(R_{\text{IDEMSL}}\right)\overline{Q}_{23}\left(E_{\text{FIDEMSL}},\lambda,n_y,n_z\right),$$

E_{FIDEMSL} is the Fermi energy in the present case and

$$Z(R_{\text{IDEMSL}}) = 2(k_BT)^{2R}(1-2^{1-2R})\xi(2R)\frac{\partial^{2R}}{\partial E_{\text{FIDEMSL}}}.$$

The field emitted current assumes the form

$$I = \frac{eg_vk_BT}{\pi\hbar}\sum_{n_y=1}^{n_{y\max}}\sum_{n_z=1}^{n_{z\max}} F_0\left(\overline{\eta}_{17}\right)\exp\left(-\overline{\theta}_{17}\right) \tag{5.47}$$

where $\overline{\eta}_{17} = \frac{E_{\text{FIDEMSL}}-\overline{E}_{15}}{k_BT}$. $\overline{E}_{15}$ is the root of the equation

$$\overline{\omega}_{19}\left(\overline{E}_{15},\lambda,n_y,n_z\right) = 0 \tag{5.48}$$

$$\overline{\theta}_{17} = \frac{4\left[\overline{\omega}_{19}\left(V_0, \lambda, n_y, n_z\right)\right]^{3/2}}{3eF_{sx}\left[\overline{\omega}_{19}\left(V_0, \lambda, n_y, n_z\right)\right]'},$$

$$[\overline{\omega}_{19}\left(V_0, \lambda, n_y, n_z\right)]' = \left[1 - \overline{f}_3^{\,2}\left(V_0, \lambda, n_y, n_z\right)\right]^{-1/2} \times \left[2[\overline{f}_3\left(V_0, \lambda, n_y, n_z\right)]'\right]\left[\cos^{-1}\left\{\overline{f}_3\left(V_0, \lambda, n_y, n_z\right)\right\}\right],$$

$$\begin{aligned}
&[\overline{f}_3(V_0, \lambda, n_y, n_z)]' \\
&= a_1 \sin\left[a_0\overline{e}_1(V_0, \lambda, E_{g_1}, \Delta_1, n_y, n_z) + b_0\overline{e}_2(V_0, \lambda, E_{g_2}, \Delta_2, n_y, n_z)\right] \\
&\quad \times \left[a_0[\overline{e}_1(V_0, \lambda, E_{g_1}, \Delta_1, n_y, n_z)]' + b_0[\overline{e}_2(V_0, \lambda, E_{g_2}, \Delta_2, n_y, n_z)]'\right] \\
&\quad - a_2 \sin\left[a_0\overline{e}_1(V_0, \lambda, E_{g_1}, \Delta_1, n_y, n_z) - b_0\overline{e}_2(V_0, \lambda, E_{g_2}, \Delta_2, n_y, n_z)\right] \\
&\quad \cdot \left[a_0[\overline{e}_1(V_0, \lambda, E_{g_1}, \Delta_1, n_y, n_z)] - b_0[\overline{e}_2(V_0, \lambda, E_{g_2}, \Delta_2, n_y, n_z)]'\right]
\end{aligned}$$

and

$$[\overline{e}_i\left(V_0, \lambda, E_{g_{0i}}, \Delta_i, n_y, n_z\right)]' = \frac{m_{ci}\beta_{i50}\left(V_0, \lambda, E_{g_{0i}}, \Delta_i\right)}{\hbar^2\overline{e}_i\left(V_0, \lambda, E_{g_{0i}}, \Delta_i, n_y, n_z\right)}.$$

In accordance with the perturbed two-band model of Kane, the electron concentration per unit length is given by

$$n_0 = \frac{2g_v}{\pi}\sum_{n_y=1}^{n_{y\max}}\sum_{n_z=1}^{n_{z\max}}\left[\overline{Q}_{25}\left(E_{\text{FIDEMSL}}, \lambda, n_y, n_z\right) + \overline{Q}_{26}\left(E_{\text{FIDEMSL}}, \lambda, n_y, n_z\right)\right] \tag{5.49}$$

Where

$$\overline{Q}_{25}\left(E_{\text{FIDEMSL}}, \lambda, n_y, n_z\right) = \left[\sqrt{\overline{\omega}_{20}\left(E_{\text{FIDEMSL}}, \lambda, n_y, n_z\right)}\right],$$

$$\overline{Q}_{26}\left(E_{\text{FIDEMSL}}, \lambda, n_y, n_z\right) = \sum_{R=1}^{R=R_0} Z\left(R_{\text{IDEMSL}}\right)\left[\overline{Q}_{25}\left(E_{\text{FIDEMSL}}, \lambda, n_y, n_z\right)\right],$$

$$\overline{\omega}_{20}\left(E, \lambda, n_y, n_z\right) = \left[\frac{1}{L_0^2}\left[\cos^{-1}\overline{f}_4\left(E, \lambda, n_y, n_z\right)\right]^2 - H\left(n_y, n_z\right)\right],$$

$$\begin{aligned}
\overline{f}_4\left(E, \lambda, n_y, n_z\right) &= \left[a_1 \cos\left[a_0\overline{g}_1\left(E, \lambda, E_{g1}, n_y, n_z\right) - b_0\overline{g}_2\left(E, \lambda, E_{g2}, n_y, n_z\right)\right]\right. \\
&\quad \left. - a_2 \cos\left[a_0\overline{g}_1\left(E, \lambda, E_{g1}, n_y, n_z\right) - b_0\overline{g}_2\left(E, \lambda, E_{g2}, n_y, n_z\right)\right]\right],
\end{aligned}$$

and

$$\overline{g}_i^2\left(E, \lambda, E_{g_{0i}}, n_y, n_z\right) = \left[\frac{2m_{ci}}{\hbar^2}\omega_{i50}\left(E, \lambda, E_{g_{0i}}\right) - H\left(n_y, n_z\right)\right].$$

The field emitted current can be written as,

$$I = \frac{eg_v k_B T}{\pi\hbar}\sum_{n_y=1}^{n_{y\max}}\sum_{n_z=1}^{n_{z\max}} F_0\left(\overline{\eta}_{18}\right)\exp\left(-\overline{\theta}_{18}\right) \tag{5.50}$$

where $\overline{\eta}_{18} = \frac{E_{\text{FIDEMSL}} - \overline{E}_{16}}{k_B T}$, $\overline{E}_{16}$ is the root of the equation

$$\overline{\omega}_{20}\left(\overline{E}_{16}, \lambda, n_y, n_z\right) = 0 \tag{5.51}$$

$$\overline{\theta}_{18} = \frac{4\left[\overline{\omega}_{20}\left(V_0, \lambda, n_y, n_z\right)\right]^{3/2\ \prime}}{3eF_{sx}[\overline{\omega}_{20}\left(V_0, \lambda, n_y, n_z\right)]},$$

$$[\overline{\omega}_{20}\left(V_0, \lambda, n_y, n_z\right)]' = \frac{2[\overline{f}_4\left(V_0, \lambda, n_y, n_z\right)]'\left[\cos^{-1}\overline{f}_4\left(V_0, \lambda, n_y, n_z\right)\right]}{\sqrt{1 - \overline{f}_4^{\,2}\left(V_0, \lambda, n_y, n_z\right)}},$$

$$\begin{aligned}
&[\overline{f}_4\left(V_0, \lambda, n_y, n_z\right)]' \\
&\quad = a_1 \sin\left[a_0\overline{g}_1\left(V_0, \lambda, E_{g1}, n_y, n_z\right) + b_0\overline{g}_2\left(V_0, \lambda, E_{g2}, n_y, n_z\right)\right] \\
&\qquad \times \left[a_0[\overline{g}_1\left(V_0, \lambda, E_{g1}, n_y, n_z\right)]' + b_0[\overline{g}_2\left(V_0, \lambda, E_{g2}, n_y, n_z\right)]'\right] \\
&\qquad - a_2 \sin\left[a_0\overline{g}_1\left(V_0, \lambda, E_{g1}, n_y, n_z\right) - b_0\overline{g}_2\left(V_0, \lambda, E_{g2}, n_y, n_z\right)\right] \\
&\qquad \times \left[a_0[\overline{g}_1\left(V_0, \lambda, E_{g1}, n_y, n_z\right)]' - b_0[\overline{g}_2\left(V_0, \lambda, E_{g2}, n_y, n_z\right)]'\right],
\end{aligned}$$

and

$$\left[\overline{g}_i\left(V_0, \lambda, E_{g0i}, n_y, n_z\right)\right]' = \frac{m_{ci}\left[\omega_{i50}\left(V_0, \lambda, E_{g0i}\right)\right]'}{\hbar^2\overline{g}_i\left(V_0, \lambda, E_{g0i}, n_y, n_z\right)}.$$

In accordance with the perturbed parabolic energy bands, the electron concentration per unit length is given by

$$n_0 = \frac{2g_v}{\pi}\sum_{n_y=1}^{n_{y\max}}\sum_{n_z=1}^{n_{z\max}}\left[\overline{Q}_{251}\left(E_{\text{FIDEMSL}}, \lambda, n_y, n_z\right) + \overline{Q}_{261}\left(E_{\text{FIDEMSL}}, \lambda, n_y, n_z\right)\right] \tag{5.52}$$

where

$$\overline{Q}_{251}\left(E_{\text{FIDEMSL}}, \lambda, n_y, n_z\right) = \left[\sqrt{\overline{\omega}_{201}\left(E_{\text{FIDEMSL}}, \lambda, n_y, n_z\right)}\right],$$

$$\overline{Q}_{261}\left(E_{\text{FIDEMSL}}, \lambda, n_y, n_z\right) = \sum_{R=1}^{R=R_0} Z\left(R_{\text{IDEMSL}}\right)\left[\overline{Q}_{251}\left(E_{\text{FIDEMSL}}, \lambda, n_y, n_z\right)\right],$$

$$\overline{\omega}_{201}\left(E, \lambda, n_y, n_z\right) = \left[\frac{1}{L_0^2}\left[\cos^{-1}\overline{f}_{41}\left(E, \lambda, n_y, n_z\right)\right]^2 - H\left(n_y, n_z\right)\right],$$

$$\overline{f}_{41}\left(E,\lambda,n_y,n_z\right)=\left[a_1\cos\left[a_0\overline{g}_{11}\left(E,\lambda,E_{g1},n_y,n_z\right)-b_0\overline{g}_{21}\left(E,\lambda,E_{g2},n_y,n_z\right)\right]\right.$$
$$\left.-a_2\cos\left[a_0\overline{g}_{11}\left(E,\lambda,E_{g1},n_y,n_z\right)-b_0\overline{g}_{21}\left(E,\lambda,E_{g2},n_y,n_z\right)\right]\right]$$

and

$$\overline{g}_{i1}^2\left(E,\lambda,E_{g_{0i}},n_y,n_z\right)=\left[\frac{2m_{ci}}{\hbar^2}\rho_{i50}\left(E,E_{g_{0i}},\lambda\right)-H\left(n_y,n_z\right)\right]$$

The field emitted current can be written as,

$$I=\frac{eg_v k_B T}{\pi\hbar}\sum_{n_y=1}^{n_{y\max}}\sum_{n_z=1}^{n_{z\max}}F_0\left(\overline{\eta}_{181}\right)\exp\left(-\overline{\theta}_{181}\right) \tag{5.53}$$

where $\overline{\eta}_{181}=\frac{E_{FIDEMSL}-\overline{E}_{161}}{k_B T}$, $\overline{E}_{161}$ is the root of the equation

$$\overline{\omega}_{201}\left(\overline{E}_{161},\lambda,n_y,n_z\right)=0 \tag{5.54}$$

$$\overline{\theta}_{181}=\frac{4\left[\overline{\omega}_{201}\left(V_0,\lambda,n_y,n_z\right)\right]^{3/2}}{3eF_{sx}\left[\overline{\omega}_{201}\left(V_0,\lambda,n_y,n_z\right)\right]'},$$

$$\left[\overline{\omega}_{201}\left(V_0,\lambda,n_y,n_z\right)\right]'=\frac{2\left[\overline{f}_{41}\left(V_0,\lambda,n_y,n_z\right)\right]'\left[\cos^{-1}\overline{f}_{41}\left(V_0,\lambda,n_y,n_z\right)\right]}{\sqrt{1-\overline{f}_{41}^2\left(V_0,\lambda,n_y,n_z\right)}},$$

$$\begin{aligned}
&\left[\overline{f}_{41}\left(V_0,\lambda,n_y,n_z\right)\right]'\\
&\quad=a_1\sin\left[a_0\overline{g}_{11}\left(V_0,\lambda,E_{g1},n_y,n_z\right)+b_0\overline{g}_{21}\left(V_0,\lambda,E_{g2},n_y,n_z\right)\right]\\
&\qquad\times\left[a_0\left[\overline{g}_{11}\left(V_0,\lambda,E_{g1},n_y,n_z\right)\right]'+b_0\left[\overline{g}_{21}\left(V_0,\lambda,E_{g2},n_y,n_z\right)\right]'\right]\\
&\qquad-a_2\sin\left[a_0\overline{g}_{11}\left(V_0,\lambda,E_{g1},n_y,n_z\right)-b_0\overline{g}_{21}\left(V_0,\lambda,E_{g2},n_y,n_z\right)\right]\\
&\qquad\times\left[a_0\left[\overline{g}_{11}\left(V_0,\lambda,E_{g1},n_y,n_z\right)\right]'-b_0\left[\overline{g}_{21}\left(V_0,\lambda,E_{g2},n_y,n_z\right)\right]'\right],
\end{aligned}$$

and

$$\left[g_{i1}\left(V_0,\lambda,E_{g_{0i}},n_y,n_z\right)\right]'=\frac{m_{ci}\left[\rho_{i50}\left(V_0,E_{g_{0i}},\lambda\right)\right]'}{\hbar^2\overline{g}_{i1}\left(V_0,\lambda,E_{g_{0i}},n_y,n_z\right)}.$$

5.2.5 *Field Emission from Superlattices of III–V Semiconductors with Graded Interfaces Under Magnetic Quantization*

The energy spectrum in superlattices of III–V compounds with graded interfaces in the presence of light waves whose constituent materials are defined by perturbed three-band model of Kane can be written following [3] as

$$\cos(L_0 k) = \frac{1}{2}\Phi_{115}(E,k_s) \tag{5.55}$$

where

$$\begin{aligned}
&\Phi_{115}(E,k_s) \\
&= \Bigg[2\cosh\{X_{215}(E,k_s)\}\cos\{Y_{215}(E,k_s)\} + \varepsilon_{215}(E,k_s)\sinh\{X_{215}(E,k_s)\} \\
&\quad \times \sin\{Y_{215}(E,k_s)\} + \Delta_{21}\left[\left(\frac{K_{215}^2(E,k_s)}{K_{225}(E,k_s)} - 3K_{225}(E,k_s)\right)\right. \\
&\quad \times \cosh\{X_{215}(E,k_s)\}\sin\{Y_{215}(E,k_s)\} \\
&\quad \left. + \left(3K_{215}(E,k_s) - \frac{\{K_{225}(E,k_s)\}^2}{K_{215}(E,k_s)}\right)\sinh\{X_{215}(E,k_s)\}\cos\{Y_{215}(E,k_s)\}\right] \\
&\quad + \Delta_{21}\Bigg[2\left(\{K_{215}(E,k_s)\}^2 - \{K_{225}(E,k_s)\}^2\right) \\
&\quad \times \cosh\{X_{215}(E,k_s)\}\cos\{Y_{215}(E,k_s)\} \\
&\quad + \frac{1}{12}\left[\frac{5\{K_{225}(E,k_s)\}^3}{K_{215}(E,k_s)} + \frac{5\{K_{215}(E,k_s)\}^3}{K_{225}(E,k_s)} - 34K_{225}(E,k_s)K_{215}(E,k_s)\right] \\
&\quad \times \sinh\{X_{215}(E,k_s)\}\sin\{Y_{215}(E,k_s)\}\Bigg]\Bigg]
\end{aligned}$$

$$X_{215}(E,k_s) = K_{215}(E,k_s)[a_0 - \Delta_{21}],$$

$$K_{215}(E,k_s) \equiv \left[-\frac{2m_{c2}}{\hbar^2}\beta_{150}(E-\overline{V_0},\lambda,E_{g02},\Delta_2) + k_s^2\right]^{1/2},$$

$$\varepsilon(E,k_s) \equiv \left[\frac{K_{215}(E,k_s)}{K_{225}(E,k_s)} - \frac{K_{225}(E,k_s)}{K_{215}(E,k_s)}\right], \quad k_s^2 = k_x^2 + k_y^2,$$

$$Y_{215}(E,k_s) = K_{225}(E,k_s)[b_0 - \Delta_{21}], \quad \text{and}$$

$$K_{225}(E,k_s) = \left[\frac{2m_{c1}\beta_{150}(E,\lambda,E_{g01},\Delta_1)}{\hbar^2} - k_s^2\right]^{1/2}.$$

In the presence of a quantizing magnetic field B along z-direction, the simplified magneto-dispersion relation can be written as

$$k_z^2 = \omega_{215}(E,\lambda,n) \tag{5.56}$$

where

$$\omega_{215}(E,\lambda,n) = \left[\frac{1}{L_0^2}\left[\cos^{-1}\left[\frac{1}{2}f_{215}(E,\lambda,n)\right]\right]^2 - \frac{2|e|B}{\hbar}\left(n+\frac{1}{2}\right)\right],$$

$$f_{215}(E,\lambda,n)$$

$$= \Bigg[2\cosh\{M_{215}(n,E)\}\cos\{N_{215}(n,E)\} + Z_{215}(n,E)\sinh\{M_{215}(n,E)\}$$

$$\times\sin\{N_{215}(n,E)\} + \Delta_{21}\left[\left(\frac{\{I_{215}(n,E)\}^2}{I_{225}(n,E)} - 3I_{225}(n,E)\right)\right.$$

$$\times\cosh\{M_{215}(n,E)\}\sin\{N_{215}(n,E)\} + \left(3I_{215}(n,E) - \frac{\{I_{225}(n,E)\}^2}{I_{215}(n,E)}\right)$$

$$\left.\times\sinh\{M_{215}(n,E)\}\cos\{N_{215}(n,E)\}\right]$$

$$+\Delta_{21}\Bigg[2\left(\{I_{215}(n,E)\}^2 - \{I_{225}(n,E)\}^2\right)\cosh\{M_{215}(n,E)\}\cos\{N_{215}(n,E)\}$$

$$+\frac{1}{12}\left(\frac{5\{I_{225}(n,E)\}^3}{I_{215}(n,E)} + \frac{5\{I_{215}(n,E)\}^3}{I_{225}(n,E)} - \{34I_{225}(n,E)\,I_{215}(n,E)\}\right)$$

$$\times\sinh\{M_{215}(n,E)\}\sin\{N_{215}(n,E)\}\Bigg]\Bigg]$$

$$Z_{215}(n,E) \equiv \left[\frac{I_{215}(n,E)}{I_{225}(n,E)} - \frac{I_{225}(n,E)}{I_{215}(n,E)}\right],$$

$$M_{215}(n,E) = I_{215}(n,E)\,[a_0 - \Delta_{21}],$$

$$I_{215}(n,E) = \left[-\frac{2m_{c2}}{\hbar^2}\beta_{250}(E-\overline{V_0},\lambda,E_{g02},\Delta_2) + \frac{2|e|B}{\hbar}\left(n+\frac{1}{2}\right)\right]^{1/2}$$

$$N_{215}(n,E) = I_{225}(n,E)\,[b_0 - \Delta_{21}]$$

and

$$I_{225}(n,E) \equiv \left[\frac{2m_{c1}}{\hbar^2}\beta_{150}(E,\lambda,E_{g01},\Delta_1) - \left\{\frac{2|e|B}{\hbar}\left(n+\frac{1}{2}\right)\right\}\right]^{1/2}.$$

The electron concentration is given by

$$n_o = \frac{g_v eB}{\pi^2\hbar}\left[\sum_{n=0}^{n_{max}}[\overline{Q}_{27}(E_{FBGISL},\lambda,n)+\overline{Q}_{28}(E_{FBGISL},\lambda,n)]\right] \tag{5.57}$$

$\overline{Q}_{27}(E_{\mathrm{FBGISL}}, \lambda, n) = [\omega_{215}(E_{\mathrm{FBGISL}}, \lambda, n)]^{1/2}$, E_{FBGISL} is the Fermi energy in the present case,

$$\overline{Q}_{28}\left(E_{\mathrm{FBGISL}}, \lambda, n\right) = \sum_{R=1}^{R=R_0} Z\left(R_{\mathrm{BGISL}}\right)\left[\overline{Q}_{27}\left(E_{\mathrm{FBGISL}}, \lambda, n\right)\right]$$

and

$$Z\left(R_{\mathrm{BGISL}}\right) = 2\left(k_{\mathrm{B}}T\right)^2\left(1-2^{1-2R}\right)\xi\left(2R\right)\frac{\partial^{2R}}{\partial E_{\mathrm{FBGISL}}}.$$

The field emitted current density is given by,

$$J = \frac{e^2 B k_{\mathrm{B}} T g_{\mathrm{v}}}{2\pi^2\hbar^2}\sum_{n=0}^{n_{\max}} F_0(\overline{\eta}_{19})\exp(-\overline{\theta}_{19}) \tag{5.58}$$

where $\overline{\eta}_{19} = \frac{E_{\mathrm{FBGISL}}-\overline{E}_{19}}{k_{\mathrm{B}}T}$, $\overline{E}_{19}$ is the root of the equation

$$\omega_{215}(\overline{E}_{19}, \lambda, n) = 0 \tag{5.59}$$

$$\overline{\theta}_{19} = \frac{4\left[\omega_{215}\left(V_0, \lambda, n\right)\right]^{3/2}}{3eF_{\mathrm{sz}}\omega'_{215}\left(V_0, \lambda, n\right)},$$

$$\omega'_{215}\left(V_0, \lambda, n\right) = \frac{1}{L_0^2}\left\{\cos^{-1}\left[\frac{1}{2}f_{215}\left(V_0, \lambda, n\right)\right]\right\}\left[1-\frac{1}{4}f_{215}^2\left(V_0, \lambda, n\right)\right]^{-1/2} \times f'_{215}\left(V_0, \lambda, n\right),$$

$$\begin{aligned}
&f'_{215}\left(V_0, \lambda, n\right)\\
&= \Big[2M'_{215}\left(n, V_0\right)\sinh\left\{M_{215}(n, V_0)\right\}\cos\left\{N_{215}(n, V_0)\right\}\\
&\quad + Z_{215}\left(n, V_0\right)M'_{215}(n, V_0)\cosh\left\{M_{215}(n, V_0)\right\}\\
&\quad \times\sin\left\{N_{215}(n, V_0)\right\} - 2N'_{215}\left(n, V_0\right)\sin\left\{N_{215}(n, V_0)\right\}\cosh\left\{M_{215}(n, V_0)\right\}\\
&\quad + Z'_{215}\left(n, V_0\right)\sinh\left\{M_{215}(n, V_0)\right\}\sin\left\{N_{215}(n, V_0)\right\}\\
&\quad + Z_{215}(n, V_0)N'_{215}(n, V_0)\cos\left\{N_{215}(n, V_0)\right\}\sinh\left\{M_{215}(n, V_0)\right\}\\
&\quad + \Delta_{21}\Bigg[\Bigg(\frac{\left\{2I_{215}\left(n, V_0\right)I'_{215}(n, V_0)\right\}}{I_{225}\left(n, V_0\right)} - \frac{\left\{I_{215}^2(n, V_0)I'_{225}\left(n, V_0\right)\right\}}{I_{225}^2(n, V_0)} - 3I'_{225}(n, V_0)\Bigg)\\
&\quad \times\cosh\left\{M_{215}\left(n, V_0\right)\right\}\sin\left\{N_{215}(n, V_0)\right\}\\
&\quad + \left(-3I_{225}(n, V_0) + \frac{\left\{I_{215}(n, V_0)\right\}^2}{I_{225}(n, V_0)}\right)\left\{M'_{215}(n, V_0)\sinh\left\{M_{215}\left(n, V_0\right)\right\}\right.\\
&\quad \times\sin\left\{N_{215}\left(n, V_0\right)\right\} + \left\{N'_{215}(n, V_0)\cosh\left\{M_{215}\left(n, V_0\right)\right\}\cos\left\{N_{215}(n, V_0)\right\}\right.
\end{aligned}$$

$$+\left(\frac{-\{2I_{225}(n,V_0)I'_{225}(n,V_0)\}}{I_{215}(n,V_0)}+\frac{\{I^2_{225}(n,V_0)I'_{215}(n,V_0)\}}{I^2_{215}(n,V_0)}+3I'_{215}(n,V_0)\right)$$

$$\times\sinh\{M_{215}(n,V_0)\}\cos\{N_{215}(n,V_0)\}$$

$$+\left(+3I_{215}(n,V_0)-\frac{\{I_{225}(n,V_0)\}^2}{I_{215}(n,V_0)}\right)\{M'_{215}(n,V_0)\cosh\{M_{215}(n,V_0)\}$$

$$\left.\times\cos\{N_{215}(n,V_0)\}-N'_{215}(n,V_0)\sin\{N_{215}(n,V_0)\}\sinh\{M_{215}(n,V_0)\}\}\right]$$

$$+\Delta_{21}\left[4\left(\{I_{215}(n,V_0)I'_{215}(n,V_0)\}-\{I_{225}(n,V_0)I'_{225}(n,V_0)\}\right)\right.$$

$$\times\cosh\{M_{215}(n,V_0)\}\cos\{N_{215}(n,V_0)\}$$

$$+2\left(\{I_{215}(n,V_0)\}^2-\{I_{225}(n,V_0)\}^2\right)\{M'_{215}(n,V_0)$$

$$\times\sinh\{M_{215}(n,V_0)\}\cos\{N_{215}(n,V_0)\}$$

$$-N'_{215}(n,V_0)\cosh\{M_{215}(n,V_0)\}\sin\{N_{215}(n,V_0)\}\}$$

$$+\frac{1}{12}\left(\frac{15\{I^2_{225}(n,V_0)\}I'_{225}(n,V_0)}{I_{215}(n,V_0)}-\frac{5\{I_{225}(n,V_0)\}^3I'_{215}(n,V_0)}{I^2_{215}(n,V_0)}\right.$$

$$+\frac{15\{I^2_{215}(n,V_0)\}I'_{215}(n,V_0)}{I_{225}(n,V_0)}-\frac{5\{I_{215}(n,V_0)\}^3I'_{225}(n,V_0)}{I^2_{225}(n,V_0)}$$

$$\left.-\{34I'_{225}(n,V_0)I_{215}(n,V_0)\}-34I_{225}(n,V_0)I'_{215}(n,V_0)\right)$$

$$\times\sinh\{M_{215}(n,V_0)\}\sin\{N_{215}(n,V_0)\}$$

$$+1/12\left(\frac{5\{I_{225}(n,V_0)\}^3}{I_{215}(n,V_0)}+\frac{5\{I_{215}(n,V_0)\}^3}{I_{225}(n,V_0)}-\{34I_{225}(n,V_0)I_{215}(n,V_0)\}\right)$$

$$\times\{M'_{215}(n,V_0)\cosh\{M_{215}(n,V_0)\}\sin\{N_{215}(n,V_0)\}+N'_{215}(n,V_0)$$

$$\left.\left.\times\sinh\{M_{215}(n,V_0)\}\cos\{N_{215}(n,V_0)\}\}\right]\right],$$

$$M'_{215}(n,V_0)=I'_{215}(n,V_0)\left[a_0-\Delta_{21}\right],$$

$$I'_{215}(V_0,n)=\frac{m_{c2}\beta'_{250}\left(V_0-\overline{V}_0,\lambda,E_{g02},\Delta_2\right)}{-\hbar^2I_{215}(V_0,n)},$$

$$I'_{225}(V_0,n)=\frac{m_{c1}\beta'_{150}\left(V_0,\lambda,E_{g01},\Delta_1\right)}{\hbar^2I_{225}(V_0,n)},$$

$$I_{215}(V_0,n)=\left[\frac{2eB}{\hbar}\left(n+\frac{1}{2}\right)-\frac{2m_{c2}}{\hbar^2}\beta_{250}\left(V_0-\overline{V}_0,\lambda,E_{g02},\Delta_2\right)\right]^{\frac{1}{2}},$$

$$N_{215}(V_0,n)=I_{225}(V_0,n)\left[b_0-\Delta_{21}\right],$$

$$I_{225}(V_0,n)=\left[\frac{2m_{c1}\beta_{150}\left(V_0,\lambda,E_{g01},\Delta_1\right)}{\hbar^2}-\frac{2eB}{\hbar}\left(n+\frac{1}{2}\right)\right]^{\frac{1}{2}},$$

$$Z'_{215}(V_0,n)$$
$$=\left[\frac{I'_{212}(V_0,n)}{I_{225}(V_0,n)}-\frac{I_{215}(V_0,n)I'_{225}(V_0,n)}{I^2_{225}(V_0,n)}-\frac{I'_{225}(V_0,n)}{I_{215}(V_0,n)}-\frac{I'_{225}(V_0,n)I'_{215}(V_0,n)}{I^2_{215}(V_0,n)}\right]$$

For perturbed two-band model of Kane, the forms of the electron concentration and field emitted current density remain same where

$$I_{215}(E,n)=\left[\frac{2eB}{\hbar}\left(n+\frac{1}{2}\right)-\frac{2m_{c2}}{\hbar^2}\omega_{250}\left(E-\overline{V}_0,\lambda,E_{g02}\right)\right]^{\frac{1}{2}},$$

$$I_{225}(E,n)=\left[\frac{2m_{c1}\omega_{150}\left(E,\lambda,E_{g01}\right)}{\hbar^2}-\frac{2eB}{\hbar}\left(n+\frac{1}{2}\right)\right]^{\frac{1}{2}},$$

$$I'_{215}(V_0,n)=\frac{m_{c2}\omega'_{250}\left(V_0-\overline{V}_0,\lambda,E_{g02}\right)}{\hbar^2 I_{215}(V_0,n)},$$

$$I'_{225}(V_0,n)=\left[\frac{m_{c1}\omega'_{150}\left(V_0,\lambda,E_{g01}\right)}{\hbar^2 I_{225}(V_0,n)}\right]$$

For perturbed parabolic energy bands, the forms of the electron concentration and field emitted current density remain same where

$$I_{215}(E,n)=\left[\frac{2eB}{\hbar}\left(n+\frac{1}{2}\right)-\frac{2m_{c2}}{\hbar^2}\rho_{250}\left(E-\overline{V}_0,\lambda,E_{g02}\right)\right]^{\frac{1}{2}},$$

$$I_{225}(E,n)=\left[\frac{2m_{c1}\rho_{150}\left(E,\lambda,E_{g01}\right)}{\hbar^2}-\frac{2eB}{\hbar}\left(n+\frac{1}{2}\right)\right]^{\frac{1}{2}},$$

$$I'_{215}(V_0,n)=\frac{m_{c2}\rho'_{250}\left(V_0-\overline{V}_0,\lambda,E_{g02}\right)}{\hbar^2 I_{215}(V_0,n)}$$

and

$$I'_{225}(V_0,n)=\left[\frac{m_{c1}\rho'_{150}\left(V_0,\lambda,E_{g01}\right)}{\hbar^2 I_{225}(V_0,n)}\right]$$

5.2.6 Field Emission from Quantum Wire Superlattices of III–V Semiconductors with Graded Interfaces

The dispersion relation in accordance with the perturbed three-band model of Kane, in this case is given by

$$k_x^2 = \omega_{225}(E,\lambda,n_y,n_z) \tag{5.60}$$

where

$$\omega_{225}\left(E,\lambda,n_y,n_z\right) = \left[\frac{1}{L_0^2}\left[\cos^{-1}\frac{1}{2}f_{135}\left(E,\lambda,n_y,n_z\right)\right]^2 - H\left(n_y,n_z\right)\right],$$

$$\begin{aligned}
&f_{135}\left(E,\lambda,n_y,n_z\right)\\
&= \Bigg[2\cosh\left\{M_{315}\left(n_y,n_z,E\right)\right\}\cos\left\{N_{315}\left(n_y,n_z,E\right)\right\}\\
&\quad + Z_{315}\left(n_y,n_z,E\right)\sinh\left\{M_{315}\left(n_y,n_z,E\right)\right\}\\
&\quad \times\sin\left\{N_{315}\left(n_y,n_z,E\right)\right\} + \Delta_{21}\Bigg[\left(\frac{\left\{I_{315}\left(n_y,n_z,E\right)\right\}^2}{I_{325}\left(n_y,n_z,E\right)} - 3I_{325}\left(n_y,n_z,E\right)\right)\\
&\quad \times\cosh\left\{M_{315}\left(n_y,n_z,E\right)\right\}\sin\left\{N_{315}\left(n_y,n_z,E\right)\right\}\\
&\quad + \left(3I_{315}\left(n_y,n_z,E\right) - \frac{\left\{I_{325}\left(n_y,n_z,E\right)\right\}^2}{I_{315}\left(n_y,n_z,E\right)}\right)\\
&\quad \times\sinh\left\{M_{315}\left(n_y,n_z,E\right)\right\}\cos\left\{N_{315}\left(n_y,n_z,E\right)\right\}\Bigg]\\
&\quad + \Delta_{21}\Bigg[2\left(\left\{I_{315}\left(n_y,n_z,E\right)\right\}^2 - \left\{I_{325}\left(n_y,n_z,E\right)\right\}^2\right)\\
&\quad \times\cosh\left\{M_{315}\left(n_y,n_z,E\right)\right\}\cos\left\{N_{315}\left(n_y,n_z,E\right)\right\}\\
&\quad + \frac{1}{12}\left(\frac{5\left\{I_{325}\left(n_y,n_z,E\right)\right\}^3}{I_{315}\left(n_y,n_z,E\right)} + \frac{5\left\{I_{315}\left(n_y,n_z,E\right)\right\}^3}{I_{325}\left(n_y,n_z,E\right)}\right.\\
&\quad \left. - \left\{34I_{325}\left(n_y,n_z,E\right)I_{315}\left(n_y,n_z,E\right)\right\}\right)\\
&\quad \times\sinh\left\{M_{315}\left(n_y,n_z,E\right)\right\}\sin\left\{N_{315}\left(n_y,n_z,E\right)\right\}\Bigg]\Bigg]
\end{aligned}$$

$$Z_{315}\left(n_y, n_z, E\right) \equiv \left[\frac{I_{315}\left(n_y, n_z, E\right)}{I_{325}\left(n_y, n_z, E\right)} - \frac{I_{325}\left(n_y, n_z, E\right)}{I_{315}\left(n_y, n_z, E\right)}\right],$$

$$M_{315}\left(n_y, n_z, E\right) = I_{315}\left(n_y, n_z, E\right)\left[a_0 - \Delta_{21}\right],$$

$$I_{315}\left(n_y, n_z, E\right) = \left[-\frac{2m_{c2}}{\hbar^2}\beta_{250}(E - \overline{V_0}, \lambda, E_{g_{02}}, \Delta_2) + H(n_y, n_z)\right]^{1/2},$$

$$N_{315}\left(n_y, n_z, E\right) = I_{325}\left(n_y, n_z, E\right)\left[b_0 - \Delta_{21}\right]$$

and

$$I_{325}\left(n_y, n_z, E\right) \equiv \left[\frac{2m_{c1}}{\hbar^2}\beta_{150}(E, \lambda, E_{g_{01}}, \Delta_1) - H(n_y, n_z)\right]^{1/2}.$$

The electron concentration per unit length is given by

$$n_0 = \frac{2g_v}{\pi}\sum_{n_y=1}^{n_{y\max}}\sum_{n_z=1}^{n_{z\max}}\left[\overline{Q}_{29}\left(E_{\text{FQWGISL}}, \lambda, n_y, n_z\right) + \overline{Q}_{30}\left(E_{\text{FQWGISL}}, \lambda, n_y, n_z\right)\right] \tag{5.61}$$

where $\overline{Q}_{29}(E_{\text{FQWGISL}}, \lambda, n_y, n_z) = [\sqrt{\omega_{225}(E_{\text{FQWGISL}}, \lambda, n_y, n_z)}]$,

$$\overline{Q}_{30}\left(E_{\text{FQWGISL}}, \lambda, n_y, n_z\right) = \sum_{R=1}^{R=R_0} Z\left(R_{\text{FQWGISL}}\right)\left[\overline{Q}_{29}\left(E_{\text{FQWGISL}}, \lambda, n_y, n_z\right)\right],$$

$Z(R_{\text{FQWGSL}}) = 2(k_BT)^{2R}(1 - 2^{1-2R})\xi(2R)\frac{\partial^{2R}}{\partial E_{\text{FQWGSL}}}$ and E_{FQWGSL} is the Fermi energy in the present case.

The field emitted current can be written as

$$I = \frac{eg_vk_BT}{\pi\hbar}\sum_{n_y=1}^{n_{y\max}}\sum_{n_z=1}^{n_{z\max}} F_0\left(\overline{\eta}_{20}\right)\exp\left(-\overline{\theta}_{20}\right) \tag{5.62}$$

$\overline{\eta}_{20} = \frac{E_{\text{FQWGISL}} - \overline{E}_{20}}{k_BT}$, $\overline{E}_{20}$ is the root of the equation

$$\omega_{225}(\overline{E}_{20}, \lambda, n_y, n_z) = 0 \tag{5.63}$$

$$\overline{\theta}_{20} = \frac{4\left[\omega_{225}\left(V_0, \lambda, n_y, n_z\right)\right]^{3/2}}{3eF_{sx}\omega'_{225}\left(V_0, \lambda, n_y, n_z\right)},$$

$$\omega'_{225}\left(V_0, \lambda, n_y, n_z\right) = \frac{2f'_{135}\left(V_0, \lambda, n_y, n_z\right)\left[\cos^{-1}\left\{\frac{1}{2}f_{135}\left(V_0, \lambda, n_y, n_z\right)\right\}\right]}{\sqrt{4 - f^2_{135}\left(V_0, \lambda, n_y, n_z\right)}},$$

$$
\begin{aligned}
& f'_{135}\left(V_0,\lambda,n_y,n_z\right) \\
& = \Bigg[2M'_{315}\left(n_y,n_z,V_0\right)\sinh\left\{M_{315}\left(n_y,n_z,V_0\right)\right\} \\
& \times \cos\left\{N_{315}\left(n_y,n_z,V_0\right)\right\} + Z_{315}\left(n_y,n_z,V_0\right)M'_{315}\left(n_y,n_z,V_0\right) \\
& \cdot \cosh\left\{M_{315}\left(n_y,n_z,V_0\right)\right\}\cdot\sin\left\{N_{315}\left(n_y,n_z,V_0\right)\right\} \\
& - 2N'_{315}\left(n_y,n_z,V_0\right)\sin\left\{N_{315}\left(n_y,n_z,V_0\right)\right\}\cosh\left\{M_{315}\left(n_y,n_z,V_0\right)\right\} \\
& + Z'_{315}\left(n_y,n_z,V_0\right)\sinh\left\{M_{315}\left(n_y,n_z,V_0\right)\right\}\sin\left\{N_{315}\left(n_y,n_z,V_0\right)\right\} \\
& + Z_{315}(n_y,n_z,V_0)N'_{315}(n_y,n_z,V_0)\cos\{N_{315}(n_y,n_z,V_0)\}\sinh\{M_{315}(n_y,n_z,V_0)\} \\
& + \Delta_{21}\Bigg[\Bigg(\frac{\{2I_{315}(n_y,n_z,V_0)I'_{315}(n_y,n_z,V_0)\}}{I_{325}(n_y,n_z,V_0)} - \frac{\{I^2_{315}(n_y,n_z,V_0)I'_{325}(n_y,n_z,V_0)\}}{I^2_{325}(n_y,n_z,V_0)} \\
& - 3I'_{325}(n_y,n_z,V_0)\Bigg) \\
& \cdot \cosh\left\{M_{315}\left(n_y,n_z,V_0\right)\right\}\sin\left\{N_{315}\left(n_y,n_z,V_0\right)\right\} \\
& + \left(-3I_{325}\left(n_y,n_z,V_0\right) + \frac{\{I^2_{315}\left(n_y,n_z,V_0\right)\}}{I_{325}\left(n_y,n_z,V_0\right)}\right)\{M'_{315}\left(n_y,n_z,V_0\right) \\
& \times \sin h\left\{M_{315}\left(n_y,n_z,V_0\right)\right\}\sin\left\{N_{315}\left(n_y,n_z,V_0\right)\right\} \\
& + \{N'_{315}\left(n_y,n_z,V_0\right)\cosh\left\{M_{315}\left(n_y,n_z,V_0\right)\right\}\cos\left\{N_{315}\left(n_y,n_z,V_0\right)\right\} \\
& + \Bigg(\frac{-\{2I_{325}(n_y,n_z,V_0)I'_{325}(n_y,n_z,V_0)\}}{I_{315}(n_y,n_z,V_0)} + \frac{\{I^2_{325}(n_y,n_z,V_0)I'_{315}(n_y,n_z,V_0)\}}{I^2_{315}(n_y,n_z,V_0)} \\
& + 3I'_{315}\left(n_y,n_z,V_0\right)\Bigg)\sinh\left\{M_{315}\left(n_y,n_z,V_0\right)\right\}\cdot\cos\left\{N_{315}\left(n_y,n_z,V_0\right)\right\} \\
& + \left(+3I_{315}\left(n_y,n_z,V_0\right) - \frac{\{I_{325}\left(n_y,n_z,V_0\right)\}^2}{I_{315}\left(n_y,n_z,V_0\right)}\right)\{M'_{315}\left(n_y,n_z,V_0\right) \\
& \times \cosh\left\{M_{315}\left(n_y,n_z,V_0\right)\right\}\cos\left\{N_{315}\left(n_y,n_z,V_0\right)\right\} \\
& - N'_{315}\left(n_y,n_z,V_0\right)\sin\left\{N_{315}\left(n_y,n_z,V_0\right)\right\}\sinh\left\{M_{315}\left(n_y,n_z,V_0\right)\right\}\}\Bigg] \\
& + \Delta_{21}\Bigg[4(\{I_{315}(n_y,n_z,V_0)I'_{315}(n_y,n_z,V_0)\} - \{I_{325}(n_y,n_z,V_0)I'_{325}(n_y,n_z,V_0)\}) \\
& \times \cosh\left\{M_{315}\left(n_y,n_z,V_0\right)\right\}\cos\left\{N_{315}\left(n_y,n_z,V_0\right)\right\} \\
& + 2\left(\left\{I_{315}\left(n_y,n_z,V_0\right)\right\}^2 - \left\{I_{325}\left(n_y,n_z,V_0\right)\right\}^2\right)\{M'_{315}\left(n_y,n_z,V_0\right)
\end{aligned}
$$

$$\times \sinh\left\{M_{315}\left(n_y,n_z,V_0\right)\right\}\cos\left\{N_{315}\left(n_y,n_z,V_0\right)\right\}$$

$$- N'_{315}\left(n_y,n_z,V_0\right)\cosh\left\{M_{315}\left(n_y,n_z,V_0\right)\right\}\sin\left\{N_{315}\left(n_y,n_z,V_0\right)\right\}\}$$

$$+ \frac{1}{12}\left(\frac{15\{I_{325}^2(n_y,n_z,V_0)\}I'_{325}(n_y,n_z,V_0)}{I_{315}(n,V_0)} - \frac{5\{I_{325}(n_y,n_z,V_0)\}^3 I'_{315}(n_y,n_z,V_0)}{I_{315}^2(n_y,n_z,V_0)}\right.$$

$$+ \frac{15\left\{I_{315}^2\left(n_y,n_z,V_0\right)\right\}I'_{315}\left(n_y,n_z,V_0\right)}{I_{325}\left(n_y,n_z,V_0\right)}$$

$$- \frac{5\left\{I_{315}\left(n_y,n_z,V_0\right)\right\}^3 I'_{325}\left(n_y,n_z,V_0\right)}{I_{325}^2\left(n_y,n_z,V_0\right)}$$

$$\left. - \left\{34I'_{325}\left(n_y,n_z,V_0\right)I_{315}\left(n_y,n_z,V_0\right)\right\} - 34I_{325}\left(n_y,n_z,V_0\right)I'_{315}\left(n_y,n_z,V_0\right)\right)$$

$$\times \sinh\left\{M_{315}\left(n_y,n_z,V_0\right)\right\}\sin\left\{N_{315}\left(n_y,n_z,V_0\right)\right\}$$

$$+ \left(\frac{5\left\{I_{325}\left(n_y,n_z,V_0\right)\right\}^3}{I_{315}\left(n_y,n_z,V_0\right)} + \frac{5\left\{I_{315}\left(n_y,n_z,V_0\right)\right\}^3}{I_{325}\left(n_y,n_z,V_0\right)}\right.$$

$$\left. - \left\{34I_{325}\left(n_y,n_z,V_0\right)I_{315}\left(n_y,n_z,V_0\right)\right\}\right)$$

$$\times \left\{M'_{315}\left(n_y,n_z,V_0\right)\cosh\left\{M_{315}\left(n_y,n_z,V_0\right)\right\}\sin\left\{N_{315}\left(n_y,n_z,V_0\right)\right\}\right.$$

$$\left. + \; N'_{315}\left(n_y,n_z,V_0\right)\sinh\left\{M_{315}\left(n_y,n_z,V_0\right)\right\}\cos\left\{N_{315}\left(n_y,n_z,V_0\right)\right\}\}\right]\Bigg],$$

$$M'_{315}\left(n_y,n_z,V_0\right) = I'_{315}\left(n_y,n_z,V_0\right)\left[a_0 - \Delta_{21}\right],$$

$$I'_{315}\left(V_0,n_y,n_z\right) = \frac{m_{c2}\beta'_{250}\left(V_0 - \overline{V}_0,\lambda,E_{g02},\Delta_2\right)}{\hbar^2 I_{315}\left(V_0,n_y,n_z\right)},$$

$$N'_{315}\left(n_y,n_z,V_0\right) = I'_{325}\left(n_y,n_z,V_0\right)\left[b_0 - \Delta_{21}\right],$$

$$I'_{325}\left(V_0,n_y,n_z\right) = \frac{m_{c1}\beta'_{150}\left(V_0,\lambda,E_{g02},\Delta_2,n_y,n_z\right)}{\hbar^2 I_{325}\left(V_0,n_y,n_z\right)},$$

$$I_{315}\left(n_y,n_z,V_0\right) = \left[-\frac{2m_{c2}}{\hbar^2}\beta_{250}(V_0 - \overline{V_0},\lambda,E_{g02},\Delta_2) + H(n_y,n_z)\right]^{1/2},$$

$$N_{315}\left(n_y,n_z,V_0\right) = I_{325}\left(n_y,n_z,V_0\right)\left[b_0 - \Delta_{21}\right],$$

$$I_{325}\left(n_y,n_z,V_0\right) = \left[\frac{2m_{c1}}{\hbar^2}\beta_{150}(V_0,\lambda,E_{g01},\Delta_1) - H(n_y,n_z)\right]^{1/2} \quad \text{and}$$

$$Z'_{315}\left(V_0,n_y,n_z\right)$$
$$=\left[\frac{I'_{315}\left(V_0,n_y,n_z\right)}{I_{325}\left(V_0,n_y,n_z\right)}-\frac{I'_{325}\left(V_0,n_y,n_z\right)}{I_{315}\left(V_0,n_y,n_z\right)}-\frac{I_{315}\left(V_0,n_y,n_z\right)I'_{325}\left(V_0,n_y,n_z\right)}{I^2_{325}\left(V_0,n_y,n_z\right)}\right.$$
$$\left.+\frac{I_{325}\left(V_0,n_y,n_z\right)I'_{315}\left(V_0,n_y,n_z\right)}{I^2_{315}\left(V_0,n_y,n_z\right)}\right]$$

For perturbed two-band model of Kane, the form of electron concentration per unit length and the field emitted current remain same where

$$I_{315}\left(n_y,n_z,E\right)=\left[H(n_y,n_z)-\frac{2m_{c2}}{\hbar^2}\omega_{250}(E-\overline{V}_0,\lambda,E_{g02})\right]^{1/2},$$

$$I'_{315}\left(V_0,n_y,n_z\right)=\frac{m_{c2}\omega'_{250}\left(V_0-\overline{V}_0,\lambda,E_{g02}\right)}{\hbar^2 I_{315}\left(V_0,n_y,n_z\right)},$$

$$I_{325}\left(n_y,n_z,V_0\right)=\left[-H(n_y,n_z)+\frac{2m_{c1}}{\hbar^2}\omega_{150}(V_0,\lambda,E_{g01})\right]^{1/2},$$

and

$$I'_{325}\left(V_0,n_y,n_z\right)=\frac{m_{c1}\omega'_{150}\left(V_0,\lambda,E_{g01}\right)}{\hbar^2 I_{325}\left(V_0,n_y,n_z\right)}.$$

For perturbed parabolic energy bands, the form of electron concentration per unit length and the field emitted current remain same where

$$I_{315}\left(n_y,n_z,E\right)=\left[H(n_y,n_z)-\frac{2m_{c2}}{\hbar^2}\rho_{250}(E-\overline{V}_0,\lambda,E_{g02})\right]^{1/2},$$

$$I'_{315}\left(V_0,n_y,n_z\right)=\frac{m_{c2}\rho'_{250}\left(V_0-\overline{V}_0,\lambda,E_{g02}\right)}{\hbar^2 I_{315}\left(V_0,n_y,n_z\right)},$$

$$I_{325}\left(n_y,n_z,V_0\right)=\left[-H(n_y,n_z)+\frac{2m_{c1}}{\hbar^2}\rho_{150}(V_0,\lambda,E_{g01})\right]^{1/2},$$

and

$$I'_{325}\left(V_0,n_y,n_z\right)=\frac{m_{c1}\rho'_{150}\left(V_0,\lambda,E_{g01}\right)}{\hbar^2 I_{325}\left(V_0,n_y,n_z\right)}.$$

5.3 Result and Discussions

Using the appropriate equations, we have plotted the field emitted current density from n-InSb under magnetic quantization as functions of $1/B$, concentration, wavelength, intensity, and electric field as shown in Figs. 5.1–5.5 in accordance with both three- and two-band models of Kane. Figures 5.6–5.8 represent the field emitted current from quantum wires of n-InSb in accordance with the three- and two-band models of Kane as functions of film thickness, concentration, and electric field, respectively. Figures 5.9 and 5.10 exhibit the field emitted current density from GaAs/AlGaAs superlattices with graded interfaces and also its effective mass counterpart under magnetic quantization as functions of $1/B$ and carrier concentration, respectively. Figures 5.11 and 5.12 exhibit the field emitted current from GaAs/AlGaAs quantum wire superlattices with graded interfaces and also its effective mass counterpart as functions of film thickness and concentration, respectively.

From Fig. 5.1, we observe that the field emitted current density from n-InSb under magnetic quantization exhibits oscillations with $1/B$, the background physics of which has already been explained. When compared with that of the corresponding Fig. 3.2 as given in Chap. 3, it appears that the effect of light waves enhances the field emitted current density to a very large extent, almost about 1,000 times. We note that although the generation Landau subbands remains same within the given bandwidth, nature of the orientation of the curves is radically different. It appears that when the wavelength of the incident light waves stays in the regime of radio wave zone and whose intensity lies within that of the solar intensity at the earth surface, the degeneracy increases. This implies that in the presence of radio waves, the Fermi energy increases leading to a decrease in the variation of the field emitted current density per subband. This was not in the case with the corresponding

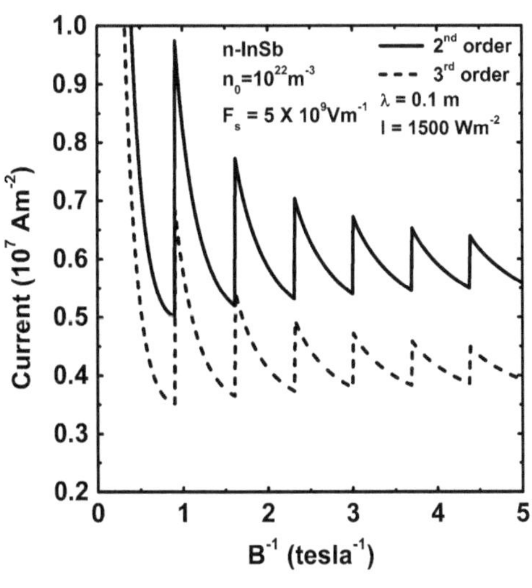

Fig. 5.1 Plot of the field emitted current density as a function of inversing magnetic field for n-InSb in the presence of light waves for both three- and two-band models of Kane

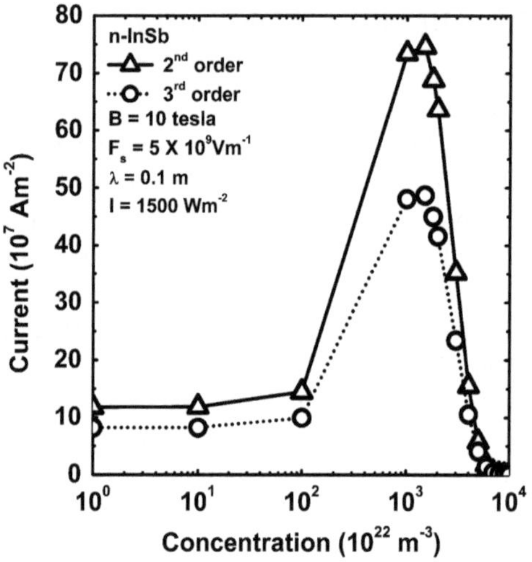

Fig. 5.2 Plot of the field emitted current density as a function of carrier concentration for n-InSb in the presence of light waves and external magnetic field for both three- and two-band models of Kane

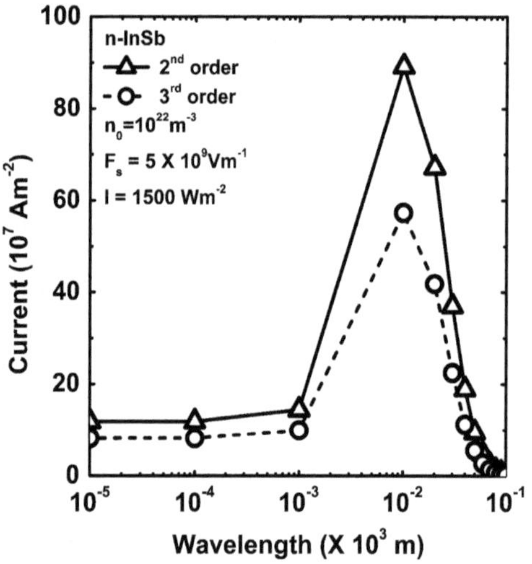

Fig. 5.3 Plot of the field emitted current density as a function of wavelength for n-InSb in the presence of light waves and external magnetic field for both three- and two-band models of Kane

figure in Chap. 3. From Fig. 5.2, exhibits the variation of the field emitted current density at magnetic quantum limit as function of carrier concentration for the said case. We observe a peak in the current density near the value of the electron concentration $10^{25}\,m^{-3}$at low temperatures for both the models. The field emitted current density remains almost constant below the degeneracy of about $10^{24}\,m^{-3}$. Besides, the said peak may alter its position with the variation of both wavelength and intensity. Similar nature of the dependence of the field emitted current density on the wavelength has been shown in Fig. 5.3. We note that the peak happens in the radio wave zone for an intensity of $1,500\,Wm^{-2}$. In Fig. 5.4, we observe an almost constant field emitted current density with respect to the light intensity up

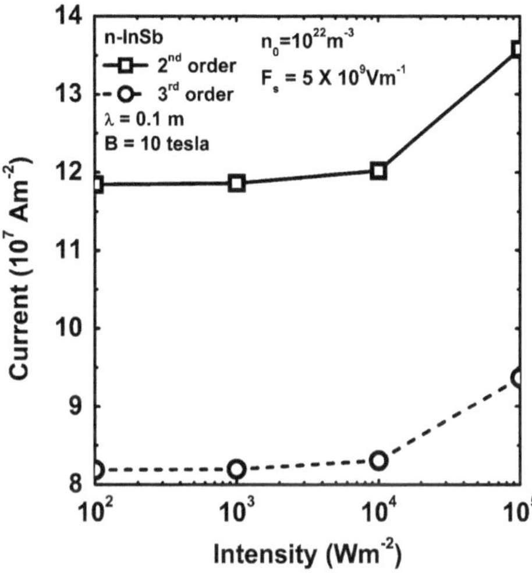

Fig. 5.4 Plot of the field emitted current density as a function of intensity for n-InSb in the presence of light waves and external magnetic field for both three- and two-band models of Kane

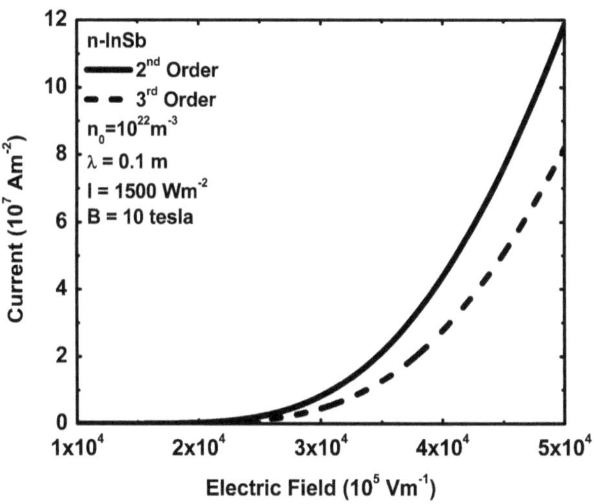

Fig. 5.5 Plot of the field emitted current density as a function of electric field for n-InSb in the presence of light waves and external magnetic field for both three- and two-band models of Kane

to 10^4 Wm^{-2}. As the intensity level increases, the current density start increasing initially slowly, while beyond 10^4 Wm^{-2} exhibits very large rise. The effect of the electric field on the field emitted current density has been plotted in Fig. 5.5, and it appears that an application of radio waves increases the cut-in field in n-InSb under magnetic quantization as exhibited in the same figure.

Composite oscillations in the field emitted current as function of film thickness in the presence of light waves has been exhibited for quantum wires of n-InSb in Fig. 5.6. Few tenths of microamperes of current has been observed in the same figure

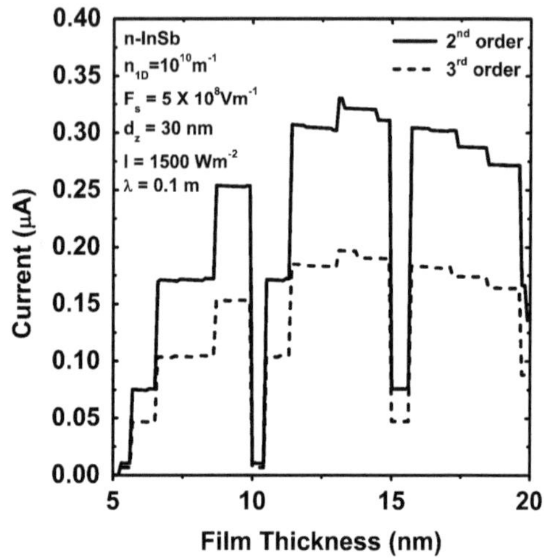

Fig. 5.6 Plot of the field emitted current as a function of film thickness from quantum wires of n-InSb in the presence of light waves for both three- and two-band models of Kane

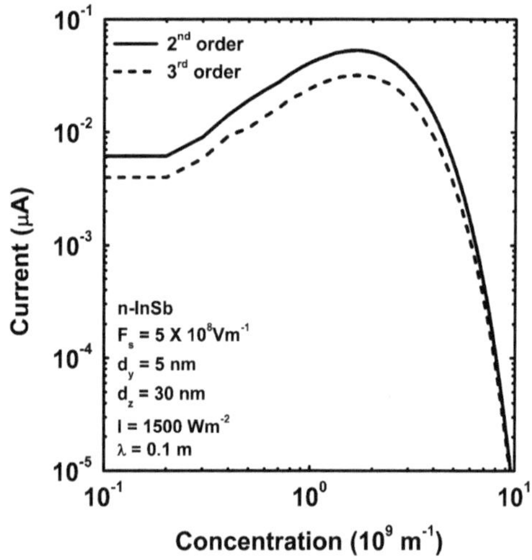

Fig. 5.7 Plot of the field emitted current as a function of electron concentration from quantum wires of n-InSb in the presence of light waves for both three- and two-band models of Kane

for 0.1 m wavelength and 1,500 Wm^{-2} for a carrier concentration of $10^{10}\,\mathrm{m}^{-1}$. In this case, we note that there exist both increment and decrement in the field emitted current, the reasons of which have already been stated in Chap. 1 and the curves can be compared with the corresponding figures there in. The influence of electron concentration on the field emitted current in this case has been shown in Fig. 5.7 for the quantum limit for the two- and three-band models of Kane in the presence of light waves. We observe that the current rises to a peak near the value of about $10^9\,\mathrm{m}^{-1}$ after which the field emitted current falls sharply. From Fig. 5.8, we note that the field emitted current from quantum wires of n-InSb increases with increasing surface electric field in the electric quantum limit. The cut-in fields in

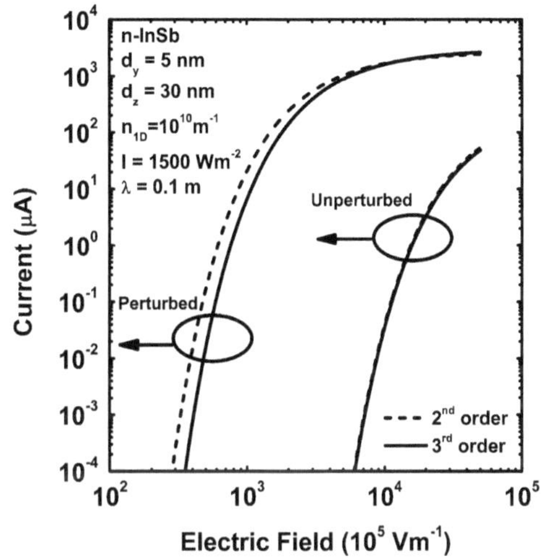

Fig. 5.8 Plot of the field emitted current as a function of electric field from quantum wires of n-InSb in the presence of light waves for both three- and two-band models of Kane

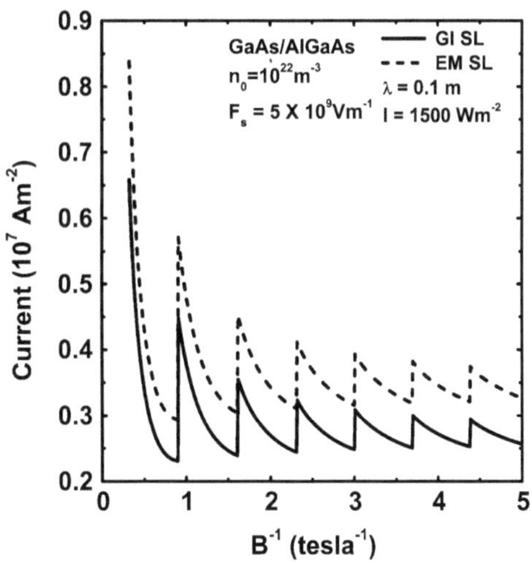

Fig. 5.9 Plot of the field emitted current as a function of inverse magnetic field from GaAs/AlGaAs effective mass superlattices (EM SL) and superlattices with graded interfaces (GI SL) in the presence of light waves, the constituent materials of which obey the unperturbed two-band model of Kane

the present perturbed case are near $10^7 Vm^{-1}$, which can be compared with the unperturbed one.

The influence of light waves increases the magnitude of the field emitted current density in both effective mass superlattices and superlattices with graded interface, respectively, which appears from Figs. 5.9 and 5.10 in the presence of quantizing magnetic field as compared with the corresponding case in Chap. 4. With the increase in the electron concentration, the field emitted current density increases with nonperiodic manner. In the case of quantum wire superlattices with graded interfaces and quantum wire effective mass superlattices, the drastic reduction of field emitted current to an order of nanoamperes is due to the high increase in the

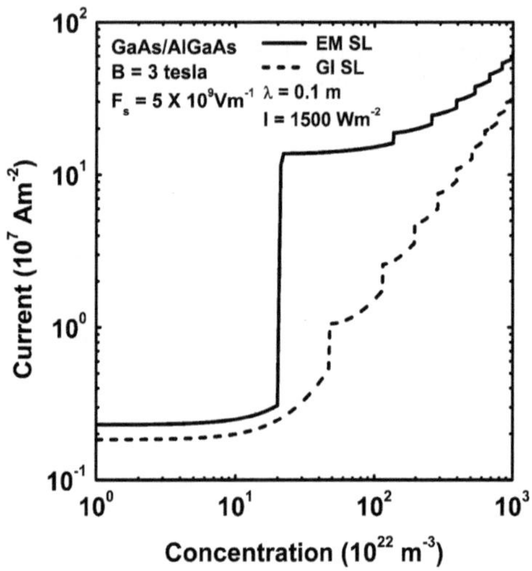

Fig. 5.10 Plot of the field emitted current as a function of carrier concentration in the presence of a magnetic field from GaAs/AlGaAs effective mass superlattices (EM SL) and superlattices with graded interfaces (GI SL) in the presence of light waves, the constituent materials of which obey the unperturbed two-band model of Kane

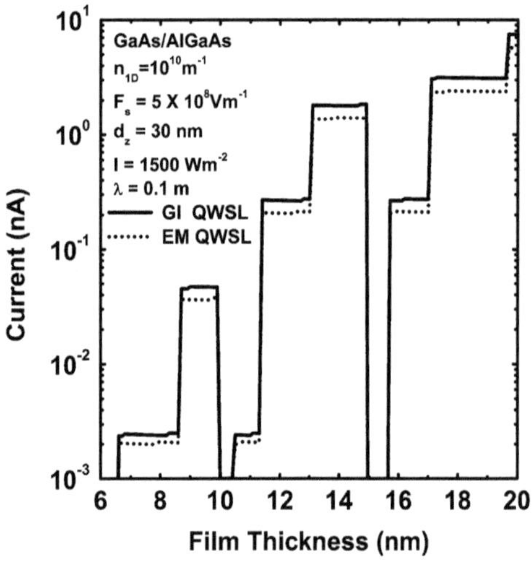

Fig. 5.11 Plot of the field emitted current as a function of film thickness from GaAs/AlGaAs quantum wire effective mass superlattices (EM QWSL) and quantum wire superlattices with graded interfaces (GI QWSL) in the presence of light waves, the constituent materials of which obey the unperturbed two-band model of Kane

Fermi energy in Fig. 5.11. Incidentally, for low values of electron concentration per unit length, from Fig. 5.12 it appears that the field emitted current increases slowly and sharply falls off above a carrier degeneracy of $10^9 \mathrm{m}^{-1}$ for both types of quantum wire superlattices as discussed in this chapter. It may be noted that although ternary and quaternary materials are known, primarily known as optoelectronic materials, their conduction electron energy band models in the absence of any field is the same as that of III–V semiconductors whose electron energy spectrum obey the three and two and models of Kane. All the results of this chapter are also equally valid for ternary and quaternary materials and only the numerical values will be different. For the purpose of condensed presentation, the specific carrier statistics for a specific

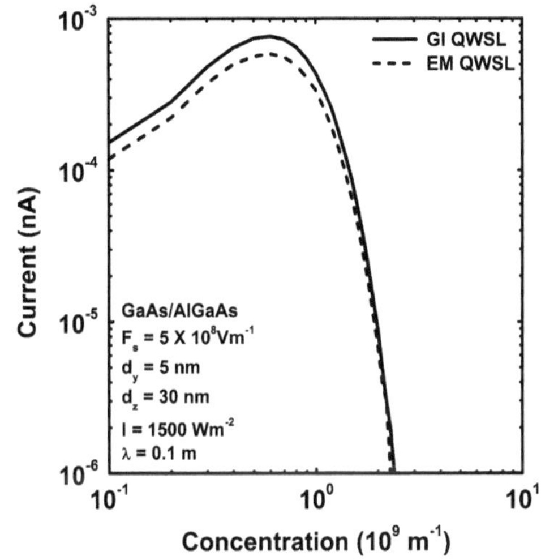

Fig. 5.12 Plot of the field emitted current as a function of carrier concentration per unit length from GaAs/AlGaAs quantum wire effective mass superlattices (EM QWSL) and quantum wire superlattices with graded interfaces (GI QWSL) in the presence of light waves, the constituent materials of which obey the unperturbed two-band model of Kane

system having a particular electron energy spectrum and the corresponding field emitted current have been presented in Table 5.1.

5.4 Open Research Problems

All the following problems should be investigated in the presence of external photoexcitation which changes the band structure in a fundamental way together with the proper inclusion of the variations of work function in appropriate cases.

(R5.1) (a) Investigate the FNFE from all the bulk semiconductors and the corresponding superlattices whose respective dispersion relations of the carriers are given in this chapter by converting the summations over the quantum numbers to the corresponding integrations by including the uniqueness conditions in the appropriate cases and considering the effect of image force in the subsequent study in each case.

(b) Investigate the FNFE for bulk specimens of all the semiconductors whose unperturbed carrier energy spectra are defined in Chap. 1 in the presence of arbitrarily oriented photoexcitation by incorporating the appropriate changes.

(R5.2) Investigate the FNFE in the presence of an arbitrarily oriented nonquantizing nonuniform electric field and photoexcitation respectively for all the cases of R5.1.

(R5.3) Investigate the FNFE in the presence of arbitrarily oriented nonquantizing alternating electric field and photoexcitation respectively for all the cases of R5.1.

Table 5.1 The carrier statistics and the Fowler–Nordheim field emission from quantum confined optoelectronic materials in the presence of light waves

The carrier statistics	The Fowler–Nordheim field emission
(1) Perturbed three- and two-band models of Kane and perturbed parabolic energy bands under magnetic quantization:	$J = \frac{e^2 B k_B T g_v}{2\pi^2 \hbar^2} \sum_{n=0}^{n_{max}} F_0(\bar{\eta}_{51,B}) \exp(-\bar{\beta}_{51,B})$ (5.13)
$n_0 = \frac{g_v e B \sqrt{2m_c}}{\pi^2 \hbar^2} \sum_{n=0}^{n_{max}} [\overline{M}_{51}(E_{FB}, \lambda, n) + \overline{N}_{51}(E_{FB}, \lambda, n)]$ (5.10)	
$n_0 = \frac{g_v e B \sqrt{2m_c}}{\pi^2 \hbar^2} \sum_{n=0}^{n_{max}} [\overline{M}_{52}(E_{FB}, \lambda, n) + \overline{N}_{52}(E_{FB}, \lambda, n)]$ (5.11)	$J = \frac{e^2 B k_B T g_v}{2\pi^2 \hbar^2} \sum_{n=0}^{n_{max}} F_0(\bar{\eta}_{52,B}) \exp(-\bar{\beta}_{52,B})$ (5.14)
$n_0 = \frac{g_v e B \sqrt{2m_c}}{\pi^2 \hbar^2} \sum_{n=0}^{n_{max}} [\overline{M}_{53}(E_{FB}, \lambda, n) + \overline{N}_{53}(E_{FB}, \lambda, n)]$ (5.12)	$J = \frac{e^2 B k_B T g_v}{2\pi^2 \hbar^2} \sum_{n=0}^{n_{max}} F_0(\bar{\eta}_{53,B}) \exp(-\bar{\beta}_{53,B})$ (5.15)
(2) Perturbed three- and two-band models of Kane and perturbed parabolic energy bands under 2D quantization:	$I = \frac{e g_v k_B T}{\pi \hbar} \sum_{n_x=1}^{n_{x max}} \sum_{n_y=1}^{n_{y max}} F_0(\bar{\eta}_{51,1D}) \exp(-\bar{\beta}_{51,1D})$ (5.25)
$n_0 = \frac{\sqrt{2m_c} 2 g_v}{\hbar \pi} \sum_{n_x=1}^{n_{x max}} \sum_{n_y=1}^{n_{y max}} [S_{51}(E_{F1D}, \lambda, n_x, n_y) + T_{51}(E_{F1D}, \lambda, n_x, n_y)]$ (5.22)	
$n_0 = \frac{\sqrt{2m_c} 2 g_v}{\hbar \pi} \sum_{n_x=1}^{n_{x max}} \sum_{n_y=1}^{n_{y max}} [S_{52}(E_{F1D}, \lambda, n_x, n_y) + T_{52}(E_{F1D}, \lambda, n_x, n_y)]$ (5.23)	$I = \frac{e g_v k_B T}{\pi \hbar} \sum_{n_x=1}^{n_{x max}} \sum_{n_y=1}^{n_{y max}} F_0(\bar{\eta}_{52,1D}) \exp(-\bar{\beta}_{52,1D})$ (5.26)
$n_0 = \frac{\sqrt{2m_c} 2 g_v}{\hbar \pi} \sum_{n_x=1}^{n_{x max}} \sum_{n_y=1}^{n_{y max}} [S_{53}(E_{F1D}, \lambda, n_x, n_y) + T_{53}(E_{F1D}, \lambda, n_x, n_y)]$ (5.24)	$I = \frac{e g_v k_B T}{\pi \hbar} \sum_{n_x=1}^{n_{x max}} \sum_{n_y=1}^{n_{y max}} F_0(\bar{\eta}_{53,1D}) \exp(-\bar{\beta}_{53,1D})$ (5.27)
(3) Effective mass superlattices of optoelectronic materials under magnetic quantization:	$J = \frac{e^2 B k_B T g_v}{2\pi^2 \hbar^2} \sum_{n=0}^{n_{max}} F_0(\bar{\eta}_{51BSL}) \exp(-\bar{\beta}_{51BSL})$ (5.35)
(a) Perturbed three-band model of Kane	
$n_0 = \frac{g_v e B}{\pi^2 \hbar L_0} \sum_{n=0}^{n_{max}} [S_{54}(E_{FB}, \lambda, n) + T_{54}(E_{FB}, \lambda, n)]$ (5.33)	
(b) Perturbed two-band model of Kane	$J = \frac{e^2 B k_B T g_v}{2\pi^2 \hbar^2} \sum_{n=0}^{n_{max}} F_0(\bar{\eta}_{52BSL}) \exp(-\bar{\beta}_{52BSL})$ (5.38)
$n_0 = \frac{g_v e B}{\pi^2 \hbar L_0} \sum_{n=0}^{n_{max}} [S_{55}(E_{FB}, \lambda, n) + T_{55}(E_{FB}, \lambda, n)]$ (5.36)	

(c) Perturbed parabolic energy bands: $n_0 = \frac{g_v eB}{\pi^2 \hbar L_0} \sum_{n=0}^{n_{\max}} [S_{56}(E_{FB}, \lambda, n) + T_{56}(E_{FB}, \lambda, n)]$ (5.40)	$J = \frac{e^2 B k_B T g_v}{2\pi^2 \hbar^2} \sum_{n=0}^{n_{\max}} F_0(\overline{\eta}_{53BSL}) \exp(-\overline{\beta}_{53B})$ (5.42)
(4) Effective mass superlattices of optoelectronic materials under 2D quantization:	
(a) Perturbed three-band model of Kane $n_0 = \frac{2g_v}{\pi} \sum_{n_y=1}^{n_{y\max}} \sum_{n_z=1}^{n_{z\max}} [\overline{Q}_{23}(E_{\text{FIDEMSL}}, \lambda, n_y, n_z) + \overline{Q}_{24}(E_{\text{FIDEMSL}}, \lambda, n_y, n_z)]$ (5.46)	$I = \frac{e g_v k_B T}{\pi \hbar} \sum_{n_y=1}^{n_{y\max}} \sum_{n_z=1}^{n_{z\max}} F_0(\overline{\eta}_{17}) \exp(-\overline{\theta}_{17})$ (5.47)
(b) Perturbed two-band model of Kane: $n_0 = \frac{2g_v}{\pi} \sum_{n_y=1}^{n_{y\max}} \sum_{n_z=1}^{n_{z\max}} [\overline{Q}_{25}(E_{\text{FIDEMSL}}, \lambda, n_y, n_z) + \overline{Q}_{26}(E_{\text{FIDEMSL}}, \lambda, n_y, n_z)]$ (5.49)	$I = \frac{e g_v k_B T}{\pi \hbar} \sum_{n_y=1}^{n_{y\max}} \sum_{n_z=1}^{n_{z\max}} F_0(\overline{\eta}_{18}) \exp(-\overline{\theta}_{18})$ (5.50)
(c) Perturbed parabolic energy bands $n_0 = \frac{2g_v}{\pi} \sum_{n_y=1}^{n_{y\max}} \sum_{n_z=1}^{n_{z\max}} [\overline{Q}_{251}(E_{\text{FIDEMSL}}, \lambda, n_y, n_z) + \overline{Q}_{261}(E_{\text{FIDEMSL}}, \lambda, n_y, n_z)]$ (5.52)	$I = \frac{e g_v k_B T}{\pi \hbar} \sum_{n_y=1}^{n_{y\max}} \sum_{n_z=1}^{n_{z\max}} F_0(\overline{\eta}_{181}) \exp(-\overline{\theta}_{181})$ (5.53)
(5) Superlattices of optoelectronic materials with graded interfaces under magnetic quantization: $n_0 = \frac{g_v eB}{\pi^2 \hbar} [\sum_{n=0}^{n_{\max}} [\overline{Q}_{27}(E_{\text{FBGISL}}, \lambda, n) + \overline{Q}_{28}(E_{\text{FBGISL}}, \lambda, n)]]$ (5.57)	$J = \frac{e^2 B k_B T g_v}{2\pi^2 \hbar^2} \sum_{n=0}^{n_{\max}} F_0(\overline{\eta}_{19}) \exp(-\overline{\theta}_{19})$ (5.58)
(6) Superlattices of optoelectronic materials with graded interfaces under 2D quantization: $n_0 = \frac{2g_v}{\pi} \sum_{n_y=1}^{n_{y\max}} \sum_{n_z=1}^{n_{z\max}} \left[\overline{Q}_{29}(E_{\text{FQWGISL}}, \lambda, n_y, n_z) + \overline{Q}_{30}(E_{\text{FQWGISL}}, \lambda, n_y, n_z)\right]$ (5.61)	$I = \frac{e g_v k_B T}{\pi \hbar} \sum_{n_y=1}^{n_{y\max}} \sum_{n_z=1}^{n_{z\max}} F_0(\overline{\eta}_{20}) \exp(-\overline{\theta}_{20})$ (5.62)

(R5.4) Investigate the FNFE for arbitrarily oriented photoexcitation from the heavily doped semiconductors in the presence of Gaussian, exponential, Kane, Halperin, Lax and Bonch–Bruevich types of band tails for all materials whose unperturbed carrier energy spectra are defined in Chap. 1.

(R5.5) Investigate the FNFE from all the semiconductors in the presence of arbitrarily oriented nonquantizing nonuniform electric field and photoexcitation for all the appropriate cases of problem R5.4.

(R5.6) Investigate the FNFE from all the semiconductors in the presence of arbitrarily oriented nonquantizing alternating electric field and photoexcitation for all the appropriate cases of problem R5.4.

(R5.7) Investigate the FNFE from negative refractive index, organic, magnetic, disordered, and other advanced materials in the presence of arbitrarily oriented photoexcitation.

(R5.8) Investigate the FNFE in the presence of arbitrarily oriented photoexcitation and alternating nonquantizing electric field for all the problems of R5.7.

(R5.9) Investigate the FNFE in the presence of arbitrarily oriented photoexcitation and nonquantizing nonuniform electric field for all the problems of R5.7.

(R5.10) Investigate the FNFE in the presence of arbitrarily oriented photoexcitation and alternating nonquantizing electric field for all the problems of R5.7.

(R5.11) Investigate the FNFE from quantum dots of all the semiconductors whose bulk dispersion relations are given in Chap. 1, in the presence of arbitrarily oriented photoexcitation and quantizing magnetic field respectively.

(R5.12) Investigate the FNFE from quantum dots of all the materials whose bulk dispersion relations are given in Chap. 1, in the presence of an arbitrarily oriented nonquantizing nonuniform electric field, photoexcitation and quantizing magnetic field, respectively.

(R5.13) Investigate the FNFE from quantum dots of all the semiconductors whose bulk dispersion relations are given in Chap. 1, in the presence of an arbitrarily oriented nonquantizing alternating electric field, photoexcitation and quantizing magnetic field respectively.

(R5.14) Investigate the FNFE from quantum dots of all the semiconductors whose bulk dispersion relations are given in Chap. 1, in the presence of an arbitrarily oriented nonquantizing alternating electric field, photoexcitation and quantizing alternating magnetic field respectively.

(R5.15) Investigate the FNFE from quantum dots of all the semiconductors whose bulk dispersion relations are given in Chap. 1, in the presence of an arbitrarily oriented photoexcitation and crossed electric and quantizing magnetic fields respectively.

(R5.16) Investigate the FNFE for arbitrarily oriented photoexcitation and quantizing magnetic field from the heavily doped semiconductors in the presence of Gaussian, exponential, Kane, Halperin, Lax, and Bonch–Bruevich types of band for all semiconductors whose unperturbed carrier energy spectra are defined in Chap. 1.

(R5.17) Investigate the FNFE for arbitrarily oriented photoexcitation and quantizing alternating magnetic field for all the cases of R5.16.

(R5.18) Investigate the FNFE for arbitrarily oriented photoexcitation and nonquantizing alternating electric field and quantizing magnetic field for all the cases of R5.16.

(R5.19) Investigate the FNFE for arbitrarily oriented photoexcitation and nonuniform alternating electric field and quantizing magnetic field for all the cases of R5.16.

(R5.20) Investigate the FNFE for arbitrarily oriented photoexcitation and crossed electric and quantizing magnetic fields for all the cases of R5.16.

(R5.21) Investigate the FNFE from negative refractive index, organic, magnetic, heavily doped, disordered, and other advanced optical materials in the presence of arbitrary oriented photoexcitation and quantizing magnetic field.

(R5.22) Investigate the FNFE in the presence of arbitrary oriented photoexcitation, quantizing magnetic field, and alternating nonquantizing electric field for all the problems of R5.21.

(R5.23) Investigate the FNFE in the presence of arbitrary oriented photoexcitation, quantizing magnetic field, and nonquantizing nonuniform electric field for all the problems of R5.21.

(R5.24) Investigate the FNFE in the presence of arbitrary oriented photoexcitation, alternating quantizing magnetic field, and crossed alternating nonquantizing electric field for all the problems of R5.21.

(R5.25) Investigate the FNFE from all the quantum confined materials (i.e., multiple quantum wells, wires and dots) whose unperturbed carrier energy spectra are defined in Chap. 1 in the presence of arbitrary oriented photoexcitation and quantizing magnetic field respectively.

(R5.26) Investigate the FNFE in the presence of arbitrary oriented photoexcitation and alternating quantizing magnetic field respectively for all the problems of R5.25.

(R5.27) Investigate the FNFE in the presence of arbitrary oriented photoexcitation, alternating quantizing magnetic field, and an additional arbitrary oriented nonquantizing nonuniform electric field respectively for all the problems of R5.25.

(R5.28) Investigate the FNFE in the presence of arbitrary oriented photoexcitation, alternating quantizing magnetic field, and additional arbitrary oriented nonquantizing alternating electric field respectively for all the problems of R5.25.

(R5.29) Investigate the FNFE in the presence of arbitrary oriented photoexcitation, and crossed quantizing magnetic and electric fields respectively for all the problems of R5.25.

(R5.30) Investigate the FNFE for arbitrarily oriented photoexcitation and quantizing magnetic field from the entire quantum confined heavily doped semiconductors in the presence of exponential, Kane, Halperin, Lax, and Bonch–Bruevich types of band tails for all semiconductors whose unperturbed carrier energy spectra are defined in Chap. 1.

(R5.31) Investigate the FNFE for arbitrarily oriented photoexcitation and alternating quantizing magnetic field for all the cases of R5.30.

(R5.32) Investigate the FNFE in the presence of arbitrarily oriented photoexcitation, alternating quantizing magnetic field, and an additional arbitrarily oriented nonquantizing nonuniform electric field for all the cases of R5.30.
(R5.33) Investigate the FNFE in the presence of arbitrary oriented photoexcitation, alternating quantizing magnetic field, and additional arbitrary oriented nonquantizing alternating electric field respectively for all the cases of R5.30.
(R5.34) Investigate the FNFE in the presence of arbitrary oriented photoexcitation, and crossed quantizing magnetic and electric fields respectively for all the cases of R5.30.
(R5.35) Investigate the FNFE for all the appropriate problems from R5.25 to R5.34 in the presence of finite potential wells.
(R5.36) Investigate the FNFE for all the appropriate problems from R5.25 to R5.34 in the presence of parabolic potential wells.
(R5.37) Investigate the FNFE for all the above appropriate problems for quantum rings.
(R5.38) Investigate the FNFE for all the above appropriate problems in the presence of elliptical Hill and quantum square rings respectively.
(R5.39) Investigate the FNFE from carbon nanotubes in the presence of arbitrary photoexcitation.
(R5.40) Investigate the FNFE from carbon nanotubes in the presence of arbitrary photoexcitation and nonquantizing alternating electric field.
(R5.41) Investigate the FNFE from carbon nanotubes in the presence of arbitrary photoexcitation and nonquantizing alternating magnetic field.
(R5.42) Investigate the FNFE from carbon nanotubes in the presence of arbitrary photoexcitation and crossed electric and quantizing magnetic fields.
(R5.43) Investigate the FNFE from heavily doped semiconductor nanotubes in the presence of arbitrary photoexcitation for all the materials whose unperturbed carrier dispersion laws are defined in Chap. 1.
(R5.44) Investigate the FNFE from heavily doped semiconductor nanotubes in the presence of nonquantizing alternating electric field and arbitrary photoexcitation for all the materials whose unperturbed carrier dispersion laws are defined in Chap. 1.
(R5.45) Investigate the FNFE from heavily doped semiconductor nanotubes in the presence of nonquantizing alternating magnetic field and arbitrary photoexcitation for all the materials whose unperturbed carrier dispersion laws are defined in Chap. 1.
(R5.46) Investigate the FNFE from heavily doped semiconductor nanotubes in the presence of arbitrary photoexcitation and nonuniform electric field for all the materials whose unperturbed carrier dispersion laws are defined in Chap. 1.
(R5.47) Investigate the FNFE from heavily doped semiconductor nanotubes in the presence of arbitrary photoexcitation and alternating quantizing magnetic fields for all the materials whose unperturbed carrier dispersion laws are defined in Chap. 1.

(R5.48) Investigate the FNFE from heavily doped semiconductor nanotubes in the presence of arbitrary photoexcitation and crossed electric and quantizing magnetic fields for all the materials whose unperturbed carrier dispersion laws are defined in Chap. 1.

(R5.49) Investigate the FNFE in the presence of arbitrary photoexcitation for all the appropriate nipi structures of the semiconductors whose unperturbed carrier energy spectra are defined in Chap. 1.

(R5.50) Investigate the FNFE in the presence of arbitrary photoexcitation for all the appropriate nipi structures of the semiconductors whose unperturbed carrier energy spectra are defined in Chap. 1 in the presence of an arbitrarily oriented nonquantizing nonuniform additional electric field.

(R5.51) Investigate the FNFE for all the appropriate nipi structures of the semiconductors whose unperturbed carrier energy spectra are defined in Chap. 1 in the presence of an arbitrarily oriented photoexcitation and nonquantizing alternating additional magnetic field.

(R5.52) Investigate the FNFE for all the appropriate nipi structures of the semiconductors whose unperturbed carrier energy spectra are defined in Chap. 1 in the presence of an arbitrarily oriented photoexcitation and quantizing alternating additional magnetic field.

(R5.53) Investigate the FNFE for all the appropriate nipi structures of the semiconductors whose unperturbed carrier energy spectra are defined in Chap. 1 in the presence of an arbitrarily oriented photoexcitation and crossed electric and quantizing magnetic fields.

(R5.54) Investigate the FNFE from heavily doped nipi structures for all the appropriate cases of all the above problems.

(R5.55) Investigate the FNFE in the presence of arbitrary photoexcitation for the appropriate inversion layers of all the materials whose unperturbed carrier energy spectra are defined in Chap. 1.

(R5.56) Investigate the FNFE in the presence of arbitrary photoexcitation for the appropriate inversion layers of all the materials whose unperturbed carrier energy spectra are defined in Chap. 1 in the presence of an arbitrarily oriented nonquantizing nonuniform additional electric field.

(R5.57) Investigate the FNFE for the appropriate inversion layers of all the materials whose unperturbed carrier energy spectra are defined in Chap. 1 in the presence of an arbitrarily oriented photoexcitation and nonquantizing alternating additional magnetic field.

(R5.58) Investigate the FNFE for the appropriate inversion layers of all the materials whose unperturbed carrier energy spectra are defined in Chap. 1 in the presence of an arbitrarily oriented photoexcitation and quantizing alternating additional magnetic field.

(R5.59) Investigate the FNFE for the appropriate inversion layers of all the materials whose unperturbed carrier energy spectra are defined in Chap. 1 in the presence of an arbitrarily oriented photoexcitation and crossed electric and quantizing magnetic fields by considering electron spin and broadening of Landau levels.

(R5.60) Investigate the FNFE in the presence of arbitrary photoexcitation for the appropriate accumulation layers of all the materials whose unperturbed carrier energy spectra are defined in Chap. 1 by modifying the above appropriate problems.

(R5.61) Investigate the FNFE in the presence of arbitrary photoexcitation from wedge shaped and cylindrical QDs of all the semiconductors whose unperturbed carrier energy spectra are defined in Chap. 1.

(R5.62) Investigate the FNFE in the presence of arbitrary photoexcitation from wedge shaped and cylindrical QDs of all the semiconductors whose unperturbed carrier energy spectra are defined in Chap. 1 in the presence of an arbitrarily oriented nonquantizing nonuniform additional electric field.

(R5.63) Investigate the FNFE from wedge shaped and cylindrical QDs of all the semiconductors whose unperturbed carrier energy spectra are defined in Chap. 1 in the presence of an arbitrarily oriented photoexcitation and nonquantizing alternating additional magnetic field.

(R5.64) Investigate the FNFE from wedge shaped and cylindrical QDs of all the semiconductors whose unperturbed carrier energy spectra are defined in Chap. 1 in the presence of an arbitrarily oriented photoexcitation and quantizing alternating additional magnetic field.

(R5.65) Investigate the FNFE from wedge shaped and cylindrical QDs of all the semiconductors whose unperturbed carrier energy spectra are defined in Chap. 1 in the presence of an arbitrarily oriented photoexcitation and crossed electric and quantizing magnetic fields.

(R5.66) Investigate the FNFE from wedge shaped and cylindrical QDs for all the appropriate cases of the above problems.

(R5.67) Investigate all the problems from R5.25 to R5.66 by removing all the mathematical approximations and establishing the respective appropriate uniqueness conditions.

(R5.68) Investigate the FNFE from quantum confined III–V, II–VI, IV–VI, HgTe/CdTe effective mass superlattices together with short period, strained layer, random, Fibonacci, polytype and sawtooth superlattices in the presence of arbitrarily oriented photoexcitation.

(R5.69) Investigate the FNFE in the presence of arbitrarily oriented photoexcitation and quantizing magnetic field respectively for all the cases of R5.68.

(R5.70) Investigate the FNFE in the presence of arbitrarily oriented photoexcitation and nonquantizing nonuniform electric field respectively for all the cases of R5.68.

(R5.71) Investigate the FNFE in the presence of arbitrarily oriented photoexcitation and nonquantizing alternating electric field respectively for all the cases of R5.68.

(R5.72) Investigate the FNFE in the presence of arbitrarily oriented photoexcitation and crossed electric and quantizing magnetic fields respectively for all the cases of R5.68.

(R5.73) Investigate the FNFE from heavily doped quantum confined superlattices for all the problems of R5.68.

(R5.74) Investigate the FNFE in the presence of arbitrarily oriented photoexcitation and quantizing magnetic field respectively for all the cases of R5.73.

(R5.75) Investigate the FNFE in the presence of arbitrarily oriented photoexcitation and nonquantizing nonuniform electric field respectively for all the cases of R5.73.

(R5.76) Investigate the FNFE in the presence of arbitrarily oriented photoexcitation and nonquantizing alternating electric field respectively for all the cases of R5.73.

(R5.77) Investigate the FNFE in the presence of arbitrarily oriented photoexcitation and crossed electric and quantizing magnetic fields respectively for all the cases of R5.73.

(R5.78) Investigate all the problems from R5.68 to R5.77 by removing all the mathematical approximations and establishing the respective appropriate uniqueness conditions.

(R5.79) Investigate the FNFE from quantum confined III–V, II–VI, IV–VI, HgTe/CdTe superlattices with graded interfaces together with short period, strained layer, random, Fibonacci, polytype and sawtooth superlattices in this context in the presence of arbitrarily oriented photoexcitation.

(R5.80) Investigate the FNFE in the presence of arbitrarily oriented photoexcitation and quantizing magnetic field respectively for all the cases of R5.79.

(R5.81) Investigate the FNFE in the presence of arbitrarily oriented photoexcitation and nonquantizing nonuniform electric field respectively for all the cases of R5.79.

(R5.82) Investigate the FNFE in the presence of arbitrarily oriented photoexcitation and nonquantizing alternating electric field respectively for all the cases of R5.79.

(R5.83) Investigate the FNFE in the presence of arbitrarily oriented photoexcitation and crossed electric and quantizing magnetic fields respectively for all the cases of R5.79.

(R5.84) Investigate the FNFE from heavily doped quantum confined superlattices for all the problems of R5.79.

(R5.85) Investigate the FNFE in the presence of arbitrarily oriented photoexcitation and quantizing magnetic field respectively for all the cases of R5.84.

(R5.86) Investigate the FNFE in the presence of arbitrarily oriented photoexcitation and nonquantizing nonuniform electric field respectively for all the cases of R5.84.

(R5.87) Investigate the FNFE in the presence of arbitrarily oriented photoexcitation and nonquantizing alternating electric field respectively for all the cases of R5.84.

(R5.88) (a) Investigate the FNFE in the presence of arbitrarily oriented photoexcitation and crossed electric and quantizing magnetic fields respectively for all the cases of R5.84.

(b) Investigate the FNFE from multiple wall carbon nanotubes in the presence of an arbitrarily oriented alternating electric field.

(c) Investigate the FNFE from heavily doped semiconductor nanotubes in the presence of an arbitrarily oriented alternating electric field for all the materials whose unperturbed carrier energy spectra are defined in R1.1 and R1.2 respectively.

(R5.89) (a) Formulate the minimum tunneling, Dwell and phase tunneling, Buttiker and Landauer and intrinsic times for all types of systems as discussed in this chapter.

(b) Investigate all the appropriate problems of this chapter for the Dirac electron.

(c) Investigate all the problems of this chapter by removing all the mathematical approximations and establishing the respective appropriate uniqueness conditions.

References

1. K.P. Ghatak, S. Bhattacharya, D. De, *Einstein Relation in Compound Semiconductors and Their nanostructures, Springer Series in Materials Science*, vol. 116 (Springer-Verlag, Berlin, 2008)
2. H. Sasaki, Phys. Rev. B **30**, 7016 (1984)
3. H.X. Jiang, J.Y. Lin, J. Appl. Phys. **61**, 624 (1987)

Part III
Fowler–Nordheim Field Emission from Quantum-Confined Optoelectronic Semiconductors in the Presence of Intense Electric Field

Chapter 6
Field Emission from Quantum-Confined Optoelectronic Semiconductors

6.1 Introduction

With the advent of modern nanodevices, there has been considerable interest in studying the electric field-induced processes in semiconductors having different band structures. It appears from the literature that the studies have been made on the assumption that the carrier dispersion laws are invariant quantities in the presence of intense electric field, which is not fundamentally true. In this chapter, we shall study the FNFE from quantum-confined optoelectronic semiconductors under strong electric field. In Sect. 6.2.1, the FNFE from the bulk specimens said compounds under strong electric field has been investigated in the presence of magnetic quantization whose unperturbed electron energy spectra are, respectively, defined by the three- and two-band models of Kane together with parabolic energy bands. Section 6.2.2 contains the investigation of the FNFE from quantum wires of optoelectronic semiconductors. In Sect. 6.2.3, the FNFE field emission from effective mass superlattices of optoelectronic semiconductors in the presence of strong electric field under magnetic quantization has been studied. In Sect. 6.2.4, we have investigated the FNFE from quantum wire effective mass superlattices of optoelectronic semiconductors. In Sect. 6.2.5, FNFE from superlattices of optoelectronic compounds with graded interfaces under magnetic quantization has been investigated. In Sect. 6.2.6, the FNFE from quantum wire superlattices of optoelectronic semiconductors with graded interfaces has been studied. Section 6.3 contains the result and discussions pertinent to this chapter. Section 6.4 presents a single challenging open research problem.

S. Bhattacharya and K.P. Ghatak, *Fowler–Nordheim Field Emission*, Springer Series in Solid-State Sciences 170, DOI 10.1007/978-3-642-20493-7_6,

6.2 Theoretical Background

6.2.1 Field Emission from Optoelectronic Semiconductors Under Magnetic Quantization

Let the wave vector $\bar{k}$ be in the direction with the polar angle (θ, ϕ) referred to the crystal symmetry axes and the spin-up and spin-down functions are written as $\uparrow$ and $\downarrow$, respectively. The $u_{\bar{k}1}(\bar{r})$ and $u_{\bar{k}2}(\bar{r})$ are the wave functions for the conduction and light-hole/spin–orbit splitting valence band and can, respectively, be expressed as [1, 2]

$$u_{\bar{k}1}(\bar{r}) = a_{k+}[(is)\downarrow'] + b_{k+}\left[\frac{X' - iY'}{\sqrt{2}}\uparrow'\right] + c_{k+}[Z'\downarrow'] \tag{6.1}$$

and

$$u_{\bar{k}2}(\bar{r}) = a_{k-}[(is)\uparrow'] - b_{k-}\left[\frac{X' + iY'}{\sqrt{2}}\downarrow'\right] + c_{k-}[Z'\uparrow'] \tag{6.2}$$

where $\bar{r}$ is the position vector of the electron,

$$a_{k\pm} = r_0\left[\frac{E_g - \gamma_{k\pm}^2(E_g - \delta')}{E_g + \delta'}\right]^{1/2} \tag{6.3}$$

$r_0 = \left\{6 \cdot \frac{\left(E_g + \frac{2\Delta}{3}\right)\cdot(E_g + \Delta)}{\chi}\right\}^{1/2}$, E_g is the band gap in the absence of any field, Δ is the spin–orbit splitting constant, $\chi = 6E_g^2 + 9E_g \cdot \Delta + 4\Delta^2$, $b_{k\pm} = s_0\gamma_{k\pm}$, $s_0 = \left[\frac{4\Delta^2}{3\chi}\right]^{1/2}$, s is the s-type atomic orbital in both unprimed and primed coordinates, X', Y', and Z' are the p-type atomic orbitals in the prime co-ordinates, $c_{k\pm} = t\gamma_{k\pm}$, $t = \left\{\frac{6\left(E_g + \frac{2\Delta}{3}\right)}{\chi}\right\}^{1/2}$, $i = \sqrt{-1}$, $\gamma_{k\pm} = \left[\frac{\eta \mp E_g}{2(\eta + \delta')}\right]^{1/2}$, $\delta' = \frac{E_g^2\Delta}{\chi}$, η is the energy difference between the conduction and the valence bands and can be written as $\eta = \{E_C(\vec{k}) - E_V(\vec{k})\} = \left[E_g^2 + \frac{E_g\hbar^2k^2}{m_r}\right]^{1/2}$, $E_C(\vec{k})$ is k-dependent energy of the electron in the conduction band (CB), $E_V(\vec{k})$ is the k-dependent energy of the heavy hole band (VB), m_r is the *reduced* effective mass and is given by $m_r = (m_c^{-1} + m_q^{-1})^{-1}$, m_c is the effective electron mass at the edge of the conduction band and m_q is the effective mass of the heavy hole at the top of the heavy hole band.

The electron energy spectrum of optoelectronic materials in the absence of any field can be expressed in accordance with the three-band model of Kane as [3]

$$\gamma(E) = \frac{\hbar^2k^2}{2m_c} \tag{6.4}$$

where $\gamma(E) = \frac{E(aE+1)(bE+1)}{(cE+1)}$; $a = 1/E_g$, $b = 1/(E_g + \Delta)$, $c = 1/\left(E_g + \frac{2}{3}\Delta\right)$.

Using (6.1) and (6.2), we can write the expression for interband transition matrix element, X_{12}, as

$$X_{12} = i \int u_{\bar{k}1}^{*}(\bar{r}) \cdot \frac{\partial}{\partial k_x} u_{\bar{k}2}(\bar{r}) \mathrm{d}^3 r \tag{6.5}$$

In the case of the presence of an external electric field, F along x-axis, the interband transition matrix element, X_{12}, has finite interaction band same band, e.g.,

$$\begin{aligned}
&\langle S|S\rangle = \langle X|X\rangle = \langle Y|Y\rangle = \langle Z|Z\rangle = 1 \\
&\langle X|Y\rangle = \langle Y|Z\rangle = \langle Z|X\rangle = 0 \\
&\langle S|X\rangle = \langle X|S\rangle = 0;\ \langle S|Y\rangle = \langle Y|S\rangle = 0 \text{ and } \langle S|Z\rangle = \langle Z|S\rangle = 0
\end{aligned}$$

We also know for the arbitrary orientation of the $\bar{k}$-vector that

$$\begin{bmatrix} \uparrow' \\ \downarrow' \end{bmatrix} = \begin{bmatrix} \mathrm{e}^{-i\phi/2} \cdot \cos\theta/2 & \mathrm{e}^{i\phi/2} \cdot \sin\theta/2 \\ -\mathrm{e}^{-i\phi/2} \cdot \sin\theta/2 & \mathrm{e}^{i\phi/2} \cdot \cos\theta/2 \end{bmatrix} \cdot \begin{bmatrix} \uparrow \\ \downarrow \end{bmatrix} \tag{6.6}$$

$$\begin{bmatrix} X' \\ Y' \\ Z' \end{bmatrix} = \begin{bmatrix} \cos\phi\cos\phi & \cos\theta \cdot \sin\phi & -\sin\theta \\ -\sin\phi & \cos\phi & 0 \\ \sin\theta\cos\phi & \sin\theta \cdot \sin\phi & \cos\theta \end{bmatrix} \cdot \begin{bmatrix} X \\ Y \\ Z \end{bmatrix} \tag{6.7}$$

The spin-vector can be expressed as

$$\vec{S} = \frac{\hbar}{2} \cdot \vec{\sigma} \tag{6.8}$$

where $\sigma_x = \begin{bmatrix} 0 & 1 \\ 1 & 0 \end{bmatrix}$, $\sigma_y = \begin{bmatrix} 0 & -i \\ i & 0 \end{bmatrix}$, and $\sigma_z = \begin{bmatrix} 1 & 0 \\ 0 & -1 \end{bmatrix}$. Using the appropriate equations we can write

$$\begin{aligned}
X_{12} &= i \int \mathrm{d}^3 r \left\{ a_{k+} [(iS)\downarrow'] + b_{k+} \left[\left(\frac{X' - iY'}{\sqrt{2}} \right) \uparrow' \right] + c_{k+} [Z'\downarrow'] \right\}^{*} \frac{\partial}{\partial k_x} \\
&= \left\{ a_{k-} [(iS)\uparrow'] - b_{k-} \left[\left(\frac{X' + iY'}{\sqrt{2}} \downarrow' \right) \right] + c_{k-} [Z'\uparrow'] \right\} \\
&\quad \times i \int \mathrm{d}^3 r \left[\left\{ \left(a_{k+} \frac{\partial}{\partial k_x} a_{k-} \right) \cdot \left([(iS)\downarrow']^{*} \cdot [(iS)\uparrow'] \right) \right.\right. \\
&\quad + \left(\frac{b_{k+}}{\sqrt{2}} \frac{\partial}{\partial k_x} a_{k-} \right) [(X' - iY')\uparrow']^{*} \cdot [(iS)\uparrow'] \\
&\quad \left.\left. + \left(c_{k+} \frac{\partial}{\partial k_x} a_{k-} \right) [(Z'\downarrow')]^{*} \cdot [(iS)\uparrow'] \right\} \right]
\end{aligned}$$

$$+i\int \mathrm{d}^3r\left[\left\{\left(-\frac{a_{k+}}{\sqrt{2}}\frac{\partial}{\partial k_x}b_{k-}\right).[(iS)\downarrow']^*\cdot[(X'+iY')\downarrow']\right.\right.$$

$$-\left(\frac{b_{k+}\frac{\partial}{\partial k_x}b_{k-}}{2}\right)[(X'-iY')\uparrow']^*\cdot[(X'+iY')\downarrow']$$

$$\left.\left.-\left(\frac{c_{k+}\frac{\partial}{\partial k_x}b_{k-}}{\sqrt{2}}\right)[(Z'\downarrow')]^*\cdot[(X'+iY')\downarrow']\right\}\right]$$

$$=i\int \mathrm{d}^3r\left[\left\{\left(a_{k+}\frac{\partial}{\partial k_x}c_{k-}\right)\cdot\left([(iS)\downarrow']^*\cdot[Z'\uparrow']\right)\right.\right.$$

$$+\left(\frac{b_{k+}}{\sqrt{2}}\frac{\partial}{\partial k_x}c_{k-}\right)[(X'-iY')\uparrow']^*\cdot[Z'\uparrow']$$

$$\left.\left.+\left(c_{k+}\frac{\partial}{\partial k_x}c_{k-}\right)[(Z'\downarrow')]^*\cdot[Z'\uparrow']\right\}\right]$$

Therefore,

$$\frac{X_{12}}{i}=\left[\left\{\left[\left(a_{k+}\frac{\partial}{\partial k_x}a_{k-}\right)\cdot\langle iS|iS\rangle\langle\downarrow'|\uparrow'\rangle\right]\right.\right.$$

$$+\left[\left(\frac{b_{k+}}{\sqrt{2}}\frac{\partial}{\partial k_x}a_{k-}\right)\langle(X'-iY')|iS\rangle\langle\uparrow'|\downarrow'\rangle\right]$$

$$\left.\left.+\left[\left(c_{k+}\frac{\partial}{\partial k_x}a_{k-}\right)\langle Z'|iS\rangle\langle\downarrow'|\uparrow'\rangle\right]\right\}\right]$$

$$-\left[\left\{\left[\left(\frac{a_{k+}}{\sqrt{2}}\frac{\partial}{\partial k_x}b_{k-}\right)\cdot\langle iS|(X'+iY')\rangle\langle\downarrow'\mid\downarrow'\rangle\right]\right.\right.$$

$$+\left[\left(\frac{b_{k+}}{2}\frac{\partial}{\partial k_x}b_{k+}\right)\langle(X'-iY')|(X'+iY')\rangle\langle\uparrow'\mid\downarrow'\rangle\right]$$

$$\left.\left.+\left[\left(\frac{c_k}{\sqrt{2}}\frac{\partial}{\partial k_x}b_{k-}\right)\langle Z'|(X'+iY')\rangle\langle\downarrow'\mid\downarrow'\rangle\right]\right\}\right]$$

$$+\left\{\left[\left(a_k\frac{\partial}{\partial k_x}c_{k-}\right)\cdot\langle iS|Z'\rangle\langle\downarrow'\mid\uparrow'\rangle\right]\right.$$

$$+\left[\left(\frac{b_{k+}}{\sqrt{2}}\frac{\partial}{\partial k_x}c_{k-}\right)\cdot\langle(X'-iY')|Z'\rangle\langle\uparrow'\mid\uparrow'\rangle\right]$$

$$\left.+\left[\left(c_k\frac{\partial}{\partial k_x}c_{k-}\right)\cdot\langle Z'|Z'\rangle\langle\downarrow'\mid\uparrow'\rangle\right]\right\}\tag{6.9}$$

From (6.9), we can write the terms

$$\langle iS|iS\rangle = -\langle S|S\rangle = -1,\ \left\langle (X' - iY')|iS\right\rangle = \left\langle X'|iS\right\rangle - \langle iY|iS\rangle .$$

From (1.7), we obtain

$$\left|X'\right\rangle = \cos\theta\cos\phi|X\rangle + \cos\theta\sin\phi\ |Y\rangle - \sin\theta\ |Z\rangle$$

Thus,

$$\left\langle X'|iS\right\rangle = \cos\theta\cos\phi\ \langle X|iS\rangle + \cos\theta\sin\phi\ \langle Y|iS\rangle - \sin\theta\ \langle Z|iS\rangle = 0$$

Since,

$$\langle X|S\rangle = \langle Y|S\rangle = \langle Z|S\rangle = 0,\quad \left|Y'\right\rangle = -\sin\phi\ |X\rangle + \cos\phi\ |Y\rangle + 0$$

Therefore,

$$\left\langle iY'|iS\right\rangle = -\left\langle Y'|S\right\rangle = +\sin\phi\ \langle X|S\rangle - \cos\phi\ \langle Y|S\rangle = 0.$$

Thus,

$$\left\langle (X' - iY')|iS\right\rangle = 0$$

$$\left\langle Z'|iS\right\rangle = \sin\theta\cos\phi\ \langle X|iS\rangle + \sin\theta\sin\phi\ \langle Y|iS\rangle + \cos\theta\ \langle Z|iS\rangle$$

$$= i\{\sin\theta\cos\phi\ \langle X|S\rangle + \sin\theta\sin\phi\ \langle Y|S\rangle + \cos\theta\ \langle Z|S\rangle\} = 0$$

$$\left\langle (X' - iY')|(X' + iY)\right\rangle = \left\langle X'|X'\right\rangle - \left\langle iY'|X'\right\rangle + \left\langle X'|iY'\right\rangle - \langle iY|iY\rangle ,$$

$$\left|X'\right\rangle = \cos\theta\cos\phi\ |X\rangle + \cos\theta\sin\phi\ |Y\rangle - \sin\theta|Z\rangle$$

Therefore, $\langle X'|X'\rangle = 1$ since $\langle X|X\rangle = \langle Y|Y\rangle = \langle Z|Z\rangle = 1$
Similarly $\langle Y'|Y'\rangle = 1$ and $\langle X'|Y'\rangle = 0$
Therefore, $\langle (X' - iY')|(X' + iY')\rangle = 1\ - 0\ + 0\ - 1 = 0$
and $\langle Z'|(X' + iY')\rangle = \langle Z'|X'\rangle + i\langle Z'|Y'\rangle = 0$
Besides $\langle Z'|Z'\rangle = \sin^2\theta\cos^2\phi\langle X|X\rangle + \sin^2\theta\sin^2\phi\langle Y|Y\rangle + \cos^2\theta\langle Z|Z\rangle = 1$
Therefore, we can write,

$$X_{12} = i\left\{-\left(a_{k+}\frac{\partial}{\partial k_x}a_{k-}\right)\langle\downarrow'\mid\uparrow'\rangle + \left(c_{k+}\frac{\partial}{\partial k_x}c_{k-}\right)\langle\downarrow'\mid\uparrow'\rangle\right\} \tag{6.10}$$

From the relation (6.6), we obtain

$$\uparrow' = \mathrm{e}^{-i\phi/2}\cos\frac{\theta}{2}\uparrow + \mathrm{e}^{i\phi/2}\sin\frac{\theta}{2}\downarrow \text{ and } \downarrow' = -\mathrm{e}^{-i\phi/2}\sin\frac{\theta}{2}\uparrow + \mathrm{e}^{i\phi/2}\cos\frac{\theta}{2}\downarrow$$

Now $\langle \downarrow' \mid \uparrow' \rangle$

$$= \left(-e^{-i\phi/2}\sin\frac{\theta}{2}\uparrow + e^{i\phi/2}\cos\frac{\theta}{2}\downarrow\right)^{*}\cdot\left(e^{-i\phi/2}\cos\frac{\theta}{2}\uparrow + e^{i\phi/2}\sin\frac{\theta}{2}\downarrow\right)$$

$$= \left(-e^{+i\phi/2}\sin\frac{\theta}{2}\uparrow^{*} + e^{-i\phi/2}\cos\frac{\theta}{2}\downarrow^{*}\right)\cdot\left(e^{-i\phi/2}\cos\frac{\theta}{2}\uparrow + e^{i\phi/2}\sin\frac{\theta}{2}\downarrow\right)$$

$$= \left(-\sin\frac{\theta}{2}\cos\frac{\theta}{2}\right)\langle \uparrow \mid \uparrow \rangle + e^{-i\phi}\cos^2\frac{\theta}{2}\langle \downarrow \mid \uparrow \rangle$$

$$-e^{i\phi}\sin^2\frac{\theta}{2}\langle \uparrow \mid \downarrow \rangle + \cos\frac{\theta}{2}\sin\frac{\theta}{2}\langle \downarrow \mid \downarrow \rangle$$

Therefore, $\langle \downarrow' \mid \uparrow' \rangle_x = \left(-\sin\frac{\theta}{2}\cos\frac{\theta}{2}\right)\langle \uparrow \mid \uparrow \rangle_x + e^{-i\phi}\cos^2\frac{\theta}{2}\langle \downarrow \mid \uparrow \rangle_x - e^{-i\phi}\sin^2\frac{\theta}{2}$ $\langle \uparrow \mid \downarrow \rangle_x + \cos\frac{\theta}{2}\sin\frac{\theta}{2}\langle \downarrow \mid \downarrow \rangle_x$

From (6.1) we can write, $\langle \uparrow \mid \uparrow \rangle_x = 0$; $\langle \downarrow \mid \uparrow \rangle_x = 1$ and $\langle \downarrow \mid \downarrow \rangle_x = 0$

Therefore, $\langle \downarrow' | \uparrow' \rangle_x = \cos\phi\cos\theta - i\sin\phi$.

Similarly, $\langle \downarrow' \mid \downarrow' \rangle_y = i\cos\phi + \sin\phi\cos\theta$ and $\langle \downarrow' \mid \downarrow' \rangle_z = -\sin\theta$

Thus, $\langle \downarrow' \mid \uparrow' \rangle = \hat{i}\langle \downarrow' \mid \uparrow' \rangle_x + \hat{j}\langle \downarrow' \mid \uparrow' \rangle_y + \hat{k}\langle \downarrow' \mid \uparrow' \rangle_z = (\hat{r}_1 + i\hat{r}_2)$, where, $\hat{r}_1$ and $\hat{r}_2$ are the unit vectors in the primed axes and are given by $\hat{r}_1 = \hat{i}\cos\theta\cos\phi + \hat{j}\cos\theta\sin\phi - \hat{k}\sin\theta$ and $\hat{r}_2 = -\hat{i}\sin\phi + \hat{j}\cos\phi - \hat{k}\cdot 0$ in which $\hat{i}$, $\hat{j}$, and $\hat{k}$ are unit vectors along x-, y- and z-axes, respectively. Considering the $\frac{1}{2}$-spin, we can write

$$\langle \downarrow' | \uparrow' \rangle = \frac{1}{2}(\hat{r}_1 + i\hat{r}_2) \tag{6.11}$$

Therefore from (6.10) and (6.11), we get

$$X_{12} = i\left\{-\left(a_{k+}\frac{\partial}{\partial k_x}a_{k-}\right) + \left(c_{k+}\frac{\partial}{\partial k_x}c_{k-}\right)\right\}\langle \downarrow' \mid \uparrow' \rangle$$

$$= -i\left\{\left(a_{k+}\frac{\partial}{\partial k_x}a_{k-}\right) - \left(c_{k+}\frac{\partial}{\partial k_x}c_{k-}\right)\right\}\cdot\frac{1}{2}(\hat{r}_1 + i\hat{r}_2) = \frac{-iA(k)}{2}(\hat{r}_1 + i\hat{r}_2) \tag{6.12}$$

where

$$A(k) = \left(a_{k+}\frac{\partial}{\partial k_x}a_{k-}\right) - \left(c_{k+}\frac{\partial}{dk_x}c_{k-}\right) \tag{6.13}$$

From (6.12), we find,

$$|X_{12}|^2 = \frac{1}{4}A^2(k)(1+1) = \frac{1}{2}A^2(k) \quad [\text{since } |\hat{r}_1| = |\hat{r}_2| = 1]]$$

Considering spin-up and spin-down, we have to multiply by 2

$$|X_{12}|^2 = 2 \times \frac{1}{2} A^2(k) = A^2(k) \tag{6.14}$$

We can evaluate X_{11} and X_{22} in the following way:

$$\begin{aligned}
X_{11} &= i \int u^*_{\bar{k}1}(\bar{r}) \cdot \frac{\partial}{\partial k_x} u_{\bar{k}1}(\bar{r}) \cdot \mathrm{d}^3 r \\
&= i \int \mathrm{d}^3 r \left\{ \left(a_{k_+} \frac{\partial}{\partial k_x} a_{k_+} \right) + \left(b_{k_+} \frac{\partial}{\partial k_x} b_{k_+} \right) + \left(c_{k_+} \frac{\partial}{\partial k_x} c_{k_+} \right) \right\} \\
&= \frac{1}{2} i \int \mathrm{d}^3 r \left\{ \frac{\partial}{\partial k_x} (a_k^2 + b_k^2 + c_k^2) \right\} \\
&= \frac{1}{2} i \int \mathrm{d}^3 r \left\{ \frac{\partial}{\partial k_x} (1) \right\} = 0, \quad \text{since } a_{k+}^2 + b_{k+}^2 + c_{k+}^2 = 1.
\end{aligned}$$

Therefore $X_{11} = 0$, and similarly we can prove $X_{22} = 0$. Thus, we conclude that intraband momentum matrix element due to external electric field (X_{CC}) is zero. From the expression of $a_{k\pm}$, we can write

$$a_{k_+}^2 = {r_0}^2 \left[\frac{E_g - \gamma_{k_+}^2 (E_g - \delta')}{E_g + \delta'} \right]^2$$

and

$$a_{k-}^2 = r_0^2 \left[\frac{E_g - \gamma_{k-}^2 (E_g - \delta')}{E_g + \delta'} \right]^2 .$$

Therefore, $2a_{k-} \frac{\partial}{\partial k_x} a_{k-} = r_0^2 \left[-\left(\frac{E_g - \delta'}{E_g + \delta'} \right) \right] \frac{\partial \gamma_{k-}^2}{\partial k_x}$ and $\frac{\partial a_{k-}}{\partial k_x} = -\frac{r_0^2}{2} \left(\frac{E_g - \delta'}{E_g + \delta'} \right) \frac{1}{a_{k-}} \frac{\partial \gamma_{k-}^2}{\partial k_x}$

Combining we can write $a_{k+} \frac{\partial a_{k-}}{\partial k_x} = -\frac{r_0^2}{2} \left(\frac{E_g - \delta\prime}{E_g + \delta\prime} \right) \frac{a_{k+}}{a_{k-}} \frac{\partial \gamma_{k-}^2}{\partial k_x}$

Similarly, $c_{k_+} = t \gamma_{k_+}$ and $c_{k_-} = t \gamma_{k_-}$. Therefore, $c_{k+} \frac{\partial}{\partial k_x} c_{k-} = \frac{t^2}{2} \frac{c_{k+}}{c_{k-}} \frac{\partial \gamma_{k-}^2}{\partial k_x}$

$$A(k) = \left\{ -\frac{r_0^2}{2} \left[\frac{E_g - \delta'}{E_g + \delta'} \right] \frac{a_{k+}}{a_{k-}} - \frac{t^2}{2} \frac{c_{k+}}{c_{k-}} \right\} \frac{\partial \gamma_{k-}^2}{\partial k_x}$$

$$\begin{aligned}
\text{Now,} \left(\frac{a_{k+}}{a_{k-}} \right)^2 &= \frac{E_g - \gamma_{k+}^2 (E_g - \delta')}{E_g - \gamma_{k-}^2 (E_g - \delta')} = \frac{E_g - \frac{\eta - E_g}{2(\eta + \delta')} (E_g - \delta')}{E_g - \frac{\eta + E_g}{2(\eta + \delta')} (E_g - \delta')} \\
&= \frac{2E_g - (\eta + \delta') - (\eta - E_g)(E_g - \delta')}{2E_g (\eta + \delta') - (\eta + E_g)(E_g - \delta')} \\
&= \frac{\eta (E_g + \delta') + E_g (E_g + \delta')}{\eta (E_g + \delta') - E_g (E_g - 3\delta')}
\end{aligned}$$

Therefore, $\left(\frac{a_{k+}}{a_{k-}}\right)^2 = \frac{\eta+E_g}{\eta-E_g\left(\frac{E_g-3\delta'}{E_g+\delta'}\right)} = \frac{\eta+E_g}{\eta-E_g'}$ where $E_g' = \frac{E_g(E_g-3\delta')}{E_g+\delta'}$. Thus, $\frac{a_{k+}}{a_{k-}} = \sqrt{\frac{\eta+E_g}{\eta-E_g'}}$ similarly, $\frac{c_{k+}}{c_{k-}} = \frac{\gamma_{k+}}{\gamma_{k-}} = \sqrt{\frac{\eta-E_g}{\eta+E_g'}}$ and thus,

$$A(k) = -\left\{P\left(\frac{\eta+E_g}{\eta-E_g'}\right)^{1/2} + Q\left(\frac{\eta-E_g}{\eta+E_g}\right)^{1/2}\right\}\frac{\partial\gamma_{k-}^2}{\partial k_x} \tag{6.15}$$

where $P = \frac{r_0^2}{2}\left(\frac{E_g-\delta'}{E_g+\delta'}\right)$ and $Q = t^2/2$.

Now $\gamma_{k-}^2 = \frac{\eta+E_g}{2(\eta+\delta')}$ so that $\frac{\partial\gamma_{k-}^2}{\partial k_x} = \frac{1}{2}\left[\frac{\partial\eta/\partial k_x}{(\eta+\delta')} - \frac{\eta+E_g}{(\eta+\delta')^2}\frac{\partial\eta}{\partial k_x}\right]$

Thus,

$$\frac{\partial\gamma_{k-}^2}{\partial k_x} = \frac{1}{2}\left[\frac{\eta+\delta'-\eta-E_g}{(\eta+\delta')}\right]\frac{\partial\eta}{\partial k_x} = -\frac{1}{2}\frac{(E_g-\delta')}{(\eta+\delta')^2}\cdot\frac{\partial\eta}{\partial k_x} \tag{6.16}$$

From (6.15) and (6.16), we get

$$A(k) = \frac{1}{2}\frac{(E_g-\delta')}{(\eta+\delta')^2}\cdot\frac{\partial\eta}{\partial k_u}\cdot\left\{P\left(\frac{\eta+E_g}{\eta-E_g'}\right)^{1/2} + Q\left(\frac{\eta-E_g}{\eta+E_g}\right)^{1/2}\right\} \tag{6.17}$$

where

$$\frac{\partial\eta}{\partial k_x} = \frac{E_g\hbar^2}{m_r}\cdot\frac{k_x}{\eta} \tag{6.18}$$

From (6.17) and (6.18), one can write

$$A(k) = \frac{E_g\hbar^2}{2m_r}\cdot\frac{k_u}{\eta}\cdot\frac{(E_g-\delta')}{(\eta+\delta')^2}\cdot\left\{P\left(\frac{\eta+E_g}{\eta-E_g'}\right)^{1/2} + Q\left(\frac{\eta-E_g}{\eta+E_g}\right)^{1/2}\right\} \tag{6.19}$$

Thus,

$$|A(k)|^2 = \frac{E_g^2(E_g-\delta')\hbar^2}{4m_r}\frac{\hbar^2k_x^2}{m_r}\frac{1}{\eta^2}\frac{1}{(\eta+\delta')^4} \times\left\{P\left(\frac{\eta+E_g}{\eta-E_g'}\right)^{1/2} + Q\left(\frac{\eta-E_g}{\eta+E_g}\right)^{1/2}\right\}^2 \tag{6.20}$$

and

$$|X_{12}|^2 = |A(\bar{k})|^2. \tag{6.21}$$

From (6.20) and (6.21), we can write the square of the magnitude of the interband transition matrix element due to external electric field ($|X_{CV}^2|$) is given by (1.20).

It is well known that the energy eigenvalue, $E_n^{(2)}(\bar{k})$ in the presence of a perturbed Hamiltonian, H', is given by [4]

$$E_n^{(2)}\left(\bar{k}\right) = E_n\left(\bar{k}\right) + \left\langle n\bar{k}|H'|n\bar{k}\right\rangle + \left\{|\left\langle n\bar{k}|H'|n\bar{k}\right\rangle|^2 / \left[E_n\left(\bar{k}\right) - E_m\left(\bar{k}\right)\right]\right\} \tag{6.22}$$

where

$$H\psi_n\left(\bar{k},\bar{r}\right) = E\psi\left(\bar{k},\bar{r}\right) \tag{6.23}$$

$$H = H_0 + H' \tag{6.24}$$

$$H_0 u_n\left(\bar{k},\bar{r}\right) = E_n\left(\bar{k}\right) u_n\left(\bar{k},\bar{r}\right) \tag{6.25}$$

in which, H is the total Hamiltonian, $\psi(\bar{k},\bar{r})$ is the wave function, where $u_n(\bar{k},\bar{r})$ is the periodic function of it, H_0 is the unperturbed Hamiltonian, n is the band index, and $E_n(\bar{k})$ is the energy of an electron in the periodic lattice.

For an external electric field (F_S) applied along the x-axis, the perturbed Hamiltonian (H') can be written as

$$H' = -F \cdot x \tag{6.26}$$

where $F(= eF_S)$

Using (6.26) in (6.22), we get

$$E_n^{(2)}\left(\bar{k}\right) = E_n\left(\bar{k}\right) - F\left\langle n\bar{k}|H'|n\bar{k}\right\rangle + F^2\left\{|\left\langle n\bar{k}|H'|n\bar{k}\right\rangle|^2 / \left[E_n\left(\bar{k}\right) - E_m\left(\bar{k}\right)\right]\right\} \tag{6.27}$$

In (6.27), the second and the third terms are due to the perturbation factor.

For

$$X_{nm}\left(\bar{k}\right) = \left\langle n\bar{k}|x|m\bar{k}\right\rangle \tag{6.28}$$

we find

$$X_{nm}\left(\bar{k}\right) = i\int u_n^*\left(\bar{k},\bar{r}\right)(\partial/\partial u)\left[u_m\left(\bar{k},\bar{r}\right)\right]\mathrm{d}^3r \tag{6.29}$$

where k_x is the x component of the $\bar{k}$ and the integration in (6.29) extends over the unit cell. From (6.27), (6.28), and (6.29) with the n corresponds to the conduction band (C) and m corresponds to the valance band (V), we get

$$E_c^{(2)}\left(\bar{k}\right) = E_c\left(\bar{k}\right) - FX_{cc} + \left\{F^2|X_{CV}|^2 / \left[E_C\left(\bar{k}\right) - E_V\left(\bar{k}\right)\right]\right\} \tag{6.30}$$

Thus, combining the appropriate equations, the dispersion relation of the conduction electrons in the presence of electric field along x-axis can be written as

$$\gamma(E) = \frac{\hbar^2 k_x^2}{2m_c} + \frac{\hbar^2 k_y^2}{2m_c} + \frac{\hbar^2 k_z^2}{2m_c} + \frac{F^2|X_{12}|^2}{\eta}$$

$$= \left[\frac{\hbar^2 k_x^2}{2m_c} + \frac{\hbar^2 k_y^2}{2m_c} + \frac{\hbar^2 k_z^2}{2m_c} + \left\{\frac{\hbar^2 k_x^2}{2m_c} \cdot \frac{2m_c}{m_r} \cdot \frac{F^2 \hbar^2 E_g^2 (E_g - \delta')^2}{4m_r}\right.\right.$$

$$\left.\left. \times \frac{1}{\eta^3} \frac{1}{(\eta + \delta')^4} \left[P\left(\frac{\eta + E_g}{\eta - E_g'}\right)^{1/2} + Q\left(\frac{\eta - E_g}{\eta + E_g}\right)^{1/2}\right]^2\right\}\right] \quad (6.31)$$

When $F \to 0$, we have from (6.31), $k^2 \to \frac{2m_c}{\hbar^2}\gamma(E)$ and $\eta_1^2 = [E_g^2 + E_g \frac{2m_c}{m_r}\gamma(E)]$. Using the method of successive approximation, one can write

$$1 = \frac{\hbar^2 k_x^2}{2m_c\gamma(E)} + \frac{\hbar^2 k_y^2}{2m_c\gamma(E)} + \frac{\hbar^2 k_z^2}{2m_c\gamma(E)} + \frac{\hbar^2 k_x^2}{2m_c\gamma(E)} \cdot \Phi(E, F) \quad (6.32)$$

where $\Phi(E, F) = \frac{2m_c}{m_r} \frac{F^2\hbar^2 E_g^2 (E_g - \delta')^2}{4m_r} \frac{1}{\eta_1^3} \frac{1}{(\eta_1 + \delta')^4} \left[P\left(\frac{\eta_1 + E_g}{\eta_1 - E_g'}\right)^{1/2} + Q\left(\frac{\eta_1 - E_g}{\eta_1 + E_g}\right)^{1/2}\right]^2$

Therefore, the E–k dispersion relation in the presence of an external electric field for III–V, ternary, and quaternary materials whose unperturbed energy band structures are defined by the three-band model of Kane can be expressed as [5]

$$\frac{k_x^2}{\frac{2m_c}{\hbar^2}\left[\frac{\gamma(E)}{1+\Phi(E,F)}\right]} + \frac{k_y^2}{\frac{2m_c}{\hbar^2}\gamma(E)} + \frac{k_z^2}{\frac{2m_c}{\hbar^2}\gamma(E)} = 1 \quad (6.33)$$

In (6.33), the coefficients of k_x, k_y, and k_z are not same and for this reason, this basic equation is "anisotropic" in nature together with the fact that the anisotropic dispersion relation is the ellipsoid of revolution in the k-space.

From (6.33), the expressions of the effective electron masses along x-, y-, and z-directions can, respectively, be written as

$$m_x^*(E, F) = \hbar^2 k_x \left.\frac{\partial k_x}{\partial E}\right|_{k_y=0, k_z=0}$$

$$= m_c [1 + \Phi(E, F)]^{-2} \big[[1 + \Phi(E, F)]\,\gamma'(E) - \gamma(E)\Phi'(E, F)\big] \quad (6.34)$$

$$m_y^*(E, F) = \hbar^2 k_y \left.\frac{\partial k_y}{\partial E}\right|_{k_x=0, k_z=0} = m_0\gamma'(E) \quad (6.35)$$

$$m_z^*(E, F) = \hbar^2 k_z \left.\frac{\partial k_z}{\partial E}\right|_{k_x=0, k_y=0} = m_0\gamma'(E) \quad (6.36)$$

where $\gamma'(E) = \frac{\partial}{\partial E}(\gamma(E))$ and $\Phi'(E, F) = \frac{\partial}{\partial E}[\Phi(E, F)]$.

It may be noted from (6.34) that the effective mass along x-direction is a function of both electron energy and electric field, respectively, whereas from (6.35) and (6.36) we can infer the expressions of the effective masses along y- and z-directions are same and they depend on the electron energy only. Thus, in the presence of an electric field, the mass anisotropy for Kane type semiconductors depends both on electron energy and electric field, respectively.

The use of the usual approximation [3]

$$k_x^2 \approx \frac{1}{3}k^2 \tag{6.37}$$

in (6.33), leads to the simplified expression of the electron energy spectrum in the present case as

$$\gamma(E) = \frac{\hbar^2 k^2}{2m_c} + \frac{F^2\hbar^2 E_g^2 (E_g - \delta')^2}{12m_r} \frac{2m_c}{m_r}\gamma(E)\frac{1}{\eta_1}\frac{1}{(\eta_1+\delta')^4} \times \left\{ P\left(\frac{\eta_1 + E_g}{\eta_1 - E_g'}\right)^{1/2} + Q\left(\frac{\eta_1 - E_g}{\eta_1 + E_g}\right)^{1/2} \right\}^2 \tag{6.38}$$

The (6.38) can be written as

$$\beta(E,F) = \frac{\hbar^2 k^2}{2m_c} \tag{6.39}$$

where

$$\beta(E,F) = \gamma(E)\left[1 - \frac{1}{3}\Phi(E,F)\right] \tag{6.40}$$

Special Cases:

I) The ***E–k*** dispersion relation of III–V, ternary, and quaternary materials in the presence of an external electric field whose unperturbed band structures are defined by the two-band model of Kane.

Under the condition $\Delta \to 0$, the (6.33) assumes the form

$$\frac{k_x^2}{\frac{2m_c}{\hbar^2}\left[\frac{\gamma_0(E)}{1+\Phi_1(E,F)}\right]} + \frac{k_y^2}{\frac{2m_c}{\hbar^2}\gamma_0(E)} + \frac{k_z^2}{\frac{2m_c}{\hbar^2}\gamma_0(E)} = 1 \tag{6.41}$$

where $\gamma_0(E) = E(1+\alpha E)$ with $\alpha = 1/E_g$ and

$$\Phi_1(E,F) = \frac{\hbar^2 F^2}{4m_r E_g^2 \gamma_0(E)}\left[1 + \frac{2m_c}{m_r}\frac{\gamma_0(E)}{E_g}\right]^{-5/2} \tag{6.42}$$

Equation (6.41) represents the electron energy spectrum of III–V, ternary, and quaternary materials in the presence of an external electric field whose unperturbed band structures are defined by the two-band model of Kane.

From (6.33) along with the substitution $\Delta \to 0$ we get

$$\beta_1(E,F) = \gamma_0(E)\left[1 - \frac{1}{3}\Phi_1(E,F)\right] = \frac{\hbar^2 k^2}{2m_c} \tag{6.43}$$

where (6.43) represents the approximate ***E*–k** dispersion relation of III–V, ternary, and quaternary compounds in the presence of an external electric field whose unperturbed band structures are defined by the two-band model of Kane.

The dispersion relation of the conduction electrons in optoelectronic materials under electric field can be written in presence of quantizing magnetic field B along x-direction whose unperturbed electron energy spectra are defined by the three- and two-band models of Kane as

$$\left(n + \frac{1}{2}\right)\hbar w_0 + \frac{\hbar^2 k_x^2}{2m_c} = \beta_{11}(E,F) \tag{6.44}$$

$$\left(n + \frac{1}{2}\right)\hbar w_0 + \frac{\hbar^2 k_x^2}{2m_c} = \overline{\beta_{12}}(E,F) \tag{6.45}$$

$$\omega_0 = \frac{eB}{m_c} \tag{6.46}$$

where,

$$\beta_{11}(E,F) = \left[\gamma(E) - \overline{C_1}\cdot\left[\frac{\gamma(E)}{\phi^3(E)}\right]\left[\frac{T_1^2(E)}{(\phi(E)+\delta')^4}\right]\right],$$

$$\overline{C_1} = \left[\frac{m_c(\hbar eFE_g)^2(E_g-\delta')^2}{6m_r^2}\right]$$

$$\phi^2(E) = \left[E_g^2 + E_g\frac{m_c}{m_r}\gamma(E)\right],$$

$$T_1(E) = \left[P\left(\frac{\phi(E)+E_g}{\phi(E)-E_g'}\right)^{1/2} + Q\left(\frac{\phi(E)-E_g}{\phi(E)+E_g}\right)^{1/2}\right],$$

$$\overline{\beta_{12}}(E,F) = \left[E(1+\alpha E) - \delta_5\left[E(1+\alpha E) + \frac{m_r E_g}{2m_c}\right]^{-\frac{5}{2}}\right],$$

$$\delta_5 = \left[\frac{\hbar^2 F^2 m_r^{3/2} E_g^{1/2}}{12(2m_c)^{5/2}}\right] \tag{6.47}$$

From (6.44) and (6.45) we get,

$$k_x^2 = w_{11}(E,F,n) \tag{6.48}$$

and

$$k_x^2 = w_{12}(E, F, n) \tag{6.49}$$

where $w_{11}(E, F, n) = \frac{2m_c}{\hbar^2}\left[\beta_{11}(E, F) - \left(n + \frac{1}{2}\right)\hbar\omega_0\right]$ and $w_{12}(E, F, n) = \frac{2m_c}{\hbar^2}\left[\overline{\beta_{12}}(E, F) - \left(n + \frac{1}{2}\right)\hbar\omega_0\right]$

The density of states function for both the cases can, respectively, be expressed as

$$N(E) = \frac{g_v eB\sqrt{2m_c}}{2\pi^2\hbar^2}\sum_{n=0}^{n_{\max}}\frac{\beta'_{11}(E, F)H(E - E_{n1})}{\sqrt{\beta_{11}(E, F)}} \tag{6.50}$$

and

$$N(E) = \frac{g_v eB\sqrt{2m_c}}{2\pi^2\hbar^2}\sum_{n=0}^{n_{\max}}\frac{\{\overline{\beta'_{12}}(E, F)\}' H(E - E_{n1})}{\sqrt{\overline{\beta_{12}}(E, F)}} \tag{6.51}$$

where g_v is the valley degeneracy, the primes denote the differentiation of the differentiable functions with respect to E, H denotes the Heaviside step function, E_{n1} is the root of the equation

$$\left(n + \frac{1}{2}\right)\hbar\omega_0 = \beta_{11}(E, F) \tag{6.52}$$

E_{n2} is the root of the equation

$$\left(n + \frac{1}{2}\right)\hbar\omega_0 = \overline{\beta_{12}}(E, F) \tag{6.53}$$

Combining (6.50) and (6.51) with the Fermi–Dirac occupation probability factor, the electron concentration can, respectively, be expressed as

$$n_0 = \frac{g_v eB}{\pi^2\hbar}\sum_{n=0}^{n_{\max}}[Q_{11}(E_{FB}, F, n) + Q_{12}(E_{FB}, F, n)] \tag{6.54}$$

and

$$n_0 = \frac{g_v eB}{\pi^2\hbar}\sum_{n=0}^{n_{\max}}[Q_{13}(E_{FB}, F, n) + Q_{14}(E_{FB}, F, n)] \tag{6.55}$$

where $Q_{11}(E_{FB}, F, n) = [\omega_{11}(E_F, F, n)]^{1/2}$, $Q_{12}(E_{FB}, F, n) = \sum_{R=1}^{R_0} Z(R)[Q_{11}(E_{FB}, F, n)]$, $Z(R) = 2(k_B T)^{2R}(1 - 2^{1-2R})\xi(2R)\frac{\partial^{2R}}{\partial E_{FB}^{2R}}$, $\xi(2R)$ is the Zeta function of order $2R$, R is the set real positive integers whose upper limit is R_0,

$$Q_{13}(E_{FB}, F, n) = [\omega_{12}(E_F, F, n)]^{1/2}$$

and

$$Q_{14}(E_{FB}, F, n) = \sum_{R=1}^{R_0} Z(R)[Q_{13}(E_{FB}, F, n)].$$

The velocity of the electron along x-axis under magnetic quantization can, respectively, be expressed from (6.48) and (6.49) as

$$v_x(E) = \frac{\omega'_{11}(E, F, n)}{(2\hbar)\sqrt{\omega_{11}(E, F, n)}} \tag{6.56}$$

So that

$$v_x(E) = \frac{\omega'_{12}(E, F, n)}{(2\hbar)\sqrt{\omega_{12}(E, F, n)}} \tag{6.57}$$

The net current density due to field emission in the x-direction is given by

$$J = \frac{e}{2}\sum_{n=0}^{n_{\max}} [v_n(E)\,(\Delta n_0)] \cdot t \tag{6.58}$$

where $(1/2)$ is introduced due to the fact that the half of the electrons which are enable to contribute the emission will migrate back into the lattice, $v_n(E)$ is the velocity of the electrons in the Landau subband characterized by the Landau quantum number n, Δn_0 is the electron concentration in that particular level, and t is the transmission coefficient. The transmission coefficient for the perturbed three- and two-band models of Kane in this case are, respectively, given by

$$t = \exp(-\theta_{11}) \tag{6.59}$$

and

$$t = \exp(-\theta_{12}) \tag{6.60}$$

where

$$\theta_{11} = \frac{4[\omega_{11}(V_0, F, n)]^{3/2}}{3eF\omega'_{11}(V_0, F)},$$

$$w_{11}(V_0, F, n) = \frac{2m_c}{\hbar^2}\left[\beta_{11}(V_0, F) - \left(n + \frac{1}{2}\right)\hbar\omega_0\right],$$

$$\omega'_{11}(V_0, F) = \frac{2m_c}{\hbar^2}\beta'_{11}(V_0, F),$$

$$\beta'_{11}(V_0, F) = \left[\gamma'(V_0) - \frac{\overline{C}_1\gamma'(V_0)}{\phi^3(V_0)} \cdot \frac{T_1^2(V_0)}{[\phi(V_0) + \delta']^4} + \frac{2\overline{C}_1\gamma(V_0)T_1(V_0)T_1'(V_0)}{\phi^3(V_0)[\phi(V_0) + \delta']^4}\right.$$
$$\left. - \frac{4\overline{C}_1\gamma(V_0)T_1^2(V_0)\phi'(V_0)}{\phi^3(V_0)[\phi(V_0) + \delta']^5} + \frac{4\overline{C}_1\gamma(V_0)T_1^2(V_0)\phi'(V_0)}{\phi^3(V_0)[\phi(V_0) + \delta']^5}\right],$$

$$\gamma'(V_0) = \gamma(V_0)\left[\frac{1}{V_0} + \frac{1}{V_0 + E_g} + \frac{1}{V_0 + E_g + \Delta} + -\frac{1}{V_0 + E_g + \frac{2}{3}\Delta}\right],$$

$$\phi'(V_0) = \left[\frac{E_g m_c \gamma'(V_0)}{2m_r\phi(V_0)}\right],$$

$$\phi'(V_0) = \left(\frac{\phi'(V_0)}{2}\right) \cdot \frac{2QE_g[\phi(V_0) - E_g]^{-1/2}}{[\phi(V_0) + E_g]^{3/2}}$$
$$- \frac{(E_g' + E_g)P[\phi(V_0) + E_g]^{-1/2}}{[\phi(V_0) - E_g']^{3/2}}$$

$$\theta_{12} = \frac{4[\omega_{12}(V_0, F, n)]^{3/2}}{3eF\omega'_{12}(V_0, F)}, \quad \omega'_{12}(V_0, F) = \frac{2m_c}{\hbar^2}[\bar{\beta}_{12}(V_0, F)]', \quad \text{and}$$

$$[\bar{\beta}_{12}(V_0, F)]' = (1 + 2\alpha(V_0) \cdot \left[1 + \frac{5}{2}\delta_5\left[V_0(1 + \alpha V_0) + \frac{m_r E_g}{2m_c}\right]^{\frac{-7}{2}}\right]$$

Using the appropriate equations, the field emitted current density in this case can be expressed as:

$$J = \frac{e^2 B k_B T g_v}{2\pi^2\hbar^2}\sum_{n=0}^{n_{\max}} F_0(\eta_{11})\exp(-\theta_{11}) \tag{6.61}$$

$$J = \frac{e^2 B k_B T g_v}{2\pi^2\hbar^2}\sum_{n=0}^{n_{\max}} F_0(\eta_{12})\exp(-\theta_{12}) \tag{6.62}$$

$\eta_{11} = \frac{E_{FB} - E_{n1}}{k_B T}$, $\eta_{12} = \frac{E_{FB} - E_{n2}}{k_B T}$, $F_j(\eta)$ is the one parameter Fermi–Dirac integral of order j which has been written in (1.10) of Chap. 1.Under the condition of extreme of degeneracy and in the absence of band nonparabolicity and also neglecting the modification of band structures of the semiconductors under intense electric field, the summation over n can be converted to the integral over n leading to the well-known result as given by (1.27) of Chap. 1.

6.2.2 Field Emission from Quantum Wires of Optoelectronic Semiconductors

From (6.44) and (6.45), the one-dimensional motion of the electron for quantum wires of optoelectronic materials can be expressed as

$$G(n_y, n_z) + \frac{\hbar^2 k_x^2}{2m_c} = \beta_{11}(E, F) \tag{6.66}$$

$$G(n_y, n_z) + \frac{\hbar^2 k_x^2}{2m_c} = \overline{\beta_{12}}(E, F) \tag{6.67}$$

where $G(n_y, n_z) = \frac{\hbar^2 \pi^2}{2m_c}[(n_y / d_y)^2 + (n_z / d_z)^2]$

The electron concentration per unit length assumes the forms

$$n_0 = \frac{2g_v}{\pi} \sum_{n_z=1}^{n_{z\max}} \sum_{n_y=1}^{n_{y\max}} [Q_{15}(E_{F1D}, F, n_y, n_z) + Q_{16}(E_{F1D}, F, n_y, n_z)] \tag{6.68}$$

$$n_0 = \frac{2g_v}{\pi} \sum_{n_z=1}^{n_{z\max}} \sum_{n_y=1}^{n_{y\max}} [Q_{17}(E_{F1D}, F, n_y, n_z) + Q_{18}(E_{F1D}, F, n_y, n_z)] \tag{6.69}$$

where $Q_{15}(E_{F1D}, F, n_y, n_z) = \sqrt{\omega_{15}(E_{F1D}, F, n_y, n_z)}, \omega_{15}(E_{F1D}, F, n_y, n_z) = \frac{2m_c}{\hbar^2}$ $[\beta_{11}(E_{F1D}, F) - G(n_y, n_z)]$, $Q_{16}(E_{F1D}, F, n_y, n_z) = \sum_{R=1}^{R_0} Z(R_{1D})[Q_{15}(E_{F1D}, F, n_y, n_z)]$, $Z(R_{1D}) = 2(k_B T)^{2R}(1-2^{1-2R})\xi(2R)\frac{\partial^{2R}}{\partial E_{F1D}^{2R}}$, E_{F1D} is the Fermi energy for one-dimensional system in the present case as measured from the edge of the conduction band in vertically upward direction in the absence of any quantization,

$$Q_{17}(E_{F1D}, F, n_y, n_z) = \sqrt{\omega_{16}(E_{F1D}, F, n_y, n_z)},\ \omega_{16}(E_{F1D}, F, n_y, n_z) = \frac{2m_c}{\hbar^2}[\bar{\beta}_{12}(E_{F1D}, F) - G(n_y, n_z)]$$

$$Q_{18}(E_{F1D}, F, n_y, n_z) = \sum_{R=1}^{R_0} Z(R_{1D})[Q_{17}(E_{F1D}, F, n_y, n_z)]$$

The transmission coefficient for both the cases can, respectively, be expressed as

$$t = \exp(-\theta_{13}) \tag{6.70}$$

$$t = \exp(-\theta_{14}) \tag{6.71}$$

where $\theta_{13} = \frac{4[\omega_{15}(V_0, F, n_y, n_z)]^{3/2}}{3eF\omega'_{15}(V_0, F)}$, $\omega_{15}(V_0, F, n_y, n_z) = \frac{2m_c}{\hbar^2}[\beta_{11}(V_0, F) - G(n_y, n_z)]$, $\omega'_{15}(V_0, F) = \frac{2m_c}{\hbar^2}[\beta'_{11}(V_0, F)]$, $\theta_{14} = \frac{4[\omega_{16}(V_0, F, n_y, n_z)]^{3/2}}{3eF\omega'_{16}(V_0, F)}$, $\omega_{16}(V_0, F, n_y, n_z) = \frac{2m_c}{\hbar^2}$ $[\bar{\beta}_{12}(V_0, F) - G(n_y, n_z)]$, and $\omega'_{16}(V_0, F) = \frac{2m_c}{\hbar^2}[\bar{\beta}_{12}(V_0, F)]'$

The field emitted current in this case can respectively be written as

$$I = \frac{eg_v k_B T}{\pi\hbar} \sum_{n_y=1}^{n_{y\max}} \sum_{n_z=1}^{n_{z\max}} F_0(\eta_{13}) \exp(-\theta_{13}) \tag{6.72}$$

$$I = \frac{eg_v k_B T}{\pi\hbar} \sum_{n_y=1}^{n_{y\max}} \sum_{n_z=1}^{n_{z\max}} F_0(\eta_{14}) \exp(-\theta_{14}) \tag{6.73}$$

$\eta_{13} = \frac{E_{F1D} - E_{1,n_y,n_z}}{k_B T}$, E_{1,n_y,n_z} is the root of the equation

$$0 = [\beta_{11}(E_{1,n_y,n_z}, F) - G(n_y, n_z)] \tag{6.74}$$

$\eta_{14} = \frac{E_{F1D} - E_{2,n_y,n_z}}{k_B T}$, E_{2,n_y,n_z} is the root of the equation

$$0 = [\bar{\beta}_{12}(E_{2,n_y,n_z}, F) - G(n_y, n_z)] \tag{6.75}$$

6.2.3 Field Emission from Effective Mass Superlattices of Optoelectronic Semiconductors Under Magnetic Quantization

The dispersion relation of the conduction electrons in effective mass superlattices of optoelectronic semiconductors can be expressed following Sasaki [5] as

$$a_1 \cdot \cos[c_1(E, F, E_{g1}, \Delta_1)a_0 + c_2(E, F, E_{g2}, \Delta_2)b_0] - a_2 \cdot \cos[c_1(E, F, E_{g1}, \Delta_1)a_0 - c_2(E, F, E_{g2}, \Delta_2)b_0] = \cos(L_0 k) \tag{6.76}$$

where $a_1 = \left[\left[1 + \sqrt{\frac{m_{c2}}{m_{c1}}}\right]^2 \cdot \left[4\left(\sqrt{\frac{m_{c2}}{m_{c1}}}\right)^{1/2}\right]^{-1}\right]$,

$a_2 = \left[\left[-1 + \sqrt{\frac{m_{c2}}{m_{c1}}}\right]^2 \cdot \left[4\left(\sqrt{\frac{m_{c2}}{m_{c1}}}\right)^{1/2}\right]^{-1}\right]$,

$$c_i^2(E, F, E_{gi}, \Delta_i) = \frac{2m_{ci}}{\hbar^2}[\beta_{1i}(E, F, E_{gi}, \Delta_i) - k_\perp^2], \quad i = 1, 2,$$

$$k_\perp^2 = k_y^2 + k_z^2, \beta_{1i}(E, F, E_{gi}, \Delta_i) = \left[\gamma(E, E_{gi}, \Delta_i) - \left[\frac{L(E_{gi}, \Delta_i, m_{ri})\gamma(E, E_{gi}, \Delta_i) T_i^2(E, E_{gi}, \Delta_i)}{\phi_i^3(E, E_{gi}, \Delta_i)[\phi_i(E, E_{gi}, \Delta_i) + \delta_i']^4}\right]\right]$$

$$\gamma\left(E, E_{gi}, \Delta_i\right) = \frac{E\left(E + E_{gi}\right)\left(E + E_{gi} + \Delta_i\right)\left(E_{gi} + \frac{2}{3}\Delta_i\right)}{E_{gi}\left(E_{gi} + \Delta_i\right)\left(E + E_{gi} + \frac{2}{3}\Delta_i\right)},$$

$$L\left(E_{gi}, \Delta_i, m_{ri}\right) = \frac{(\hbar eF)^2\left(E_{gi} - \delta_i'\right)^2 m_{ci}}{6m_{ri}^2},$$

$$\delta_i' = \frac{(E_{gi})^2\Delta_i}{\chi_i}, \quad \chi_i = [6\left(E_{gi}\right)^2 + 9E_{gi}\cdot\Delta_i + 4\Delta_i^2], \quad \frac{1}{m_{ri}} = \left(\frac{1}{m_{ci}} + \frac{1}{m_{vi}}\right),$$

$$T_i(E, E_{gi}, \Delta_i) = \left[P_i\left[\frac{\{\phi_i(E, E_{gi}, \Delta_i) + E_{gi}\}}{\{\phi_i(E, E_{gi}, \Delta_i) - E_{gi}'\}}\right]^{1/2}\right.$$
$$\left. + Q_i\left[\frac{\{\phi_i(E, E_{gi}, \Delta_i) - E_{gi}\}}{\{\phi_i(E, E_{gi}, \Delta_i) + E_{gi}\}}\right]^{1/2}\right]$$

$$P_i = \frac{r_{0i}^2}{2}\left(\frac{E_{gi} - \delta_i'}{E_{gi} + \delta_i'}\right), \quad r_{0i}^2 = \left[6\left(E_{gi} + \frac{2}{3}\Delta_i\right)(E_{gi} + \Delta_i)\right][\chi_i]^{-1}, \phi_i^2(E, E_{gi}, \Delta_i)$$
$$= [E_{gi}^2 + E_{gi}(m_{ci}/m_{ri})\gamma(E, E_{gi}, \Delta_i)],$$

$$E_{gi}' = \left[\frac{E_{gi}(E_{gi} - 3\delta_i')}{(E_{gi} + \delta_i')}\right], \quad Q_i = \frac{t_i^2}{2} \quad \text{and} \quad t_i^2 = \left[\frac{6\left(E_{g_i} + \frac{2}{3}\Delta_i\right)}{\chi_i}\right].$$

In the presence of a quantizing magnetic field B, along z-direction the magneto-energy spectrum assumes the form

$$k_x^2 = \omega_{17}(E, F, n) \tag{6.77}$$

where $\omega_{17}(E, F, n) = \frac{1}{L_0^2}[\cos^{-1}\{f_1(E, F, n)\}]^2 - \frac{2eB}{\hbar}\left(n + \frac{1}{2}\right)$,

$$f_1(E, F, n) = [a_1\cos[c_1(E, F, E_{g_1}, \Delta_1, n)a_0 + b_0c_2(E, F, E_{g_2}, \Delta_2, n)]$$
$$- a_2\cos[c_1(E, F, E_{g_1}, \Delta_1, n)a_0 - b_0c_2(E, F, E_{g_2}, \Delta_2, n)]]$$

and

$$c_i^2(E, F, E_{g_i}, \Delta_i, n) = \left[\left(\frac{2m_{c_i}}{\hbar^2}\right)[\beta_{1_i}(E, F, E_{g_i}, \Delta_i)] - \frac{2eB}{\hbar}\left(n + \frac{1}{2}\right)\right].$$

The electron concentration assumes the form

$$n_o = \frac{g_v eB}{\pi^2\hbar}\left[\sum_{n=0}^{n_{\max}}[Q_{19}(E_{FB}, F, n) + Q_{20}(E_{FB}, F, n)]\right] \tag{6.78}$$

where $Q_{19}(E_{FB}, F, n) = [\omega_{17}(E_{FB}, F, n)]^{1/2}$ and

$$Q_{20}(E_{FB}, F, n) = \sum_{R=1}^{R=R_0} Z(R)[Q_{19}(E_{FB}, F, n)]$$

The transmission coefficient in this case is given by

$$t = \exp(-\theta_{15}) \tag{6.79}$$
$$\theta_{15} = \frac{4[\omega_{17}(V_0, F, n)]^{3/2}}{3eF\omega'_{17}(V_0, F, n)},$$

where

$$\omega'_{17}(V_0, F, n) = [(2/L_0^2) f_1'(V_0, F, n)\cos^{-1}[f_1(V_0, F, n)][1 - f_1^2(V_0, F, n)]^{-1/2}]$$

$$\begin{aligned} f_1'(V_0, F, n) = {} & [a_1 \sin[c_1(V_0, F, E_{g_1}, \Delta_1, n)a_0 + b_0 c_2(V_0, F, E_{g_2}, \Delta_2, n)] \\ & \cdot [c_1'(V_0, F, E_{g_1}, \Delta_1, n)a_0 + b_0 c_2'(V_0, F, E_{g_2}, \Delta_2, n)] \\ & - a_2 \sin[c_1(E, F, E_{g_1}, \Delta_1, n)a_0 - b_0 c_2(E, F, E_{g_2}, \Delta_2, n_0)] \\ & \cdot [c_1'(E, F, E_{g_1}, \Delta_1, n)a_0 - b_0 c_2'(E, F, E_{g_2}, \Delta_2, n_0)]] \end{aligned}$$

$$c_i'(V_0, F, E_{g_i}, \Delta_i) = \left(\frac{m_{c_i}}{\hbar^2}\right)[c_i(V_0, F, E_{g_i}, \Delta_i)]^{-1}[\beta'_{1i}(V_0, F, E_{g_i}, \Delta_i)]$$

$$\begin{aligned} & \beta'_{1i}(V_0, F, E_{g_i}, \Delta_i) \\ & = \left[\gamma'(V_0, E_{g_i}, \Delta_i) - \frac{L(E_{g_i}, \Delta_i, m_{r_i})\gamma'(V_0, E_{g_i}, \Delta_i)T_i^2(V_0, E_{g_i}, \Delta_i)}{\phi_i^3(V_0, E_{g_i}, \Delta_i)[\phi_i(V_0, E_{g_i}, \Delta_i) + \delta_i']^4}\right. \\ & \quad - \frac{2L(E_{g_i}, \Delta_i, m_{r_i})\gamma(V_0, E_{g_i}, \Delta_i)T_i(V_0, E_{g_i}, \Delta_i)T_i'(V_0, E_{g_i}, \Delta_i)}{\phi_i^3(V_0, E_{g_i}, \Delta_i)[\phi_i(V_0, E_{g_i}, \Delta_i) + \delta_i']^4} \\ & \quad + \frac{3L(E_{g_i}, \Delta_i, m_{r_i})\gamma(V_0, E_{g_i}, \Delta_i)T_i^2(V_0, E_{g_i}, \Delta_i)\phi_i'(V_0, E_{g_i}, \Delta_i)}{\phi_i^4(V_0, E_{g_i}, \Delta_i)[\phi_i(V_0, E_{g_i}, \Delta_i) + \delta_i']^4} \\ & \quad + \frac{4L(E_{g_i}, \Delta_i, m_{r_i})\gamma(V_0, E_{g_i}, \Delta_i)T_i^2(V_0, E_{g_i}, \Delta_i)\phi_i'(V_0, E_{g_i}, \Delta_i)}{\phi_i^3(V_0, E_{g_i}, \Delta_i)[\phi_i(V_0, E_{g_i}, \Delta_i) + \delta_i']^5} \end{aligned}$$

$$\begin{aligned} \gamma'(V_0, E_{g_i}, \Delta_i) = \gamma(V_0, E_{g_i}, \Delta_i)\Bigg[& \frac{1}{V_0} + \frac{1}{V_0 + E_{g_i}} + \frac{1}{V_0 + E_{g_i} + \Delta_i} \\ & - \frac{1}{V_0 + E_{g_i} + \frac{2}{3}\Delta_i}\Bigg] \end{aligned}$$

$$\phi_i'(V_0, E_{g_i}, \Delta_i) = \frac{E_{g_i} m_{c_i}\gamma'(V_0, E_{g_i}, \Delta_i)}{2m_{r_i}\phi_i(V_0, E_{g_i}, \Delta_i)}$$

$$T_i'(V_0, E_{g_i}, \Delta_i) = \left[\frac{\phi_i'(V_0, E_{g_i}, \Delta_i)}{2}\right]\left[\frac{2E_{g_i} Q_i[\phi_i(V_0, E_{g_i}, \Delta_i) - E_{g_i}]^{-1/2}}{[\phi_i(V_0, E_{g_i}, \Delta_i) - E_{g_i}]^{3/2}} - (E'_{g_i} + E_{g_i})P_i[\phi_i(V_0, E_{g_i}, \Delta_i) - E_{g_i}]^{-1/2}[\phi_i(V_0, E_{g_i}, \Delta_i) - E'_{g_i}]^{-3/2}\right]$$

The photo-emitted current density in this case is given by

$$J = \frac{e^2 B k_B T g_v}{2\pi^2 \hbar^2} \sum_{n=0}^{n_{max}} F_0(\eta_{15}) \exp(-\theta_{15}) \tag{6.80}$$

where $\eta_{15} = \frac{E_{FB} - E_{n3}}{k_B T}$, E_{n3} is the root of the equation

$$\omega_{17}(E_{n3}, F, n) = 0 \tag{6.81}$$

The electron concentration and the field emitted current density in this case when the dispersion relations of the constituent materials are defined by the perturbed two-band model of Kane, can respectively be expressed as

$$n_0 = \frac{g_v e B}{\pi^2 \hbar}\left[\sum_{n=0}^{n_{max}} [Q_{21}(E_{FB}, F, n) + Q_{22}(E_{FB}, F, n)]\right] \tag{6.82}$$

and

$$J = \frac{e^2 B k_B T g_v}{2\pi^2 \hbar^2} \sum_{n=0}^{n_{max}} F_0(\eta_{16}) \exp(-\theta_{16}) \tag{6.83}$$

where $Q_{21}(E_{FB}, F, n) = [\omega_{18}(E_{FB}, F, n)]^{1/2}$, $Q_{22}(E_{FB}, F, n) = \sum_{R=1}^{R=R_o} Z(R)[Q_{21}(E_{FB}, F, n)]$

$$\omega_{18}(E, F, n) = \left[\frac{1}{L_o^2}[\cos^{-1}\{f_2(E, F, n)\}]^2 - \frac{2eB}{\hbar}\left(n + \frac{1}{2}\right)\right]$$

$$f_2(E, F, n) = [a_1 \cos[D_1(E, F, E_{g_1}, n)a_o + b_o D_2(E, F, E_{g_2}, n)] - a_2 \cos[D_1(E, F, E_{g_1}, n)a_o - b_o D_2(E, F, E_{g_2}, n)]]$$

$$D_i^2(E, F, E_{gi}, n) = \left[\frac{2m_{ci}}{\hbar^2}\rho_{1i}(E, F, E_{gi}) - \frac{2eB}{\hbar}\left(n + \frac{1}{2}\right)\right],$$

$$\rho_{1i}(E, F, E_{gi}) = \left[E(1 + \alpha_i E) - \delta_{5i}\left[E(1 + \alpha_i E) + \frac{m_{ri} E_{gi}}{2m_{ci}}\right]^{-\frac{5}{2}}\right],$$

$$\delta_{5i} = \left[\frac{(\hbar e F)^2 m_{ri}^{3/2} (E_{gi})^{1/2}}{12(2m_{ci})^{5/2}}\right]$$

$\eta_{16} = \frac{E_{FB}-E_{n4}}{k_B T}$, E_{n4} is the root of the equation

$$\omega'_{18}(V_0, F, n) = [(2/L_0^2) f'_2(V_0, F, n) \cos^{-1}[f_2(V_0, F, n)] \times [1 - f_2^2(V_0, F, n)]^{-1/2}] \tag{6.84}$$

$$f'_2(V_0, F, n) = [a_1 \sin[D_1(V_0, F, E_{g_1}, n)a_0 + b_0 D_2(V_0, F, E_{g_2}, n)] \cdot [D'_1(V_0, F, E_{g_1}, n)a_0 + b_0 D'_2(V_0, F, E_{g_2}, n)] - a_2 \sin[D_1(E, F, E_{g_1}, n)a_0 - b_0 D_2(E, F, E_{g_2}, n_0)] \cdot [D'_1(E, F, E_{g_1}, n)a_0 - b_0 D'_2(E, F, E_{g_2}, n_0)]]$$

$$D'_i(V_0, F, E_{gi}, n) = \left[\frac{m_{ci}\rho'_{1i}(V_0, F, E_{gi})}{\hbar^2 D_i(V_0, F, E_{gi}, n)}\right]$$

and

$$\rho'_{1i}(V_0, F, E_{gi}) = \left[1 + \frac{2V_0}{E_{gi}}\right]\left[1 + \frac{5}{2}\delta_{5i}\left[V_0(1 + \alpha_i V_0) + \frac{m_{ri}E_{gi}}{2m_{ci}}\right]^{-7/2}\right]$$

6.2.4 *Field Emission from Quantum Wire Effective Mass Superlattices of Optoelectronic Semiconductors*

The dispersion relation of the conduction electrons for quantum wire effective mass superlattices in accordance with the perturbed three-band model of Kane is given by

$$k_x^2 = \omega_{19}(E, F, n_y, n_z) \tag{6.85}$$

where, $\omega_{19}(E, F, n_y, n_z) = \left[\frac{1}{L_0^2}\left[\cos^{-1}\{f_3(E, F, n_y, n_z)\}\right]^2 - H(n_y, n_z)\right]$

$$f_3(E, F, n_y, n_z) = [a_1 \cos[e_1(E, F, E_{g_1}, \Delta_1, n_y, n_z)a_o + b_o e_2(E, F, E_{g_2}, \Delta_2, n_y, n_z)] - a_2 \cos[e_1(E, F, E_{g_1}, \Delta_1, n_y, n_z)a_o - b_o e_2(E, F, E_{g_2}, \Delta_2, n_y, n_z)]], \tag{6.86}$$

$$e_i^2(E, F, E_{g_i}, \Delta_i, n_y, n_z) = \left[\left(\frac{2m_{c_i}}{\hbar^2}\right)[\beta_{1_i}(E, F, E_{g_i}, \Delta_i)] - H\left(n_y, n_z\right)\right] \tag{6.87}$$

where $H\left(n_y, n_z\right) = \left[\left(\frac{n_y\pi}{d_y}\right)^2 + \left(\frac{n_z\pi}{d_z}\right)^2\right]$

The expression of the electron concentration in this case can be written as

$$n_0 = \frac{2g_v}{\pi} \sum_{n_y=1}^{n_{y\max}} \sum_{n_z=1}^{n_{z\max}} \left[Q_{23}\left(E_{\text{FIDEMSL}}, F, n_y, n_z\right) + Q_{24}\left(E_{\text{FIDEMSL}}, F, n_y, n_z\right)\right] \tag{6.88}$$

where $Q_{23}\left(E_{\text{FIDEMSL}}, F, n_y, n_z\right) = \sqrt{\omega_{19}\left(E_{\text{FIDEMSL}}, F, n_y, n_z\right)}$,

$$Q_{24}\left(E_{\text{FIDEMSL}}, F, n_y, n_z\right) = \sum_{R=1}^{R=R_0} Z\left(R_{\text{IDEMSL}}\right) Q_{23}\left(E_{\text{FIDEMSL}}, F, n_y, n_z\right),$$

E_{FIDEMSL} is the Fermi energy in the present case and

$$Z\left(R_{\text{IDEMSL}}\right) = 2\left(k_B T\right)^{2R}\left(1 - 2^{1-2R}\right)\xi\left(2R\right)\frac{\partial^{2R}}{\partial E_{\text{FIDEMSL}}}.$$

The field emitted current assumes the form

$$I = \frac{e g_v k_B T}{\pi \hbar} \sum_{n_y=1}^{n_{y\max}} \sum_{n_z=1}^{n_{z\max}} F_0\left(\eta_{17}\right)\exp\left(-\theta_{17}\right) \tag{6.89}$$

where $\eta_{17} = \frac{E_{\text{FIDEMSL}} - E_{15}}{k_B T}$. E_{15} is the root of the equation

$$\omega_{17}\left(E_{15}, F, n_y, n_z\right) = 0 \tag{6.90}$$

$$\theta_{17} = \frac{4\left[\omega_{19}\left(V_0, F, n_y, n_z\right)\right]^{3/2}}{3eF\left[\omega_{19}'\left(V_0, F, n_y, n_z\right)\right]},$$

$$\omega_{19}'\left(V_0, F, n_y, n_z\right) = \left[1 - f_3^2\left(V_0, F, n_y, n_z\right)\right]^{-1/2} \times \left[\left(2/L_0^2\right) f_3'\left(V_0, F, n_y, n_z\right)\right]\left[\cos^{-1}\left\{f_3\left(V_0, F, n_y, n_z\right)\right\}\right],$$

$$\begin{aligned} &f_3'(V_0, F, n_y, n_z) \\ &= a_1 \sin\left[a_0 e_1(V_0, F, E_{g_1}, \Delta_1, n_y, n_z) + b_o e_2(V_0, F, E_{g_2}, \Delta_2, n_y, n_z)\right] \\ &\cdot\left[a_0 e_1'(V_0, F, E_{g_1}, \Delta_1, n_y, n_z) + b_o e_2'(V_0, F, E_{g_2}, \Delta_2, n_y, n_z)\right] \\ &- a_2 \sin\left[a_0 e_1(V_0, F, E_{g_1}, \Delta_1, n_y, n_z) - b_o e_2(V_0, F, E_{g_2}, \Delta_2, n_y, n_z)\right] \\ &\cdot\left[a_0 e_1'(V_0, F, E_{g_1}, \Delta_1, n_y, n_z) - b_o e_2'(V_0, F, E_{g_2}, \Delta_2, n_y, n_z)\right] \end{aligned}$$

and

$$e_i'\left(V_0, F, E_{gi}, \Delta_i, n_y, n_z\right) = \frac{m_{ci}\beta_{1i}\left(V_0, F, E_{gi}, \Delta_i\right)}{\hbar^2 e_i\left(V_0, F, E_{gi}, \Delta_i, n_y, n_z\right)}.$$

In accordance with the perturbed two-band model of Kane, the electron concentration per unit length is given by,

$$n_0 = \frac{2g_v}{\pi}\sum_{n_y=1}^{n_{y\max}}\sum_{n_z=1}^{n_{z\max}}\left[Q_{25}\left(E_{\text{FIDEMSL}}, F, n_y, n_z\right) + Q_{26}\left(E_{\text{FIDEMSL}}, F, n_y, n_z\right)\right] \tag{6.91}$$

where, $Q_{25}\left(E_{\text{FIDEMSL}}, F, n_y, n_z\right) = \left[\sqrt{\omega_{20}\left(E_{\text{FIDEMSL}}, F, n_y, n_z\right)}\right]$, $Q_{26}\big(E_{\text{FIDEMSL}}, F, n_y, n_z\big) = \sum_{R=1}^{R=R_0} Z\left(R_{\text{IDEMSL}}\right)\left[Q_{25}\left(E_{\text{FIDEMSL}}, F, n_y, n_z\right)\right]$,

$$\omega_{20}\left(E, F, n_y, n_z\right) = \left[\frac{1}{L_0^2}\left[\cos^{-1} f_4\left(E, F, n_y, n_z\right)\right]^2 - H\left(n_y, n_z\right)\right],$$

$$\begin{aligned} f_4\left(E, F, n_y, n_z\right) = \big[&a_1\cos\left[a_0 g_1\left(E, F, E_{g1}, n_y, n_z\right) + b_0 g_2\left(E, F, E_{g2}, n_y, n_z\right)\right] \\ &- a_2\cos\left[a_0 g_1\left(E, F, E_{g1}, n_y, n_z\right) - b_0 g_2\left(E, F, E_{g2}, n_y, n_z\right)\right]\big] \end{aligned}$$

and

$$g_i^2\left(E, F, E_{gi}, n_y, n_z\right) = \left[\frac{2m_{ci}}{\hbar^2}\theta_{1i}\left(E, F, E_{gi}\right) - H\left(n_y, n_z\right)\right]$$

The field emitted current can be written as,

$$I = \frac{eg_v k_B T}{\pi\hbar}\sum_{n_y=1}^{n_{y\max}}\sum_{n_z=1}^{n_{z\max}} F_0\left(\eta_{18}\right)\exp\left(-\theta_{18}\right) \tag{6.92}$$

where, $\eta_{18} = \frac{E_{\text{FIDEMSL}} - E_{16}}{k_B T}$, E_{16} is the root of the equation

$$\omega_{20}\left(E_{16}, F, n_y, n_z\right) = 0 \tag{6.93}$$

$$\theta_{18} = \frac{4\left[\omega_{20}\left(V_0, F, n_y, n_z\right)\right]^{3/2}}{3eF\omega_{20}'\left(V_0, F, n_y, n_z\right)},$$

$$\omega_{20}'\left(V_0, F, n_y, n_z\right) = \frac{\left(2/L_0^2\right) f_4'\left(V_0, F, n_y, n_z\right)\left[\cos^{-1} f_4\left(V_0, F, n_y, n_z\right)\right]}{\sqrt{1 - f_4^2\left(V_0, F, n_y, n_z\right)}},$$

$$f_4'(V_0, F, n_y, n_z) = a_1 \sin\left[a_0 g_1(V_0, F, E_{g1}, n_y, n_z) + b_0 g_2(V_0, F, E_{g2}, n_y, n_z)\right] \\ \times \left[a_0 g_1'(V_0, F, E_{g1}, n_y, n_z) + b_0 g_2'(V_0, F, E_{g2}, n_y, n_z)\right] \\ - a_2 \sin\left[a_0 g_1(V_0, F, E_{g1}, n_y, n_z) - b_0 g_2(V_0, F, E_{g2}, n_y, n_z)\right] \\ \times \left[a_0 g_1(V_0, F, E_{g1}, n_y, n_z) - b_0 g_2(V_0, F, E_{g2}, n_y, n_z)\right],$$

and

$$g_i'(V_0, F, n_y, n_z) = \frac{m_{ci}\theta_{1i}'(V_0, F, E_{gi})}{\hbar^2 g_i(V_0, F, E_{gi}, n_y, n_z)}.$$

6.2.5 Field Emission from Superlattices of Optoelectronic Semiconductors with Graded Interfaces Under Magnetic Quantization

The energy spectrum in superlattices of optoelectronic compounds with graded interfaces in the presence of electric field whose constituent materials are defined by perturbed three-band model of Kane can be written following [5] as

$$\cos(L_0 k) = \frac{1}{2}\Phi_{11}(E, k_s) \tag{6.94}$$

where

$$\Phi_{11}(E, k_s) = \Bigg[2\cosh\{X_{21}(E, k_s)\}\cos\{Y_{21}(E, k_s)\} \\ + \varepsilon_{21}(E, k_s)\sinh\{X_{21}(E, k_s)\}\sin\{Y_{21}(E, k_s)\} \\ + \Delta_{21}\left[\left(\frac{K_{21}^2(E, k_s)}{K_{22}(E, k_s)} - 3K_{22}(E, k_s)\right)\cosh\{X_{21}(E, k_s)\}\right. \\ \times \sin\{Y_{21}(E, k_s)\} + \left(3K_{21}(E, k_s) - \frac{\{K_{22}(E, k_s)\}^2}{K_{21}(E, k_s)}\right) \\ \times \sinh\{X_{21}(E, k_s)\}\cos\{Y_{21}(E, k_s)\}] \\ + \Delta_{21}\Bigg[2\left(\{K_{21}(E, k_s)\}^2 - \{K_{22}(E, k_s)\}^2\right)\cos h\{X_{21}(E, k_s)\} \\ \times \cos\{Y_{21}(E, k_s)\} + \frac{1}{12}\left[\frac{5\{K_{22}(E, k_s)\}^3}{K_{21}(E, k_s)} + \frac{5\{K_{21}(E, k_s)\}^3}{K_{22}(E, k_s)}\right. \\ \left. - 34K_{22}(E, k_s)K_{21}(E, k_s)\right]\sinh\{X_{21}(E, k_s)\}\sin\{Y_{21}(E, k_s)\}\Bigg]\Bigg],$$

$$X_{21}(E,k_s) = K_{21}(E,k_s)[a_0 - \Delta_{21}],$$

$$K_{21}(E,k_s) \equiv \left[-\frac{2m_{c2}}{\hbar^2}\beta_{012}(E-\overline{V_0},F,E_{g2},\Delta_2)+k_s^2\right]^{1/2},$$

$$\beta_{012}\left(E-\overline{V_0},F,E_{g2},\Delta_2\right)$$

$$= \left[\gamma(E-\overline{V_0},E_{g2},\Delta_2) - \frac{L(E_{g2},\Delta_2,m_{r2})\gamma(E-\overline{V_0},E_{g2},\Delta_2)T_2^2(E-\overline{V_0},E_{g2},\Delta_2)}{\phi_2^3(E-\overline{V_0},E_{g2},\Delta_2)[\phi_2(E-\overline{V_0},E_{g2},\Delta_2)+\delta_2']^4}\right],$$

$$\varepsilon(E,k_s) \equiv \left[\frac{K_1(E,k_s)}{K_2(E,k_s)} - \frac{K_2(E,k_s)}{K_1(E,k_s)}\right],$$

$k_s^2 = k_x^2 + k_y^2$, $Y_{21}(E,k_s) = K_{22}(E,k_s)[b_0 - \Delta_{21}]$ and

$$K_{22}(E,k_s) = \left[\frac{2m_{c1}\beta_{11}(E,F,E_{g1},\Delta_1)}{\hbar^2} - k_s^2\right]^{1/2}.$$

In the presence of a quantizing magnetic field B along z-direction, the simplified magneto-dispersion relation can be written as

$$k_z^2 = \omega_{21}(E,F,n) \tag{6.95}$$

where $\omega_{21}(E,F,n) = \left[\frac{1}{L_0^2}[\cos^{-1}[\frac{1}{2}f_{11}(E,F,n)]]^2 - \frac{2|e|B}{\hbar}\left(n+\frac{1}{2}\right)\right]$,

$$f_{11}(E,F,n) = \Bigg[2\cosh\{M_{21}(n,E)\}\cos\{N_{21}(n,E)\}$$
$$+ Z_{21}(n,E)\sinh\{M_{21}(n,E)\}\sin\{N_{21}(n,E)\}$$
$$+ \Delta_{21}\left[\left(\frac{\{I_{21}(n,E)\}^2}{I_{22}(n,E)} - 3I_{22}(n,E)\right)\right.$$
$$\times \cosh\{M_{21}(n,E)\}\sin\{N_{21}(n,E)\}$$
$$\left.+\left(3I_{21}(n,E) - \frac{\{I_{22}(n,E)\}^2}{I_{21}(n,E)}\right)\sinh\{M_{21}(n,E)\}\cos\{N_{21}(n,E)\}\right]$$
$$+ \Delta_{21}\Bigg[2\left(\{I_{21}(n,E)\}^2 - \{I_{22}(n,E)\}^2\right)$$
$$\times \cosh\{M_{21}(n,E)\}\cos\{N_{21}(n,E)\}$$
$$+ \frac{1}{12}\left(\frac{5\{I_{22}(n,E)\}^3}{I_{21}(n,E)} + \frac{5\{I_{21}(n,E)\}^3}{I_{22}(n,E)} - \{34I_{22}(n,E)I_{21}(n,E)\}\right)$$
$$\times \sinh\{M_{21}(n,E)\}\sin\{N_{21}(n,E)\}\Bigg]\Bigg]$$

$$Z_{21}(n,E) \equiv \left[\frac{I_{21}(n,E)}{I_{22}(n,E)} - \frac{I_{22}(n,E)}{I_{21}(n,E)}\right], \quad M_{21}(n,E) = I_{21}(n,E)\left[a_0 - \Delta_{21}\right],$$

$$I_{21}(n,E) = \left[-\frac{2m_{c2}}{\hbar^2}\beta_{012}(E-\overline{V_0}, F, E_{g2}, \Delta_2) + \frac{2|e|B}{\hbar}\left(n+\frac{1}{2}\right)\right]^{1/2}$$

$$N_{21}(n,E) = I_{22}(n,E)\left[b_0 - \Delta_{21}\right]$$

and

$$I_{22}(n,E) \equiv \left[\frac{2m_{c1}}{\hbar^2}\beta_{11}(E,F,E_{g1},\Delta_1) - \left\{\frac{2|e|B}{\hbar}\left(n+\frac{1}{2}\right)\right\}\right]^{1/2}.$$

The electron concentration is given by

$$n_0 = \frac{g_v eB}{\pi^2\hbar}\left[\sum_{n=0}^{n_{\max}}\left[Q_{27}(E_{\text{FBGISL}},F,n) + Q_{28}(E_{\text{FBGISL}},F,n)\right]\right] \tag{6.96}$$

$Q_{27}(E_{\text{FBGISL}},F,n) = \left[\omega_{21}(E_{\text{FBGISL}},F,n)\right]^{1/2}$, E_{FBGISL} is the Fermi energy in the present case,

$$Q_{28}\left(E_{\text{FBGISL}},F,n\right) = \sum_{R=1}^{R=R_0} Z\left(R_{\text{BGISL}}\right)\left[Q_{27}\left(E_{\text{FBGISL}},F,n\right)\right]$$

and

$$Z\left(R_{\text{BGISL}}\right) = 2\left(k_B T\right)^2\left(1-2^{1-2R}\right)\xi\left(2R\right)\frac{\partial^{2R}}{\partial E_{\text{FBGISL}}}.$$

The field emitted current density is given by,

$$J = \frac{e^2 B k_B T g_v}{2\pi^2\hbar^2}\sum_{n=0}^{n_{\max}} F_0(\eta_{19})\exp(-\theta_{19}) \tag{6.97}$$

where $\eta_{19} = \frac{E_{\text{FBGISL}} - E_{19}}{k_B T}$, E_{19} is the root of the equation

$$\omega_{21}(E_{19},F,n) = 0 \tag{6.98}$$

$$\theta_{19} = \frac{4\left[\omega_{21}(V_0,F,n)\right]^{3/2}}{3eF\omega_{21}'\left(V_0,F,n\right)},$$

$$\begin{aligned} f_{11}'\left(V_0,F,n\right) = \Big[& 2M_{21}'(n,V_0)\sinh\left\{M_{21}(n,V_0)\right\}\cos\left\{N_{21}(n,V_0)\right\} \\ & + Z_{21}(n,V_0)M_{21}'(n,V_0)\cosh\left\{M_{21}(n,V_0)\right\}\sin\left\{N_{21}(n,V_0)\right\} \\ & - 2N_{21}'(n,V_0)\sin\left\{N_{21}(n,V_0)\right\}\cosh\left\{M_{21}(n,V_0)\right\} \\ & + Z_{21}'(n,V_0)\sinh\left\{M_{21}(n,V_0)\right\}\sin\left\{N_{21}(n,V_0)\right\} \end{aligned}$$

$$+ Z_{21}(n, V_0)N'_{21}(n, V_0)\cos\{N_{21}(n, V_0)\}\sinh\{M_{21}(n, V_0)\}$$

$$+ \Delta_{21}\left[\left(\frac{\{2I_{21}(n, V_0)I'_{21}(n, V_0)\}}{I_{22}(n, V_0)} - \frac{\{I_{21}^2(n, V_0)I'_{22}(n, V_0)\}}{I_{22}^2(n, V_0)}\right.\right.$$

$$\left. - 3I'_{22}(n, V_0)\right)\cosh\{M_{21}(n, V_0)\}\sin\{N_{21}(n, V_0)\}$$

$$+ \left(-3I_{22}(n, V_0) + \frac{\{I_{21}(n, V_0)\}^2}{I_{22}(n, V_0)}\right)\{M'_{21}(n, V_0)\}\sinh\{M_{21}(n, V_0)\}$$

$$\times \sin\{N_{21}(n,V_0)\} + \{N'_{21}(n,V_0)\cosh\{M_{21}(n,V_0)\}\cos\{N_{21}(n,V_0)\}$$

$$+ \left(\frac{-\{2I_{22}(n, V_0)I'_{22}(n, V_0)\}}{I_{21}(n, V_0)} + \frac{\{I_{22}^2(n, V_0)I'_{21}(n, V_0)\}}{I_{21}^2(n, V_0)}\right.$$

$$\left. + 3I'_{21}(n, V_0)\right)\sinh\{M_{21}(n, V_0)\}\cos\{N_{21}(n, V_0)\}$$

$$+ \left(+3I_{21}(n, V_0) - \frac{\{I_{22}(n, V_0)\}^2}{I_{21}(n, V_0)}\right)\{M'_{21}(n, V_0)\cosh\{M_{21}(n, V_0)\}$$

$$\left.\times \cos\{N_{21}(n,V_0)\} - N'_{21}(n,V_0)\sin\{N_{21}(n,V_0)\}\sinh\{M_{21}(n,V_0)\}\}\right]$$

$$+ \Delta_{21}\left[4\left(\{I_{21}(n, V_0)I'_{21}(n, V_0)\} - \{I_{22}(n, V_0)I'_{22}(n, V_0)\}\right)\right.$$

$$\times \cosh\{M_{21}(n, V_0)\}\cos\{N_{21}(n, V_0)\}$$

$$+ 2\left(\{I_{21}(n,V_0)\}^2 - \{I_{22}(n,V_0)\}^2\right)\{M'_{21}(n,V_0)\sinh\{M_{21}(n,V_0)\}$$

$$\times \cos\{N_{21}(n,V_0)\} - N'_{21}(n,V_0)\cosh\{M_{21}(n,V_0)\}\sin\{N_{21}(n,V_0)\}\}$$

$$+ \frac{1}{12}\left(\frac{15\{I_{22}^2(n, V_0)\}I'_{22}(n, V_0)}{I_{21}(n, V_0)} - \frac{5\{I_{22}(n, V_0)\}^3 I'_{21}(n, V_0)}{I_{21}^2(n, V_0)}\right.$$

$$+ \frac{15\{I_{21}^2(n, V_0)\}I'_{21}(n, V_0)}{I_{22}(n, V_0)} - \frac{5\{I_{21}(n, V_0)\}^3 I'_{22}(n, V_0)}{I_{22}^2(n, V_0)}$$

$$\left. - \{34I'_{22}(n, V_0)I_{21}(n, V_0)\} - 34I_{22}(n, V_0)I'_{21}(n, V_0)\right)$$

$$\times \sinh\{M_{21}(n, V_0)\}\sin\{N_{21}(n, V_0)\}$$

$$+ \left(\frac{5\{I_{22}(n,V_0)\}^3}{I_{21}(n, V_0)} + \frac{5\{I_{21}(n,V_0)\}^3}{I_{22}(n,V_0)} - \{34I_{22}(n,V_0)I_{21}(n,V_0)\}\right)$$

$$\times\left\{M'_{21}(n,V_0)\cos h\left\{M_{21}(n,V_0)\right\}\sin\left\{N_{21}(n,V_0)\right\}\right.$$

$$\left.\left.\left.+N'_{21}(n,V_0)\sin h\left\{M_{21}(n,V_0)\right\}\cos\left\{N_{21}(n,V_0)\right\}\right\}\right]\right],$$

$$M'_{21}(n,V_0)=I'_{21}(n,V_0)\left[a_0-\Delta_{21}\right],\quad I'_{21}\left(V_0,n\right)=\frac{m_{c_2}\beta'_{012}\left(V_0-\overline{V}_0,F,E_{g2},\Delta_2\right)}{-\hbar^2 I_{21}\left(V_0,n\right)},$$

$$\beta'_{012}(V_0-\overline{V}_0,F,E_{g2},\Delta_2)=\left[(\gamma'(V_0-\overline{V}_0,E_{g2},\Delta_2))\right.$$

$$-\frac{L(E_{g_2},\Delta_2,m_{r2})\gamma'(V_0-\overline{V}_0,E_{g2},\Delta_i)T_i^2(V_0-\overline{V}_0,E_{g2},\Delta_2)}{\phi_i^3(V_0-\overline{V}_0,E_{g_2},\Delta_2)[\phi_2(V_0-\overline{V}_0,E_{g_2},\Delta_2)+\delta'_2]^4}$$

$$-\frac{2L(E_{g_2},\Delta_2,m_{r_2})\gamma(V_0-\overline{V}_0,E_{g_2},\Delta_2)T_2(V_0-\overline{V}_0,E_{g_2},\Delta_2)T_2{}'(V_0-\overline{V}_0,E_{g_2},\Delta_2)}{\phi_2^3(V_0-\overline{V}_0,E_{g_2},\Delta_2)[\phi_2(V_0-\overline{V}_0,E_{g_2},\Delta_2)+\delta'_2]^4}$$

$$+\frac{3L(E_{g_2},\Delta_2,m_{r_2})\gamma(V_0-\overline{V}_0,E_{g_2},\Delta_2)T_2^2(V_0-\overline{V}_0,E_{g_2},\Delta_2)\phi'_2(V_0-\overline{V}_0,E_{g_2},\Delta_2)}{\phi_i^4(V_0-\overline{V}_0,E_{g_2},\Delta_2)[\phi_2(V_0-\overline{V}_0,E_{g_2},\Delta_2)+\delta'_2]^4}$$

$$\left.+\frac{4L(E_{g_2},\Delta_2,m_{r2})\gamma(V_0-\overline{V}_0,E_{g_2},\Delta_2)T_2^2(V_0-\overline{V}_0,E_{g_2},\Delta_2)\phi'_2(V_0-\overline{V}_0,E_{g_2},\Delta_2)}{\phi_2^3(V_0-\overline{V}_0,E_{g_2},\Delta_2)[\phi_2(V_0-\overline{V}_0,E_{g_2},\Delta_2)+\delta'_2]^5}\right],$$

$$\gamma'(V_0-\overline{V}_0,E_{g_2},\Delta_2)=\gamma(V_0-\overline{V}_0,E_{g_2},\Delta_2)\left[\frac{1}{V_0-\overline{V}_0}+\frac{1}{V_0-\overline{V}_0+E_{g2}}\right.$$

$$\left.+\frac{1}{V_0-\overline{V}_0+E_{g2}+\Delta_2}-\frac{1}{V_0-\overline{V}_0+E_{g_2}+\frac{2}{3}\Delta_2}\right],$$

$$\gamma\left(V_0-\overline{V}_0,E_{g_2},\Delta_2\right)=\frac{(V_0-\overline{V}_0)(V_0-\overline{V}_0+E_{g2})(V_0-\overline{V}_0+E_{g2}+\Delta_2)\left(E_{g_2}+\frac{2}{3}\Delta_2\right)}{E_{g2}\left(E_{g_2}+\Delta_2\right)\left(V_0-\overline{V}_0+E_{g_2}+\frac{2}{3}\Delta_2\right)},$$

$$L\left(E_{g_2},\Delta_2,m_{r_2}\right)=\frac{(\hbar eF)^2\left(E_{g_2}-\delta'_2\right)^2 m_{c2}}{6m_{r_2}^2},\quad \delta'_2=\frac{\left(E_{g_2}\right)^2\Delta_2}{\chi_2},$$

$$\chi_2=\left[6\left(E_{g_2}\right)^2+9E_{g_2}.\Delta_2+4\Delta_2^2\right],\quad \frac{1}{m_{r_2}}=\left(\frac{1}{m_{c2}}+\frac{1}{m_{v_2}}\right),$$

$$T_2(V_0-\overline{V}_0,E_{g_2},\Delta_2)=\left[P_2\left[\frac{\{\phi_2(V_0-\overline{V}_0,E_{g_2},\Delta_2)+E_{g2}\}}{\{\phi_i(V_0-\overline{V}_0,E_{g_2},\Delta_2)-E'_{g_2}\}}\right]^{1/2}\right.$$

$$\left.+Q_2\left[\frac{\{\phi_2(V_0-\overline{V}_0,E_{g_2},\Delta_2)-E_{g2}\}}{\{\phi_2(V_0-\overline{V}_0,E_{g_2},\Delta_2)+E_{g2}\}}\right]^{1/2}\right]$$

$$P_2=\frac{r_{02}^2}{2}\left(\frac{E_{g2}-\delta_2'}{E_{g_2}+\delta_2'}\right),\quad r_{o2}^2=\left[6\left(E_{g_2}+\frac{2}{3}\Delta_2\right)(E_{g_2}+\Delta_2)\right][\chi_2]^{-1},$$

$$\phi_2^2(V_0-\overline{V}_0,E_{g_2},\Delta_2)=\left[E_{g_2}^2+E_{g_2}(m_{c_2}/m_{r_2})\gamma(V_0-\overline{V}_0,E_{g_2},\Delta_2)\right],$$

$$E_{g_2}'=\left[\frac{E_{g_2}(E_{g_2}-3\delta_2')}{(E_{g_2}+\delta_2')}\right],\quad Q_2=\frac{t_2^2}{2},\quad t_2^2=\left[\frac{6\left(E_{g_2}+\frac{2}{3}\Delta_2\right)}{\chi_2}\right],$$

$$\phi_2'(V_0-\overline{V}_0,E_{g_2},\Delta_2)=\frac{E_{g_2}m_{c_2}\gamma'(V_0-\overline{V}_0,E_{g_2},\Delta_2)}{2m_{r_2}\phi_2(V_0-\overline{V}_0,E_{g_2},\Delta_2)},$$

$$\begin{aligned}T_2'(V_0-\overline{V}_0,E_{g_2},\Delta_2)=&\left[\frac{\phi_2'(V_0-\overline{V}_0,E_{g_2},\Delta_2)}{2}\right]\\&\times\left[\frac{2E_{g_2}Q_2[\phi_2(V_0-\overline{V}_0,E_{g_2},\Delta_2)-E_{g2}]^{-1/2}}{[\phi_2(V_0-\overline{V}_0,E_{g_2},\Delta_2)-E_{g_2}]^{3/2}}\right.\\&-(E_{g_2}'+E_{g_2})P_2[\phi_2(V_0-\overline{V}_0,E_{g_2},\Delta_2)-E_{g_2}]^{-1/2}\\&\left.\times[\phi_2(V_0-\overline{V}_0,E_{g_2},\Delta_2)-E_{g_2}']^{-3/2}\right],\end{aligned}$$

$$I_{22}'(V_0,n)=\frac{m_{c1}\beta_{11}'\left(V_0,F,E_{g1},\Delta_1\right)}{\hbar^2 I_{22}(V_0,n)},$$

$$I_{21}(V_0,n)=\left[\frac{2eB}{\hbar}\left(n+\frac{1}{2}\right)-\frac{2m_{c2}}{\hbar^2}\beta_{012}\left(V_0-\overline{V}_0,F,E_{g2,}\Delta_2\right)\right]^{\frac{1}{2}},$$

$$\begin{aligned}N_{21}(V_0,n)&=I_{22}(V_0,n)\left[b_0-\Delta_{21}\right],\ I_{22}(V_0,n)\\&=\left[\frac{2m_{c1}\beta_{11}\left(V_0,F,E_{g1},\Delta_1\right)}{\hbar^2}-\frac{2eB}{\hbar}\left(n+\frac{1}{2}\right)\right]^{\frac{1}{2}},\end{aligned}$$

$$\begin{aligned}Z_{21}'(V_0,n)=&\left[\frac{I_{21}'(V_0,n)}{I_{22}(V_0,n)}-\frac{I_{21}(V_0,n)I_{22}'(V_0,n)}{I_{22}^2(V_0,n)}\right.\\&\left.-\frac{I_{22}'(V_0,n)}{I_{21}(V_0,n)}+\frac{I_{22}(V_0,n)I_{21}'(V_0,n)}{I_{21}^2(V_0,n)}\right]\end{aligned}$$

For perturbed two-band model of Kane, the forms of the electron concentration and field emitted current density remain same where

$$I_{21}(E,n)=\left[\frac{2eB}{\hbar}\left(n+\frac{1}{2}\right)-\frac{2m_{c2}}{\hbar^2}\rho_{012}\left(E-\overline{V}_0,F,E_{g2}\right)\right]^{\frac{1}{2}},$$

$$\rho_{012}\left(E-\overline{V}_0,F,E_{g2}\right)=\left[\left(E-\overline{V}_0\right)\left[1+\alpha_2\left(E-\overline{V}_0\right)\right]\right.$$
$$\left.-\delta_{52}\left[\left(E-V_0\right)\left[1+\alpha_2\left(E-\overline{V}_0\right)\right]+\frac{m_{r2}E_{g2}}{2m_{c2}}\right]^{-5/2}\right]$$

$$\delta_{52}=\left[\frac{(\hbar eF)^2 m_{r2}^{3/2}\left(E_{g2}\right)^{1/2}}{12\left(2m_{c2}\right)^{5/2}}\right],$$

$$I_{22}(E,n)=\left[\frac{2m_{c1}\rho_{11}\left(E,F,E_{g1}\right)}{\hbar^2}-\frac{2eB}{\hbar}\left(n+\frac{1}{2}\right)\right]^{\frac{1}{2}},$$

$$\rho_{11}\left(E,F,E_{g1}\right)=\left[E(1+\alpha_1E)-\delta_{51}\left[E(1+\alpha_1E)+\frac{m_{r1}E_{g1}}{2m_{c1}}\right]^{-\frac{5}{2}}\right],\quad \alpha_i=\frac{1}{E_{gi}},$$

$$\delta_{51}=\left[\frac{(\hbar eF)^2 m_{r1}^{3/2}\left(E_{g1}\right)^{1/2}}{12\left(2m_{c1}\right)^{5/2}}\right],\quad I'_{21}(V_0,n)=\frac{m_{c2}\rho'_{012}\left(V_0-\overline{V}_0,F,E_{g2}\right)}{-\hbar^2 I_{21}(V_0,n)},$$

$$\rho'_{012}\left(V_0-\overline{V}_0,F,E_{g2}\right)=\left[\left[1+2\alpha_2\left(V_0-\overline{V}_0\right)\right]\left[1+\frac{5}{2}\delta_{52}\left(V_0-\overline{V}_0\right)\right.\right.$$
$$\left.\left.\times\left[1+\alpha_2\left(V_0-\overline{V}_0\right)\right]+\frac{m_{r2}E_{g2}}{2m_{c2}}\right]^{-7/2}\right]$$

$$I'_{22}(V_0,n)=\left[\frac{2m_{c1}\rho'_{11}\left(V_0,F,E_{g1}\right)}{\hbar^2 I_{22}(V_0,n)}\right]$$

and

$$\rho'_{11}\left(V_0,F,E_{g1}\right)=\left[\left[1+2\alpha_2\left(V_0\right)\right]\left[1+\frac{5}{2}\delta_{51}\left(V_0\right)\left[1+\alpha_1\left(V_0\right)\right]+\frac{m_{r1}E_{g1}}{2m_{c1}}\right]^{-7/2}\right].$$

6.2.6 *Field Emission from Quantum Wire Superlattices of Optoelectronic Semiconductors with Graded Interfaces*

The dispersion relation in accordance with the perturbed three-band model of Kane, in this case is given by

$$k_x^2=\omega_{22}(E,F,n_y,n_z) \tag{6.99}$$

where $\omega_{22}\left(E,F,n_y,n_z\right)=\left[\frac{1}{L_0^2}\left[\cos^{-1}\frac{1}{2}f_{13}\left(E,F,n_y,n_z\right)\right]^2-H\left(n_y,n_z\right)\right]$,

$$
\begin{aligned}
f_{13}\left(E,F,n_y,n_z\right) = \Bigg[& 2\cosh\left\{M_{31}\left(n_y,n_z,E\right)\right\}\cos\left\{N_{31}\left(n_y,n_z,E\right)\right\} \\
& + Z_{31}\left(n_y,n_z,E\right)\sinh\{M_{31}(n_y,n_z,E)\}\sin\{N_{31}(n_y,n_z,E)\} \\
& + \Delta_{21}\left[\left(\frac{\left\{I_{31}\left(n_y,n_z,E\right)\right\}^2}{I_{32}\left(n_y,n_z,E\right)} - 3I_{32}\left(n_y,n_z,E\right)\right)\right. \\
& \times \cosh\left\{M_{31}\left(n_y,n_z,E\right)\right\}\sin\left\{N_{31}\left(n_y,n_z,E\right)\right\} \\
& + \left(3I_{31}\left(n_y,n_z,E\right) - \frac{\left\{I_{32}\left(n_y,n_z,E\right)\right\}^2}{I_{31}\left(n_y,n_z,E\right)}\right) \\
& \times \sinh\left\{M_{31}\left(n_y,n_z,E\right)\right\}\cos\left\{N_{31}\left(n_y,n_z,E\right)\right\}] \\
& + \Delta_{21}\left[2\left(\left\{I_{31}\left(n_y,n_z,E\right)\right\}^2 - \left\{I_{32}\left(n_y,n_z,E\right)\right\}^2\right)\right. \\
& \times \cosh\left\{M_{31}\left(n_y,n_z,E\right)\right\}\cos\left\{N_{31}\left(n_y,n_z,E\right)\right\} \\
& + \frac{1}{12}\left(\frac{5\left\{I_{32}\left(n_y,n_z,E\right)\right\}^3}{I_{31}\left(n_y,n_z,E\right)} + \frac{5\left\{I_{31}\left(n_y,n_z,E\right)\right\}^3}{I_{32}\left(n_y,n_z,E\right)}\right. \\
& \qquad \left. - \left\{34I_{32}\left(n_y,n_z,E\right)I_{31}\left(n_y,n_z,E\right)\right\}\right) \\
& \left.\times \sinh\left\{M_{31}\left(n_y,n_z,E\right)\right\}\sin\left\{N_{31}\left(n_y,n_z,E\right)\right\}\right]\Bigg]
\end{aligned}
$$

$$
Z_{31}\left(n_y,n_z,E\right) \equiv \left[\frac{I_{31}\left(n_y,n_z,E\right)}{I_{32}\left(n_y,n_z,E\right)} - \frac{I_{32}\left(n_y,n_z,E\right)}{I_{31}\left(n_y,n_z,E\right)}\right],
$$

$$
M_{31}\left(n_y,n_z,E\right) = I_{31}\left(n_y,n_z,E\right)\left[a_0 - \Delta_{21}\right],
$$

$$
I_{31}\left(n_y,n_z,E\right) = \left[-\frac{2m_{c2}}{\hbar^2}\beta_{012}(E-\overline{V_0},F,E_{g2},\Delta_2) + H(n_y,n_z)\right]^{1/2},
$$

$$
N_{31}\left(n_y,n_z,E\right) = I_{32}\left(n_y,n_z,E\right)\left[b_0 - \Delta_{21}\right]
$$

and

$$
I_{32}\left(n_y,n_z,E\right) \equiv \left[\frac{2m_{c1}}{\hbar^2}\beta_{11}(E,F,E_{g1},\Delta_1) - H(n_y,n_z)\right]^{1/2}.
$$

The electron concentration per unit length is given by

$$
n_0 = \frac{2g_v}{\pi}\sum_{n_y=1}^{n_{y\max}}\sum_{n_z=1}^{n_{z\max}}\left[Q_{29}\left(E_{\mathrm{FQWGISL}},F,n_y,n_z\right) + Q_{30}\left(E_{\mathrm{FQWGISL}},F,n_y,n_z\right)\right] \tag{6.100}
$$

where $Q_{29}(E_{\mathrm{FQWGISL}}, F, n_y, n_z) = [\sqrt{\omega_{22}(E_{\mathrm{FQWGISL}}, F, n_y, n_z)}]$,

$$Q_{30}\left(E_{\mathrm{FQWGISL}}, F, n_y, n_z\right) = \sum_{R=1}^{R=R_0} Z\left(R_{\mathrm{FQWGISL}}\right)\left[Q_{29}\left(E_{\mathrm{FQWGISL}}, F, n_y, n_z\right)\right],$$

$Z(R_{\mathrm{FQWGSL}}) = 2(k_B T)^{2R}(1-2^{1-2R})\xi(2R)\frac{\partial^{2R}}{\partial E_{\mathrm{FQWGSL}}}$ and E_{FQWGSL} is the Fermi energy in the present case. The field emitted current can be written as

$$I = \frac{e g_v k_B T}{\pi \hbar} \sum_{n_y=1}^{n_{y\max}} \sum_{n_z=1}^{n_{z\max}} F_0\left(\eta_{20}\right) \exp\left(-\theta_{20}\right) \tag{6.101}$$

$\eta_{20} = \frac{E_{\mathrm{FQWGISL}} - E_{20}}{k_B T}$, E_{20} is the root of the equation

$$\omega_{22}(E_{20}, F, n_y, n_z) = 0 \tag{6.102}$$

$$\theta_{20} = \frac{4\left[\omega_{22}\left(V_0, F, n_y, n_z\right)\right]^{3/2}}{3eF\omega'_{22}\left(V_0, F, n_y, n_z\right)},$$

$$\omega'_{22}\left(V_0, F, n_y, n_z\right) = \frac{2f'_{13}\left(V_0, F, n_y, n_z\right)\left[\cos^{-1}\{\frac{1}{2}f_{13}\left(V_0, F, n_y, n_z\right)\}\right]}{L_0^2\sqrt{4 - f_{13}^2\left(V_0, F, n_y, n_z\right)}},$$

$$\begin{aligned}
&f'_{13}(V_0, F, n_y, n_z)\\
&= \Bigg[2M'_{31}(n_y, n_z, V_0)\sinh\{M_{31}(n_y, n_z, V_0)\}\cos\{N_{31}(n_y, n_z, V_0)\}\\
&\quad + Z_{31}(n_y, n_z, V_0)\, M'_{31}(n_y, n_z, V_0)\cosh\{M_{31}(n_y, n_z, V_0)\}\\
&\quad \times \sin\{N_{31}(n_y, n_z, V_0)\} - 2N'_{31}(n_y, n_z, V_0)\sin\{N_{31}(n_y, n_z, V_0)\}\\
&\quad \times \cosh\{M_{31}(n_y, n_z, V_0)\} + Z'_{31}(n_y, n_z, V_0)\sinh\{M_{31}(n_y, n_z, V_0)\}\\
&\quad \times \sin\{N_{31}(n_y, n_z, V_0)\} + Z_{31}(n_y, n_z, V_0)\, N'_{31}(n_y, n_z, V_0)\\
&\quad \times \cos\{N_{31}(n_y, n_z, V_0)\}\sinh\{M_{31}(n_y, n_z, V_0)\}\\
&\quad + \Delta_{21}\Bigg[\Bigg(\frac{\{2I_{31}(n_y, n_z, V_0)\, I'_{31}(n_y, n_z, V_0)\}}{I_{32}(n_y, n_z, V_0)}\\
&\quad - \frac{\{I_{31}^2(n_y, n_z, V_0)\, I'_{32}(n_y, n_z, V_0)\}}{I_{32}^2(n_y, n_z, V_0)} - 3I'_{32}(n_y, n_z, V_0)\Bigg)\\
&\quad \cdot \cosh\{M_{31}(n_y, n_z, V_0)\}\sin\{N_{31}(n_y, n_z, V_0)\}\\
&\quad + \left(-3I_{32}(n_y, n_z, V_0) + \frac{\{I_{31}^2(n_y, n_z, V_0)\}}{I_{32}(n_y, n_z, V_0)}\right)\{M'_{31}(n_y, n_z, V_0)\\
&\quad \times \sinh\{M_{31}(n_y, n_z, V_0)\}\sin\{N_{31}(n_y, n_z, V_0)\}
\end{aligned}$$

$$
\begin{aligned}
&+\{N'_{31}(n_y,n_z,V_0)\cosh\{M_{31}(n_y,n_z,V_0)\}\cos\{N_{31}(n_y,n_z,V_0)\}\\
&+\left(\frac{-\{2I_{32}(n_y,n_z,V_0)\,I'_{32}(n_y,n_z,V_0)\}}{I_{31}(n_y,n_z,V_0)}\right.\\
&\left.+\frac{\{I^2_{32}(n_y,n_z,V_0)\,I'_{31}(n_y,n_z,V_0)\}}{I^2_{31}(n_y,n_z,V_0)}+3I'_{31}(n_y,n_z,V_0)\right)\\
&\times\sinh\{M_{31}(n_y,n_z,V_0)\}\cos\{N_{31}(n_y,n_z,V_0)\}\\
&+\left(+3I_{31}(n_y,n_z,V_0)-\frac{\{I_{32}(n_y,n_z,V_0)\}^2}{I_{31}(n_y,n_z,V_0)}\right)\{M'_{31}(n_y,n_z,V_0)\\
&\times\cosh\{M_{31}(n_y,n_z,V_0)\}\cos\{N_{31}(n_y,n_z,V_0)\}-N'_{31}(n_y,n_z,V_0)\\
&\left.\times\sin\{N_{31}(n_y,n_z,V_0)\}\sinh\{M_{31}(n_y,n_z,V_0)\}\}\right]\\
&+\Delta_{21}\left[4(\{I_{31}(n_y,n_z,V_0)\,I'_{31}(n_y,n_z,V_0)\}\right.\\
&-\{I_{32}(n_y,n_z,V_0)\,I'_{32}(n_y,n_z,V_0)\})\\
&\times\cosh\{M_{31}(n_y,n_z,V_0)\}\cos\{N_{31}(n_y,n_z,V_0)\}\\
&+2\left(\{I_{31}(n_y,n_z,V_0)\}^2-\{I_{32}(n_y,n_z,V_0)\}^2\right)\\
&\times\{M'_{31}(n_y,n_z,V_0)\sin h\{M_{31}(n_y,n_z,V_0)\}\cos\{N_{31}(n_y,n_z,V_0)\}\\
&-N'_{31}(n_y,n_z,V_0)\cosh\{M_{31}(n_y,n_z,V_0)\}\sin\{N_{31}(n_y,n_z,V_0)\}\}\\
&+\frac{1}{12}\left(\frac{15\{I^2_{32}(n_y,n_z,V_0)\}\,I'_{32}(n_y,n_z,V_0)}{I_{21}(n,V_0)}\right.\\
&-\frac{5\{I_{32}(n_y,n_z,V_0)\}^3\,I'_{31}(n_y,n_z,V_0)}{I^2_{31}(n_y,n_z,V_0)}\\
&+\frac{15\{I^2_{31}(n_y,n_z,V_0)\}\,I'_{31}(n_y,n_z,V_0)}{I_{32}(n_y,n_z,V_0)}\\
&-\frac{5\{I_{31}(n_y,n_z,V_0)\}^3\,I'_{32}(n_y,n_z,V_0)}{I^2_{32}(n_y,n_z,V_0)}\\
&-\{34I'_{32}(n_y,n_z,V_0)\,I_{31}(n_y,n_z,V_0)\}-34I_{32}(n_y,n_z,V_0)\\
&\left.\times I'_{31}(n_y,n_z,V_0)\right)\sinh\{M_{31}(n_y,n_z,V_0)\}\sin\{N_{31}(n_y,n_z,V_0)\}
\end{aligned}
$$

$$
+ (1/12)\left(\frac{5\left\{I_{32}\left(n_y, n_z, V_0\right)\right\}^3}{I_{31}\left(n_y, n_z, V_0\right)} + \frac{5\left\{I_{31}\left(n_y, n_z, V_0\right)\right\}^3}{I_{32}\left(n_y, n_z, V_0\right)}\right.
$$

$$
\left. - \left\{34 I_{32}\left(n_y, n_z, V_0\right) I_{31}\left(n_y, n_z, V_0\right)\right\}\right)\left\{M'_{31}\left(n_y, n_z, V_0\right)\right.
$$

$$
\times \cosh\left\{M_{31}\left(n_y, n_z, V_0\right)\right\} \sin\left\{N_{31}\left(n_y, n_z, V_0\right)\right\} + N'_{31}\left(n_y, n_z, V_0\right)
$$

$$
\left.\left. \times \sinh\left\{M_{31}\left(n_y, n_z, V_0\right)\right\} \cos\left\{N_{31}\left(n_y, n_z, V_0\right)\right\}\right\}\right]\Bigg],
$$

$$
\begin{aligned}
M'_{31}\left(n_y, n_z, V_0\right) &= I'_{31}\left(n_y, n_z, V_0\right)\left[a_0 - \Delta_{21}\right], I'_{31}\left(V_0, n_y, n_z\right) \\
&= \frac{m_{c_2} \beta'_{012}\left(V_0 - \overline{V}_0, F, E_{g2}, \Delta_2\right)}{-\hbar^2 I_{31}\left(V_0, n_y, n_z\right)}, \\
N'_{31}\left(n_y, n_z, V_0\right) &= I'_{32}\left(n_y, n_z, V_0\right)\left[b_0 - \Delta_{21}\right], I'_{32}\left(V_0, n_y, n_z\right) \\
&= \frac{m_{c_1} \beta'_{11}\left(V_0, F, E_{g2}, \Delta_2, n_y, n_z\right)}{\hbar^2 I_{32}\left(V_0, n_y, n_z\right)},
\end{aligned}
$$

$$
\begin{aligned}
I_{31}\left(n_y, n_z, V_0\right) &= \left[-\frac{2m_{c2}}{\hbar^2}\beta_{012}(V_0 - \overline{V_0}, F, E_{g2}, \Delta_2) + H(n_y, n_z)\right]^{1/2}, \\
N_{31}\left(n_y, n_z, V_0\right) &= I_{32}\left(n_y, n_z, V_0\right)\left[b_0 - \Delta_{21}\right], I_{32}\left(n_y, n_z, V_0\right) \\
&= \left[\frac{2m_{c1}}{\hbar^2}\beta_{012}(V_0, F, E_{g1}, \Delta_1) - H(n_y, n_z)\right]^{1/2}
\end{aligned}
$$

and

$$
\begin{aligned}
& Z'_{31}\left(V_0, n_y, n_z\right) \\
&= \left[\frac{I'_{31}\left(V_0, n_y, n_z\right)}{I_{32}\left(V_0, n_y, n_z\right)} - \frac{I'_{32}\left(V_0, n_y, n_z\right)}{I_{31}\left(V_0, n_y, n_z\right)} - \frac{I_{31}\left(V_0, n_y, n_z\right) I'_{32}\left(V_0, n_y, n_z\right)}{I^2_{32}\left(V_0, n_y, n_z\right)}\right. \\
&\quad \left. + \frac{I_{32}\left(V_0, n_y, n_z\right) I'_{31}\left(V_0, n_y, n_z\right)}{I^2_{31}\left(V_0, n_y, n_z\right)}\right]
\end{aligned}
$$

For perturbed two-band model of Kane, the form of electron concentration per unit length and the field emitted current remain same where

$$
I_{31}\left(n_y, n_z, E\right) = \left[H(n_y, n_z) - \frac{2m_{c2}}{\hbar^2}\rho_{012}(E - \overline{V}_0, F, E_{g2})\right]^{1/2},
$$

$$I'_{31}\left(V_0, n_y, n_z\right) = \frac{m_{c2}\rho'_{012}\left(V_0 - \overline{V}_0, F, E_{g2}\right)}{-\hbar^2 I_{31}\left(V_0, n_y, n_z\right)}, I_{32}\left(n_y, n_z, V_0\right)$$

$$= \left[-H(n_y, n_z) + \frac{2m_{c1}}{\hbar^2}\rho_{11}(V_0, F, E_{g1})\right]^{1/2},$$

and

$$I'_{32}(V_0, n_y, n_z) = \frac{m_{c1}\rho'_{11}(V_0, F, E_{g1})}{\hbar^2 I_{32}(V_0, n_y, n_z)}.$$

6.3 Results and Discussion

Using (6.54 and 6.61) and (6.55 and 6.62) together with the energy band constants as given in Table 1.1, the field emitted current density under magnetic quantization have been plotted with inverse quantizing magnetic field in accordance with the perturbed three and two-band models of Kane in the presence of electric field for InSb, InAs, $Hg_{1-x}Cd_xTe$ and $In_{1-x}Ga_xAs_yP_{1-y}$ lattice matched to InP as shown in Figs. 6.1 and 6.2 respectively. The band gap of InSb and the lattice constant ($\bar{a}$) of the said material will determine the maximum allowable value of the electric field, where $F_m = E_g/e\bar{a}$.

The use of the said energy band constants leads to the numerical value of the maximum allowable electric field as $F_m = 3.63 \times 10^8$ V/m. It is important to note that one cannot use arbitrary values of electric field F. This is because of the fact that if F is greater than F_m, the electrical breakdown of the aforementioned compound will occur. Therefore, one has to select those values of F for various materials, so that the electrical breakdown does not occur. As a result, for different compounds, the values of the electric field are different since the values of the band gap and the lattice constants are different for different materials. From Table 1.1, we can write that the values of F_m for InAs, $Hg_{1-x}Cd_xTe(x = 0.2)$ and $In_{1-x}Ga_x As_yP_{1-y}(x = 0.2$ and $y = 0.9)$ lattice matched to InP are 7×10^8 V/m, 1×10^8 V/m, and 1.5×10^9 V/m, respectively. The value of the electric field should be less than F_m and since the numerical values of F_m are different for different materials, the value of F will also be selected accordingly to maintain the constraint.

Figure 6.1 shows that the difference between the perturbed three- and two-band models of Kane at high values of the magnetic field is less. It appears that with the increase of the magnetic field strength, the effect of band nonparabolicity merges with each other due to the suppression of the spin–orbit splitting constant and the energy band gap by the respective high value of the Fermi energy. It also appears that the periods of oscillations are constant, together with the fact that with the decrease in the magnetic field strength, the magnitude of the peak of oscillation decreases. Figure 6.2 also exhibits the same phenomena for n-InSb and GaAs for both perturbed three- and two-band models of Kane. The oscillatory dependence

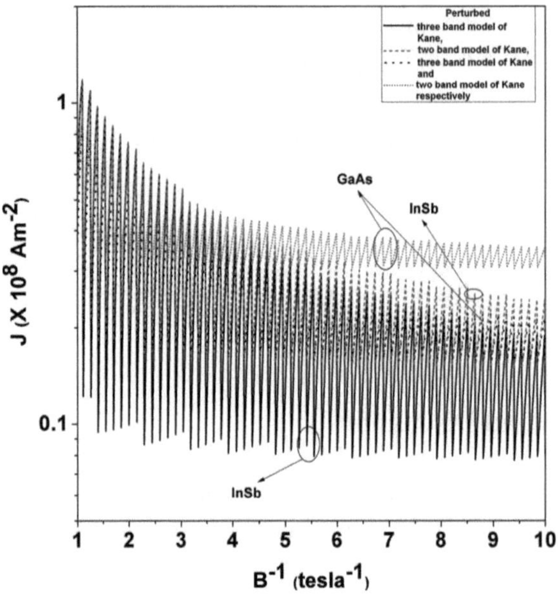

Fig. 6.1 Plot of the field emitted current density as a function of inverse quantizing magnetic field in accordance with the perturbed three- and two-band models of Kane in the presence of electric field for n-InSb and n-GaAs at field strength of 10^6 V m^{-1} and electron concentration of 10^{22} m^{-3}respectively

of the field emitted current density is due to the SdH effect as discussed already. The nature of variation of the curves is due to the fact that for large values of the Fermi energies, the exponential part of the field emitted current density as given by (6.61) and (6.62) dominates as compared with the $F_0(\eta_{11})$ and $F_0(\eta_{12})$ respectively. This tends to make an exponential fall of the magnitude of J for each crossing of the Landau subbands. Besides for low values of the Fermi energies, the opposite phenomenon happens and, in turn, the variations become more sharper near the discontinuities. It appears from the Figs. 6.1 and 6.2 that at high magnetic field, both the materials InSb and InGaAsP are more prone to exhibit quantized oscillations in the field emitted current density with higher magnitude than that of the GaAs and HgCdTe. The reason behind this behavior is the very low effective masses of InSb and InGaAsP, which marks significant peak in oscillations than that of the other heavy effective electron mass systems.

Figures 6.3 and 6.4 exhibit the variation in the field emitted current density as function of the carrier concentration for aforementioned materials. Sharp drop in the field current density is due to the high value of the Fermi energy which makes the exponential part to dominate. From the said figures, it appears that with the increase in the carrier concentration, both the perturbed three- and two-band models of Kane merge with each other.

Figure 6.5 exhibits the field emitted current density as function of electric field for the said materials. It appears that the field current saturates above 10^6 V m^{-1} for

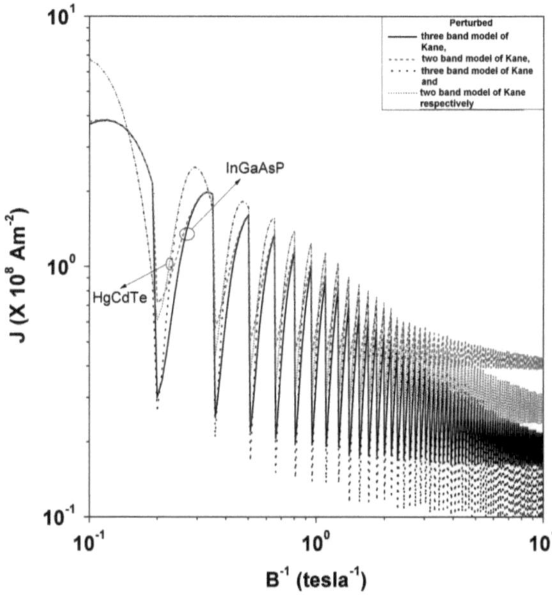

Fig. 6.2 Plot of the field emitted current density as a function of inverse quantizing magnetic field in accordance with the perturbed three- and two-band models of Kane in the presence of electric field for n-HgCdTe and n-InGaAsP at field strength of $10^6\ \mathrm{V\ m^{-1}}$ and electron concentration of $10^{22}\ \mathrm{m^{-3}}$ respectively

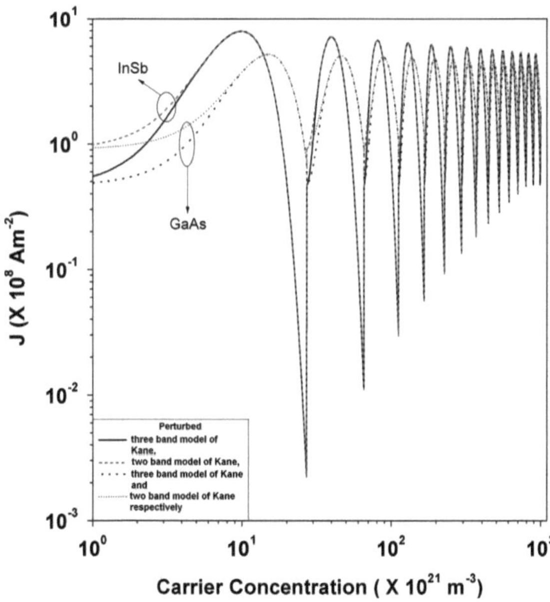

Fig. 6.3 Plot of the field emitted current density as a function of carrier concentration in accordance with the perturbed three- and two-band models of Kane in the presence of electric field for n-InSb and n-GaAs at field strength of $10^6\ \mathrm{V\ m^{-1}}$ where $B = 10\,\mathrm{T}$

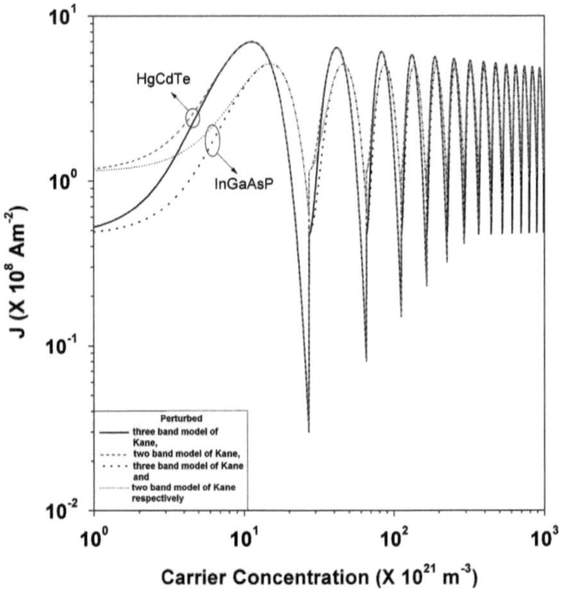

Fig. 6.4 Plot of the field emitted current density as a function of carrier concentration in accordance with the perturbed three- and two-band models of Kane in the presence of electric field for n-HgCdTe and n-InGaAsP at field strength of $10^6\ \mathrm{V\ m^{-1}}$ where $B = 10\,\mathrm{T}$

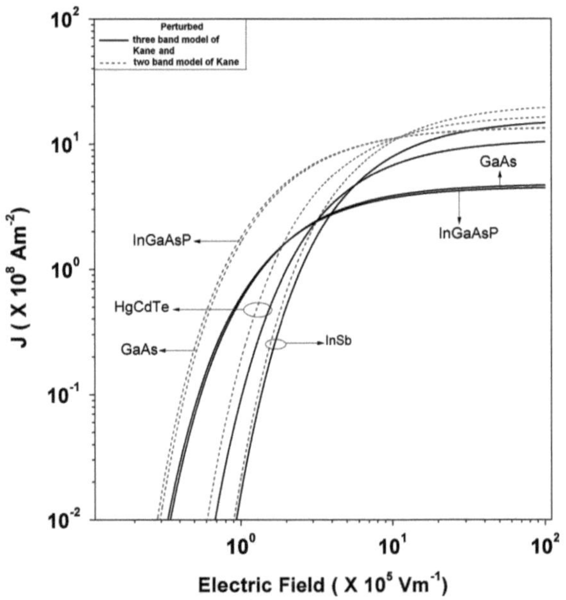

Fig. 6.5 Plot of the field emitted current density as a function of electric field in accordance with the perturbed three- and two-band models of Kane in the presence of electric field for n-InSb, n-GaAs, n-HgCdTe and n-InGaAsP at magnetic field strength of 10 T and carrier concentration of $10^{22}\ \mathrm{m^{-3}}$ respectively

all the materials. Thus, we observe the effect of saturation of electron velocity. Since the graphs are presented for the lowest occupied subband, no oscillations are being observed in this case. Incorporating several higher subbands, one would expect the oscillatory velocity saturation curves. At this point, we note that the velocity of the carriers at the Fermi level can also be a function of the magnetic field, which in general, will oscillate if the field is varied. We have not shown such curves in this case. One can generate this by numerically solving the (6.54) and (6.55) of carrier statistics respectively for both the bands. It should be noted that the effect of electron spin has not been considered in obtaining the oscillatory plots. The peaks in all the figures would increase in number with decrease in amplitude if spin splitting term is included in the respective numerical computations. Although in a more rigorous treatment, the self-consistent procedure should be used, the simplified analysis as presented here exhibits the basic qualitative features of the field emitted current density in nonparabolic materials having perturbed three- and two-band model of Kane in the presence of electric field under the magnetic quantization with reasonable accuracy. We also do not increase the field strength beyond $10^7\,\mathrm{V\,m^{-1}}$, since many other high field transport effect such as hot phonons and hot electrons will arise signifying the issues of transverse negative differential resistance, etc., and transforming the mathematical analysis into a formidable one for the generalized systems as considered here and is clearly beyond the scope of the present literature.

Using (6.69) and (6.70), the field emitted current as function of electric field has been plotted for quantum wires of n-InSb and n-InGaAsP as shown in Fig. 6.6 which exhibits the fact that the field emitted current is more for InSb than that of InGaAsP. Spatial oscillations in the current are clearly exhibited in Fig. 6.6 due to the existence of van-Hove singularity in the density of states function in such 1D structures. On comparing Fig. 6.6 with Fig. 6.5, we observe that the cut-in value of the field current in quantum wires are more than that of the bulk case under magnetic quantization. Physically, this is due to the spatial confinement of the carriers along the two orthogonal directions. It appears that the field current increases with the increase in the well width because of the generation of the subband levels in the presence of size quantization. The close inspection signifies that for a particular subband, the field current actually decreases. This is logical, since with the increase in the film thickness the Fermi energy reduces, which consequently reduces the field current. With the increase in the film thickness along both the directions, the current starts to saturate at their respective bulk value. The field emitted current as function of film thickness in quantum wires has been plotted in Fig. 6.7 in accordance with the perturbed three- and two-band models of Kane in the presence of electric field for n-InSb and n-InGaAsP respectively. One immediate conclusion about the tunneling conductance can also be drawn from the Fig. 6.7. It appears that with the increase of lateral dimension, the transmission of the carriers decays, although the carriers present at the different higher sub-bands participate in the conduction process. This tends to increase the transverse conductance due to tunneling, although for each subband, the conductance decreases. It appears that the magnitudes of the quantum jumps are not of same height indicating the signature of the band structure of the material concerned.

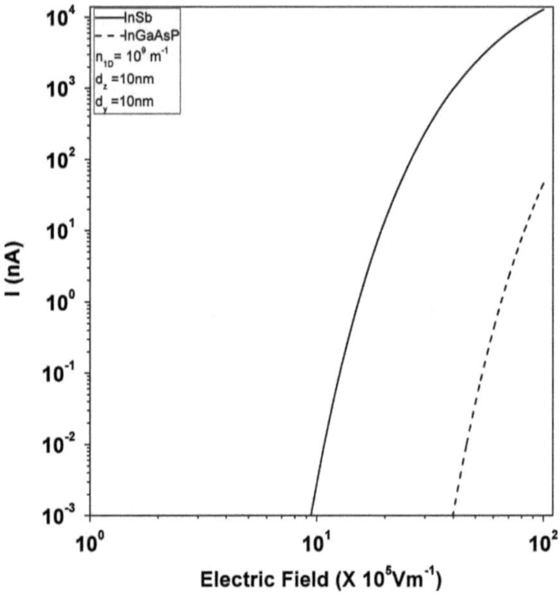

Fig. 6.6 Plot of the field emitted current as a function of electric field in quantum wires in accordance with the perturbed three-band model of Kane, where the materials are quantum wires of n-InSb, and n-InGaAsP respectively

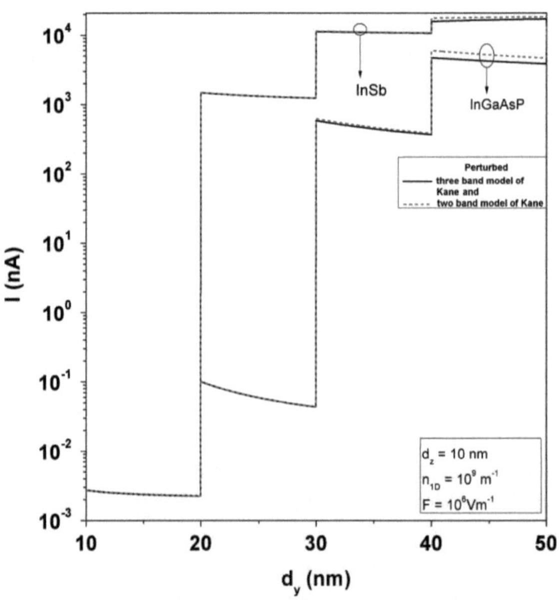

Fig. 6.7 Plot of the field emitted current as a function of film thickness in quantum wires in accordance with the perturbed three- and two-band models of Kane in the presence of electric field for n-InSb and n-InGaAsP respectively

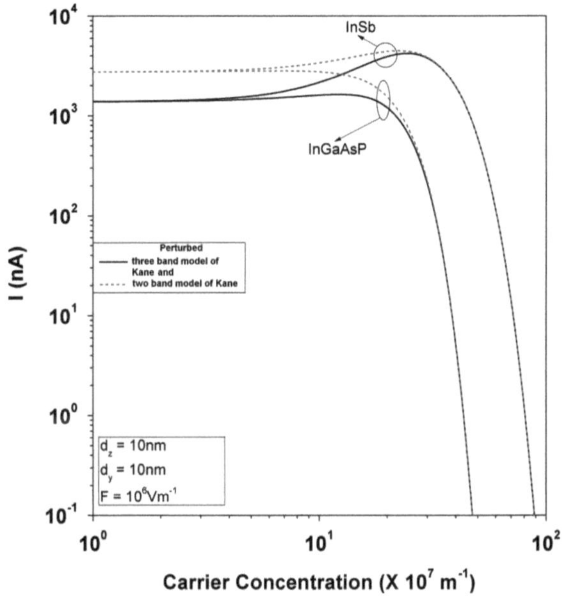

Fig. 6.8 Plot of the field emitted current as a function of carrier concentration in quantum wires in accordance with the perturbed three- and two-band models of Kane in the presence of electric field for n-InSb and n-InGaAsP

Figure 6.8 exhibits the variation of the filed current as function of the carrier density per unit length in this case. Similar trend of variation has been observed when one compares Fig. 6.8 with Fig. 6.5. Since, we have taken the lowest subband, it is logical to assume that almost all the carriers at low temperatures occupy the lowest energy subband, we cannot observe any oscillatory behavior in the field current. We find that a smooth increase and a sharp decrease in the field current with carrier density are the characteristic behavior of the FNFE. With large carrier distribution, we observe negligible deviation between the two types of band nonparabolicity. It also appears from the Fig. 6.7 that the field current in quantum wires exhibits a step like manner as considered here although the numerical values vary widely and determined by the constants of the energy spectra. The step dependence is due to the crossing over of the Fermi level by the size quantized levels. For each coincidence of a quantized level with the Fermi level, there would be a discontinuity in the density of states function resulting in a peak. With large values of film thickness, the height of the steps decreases and the current decreases with increasing film thickness in nonoscillatory manner and exhibit monotonic decreasing dependence. The height of step size and the rate of decrement are totally dependent on the band structure. The numerical values of the field current in accordance with the perturbed three-band model of Kane is lower than that of the corresponding two-band model, which reflects that fact that the presence of the spin orbit splitting constant decreases the magnitude of the tunneling field current.

It may be noted that the presence of the band nonparabolicity in accordance with the perturbed two-band model of Kane enhances the peaks of the ladder type field current for all cases of quantum confinements. The appearance of the humps of the respective curves is due to the redistribution of the electrons among the quantized energy levels when the quantum numbers corresponding to the highest occupied level changes from one fixed value to the others. Although the field current varies in various manners with all the variables in all the limiting cases as evident from all the curves, the rates of variations are totally band-structure dependent.

The variation of field current density for the effective mass superlattices for InGaAs/InGaAsP and InAlSb/InSb as function of inverse magnetic field, carrier density and electric field have been plotted in Figs. 6.9–6.11 respectively. It appears that the variation in the field currents for all the present cases resembles a similar shape of their corresponding bulk part under magnetic quantization. The decaying rates of the currents in all the aforementioned cases are sharper than that of their bulk counterpart. We also observe a significant increase in the magnitude of the current in the aforementioned figures when compared with their bulk. This is due to the very narrow value of the barrier width which makes a high transmission of carriers. With the increase in the width, the current reduces and tends to its bulk counterpart. Figures 6.12–6.14 exhibit the variation in the field current as function of film thickness, carrier density and electric field respectively for quantum wire effective mass superlattices and quantum wire superlattices with graded interfaces. High suppression of the field current is exhibited with the increase in the lateral

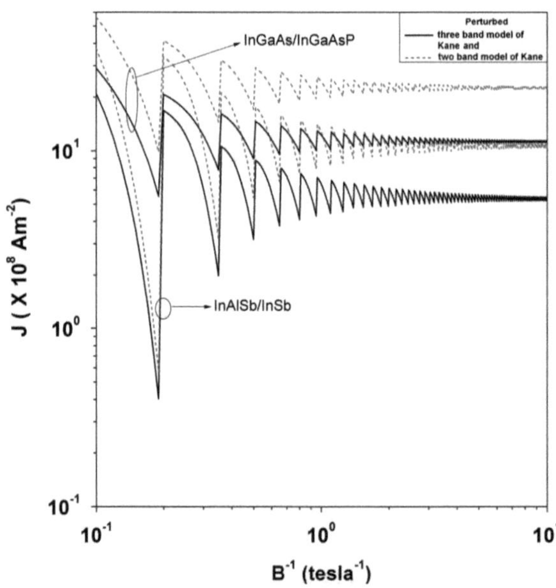

Fig. 6.9 Plot of the field emitted current density as a function of inverse magnetic field in accordance with the perturbed three- and two-band models of Kane in the presence of electric field for InAlSb/InSb and InGaAs/InGaAsP effective mass superlattices

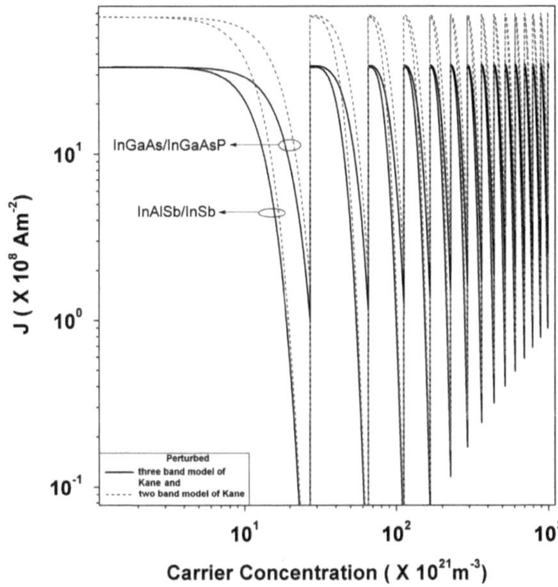

Fig. 6.10 Plot of the field emitted current density in the presence of magnetic field as a function of carrier concentration for InAlSb/InSb and InGaAs/InGaAsP effective mass superlattices respectively in accordance with the perturbed three- and two-band models of Kane in the presence of electric field

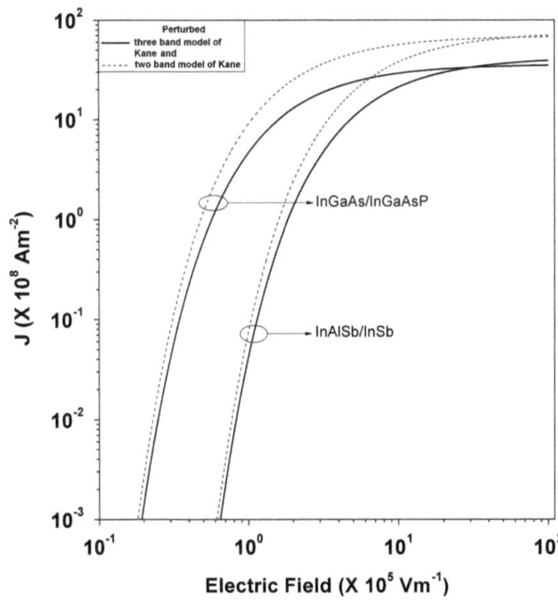

Fig. 6.11 Plot of the field emitted current density as a function of electric field for the InAlSb/InSb and InGaAs/InGaAsP effective mass superlattices respectively in accordance with the perturbed three- and two-band models of Kane in the presence of electric field

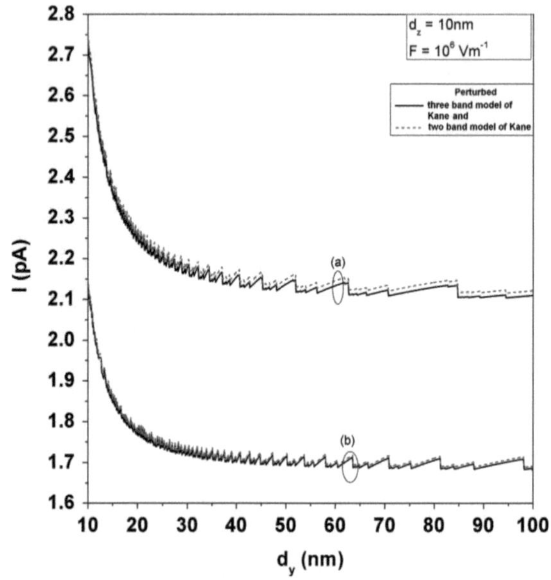

Fig. 6.12 Plot of the field emitted current as a function of film thickness in accordance with the perturbed three- and two-band models of Kane in the presence of electric field for (**a**) quantum wire effective mass superlattices and (**b**) quantum wire superlattices with graded interface.The superlattices taken are InAlSb/InSb and InGaAs/InGaAsP respectively

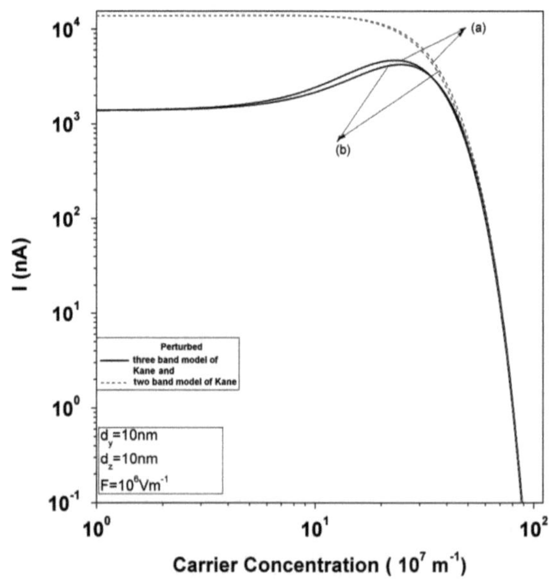

Fig. 6.13 Plot of the field emitted current as a function of carrier density in accordance with the perturbed three- and two-band models of Kane in the presence of electric field for (**a**) quantum wire effective mass superlattices and (**b**) quantum wire superlattices with graded interfaces. The superlattices taken are InAlSb/InSb and InGaAs/InGaAsP respectively

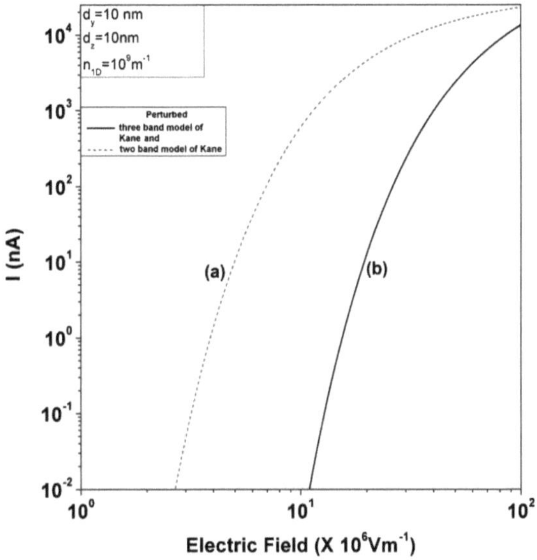

Fig. 6.14 Plot of the field emitted current as function of electric field in accordance with the perturbed (**a**) three and (**b**) two band models of Kane in the presence of electric field for InAlSb/InSb quantum wire effective mass superlattices and InGaAs/InGaAsP quantum wire superlattices with graded interface respectively

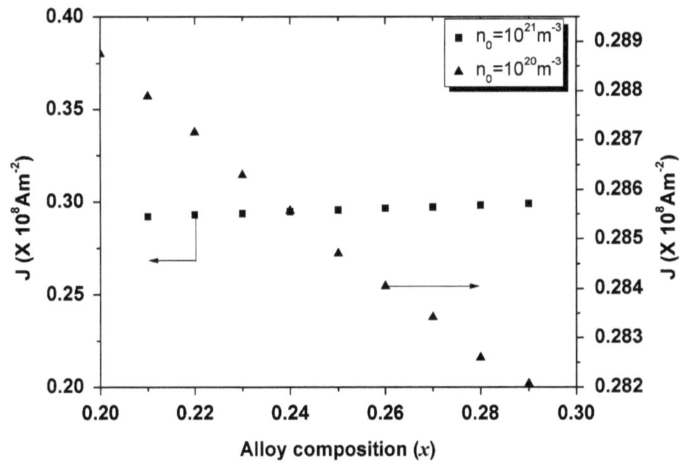

Fig. 6.15 Plot of the field emitted current density as a function of alloy composition in accordance with the perturbed three band model of Kane in the presence of electric field for InGaAs/InGaAsP superlattices with graded interfaces where $B = 10\,\mathrm{T}$ and $F = 10^5\ \mathrm{V\ m^{-1}}$ respectively

dimension. Although, many higher subbands levels are considered, still, we do not observe such high quantum jumps as shown for the quantum wire systems. In the present case, we see composite spatial oscillations in the current due to the selection rules in the quantum numbers along the three confined directions. Figure 6.15

Table 6.1 The carrier statistics and the FNFE from different quantized of optoelectronic materials under intense electric field

Type of materials	The carrier statistics	FNFE
Opto-electronic materials under magnetic quantization:		
(a) Perturbed three-band model of Kane	$n_0 = \frac{g_v eB}{\pi^2\hbar}\sum_{n=0}^{n_{\max}}[Q_{11}(E_{FB},F,n)+Q_{12}(E_{FB},F,n)]$ (6.54)	$J = \frac{e^2 Bk_B Tg_v}{2\pi^2\hbar^2}\sum_{n=0}^{n_{\max}}F_0(\eta_{11})\exp(-\theta_{11})$ (6.61)
(b) Perturbed two-band model of Kane	$n_0 = \frac{g_v eB}{\pi^2\hbar}\sum_{n=0}^{n_{\max}}[Q_{13}(E_{FB},F,n)+Q_{14}(E_{FB},F,n)]$ (6.55)	$J = \frac{e^2 Bk_B Tg_v}{2\pi^2\hbar^2}\sum_{n=0}^{n_{\max}}F_0(\eta_{12})\exp(-\theta_{12})$ (6.62)
Quantum wires of opto electronic materials		
(a) Perturbed three band model of Kane	$n_0 = \frac{2g_v}{\pi}\sum_{n_z=1}^{n_{z\max}}\sum_{n_y=1}^{n_{y\max}}[Q_{15}(E_{F1D},F,n_y,n_z)+Q_{16}(E_{F1D},F,n_y,n_z)]$ (6.68)	$I = \frac{eg_v k_B T}{\pi\hbar}\sum_{n_y=1}^{n_{y\max}}\sum_{n_z=1}^{n_{z\max}}F_0(\eta_{13})\exp(-\theta_{13})$ (6.72)
(b) Perturbed two-band model of Kane	$n_0 = \frac{2g_v}{\pi}\sum_{n_z=1}^{n_{z\max}}\sum_{n_y=1}^{n_{y\max}}[Q_{17}(E_{F1D},F,n_y,n_z)+Q_{18}(E_{F1D},F,n_y,n_z)]$ (6.69)	$I = \frac{eg_v k_B T}{\pi\hbar}\sum_{n_y=1}^{n_{y\max}}\sum_{n_z=1}^{n_{z\max}}F_0(\eta_{14})\exp(-\theta_{14})$ (6.73)
Effective mass superlattices of optoelectronic materials under magnetic quantization		
(a) Perturbed three-band model of Kane	$n_0 = \frac{g_v eB}{\pi^2\hbar}[\sum_{n=0}^{n_{\max}}[Q_{19}(E_{FB},F,n)+Q_{20}(E_{FB},F,n)]]$ (6.78)	$J = \frac{e^2 Bk_B Tg_v}{2\pi^2\hbar^2}\sum_{n=0}^{n_{\max}}F_0(\eta_{15})\exp(-\theta_{15})$ (6.80)
(b) Perturbed two-band model of Kane	$n_0 = \frac{g_v eB}{\pi^2\hbar}[\sum_{n=0}^{n_{\max}}[Q_{21}(E_{FB},F,n)+Q_{22}(E_{FB},F,n)]]$ (6.82)	$J = \frac{e^2 Bk_B Tg_v}{2\pi^2\hbar^2}\sum_{n=0}^{n_{\max}}F_0(\eta_{16})\exp(-\theta_{16})$ (6.83)

Quantum wires effective mass superlattices of Optoelectronic materials		
(a) Perturbed three-band model of Kane	$n_0 = \frac{2g_v}{\pi} \sum_{n_y=1}^{n_{y\max}} \sum_{n_z=1}^{n_{z\max}} \left[Q_{23}\left(E_{\mathrm{FIDEMSL}}, F, n_y, n_z\right) + Q_{24}\left(E_{\mathrm{FIDEMSL}}, F, n_y, n_z\right)\right]$ (6.88)	$I = \frac{eg_v k_B T}{\pi \hbar} \sum_{n_y=1}^{n_{y\max}} \sum_{n_z=1}^{n_{z\max}} F_0(\eta_{17}) \exp(-\theta_{17})$ (6.89)
(b) Perturbed two-band model of Kane	$n_0 = \frac{2g_v}{\pi} \sum_{n_y=1}^{n_{y\max}} \sum_{n_z=1}^{n_{z\max}} \left[Q_{25}\left(E_{\mathrm{FIDEMSL}}, F, n_y, n_z\right) + Q_{26}\left(E_{\mathrm{FIDEMSL}}, F, n_y, n_z\right)\right]$ (6.91)	$I = \frac{eg_v k_B T}{\pi \hbar} \sum_{n_y=1}^{n_{y\max}} \sum_{n_z=1}^{n_{z\max}} F_0(\eta_{18}) \exp(-\theta_{18})$ (6.92)
Superlattices of Optoelectronic materials with graded interfaces under magnetic quantization Perturbed three-band model of Kane	$n_o = \frac{g_v eB}{\pi^2 \hbar}\left[\sum_{n=0}^{n_{\max}} \left[Q_{27}(E_{\mathrm{FBGISL}}, F, n) + Q_{28}(E_{\mathrm{FBGISL}}, F, n)\right]\right]$ (6.96)	$J = \frac{e^2 B k_B T g_v}{2\pi^2 \hbar^2} \sum_{n=0}^{n_{\max}} F_0(\eta_{19}) \exp(-\theta_{19})$ (6.97)
Quantum wire superlattices of optoelectronic materials with graded interfaces Perturbed three-band model of Kane	$n_0 = \frac{2g_v}{\pi} \sum_{n_y=1}^{n_{y\max}} \sum_{n_z=1}^{n_{z\max}} \left[Q_{29}\left(E_{\mathrm{FQWGISL}}, F, n_y, n_z\right) + Q_{30}\left(E_{\mathrm{FQWGISL}}, F, n_y, n_z\right)\right]$ (6.100)	$I = \frac{eg_v k_B T}{\pi \hbar} \sum_{n_y=1}^{n_{y\max}} \sum_{n_z=1}^{n_{z\max}} F_0(\eta_{20}) \exp(-\theta_{20})$ (6.101)

exhibits the field emitted current density as function of alloy composition for the perturbed three-band model of Kane electron dispersion relation of the constituent materials in superlattices with graded interfaces. It appears that with the increase in the alloy composition, the current density decreases for low carrier degeneracy, while for higher values of the electron concentration, the field current density is relatively invariant of alloy composition. It may be noted that the experimental results for the verification of the theoretical analyses of this chapter are still not available in the literature. For the last time, the carrier statistics and the FNFE from different quantized optoelectronic materials have been presented in Table 6.1.

6.4 Open Research Problem

(R.6.1) Investigate the FNFE for all the systems as discussed in this book in the presence of very strong electric field which changes the original band structure and consider the effect of image force in the subsequent: study in each case.

References

1. E. Haga, H. Kimura, J. Phys. Soc. Japan **18(6),** 777 (1963)
2. E. Haga, H. Kimura, J. Phys. Soc. Japan **19(4)**, 471 (1964)
3. B.R. Nag, *Electron Transport in Compound Semiconductors* (Springer-Verlag, Berlin, 1980)
4. L.I. Schiff, *Quantum Mechanics* (McGraw-Hill, New York, 1968)
5. P.K. Chakraborty, S. Choudhury, K.P. Ghatak, Phys. B **387**, 333 (2007)

Chapter 7
Applications and Brief Review of Experimental Results

7.1 Introduction

In this monograph, we have investigated many aspects of FNFE based on the dispersion relations of the semiconductor nanostructures of different technologically important quantum-confined materials having different band structures. In this chapter, we shall discuss few applications in this context in Sect. 7.2 and also present a very brief review of the experimental investigations in Sect. 7.3 which is a sea in itself. Section 7.4 contains the single experimental open research problem.

7.2 Applications

The investigations as presented in this monograph find nine different applications in the realm of modern quantum effect devices.

7.2.1 Debye Screening Length

The Debye screening length (DSL) of the carriers in the semiconductors is a fundamental quantity, characterizing the screening of the Coulomb field of the ionized impurity centers by the free carriers. It affects many special features of the modern semiconductor devices, the carrier mobility under different mechanisms of scattering, and the carrier plasmas in semiconductors [1–14]. The DSL (L_D) can, in general, be written as [4–14]

$$L_{\mathrm{D}} = \left(\frac{|e|^2}{\varepsilon_{\mathrm{sc}}} \frac{\partial n_0}{\partial E_{\mathrm{F}}} \right)^{-1/2} \tag{7.1}$$

where n_0 and E_F are applicable for bulk samples.

S. Bhattacharya and K.P. Ghatak, *Fowler–Nordheim Field Emission*, Springer Series in Solid-State Sciences 170, DOI 10.1007/978-3-642-20493-7_7,

It is well known that the thermoelectric power of the carriers in semiconductors in the presence of a classically large magnetic field is independent of scattering mechanisms and is determined only by their energy band spectra [15]. The magnitude of the thermoelectric power G can be written under the condition of carrier degeneracy [15] as

$$G = \left(\frac{\pi^2 k_B^2 T}{3|e|n_0}\right)\left(\frac{\partial n_0}{\partial E_F}\right) \tag{7.2}$$

Using (7.1) and (7.2), one obtains

$$L_D = \left(3|e|^3 n_0 G / \varepsilon_{sc}\pi^2 k_B^2 T\right)^{-1/2} \tag{7.3}$$

Therefore, we can experimentally determine L_D by knowing the experimental curve of G versus carrier concentration at a fixed temperature. It is evident that the DSL for a system can be investigated if the functional dependence between the electron concentration and the Fermi energy of that particular material is known. For the purpose of completeness, we present a few results of DSL as written below:

1. In the presence of external light waves, the DSL in optoelectronic materials whose unperturbed conduction electrons obey the three- and two-band models of Kane together with parabolic energy bands can, respectively, be expressed as

$$L_D = \left[\left(\frac{e^2}{3\pi^2\varepsilon_{sc}}\right)\left(\frac{2m_c}{\hbar^2}\right)^{3/2}\right]^{-1/2} \left[G'_{70}\left(E_{Fl}, \lambda, E_{g0}, \Delta\right) + H'_{70}\left(E_{Fl}, \lambda, E_{g0}, \Delta\right)\right]^{-1/2} \tag{7.4}$$

$$L_D = \left[\left(\frac{e^2}{3\pi^2\varepsilon_{sc}}\right)\left(\frac{2m_c}{\hbar^2}\right)^{3/2}\right]^{-1/2} \left[G'_{71}\left(E_{Fl}, \lambda, E_{g0}\right) + H'_{71}\left(E_{Fl}, \lambda, E_{g0}\right)\right]^{-1/2} \tag{7.5}$$

$$L_D = \left[\left(\frac{e^2}{3\pi^2\varepsilon_{sc}}\right)\left(\frac{2m_c}{\hbar^2}\right)^{3/2}\right]^{-1/2} \left[G'_{72}\left(E_{Fl}, \lambda, E_{g0}\right) + H'_{72}\left(E_{Fl}, \lambda, E_{g0}\right)\right]^{-1/2} \tag{7.6}$$

where the primes indicate the differentiation of the differentiable functions with respect to the Fermi energy, $G_{70}(E_{Fl}, \lambda, E_{g0}, \Delta) = [\beta_{50}(E_{Fl}, \lambda, E_{g0}, \Delta)]^{3/2}$, $H_{70}(E_{Fl}, \lambda, E_{g0}, \Delta) = \sum_{r=1}^{s} z_t(r) G_{70}(E_{Fl}, \lambda, E_{g0}, \Delta) z_t(r) = 2(k_B T)^{2r}(1 - 2^{1-2r})$ $\xi(2r)\frac{\partial^{2r}}{\partial E_{Fl}^{2r}}$, $t = l$ or F_s, E_{Fl} is the Fermi energy as measured in the presence of light waves as measured from the edge of the conduction band in the vertically upward direction in the absence of any field, $G_{71}(E_{Fl}, \lambda, E_{g0}) = [\omega_{50}(E_{Fl}, \lambda, E_{g0})]^{3/2}$,

$$H_{71}\left(E_{Fl}, \lambda, E_{g0}\right) = \sum_{r=1}^{s} z_t(r) G_{71}\left(E_{Fl}, \lambda, E_{g0}\right), G_{72}\left(E_{Fl}, \lambda, E_{g0}\right)$$

$$= \left[\rho_{50}\left(E_{Fl}, \lambda, E_{g0}\right)\right]^{3/2}, H_{72}\left(E_{Fl}, \lambda, E_{g0}\right) = \sum_{r=1}^{s} z_t(r) G_{72}\left(E_{Fl}, \lambda, E_{g0}\right)$$

$$n_0 = \frac{1}{3\pi^2}\left(\frac{2m_c}{\hbar^2}\right)^{3/2}\left[G_{70}(E_{Fl}, \lambda, E_{g0}, \Delta) + H_{70}(E_{Fl}, \lambda, E_{g0}, \Delta)\right],$$

$$n_0 = \frac{1}{3\pi^2}\left(\frac{2m_c}{\hbar^2}\right)^{3/2}\left[G_{71}(E_{Fl}, \lambda, E_{g0}) + H_{71}(E_{Fl}, \lambda, E_{g0})\right]$$

and

$$n_0 = \frac{1}{3\pi^2}\left(\frac{2m_c}{\hbar^2}\right)^{3/2}\left[G_{72}(E_{Fl}, \lambda, E_{g0}) + H_{72}(E_{Fl}, \lambda, E_{g0})\right]$$

2. In the presence of intense electric field, the DSL in optoelectronic semiconductors in accordance with the perturbed three- and two-band models of Kane can, respectively, be expressed as

$$L_D = \left[\left(\frac{e^2}{3\pi^2\varepsilon_{sc}}\right)\left(\frac{2m_c}{\hbar^2}\right)^{3/2}\right]^{-1/2}\left[g'_{70}\left(E_{Fs}, F\right) + h'_{70}\left(E_{Fs}, F\right)\right]^{-1/2} \quad (7.7)$$

$$L_D = \left[\left(\frac{e^2}{3\pi^2\varepsilon_{sc}}\right)\left(\frac{2m_c}{\hbar^2}\right)^{3/2}\right]^{-1/2}\left[g'_{71}\left(E_{Fs}, F\right) + h'_{71}\left(E_{Fs}, F\right)\right]^{-1/2} \quad (7.8)$$

$$g_{70}\left(E_{Fs}, F\right) = \left[\beta\left(E_{Fs}, F\right)\right]^{3/2}, h_{70}\left(E_{Fs}, F\right) = \sum_{r=1}^{s} z_t(r) g_{70}\left(E_{Fs}, F\right)$$

E_{Fs} is the Fermi energy as measured in the presence of intense electric field as measured from the edge of the conduction band in the vertically upward direction in the absence of any field $g_{71}(E_{Fs}, F) = [\beta_1(E_{Fs}, F)]^{3/2}$, $h_{71}(E_{Fs}, F) = \sum_{r=1}^{s} z_t(r) g_{71}(E_{Fs}, F)$,

$$n_0 = \frac{1}{3\pi^2}\left(\frac{2m_c}{\hbar^2}\right)^{3/2}\left[g_{70}(E_{Fs}, F) + h_{70}(E_{Fs}, F)\right],$$

$$n_0 = \frac{1}{3\pi^2}\left(\frac{2m_c}{\hbar^2}\right)^{3/2}\left[g_{71}(E_{Fs}, F) + h_{71}(E_{Fs}, F)\right]$$

In the absence of any field, the expressions for the DSL and the electron concentration for optoelectronic semiconductors whose energy band structures are defined by the unperturbed two-band model of Kane, under the condition $(E_F E_{g0}^{-1}) << 1$, assume the well-known forms as [8]

$$L_D = \left[\frac{e^2}{\varepsilon_{sc}} N_c k_B T \left[F_{-1/2}(\eta) + \frac{15\alpha k_B T}{4} F_{1/2}(\eta)\right]\right]^{-1/2} \tag{7.9}$$

$$n_0 = N_c \left[F_{1/2}(\eta) + \frac{15\alpha k_B T}{4} F_{3/2}(\eta)\right] \tag{7.10}$$

where $\eta = \frac{E_F}{k_B T}$.

7.2.2 Carrier Contribution to the Elastic Constants

The knowledge of the carrier contribution to the elastic constants is important in studying the mechanical properties of the materials and has been investigated in the literature [15–37]. The electronic contribution to the second- and third-order elastic constants can be written as [15–37]

$$\Delta C_{44} = -\frac{\left(\overline{G_0}\right)^2}{9} \frac{\partial n_0}{\partial E_F}, \tag{7.11}$$

and

$$\Delta C_{456} = \frac{\left(\overline{G_0}\right)^3}{27} \frac{\partial^2 n_0}{\partial E_F^2}, \tag{7.12}$$

where $\overline{G_0}$ is the deformation potential constant. Thus, using (7.11), (7.12), and (7.2), we can write

$$\Delta C_{44} = \left[-n_0 \left(\overline{G_0}\right)^2 |e| \, G_0 / \left(3\pi^2 k_B^2 T\right)\right] \tag{7.13}$$

and

$$\Delta C_{456} = \left(n_0 \, |e| \left(\overline{G_0}\right)^3 G_0^2 / (3\pi^4 k_B^3 T)\right) \left(1 + \frac{n_0}{G_0} \frac{\partial G_0}{\partial n_0}\right) \tag{7.14}$$

Thus, again the experimental graph of G_0 versus n_0 allows us to determine the electronic contribution to the elastic constants for materials having arbitrary spectra. We present a few results in this context:

i. The expressions for ΔC_{44} and ΔC_{456} in quantum wires of nonlinear optical materials, III–V, II–VI, bismuth, IV–VI, stressed semiconductors, Te, n-GaP, $PtSb_2$, Bi_2Te_3, n-Ge, and II–V can, respectively, be expressed as

(a) Nonlinear optical materials:

$$\Delta C_{44} = -\left(\frac{2(\overline{G}_0)^2 g_v}{9\pi}\right)\sum_{n_x=1}^{n_{x\max}}\sum_{n_y=1}^{n_{y\max}}[B'_{11}(E_{F1D}, n_x, n_y)$$

$$+ B'_{12}(E_{F1D}, n_x, n_y)] \tag{7.15}$$

$$\Delta C_{456} = \left(\frac{2(\overline{G}_0)^3 g_v}{27\pi}\right)\sum_{n_x=1}^{n_{x\max}}\sum_{n_y=1}^{n_{y\max}}[B''_{11}(E_{F1D}, n_x, n_y)$$

$$+B''_{12}(E_{F1D}, n_x, n_y)] \tag{7.16}$$

(b) *III–V materials:*

1. Three-band model of Kane:

$$\Delta C_{44} = -\left(\frac{2(\overline{G}_0)^2 g_v}{9\pi}\right)\sum_{n_x=1}^{n_{x\max}}\sum_{n_y=1}^{n_{y\max}}[B'_{13}(E_{F1D}, n_x, n_y)$$

$$+ B'_{14}(E_{F1D}, n_x, n_y)] \tag{7.17}$$

$$\Delta C_{456} = \left(\frac{2(\overline{G}_0)^3 g_v}{27\pi}\right)\sum_{n_x=1}^{n_{x\max}}\sum_{n_y=1}^{n_{y\max}}[B''_{13}(E_{F1D}, n_x, n_y)$$

$$+ B''_{14}(E_{F1D}, n_x, n_y)] \tag{7.18}$$

2. Two-band model of Kane:

$$\Delta C_{44} = -(\frac{2(\overline{G}_0)^2 g_v}{9\pi})\sum_{n_x=1}^{n_{x\max}}\sum_{n_y=1}^{n_{y\max}}[B'_{15}(E_{F1D}, n_x, n_y)$$

$$+ B'_{16}(E_{F1D}, n_x, n_y)] \tag{7.19}$$

$$\Delta C_{456} = \left(\frac{2(\overline{G}_0)^3 g_v}{27\pi}\right)\sum_{n_x=1}^{n_{x\max}}\sum_{n_y=1}^{n_{y\max}}[B''_{15}(E_{F1D}, n_x, n_y)$$

$$+ B''_{16}(E_{F1D}, n_x, n_y)] \tag{7.20}$$

3. The model of Stillman et al.:

$$\Delta C_{44} = -\left(\frac{2(\overline{G}_0)^2 g_v}{9\pi}\right)\sum_{n_x=1}^{n_{x\max}}\sum_{n_y=1}^{n_{y\max}}[B'_{17}(E_{F1D}, n_x, n_y)$$

$$+ B'_{18}(E_{F1D}, n_x, n_y)] \tag{7.21}$$

$$\Delta C_{456} = \left(\frac{2(\overline{G}_0)^3 g_v}{27\pi}\right) \sum_{n_x=1}^{n_{x\max}} \sum_{n_y=1}^{n_{y\max}} [B''_{17}(E_{F1D}, n_x, n_y) + B''_{18}(E_{F1D}, n_x, n_y)] \tag{7.22}$$

4. The model of Newson and Kurobe:

$$\Delta C_{44} = -\left(\frac{2(\overline{G}_0)^2 g_v}{9\pi}\right) \sum_{n_x=1}^{n_{x\max}} \sum_{n_y=1}^{n_{y\max}} [B'_{19}(E_{F1D}, n_x, n_y) + B'_{20}(E_{F1D}, n_x, n_y)] \tag{7.23}$$

$$\Delta C_{456} = \left(\frac{2(\overline{G}_0)^3 g_v}{27\pi}\right) \sum_{n_x=1}^{n_{x\max}} \sum_{n_y=1}^{n_{y\max}} [B''_{19}(E_{F1D}, n_x, n_y) + B''_{20}(E_{F1D}, n_x, n_y)] \tag{7.24}$$

5. The model of Palik et al.:

$$\Delta C_{44} = -\left(\frac{2(\overline{G}_0)^2 g_v}{9\pi}\right) \sum_{n_x=1}^{n_{x\max}} \sum_{n_y=1}^{n_{y\max}} [B'_{21}(E_{F1D}, n_x, n_y) + B'_{22}(E_{F1D}, n_x, n_y)] \tag{7.25}$$

$$\Delta C_{456} = \left(\frac{2(\overline{G}_0)^3 g_v}{27\pi}\right) \sum_{n_x=1}^{n_{x\max}} \sum_{n_y=1}^{n_{y\max}} [B''_{21}(E_{F1D}, n_x, n_y) + B''_{22}(E_{F1D}, n_x, n_y)] \tag{7.26}$$

(c) *II–VI materials:*

$$\Delta C_{44} = -\frac{(\overline{G}_0)^2 g_v}{9\pi\sqrt{B_0}} \sum_{n_x=1}^{n_{x\max}} \sum_{n_y=1}^{n_{y\max}} \left[t'_7\left(E_{F1D}, n_x, n_y\right) + t'_8\left(E_{F1D}, n_x, n_y\right)\right] \tag{7.27}$$

$$\Delta C_{456} = \frac{(\overline{G}_0)^3 g_v}{27\pi\sqrt{B_0}} \sum_{n_x=1}^{n_{x\max}} \sum_{n_y=1}^{n_{y\max}} \left[t''_7\left(E_{F1D}, n_x, n_y\right) + t''_8\left(E_{F1D}, n_x, n_y\right)\right] \tag{7.28}$$

(d) *Bismuth:*

1. The model of McClure and Choi:

$$\Delta C_{44} = -\frac{2(\overline{G}_0)^2 g_v}{9\pi}\frac{\sqrt{2m_1}}{\hbar}\sum_{n_y=1}^{n_{y\max}}\sum_{n_z=1}^{n_{z\max}}\left[t'_{27}\left(E_{F1D}, n_y, n_z\right) + t'_{28}\left(E_{F1D}, n_y, n_z\right)\right] \tag{7.29}$$

$$\Delta C_{456} = \frac{2(\overline{G}_0)^3 g_v}{27\pi}\frac{\sqrt{2m_1}}{\hbar}\sum_{n_y=1}^{n_{y\max}}\sum_{n_z=1}^{n_{z\max}}\left[t''_{27}\left(E_{F1D}, n_y, n_z\right) + t''_{28}\left(E_{F1D}, n_y, n_z\right)\right] \tag{7.30}$$

2. Hybrid model:

$$\Delta C_{44} = -\frac{2(\overline{G}_0)^2 g_v}{9\pi}\frac{\sqrt{2m_1}}{\hbar}\sum_{n_y=1}^{n_{y\max}}\sum_{n_z=1}^{n_{z\max}}\left[t'_{31}\left(E_{F1D}, n_y, n_z\right) + t'_{32}\left(E_{F1D}, n_y, n_z\right)\right] \tag{7.31}$$

$$\Delta C_{456} = \frac{2(\overline{G}_0)^3 g_v}{27\pi}\frac{\sqrt{2m_1}}{\hbar}\sum_{n_y=1}^{n_{y\max}}\sum_{n_z=1}^{n_{z\max}}\left[t''_{31}\left(E_{F1D}, n_y, n_z\right) + t''_{32}\left(E_{F1D}, n_y, n_z\right)\right] \tag{7.32}$$

3. Cohen model:

$$\Delta C_{44} = -\frac{2(\overline{G}_0)^2 g_v}{9\pi}\frac{\sqrt{2m_1}}{\hbar}\sum_{n_y=1}^{n_{y\max}}\sum_{n_z=1}^{n_{z\max}}\left[t'_{35}\left(E_{F1D}, n_y, n_z\right) + t'_{36}\left(E_{F1D}, n_y, n_z\right)\right] \tag{7.33}$$

$$\Delta C_{456} = \frac{2(\overline{G}_0)^3 g_v}{27\pi}\frac{\sqrt{2m_1}}{\hbar}\sum_{n_y=1}^{n_{y\max}}\sum_{n_z=1}^{n_{z\max}}\left[t''_{35}\left(E_{F1D}, n_y, n_z\right) + t''_{36}\left(E_{F1D}, n_y, n_z\right)\right] \tag{7.34}$$

4. Lax model:

$$\Delta C_{44} = -\frac{2(\overline{G}_0)^2 g_v}{9\pi}\frac{\sqrt{2m_1}}{\hbar}\sum_{n_y=1}^{n_{y\max}}\sum_{n_z=1}^{n_{z\max}}\left[t'_{37}\left(E_{F1D}, n_y, n_z\right) + t'_{38}\left(E_{F1D}, n_y, n_z\right)\right] \tag{7.35}$$

$$\Delta C_{456} = \frac{2(\overline{G}_0)^3 g_v}{27\pi} \frac{\sqrt{2m_1}}{\hbar} \sum_{n_y=1}^{n_{y\max}} \sum_{n_z=1}^{n_{z\max}} \left[t''_{37}\left(E_{F1D}, n_y, n_z\right) + t''_{38}\left(E_{F1D}, n_y, n_z\right) \right] \tag{7.36}$$

(e) *IV–VI materials:*
Dimmock model:

$$\Delta C_{44} = -\frac{2(\overline{G}_0)^2 g_v}{9\pi} \sum_{n_x=1}^{n_{x\max}} \sum_{n_y=1}^{n_{y\max}} [B'_{32}(E_{F1D}, n_x, n_y) + B'_{33}(E_{F1D}, n_x, n_y)] \tag{7.37}$$

$$\Delta C_{456} = \frac{2(\overline{G}_0)^3 g_v}{27\pi} \sum_{n_x=1}^{n_{x\max}} \sum_{n_y=1}^{n_{y\max}} [B''_{32}(E_{F1D}, n_x, n_y) + B''_{33}(E_{F1D}, n_x, n_y)] \tag{7.38}$$

(f) *Stressed materials:*

$$\Delta C_{44} = -\frac{2(\overline{G}_0)^2 g_v}{9\pi} \sum_{n_y=1}^{n_{y\max}} \sum_{n_z=1}^{n_{z\max}} \left[B'_{34}\left(E_{F1D}, n_y, n_z\right) + B'_{35}\left(E_{F1D}, n_y, n_z\right) \right] \tag{7.39}$$

$$\Delta C_{456} = \frac{2(\overline{G}_0)^3 g_v}{27\pi} \sum_{n_y=1}^{n_{y\max}} \sum_{n_z=1}^{n_{z\max}} \left[B''_{34}\left(E_{F1D}, n_y, n_z\right) + B''_{35}\left(E_{F1D}, n_y, n_z\right) \right] \tag{7.40}$$

(g) *Tellurium:*

$$\Delta C_{44} = -\frac{(\overline{G}_0)^2 g_v}{9\pi} \sum_{n_x=1}^{n_{x\max}} \sum_{n_y=1}^{n_{y\max}} \left[B'_{36,\pm}\left(E_{F1D}, n_x, n_y\right) + B'_{37,\pm}\left(E_{F1D}, n_x, n_y\right) \right] \tag{7.41}$$

$$\Delta C_{456} = \frac{(\overline{G}_0)^3 g_v}{27\pi} \sum_{n_x=1}^{n_{x\max}} \sum_{n_y=1}^{n_{y\max}} \left[B''_{36,\pm}\left(E_{F1D}, n_x, n_y\right) + B''_{37,\pm}\left(E_{F1D}, n_x, n_y\right) \right] \tag{7.42}$$

(h) *Gallium phosphide:*

$$\Delta C_{44} = -\frac{2(\overline{G}_0)^2 g_v}{9\pi} \sum_{n_x=1}^{n_{x\max}} \sum_{n_y=1}^{n_{y\max}} \left[B'_{38}\left(E_{F1D}, n_x, n_y\right) + B'_{39}\left(E_{F1D}, n_x, n_y\right)\right] \quad (7.43)$$

$$\Delta C_{456} = \frac{2(\overline{G}_0)^3 g_v}{27\pi} \sum_{n_x=1}^{n_{x\max}} \sum_{n_y=1}^{n_{y\max}} \left[B''_{38}\left(E_{F1D}, n_x, n_y\right) + B''_{39}\left(E_{F1D}, n_x, n_y\right)\right] \quad (7.44)$$

(i) *Platinum Antimonide:*

$$\Delta C_{44} = -\frac{2(\overline{G}_0)^2 g_v}{9\pi} \sum_{n_x=1}^{n_{x\max}} \sum_{n_y=1}^{n_{y\max}} \left[B'_{40}\left(E_{F1D}, n_x, n_y\right) + B'_{41}\left(E_{F1D}, n_x, n_y\right)\right] \quad (7.45)$$

$$\Delta C_{456} = \frac{2(\overline{G}_0)^3 g_v}{27\pi} \sum_{n_x=1}^{n_{x\max}} \sum_{n_y=1}^{n_{y\max}} \left[B''_{40}\left(E_{F1D}, n_x, n_y\right) + B''_{41}\left(E_{F1D}, n_x, n_y\right)\right] \quad (7.46)$$

(j) *Bismuth Telluride:*

$$\Delta C_{44} = -\frac{2(\overline{G}_0)^2 g_v}{9\pi} \sum_{n_z=1}^{n_{z\max}} \sum_{n_y=1}^{n_{y\max}} \left[B'_{42}\left(E_{F1D}, n_z, n_y\right) + B'_{43}\left(E_{F1D}, n_z, n_y\right)\right] \quad (7.47)$$

$$\Delta C_{456} = \frac{2(\overline{G}_0)^3 g_v}{27\pi} \sum_{n_z=1}^{n_{z\max}} \sum_{n_y=1}^{n_{y\max}} \left[B''_{42}\left(E_{F1D}, n_z, n_y\right) + B''_{43}\left(E_{F1D}, n_z, n_y\right)\right] \quad (7.48)$$

(k) *Germanium:*

1. The model of Cardona et al:

$$\Delta C_{44} = -\frac{2(\overline{G}_0)^2 g_v}{9\pi} \sum_{n_x=1}^{n_{x\max}} \sum_{n_z=1}^{n_{z\max}} \left[B'_{44}\left(E_{F1D}, n_x, n_z\right) + B'_{45}\left(E_{F1D}, n_x, n_z\right)\right] \quad (7.49)$$

$$\Delta C_{456} = \frac{2(\overline{G}_0)^3 g_v}{27\pi} \sum_{n_x=1}^{n_{x\max}} \sum_{n_z=1}^{n_{z\max}} [B''_{44}(E_{F1D}, n_x, n_z) + B''_{45}(E_{F1D}, n_x, n_z)] \quad (7.50)$$

2. The model of Wang and Ressler:

$$\Delta C_{44} = -\frac{2(\overline{G}_0)^2 g_v}{9\pi} \sum_{n_x=1}^{n_{x\max}} \sum_{n_z=1}^{n_{z\max}} [B'_{46}(E_{F1D}, n_x, n_z) + B'_{47}(E_{F1D}, n_x, n_z)] \quad (7.51)$$

$$\Delta C_{456} = \frac{2(\overline{G}_0)^3 g_v}{27\pi} \sum_{n_x=1}^{n_{x\max}} \sum_{n_z=1}^{n_{z\max}} [B''_{46}(E_{F1D}, n_x, n_z) + B''_{47}(E_{F1D}, n_x, n_z)] \quad (7.52)$$

(l) *Gallium Antimonide:*

$$\Delta C_{44} = -\frac{2(\overline{G}_0)^2 g_v}{9\pi} \sum_{n_x=1}^{n_{x\max}} \sum_{n_y=1}^{n_{y\max}} \left[B'_{48}\left(E_{F1D}, n_x, n_y\right) + B'_{49}\left(E_{F1D}, n_x, n_y\right)\right] \quad (7.53)$$

$$\Delta C_{456} = \frac{2(\overline{G}_0)^3 g_v}{27\pi} \sum_{n_x=1}^{n_{x\max}} \sum_{n_y=1}^{n_{y\max}} \left[B''_{48}\left(E_{F1D}, n_x, n_y\right) + B''_{49}\left(E_{F1D}, n_x, n_y\right)\right] \quad (7.54)$$

(m) *II–V materials:*

$$\Delta C_{44} = -\frac{(\overline{G}_0)^2 g_v}{9\pi} \sum_{n_x=1}^{n_{x\max}} \sum_{n_y=1}^{n_{y\max}} \left[B'_{49}\left(E_{F1D}, n_x, n_y\right) + B'_{50}\left(E_{F1D}, n_x, n_y\right)\right] \quad (7.55)$$

$$\Delta C_{456} = \frac{(\overline{G}_0)^3 g_v}{27\pi} \sum_{n_x=1}^{n_{x\max}} \sum_{n_y=1}^{n_{y\max}} \left[B''_{49}\left(E_{F1D}, n_x, n_y\right) + B''_{50}\left(E_{F1D}, n_x, n_y\right)\right] \quad (7.56)$$

7.2.3 Effective Electron Mass

The concept of effective mass of the carriers in different materials, being connected with the mobility, is important in the whole field of solid state and related sciences and is used for the analysis of the semiconductor devices under different operating conditions in general [38]. Among the various definitions of the effective electron mass [39] (e.g., density of states mass, conductivity mass, acceleration effective mass, Faraday rotation effective mass, concentration effective mass), it is the effective momentum mass that should be regarded as the basic quantity [40]. This is due to the fact that it is this mass which appears in the description of transport phenomena and all other properties of the conduction electrons of the semiconductors having arbitrary dispersion laws [41]. It is the effective momentum mass which enters in various transport coefficients and plays the most dominant role in explaining the experimental results under different scattering mechanisms [41, 42]. The carrier degeneracy in semiconductors influences the effective mass when it is energy dependent. Under degenerate conditions, only the electrons at the Fermi surface of n-type semiconductors participate in the conduction process and hence the effective momentum mass of the electrons (EMM) corresponding to the Fermi level would be of interest in electron transport under such conditions. The Fermi energy is again determined by the carrier energy spectrum and the carrier concentration and therefore these two features would determine the dependence of the EMM in degenerate materials on the degree of carrier degeneracy. In recent years, the EMM in such materials under different external conditions has been studied extensively [43–62]. It has different values in different materials and varies with electron concentration, with the magnitude of the reciprocal quantizing magnetic field under magnetic quantization, with the quantizing electric field as in inversion layers, with the nanothickness as in quantum wells and quantum well wires and with superlattice period as in the quantum confined superlattices having various carrier energy spectra.

The expression of the EMM in the ith direction is given by

$$m_i^*(E_F) = \hbar^2 \left[k_{i_0} \left(\frac{\partial k_{i_0}}{\partial E} \right) \right] \Bigg|_{E=E_F} \tag{7.57}$$

where $i_0 = x, y$, and z.

For the purpose of condensed presentation, we present a few results of the EMM under different external conditions in this context:

1. The expressions of EMMs in bulk specimens of optoelectronic materials in the presence of light waves whose unperturbed conduction electrons obey the three- and two-band models of Kane together with parabolic energy bands can, respectively, be written as

$$m^*(E_{Fl}, \lambda, E_{g0}, \Delta) = m_c \beta'_{50}(E_{Fl}, \lambda, E_{g0}, \Delta) \tag{7.58}$$

$$m^*(E_{Fl}, \lambda, E_{g0}) = m_c \omega'_{50}(E_{Fl}, \lambda, E_{g0}) \tag{7.59}$$

$$m^*(E_{Fl}, \lambda, E_{g0}) = m_c \rho'_{50}(E_{Fl}, \lambda, E_{g0}) \tag{7.60}$$

2. The expressions of EMMs in bulk specimens of optoelectronic materials in the presence of intense electric field whose unperturbed conduction electrons obey the three- and two-band models of Kane can, respectively, be written as

$$m^*(E_{Fs}, F) = m_c \beta'(E_{Fs}, F) \tag{7.61}$$

$$m^*(E_{Fs}, F) = m_c \beta'_1(E_{Fs}, F) \tag{7.62}$$

In the absence of any fields, the EMM in bulk specimens of optoelectronic materials whose unperturbed conduction electrons obey the three- and two-band models of Kane together with parabolic energy bands can, respectively, be written as

$$m^*(E_F) = m_c \gamma'(E_F) \tag{7.63}$$

$$m^*(E_F) = m_c(1 + 2\alpha E_F) \tag{7.64}$$

$$m^*(E_F) = m_c \tag{7.65}$$

Comparing (7.58) with (7.63), we observe that the presence of light waves makes the mass wavelength dependent and again comparing (7.61) with (7.63) we can write that in the presence of intense electric field generates electric field dependent EMM in accordance with three-band model of Kane. Besides, the comparison among (7.65) and (7.60) attribute the fact that the EMM for materials, whose conduction electrons obey the perfect parabolic energy bands in the absence of any fields, in the presence of light waves is a function of Fermi energy, wavelength, and the band gap instead of well-known constant independent of any variable.

3. The expressions of the EMMs in nonlinear optical, III–V, II–VI, bismuth, IV–VI, stressed materials, Te, n-GaP, $PtSb_2$, Bi_2Te_3, n-Ge, GaSb, and II–V in the presence of quantizing magnetic field are given below:

   (a) Nonlinear optical materials:

$$m^*_{\pm}(E_{FB}, n) = \frac{\hbar^2}{2} A'_{31,\pm}(E_{FB}, n) \tag{7.66}$$

   (b) III–V materials:

      1. Three-band model of Kane:

$$m^*_{\pm}(E_{FB}, n) = m_c A'_{32,\pm}(E_{FB}, n) \tag{7.67}$$

2. Two-band model of Kane:

$$m^*(E_{FB}) = m_c(1 + 2\alpha E_{FB}) \tag{7.68}$$

3. The model of Stillman et al:

$$m^*(E_{FB}, n) = m_c A'_{33}(E_{FB}, n) \tag{7.69}$$

4. The model of Palik et al:

$$m^*_\pm(E_{FB}, n) = \frac{\hbar^2}{2} A'_{35,\pm}(E_{FB}, n) \tag{7.70}$$

(c) II–VI materials:

$$m^*_\pm(E_{FB}, n) = m^*_\parallel \tag{7.71}$$

(d) Bismuth:

1. The McClure and Choi model:

$$m^*_\pm(E_{FB}, n) = \frac{\hbar^2}{2} A'_{36,\pm}(E_{FB}, n) \tag{7.72}$$

2. The Cohen model:

$$m^*_\pm(E_{FB}, n) = m_3 A'_{38,\pm}(E_{FB}, n) \tag{7.73}$$

3. The Lax model:

$$m^*(E_{FB}) = m_3(1 + 2\alpha E_{FB}) \tag{7.74}$$

4. The Ellipsoidal energy bands:

$$m^*(E_{FB}) = m_3 \tag{7.75}$$

(e) IV–VI materials:

1. The Dimmock model:

$$m^*(E_{FB}, n) = \frac{\hbar^2}{2} A'_{42}(E_{FB}, n) \tag{7.76}$$

2. The Model of Bangert and Kastner:

$$m^*(E_{FB}, n) = \frac{\hbar^2}{2} A'_{44}(E_{FB}, n) \tag{7.77}$$

3. The Model of Foley and Landenberg:

$$m^*(E_{FB}, n) = \frac{\hbar^2}{2} A'_{46}(E_{FB}, n) \tag{7.78}$$

(f) Stressed materials:

$$m^*(E_{FB}, n) = \frac{\hbar^2}{2} A'_{48}(E_{FB}, n) \tag{7.79}$$

(g) Tellurium:

$$m^*_{\pm}(E_{FB}, n) = \frac{\hbar^2}{2} A'_{50,\pm}(E_{FB}, n) \tag{7.80}$$

(h) Gallium Phosphide:

$$m^*_{\pm}(E_{FB}, n) = \frac{\hbar^2}{2} A'_{52,\pm}(E_{FB}, n) \tag{7.81}$$

(i) Platinum Antimonide:

$$m^*_{\pm}(E_{FB}, n) = \frac{\hbar^2}{2} A'_{55,\pm}(E_{FB}, n) \tag{7.82}$$

(j) Bismuth Telluride:

$$m^*(E_{FB}) = \frac{\hbar^2}{2\bar{\omega}_1}(1 + 2\alpha E_{FB}) \tag{7.83}$$

(k) Germanium:

1. The model of Cardona et al.:

$$m^*(E_{FB}, n) = m^*_{\parallel} A'_{69}(E_{FB}, n) \tag{7.84}$$

2. The model of Wang and Ressler:

$$m^*(E_{FB}, n) = m^*_{\parallel} A'_{71}(E_{FB}, n) \tag{7.85}$$

(l) Gallium Antimonide:

$$m^*(E_{FB}) = m_c I'_{16}(E_{FB}) \tag{7.86}$$

(m) II–V compounds:

$$m^*_{\pm}(E_{FB}, n) = \frac{\hbar^2}{2} A'_{75,\pm}(E_{FB}, n) \tag{7.87}$$

From the different chapters of this monograph, the EMM can be formulated by using the respective dispersion relation and their dependences with respect to various variables can also be studied. In many cases, in addition to Fermi energy and other system constraints, the effective mass will depend on the quantum numbers depending on particular band structure under different physical conditions.

7.2.4 Diffusivity–Mobility Ratio

The diffusivity (D) to mobility (μ) ratio (DMR) of the carriers in semiconductor devices is known to be very useful [63] since the diffusion constant (a quantity often used in device analysis but whose exact experimental determination is rather difficult) can be obtained from this ratio by knowing the experimental values of the mobility. In addition, it is more accurate than any of the individual relation for the diffusivity or the mobility, which are the two widely used quantities of carrier transport of modern nanostructured materials and devices. The classical DMR equation is valid for both types of carriers. In its conventional form, it appears that the DMR increases linearly with the temperature T being independent of the carrier concentration. This relation holds only under the condition of carrier nondegeneracy although its validity has been suggested erroneously for degenerate materials [64]. The performance of the electron devices at the device terminals and the speed of operation of modern switching transistors are significantly influenced by the degree of carrier degeneracy present in these devices [65]. The simplest way of analyzing them under degenerate condition is to use the appropriate DMR to express the performance of the devices at the device terminals and the switching speed in terms of the carrier concentration [65].

It is well known from the fundamental work of Landsberg [66–68] that the DMR for electronic materials having degenerate electron concentration is essentially determined by their respective energy band structures. This relation is useful for semiconductor homostructures [69,70], semiconductor–semiconductor heterostructures [71, 72], metals–semiconductor heterostructures [73–77], and insulator–semiconductor heterostructures [78–81]. It has different values in different materials and varies with the doping, with the magnitude of the reciprocal quantizing magnetic field under magnetic quantization, with the quantizing electric field as in inversion layers, with the nanothickness as in quantum wells and quantum well wires and with superlattice period as in the quantum-confined superlattices of small gap semiconductors with graded interfaces having various carrier energy spectra [82–94]. It can, in general, be proved that for bulk specimens the DMR is given by [82]

$$\frac{D}{\mu} = \left(\frac{n_0}{|e|}\right) \Big/ \left(\frac{\partial n_0}{\partial E_F}\right) \tag{7.88}$$

The electric quantum limit as in inversion layers and nipi structures refers to the lowest electric subband and (7.88) assumes the form [82]

$$\frac{D}{\mu} = \left(\frac{\bar{n}_0}{|e|}\right) \Big/ \left(\frac{\partial \bar{n}_0}{\partial \left(\overline{E}_{F0} - \overline{E}_0\right)}\right) \tag{7.89}$$

where $\bar{n}_0$, $\overline{E}_{F0}$, and $\overline{E}_0$ are the electron concentration, the energy of the electric subband, and the Fermi energy in the electric quantum limit.

For inversion layers and the nipi structures, under the condition of electric quantum limit, (7.2) assumes the form [15]

$$G = \left(\frac{\pi^2 k_B^2 T}{3|e|\bar{n}_0}\right) \left[\frac{d\bar{n}_0}{d\left(\overline{E}_{F0} - \overline{E}_0\right)}\right] \tag{7.90}$$

Using the appropriate equations one obtains

$$\frac{D}{\mu} = \left(\frac{\pi^2 k_B^2 T}{3|e|^2 G}\right) \tag{7.91}$$

Thus, the DMR for degenerate materials can be determined by knowing the experimental values of G.

The suggestion for the experimental determination of the DMR for degenerate semiconductors having arbitrary dispersion laws as given by (7.91) does not contain any energy band constants. For a fixed temperature, the DMR varies inversely as G. Only the experimental values of G for any material as a function of electron concentration will generate the experimental values of the DMR for that range of n_0 for that system. Since G decreases with increasing n_0, from (7.91) one can infer that the DMR will increase with increase in n_0. This statement is the compatibility test so far as the suggestion for the experimental determination of DMR for degenerate materials is concerned.

Although the DMR has extensively been investigated in the literature [82–89], it appears that the influence of electric field on the DMR in optoelectronic semiconductors together with its various quantum-confined counterpart has yet to be reported. We present few results in this context.

(a) In the presence of intense electric field, the DMR in III–V, ternary, and quaternary materials in accordance with perturbed three- and two-band models of Kane can, respectively, be expressed as

$$\frac{D}{\mu} = \frac{1}{e}\left[\frac{g_{70}(E_{Fs}, F) + h_{70}(E_{Fs}, F)}{g'_{70}(E_{Fs}, F) + h'_{70}(E_{Fs}, F)}\right] \tag{7.92}$$

$$\frac{D}{\mu} = \frac{1}{e}\left[\frac{g_{71}(E_{Fs}, F) + h_{71}(E_{Fs}, F)}{g'_{71}(E_{Fs}, F) + h'_{71}(E_{Fs}, F)}\right] \tag{7.93}$$

In the absence of any field, the expressions for the DMR for optoelectronic materials whose energy band structures are defined by the unperturbed two-band model of Kane, under the condition $(E_F E_{g0}^{-1}) \ll 1$ assume the well-known forms as [82]

$$\frac{D}{\mu} = \frac{k_B T}{e} \left[\frac{F_{1/2}(\eta) + \frac{15\alpha k_B T}{4} F_{3/2}(\eta)}{F_{-1/2}(\eta) + \frac{15\alpha k_B T}{4} F_{1/2}(\eta)} \right] \tag{7.94}$$

(b) In the presence of intense electric field, the DMR in III–V, ternary, and quaternary materials in accordance with perturbed three- and two-band models of Kane can, respectively, be expressed under magnetic quantization as

$$\frac{D}{\mu} = \frac{1}{e} \left[\frac{\sum_{n=0}^{n_{\max}} [Q_{11}(E_{FB}, F, n) + Q_{12}(E_{FB}, F, n)]}{\sum_{n=0}^{n_{\max}} [Q'_{11}(E_{FB}, F, n) + Q'_{12}(E_{FB}, F, n)]} \right] \tag{7.95}$$

and

$$\frac{D}{\mu} = \frac{1}{e} \left[\frac{\sum_{n=0}^{n_{\max}} [Q_{13}(E_{FB}, F, n) + Q_{14}(E_{FB}, F, n)]}{\sum_{n=0}^{n_{\max}} [Q'_{13}(E_{FB}, F, n) + Q'_{14}(E_{FB}, F, n)]} \right] \tag{7.96}$$

(c) In the presence of intense electric field, the DMR in quantum wires of III–V, ternary, and quaternary materials in accordance with perturbed three- and two-band models of Kane can, respectively, be expressed as

$$\frac{D}{\mu} = \frac{1}{e} \left[\frac{\sum_{n_z=1}^{n_{z\max}} \sum_{n_y=1}^{n_{y\max}} [Q_{15}(E_{F1D}, F, n_y, n_z) + Q_{16}(E_{F1D}, F, n_y, n_z)]}{\sum_{n_z=1}^{n_{z\max}} \sum_{n_y=1}^{n_{y\max}} [Q'_{15}(E_{F1D}, F, n_y, n_z) + Q'_{16}(E_{F1D}, F, n_y, n_z)]} \right] \tag{7.97}$$

$$\frac{D}{\mu} = \frac{1}{e} \left[\frac{\sum_{n_z=1}^{n_{z\max}} \sum_{n_y=1}^{n_{y\max}} [Q_{17}(E_{F1D}, F, n_y, n_z) + Q_{18}(E_{F1D}, F, n_y, n_z)]}{\sum_{n_z=1}^{n_{z\max}} \sum_{n_y=1}^{n_{y\max}} [Q'_{17}(E_{F1D}, F, n_y, n_z) + Q'_{18}(E_{F1D}, F, n_y, n_z)]} \right] \tag{7.98}$$

(d) In the presence of intense electric field, the DMR in effective mass superlattices of optoelectronic materials in accordance with perturbed three- and two-band models of Kane can, respectively, be expressed under magnetic quantization as

$$\frac{D}{\mu} = \frac{1}{e} \left[\frac{\sum_{n=0}^{n_{\max}} [Q_{19}(E_{FB}, F, n) + Q_{20}(E_{FB}, F, n)]}{\sum_{n=0}^{n_{\max}} [Q'_{19}(E_{FB}, F, n) + Q'_{20}(E_{FB}, F, n)]} \right] \tag{7.99}$$

and

$$\frac{D}{\mu} = \frac{1}{e}\left[\frac{\sum_{n=0}^{n_{\max}}[Q_{21}(E_{FB}, F, n) + Q_{22}(E_{FB}, F, n)]}{\sum_{n=0}^{n_{\max}}[Q'_{21}(E_{FB}, F, n) + Q'_{22}(E_{FB}, F, n)]}\right] \tag{7.100}$$

(e) In the presence of intense electric field, the DMR in quantum wire effective mass superlattices of optoelectronic materials in accordance with perturbed three- and two-band models of Kane can, respectively, be expressed as

$$\frac{D}{\mu} = \frac{1}{e}\left[\frac{\sum_{n_z=1}^{n_{z\max}}\sum_{n_y=1}^{n_{y\max}}[Q_{23}(E_{\text{FIDEMSL}}, F, n_y, n_z) + Q_{24}(E_{\text{FIDEMSL}}, F, n_y, n_z)]}{\sum_{n_z=1}^{n_{z\max}}\sum_{n_y=1}^{n_{y\max}}[Q'_{23}(E_{\text{FIDEMSL}}, F, n_y, n_z) + Q'_{24}(E_{\text{FIDEMSL}}, F, n_y, n_z)]}\right] \tag{7.101}$$

$$\frac{D}{\mu} = \frac{1}{e}\left[\frac{\sum_{n_z=1}^{n_{z\max}}\sum_{n_y=1}^{n_{y\max}}[Q_{25}(E_{\text{FIDEMSL}}, F, n_y, n_z) + Q_{26}(E_{\text{FIDEMSL}}, F, n_y, n_z)]}{\sum_{n_z=1}^{n_{z\max}}\sum_{n_y=1}^{n_{y\max}}[Q'_{25}(E_{\text{FIDEMSL}}, F, n_y, n_z) + Q'_{26}(E_{\text{FIDEMSL}}, F, n_y, n_z)]}\right] \tag{7.102}$$

(f) In the presence of intense electric field, the DMR in superlattices of optoelectronic materials with graded interfaces in accordance with perturbed three-band model of Kane can be expressed under magnetic quantization as

$$\frac{D}{\mu} = \frac{1}{e}\left[\frac{\sum_{n=0}^{n_{\max}}[Q_{27}(E_{\text{FBGISL}}, F, n)+Q_{28}(E_{\text{FBGISL}}, F, n)]}{\sum_{n=0}^{n_{\max}}[Q'_{27}(E_{\text{FBGISL}}, F, n)+Q'_{28}(E_{\text{FBGISL}}, F, n)]}\right] \tag{7.103}$$

(g) In the presence of intense electric field, the DMR in quantum wire superlattices of optoelectronic materials with graded interfaces in accordance with perturbed three-band model of Kane can be expressed as

$$\frac{D}{\mu} = \frac{1}{e}\left[\frac{\sum_{n_z=1}^{n_{z\max}}\sum_{n_y=1}^{n_{y\max}}[Q_{29}\left(E_{\text{FQWGISL}}, F, n_y, n_z\right) + Q_{30}\left(E_{\text{FQWGISL}}, F, n_y, n_z\right)]}{\sum_{n_z=1}^{n_{z\max}}\sum_{n_y=1}^{n_{y\max}}[Q'_{29}\left(E_{\text{FQWGISL}}, F, n_y, n_z\right) + Q'_{30}\left(E_{\text{FQWGISL}}, F, n_y, n_z\right)]}\right] \tag{7.104}$$

With the advent of ultra-small devices, the influence of electric field is of crucial importance in the whole spectrum of nanoscience and technology. In this particular section, we have formulated the DMR in optoelectronic semiconductors and their nanostructures in the presence of intense electric field.

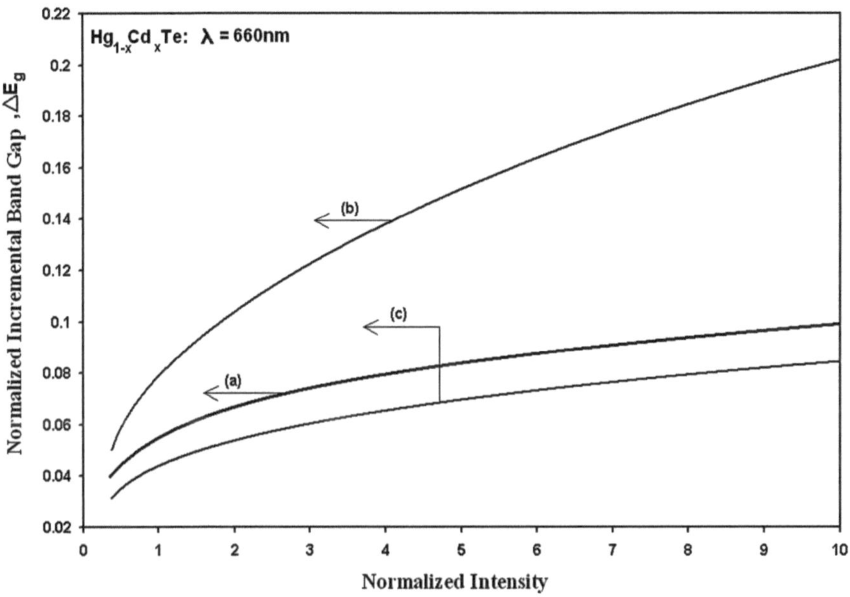

Fig. 7.1 Plots of the normalized incremental band gap (ΔE_g) for $n - Hg_{1-x}Cd_xTe$ as a function of normalized light intensity in which the curves (a) and (b) represent the perturbed three- and two-band models of Kane respectively. The curve (c) represents the same variation in $n - Hg_{1-x}Cd_xTe$ in accordance with the perturbed parabolic energy bands

7.2.5 *Measurement of Bandgap in the Presence of Light Waves*

Using (5.1), (5.2), and (5.3), the normalized incremental band gap (ΔE_g) has been plotted as a function of normalized I_0 (for a given wavelength and considering red light for which $\lambda = 660\,\text{nm}$) at $T = 4.2\,\text{K}$ in Figs. 7.1 and 7.2 for $n - Hg_{1-x}Cd_xTe$ and n-$In_{1-x}Ga_xAs_yP_{1-y}$ lattice matched to InP in accordance with the perturbed three- and two-band models of Kane and that of perturbed parabolic energy bands, respectively. In Figs. 7.3 and 7.4, the normalized incremental band gap has been plotted for the aforementioned optoelectronic compounds as a function of λ. It is worth remarking that the influence of an external photoexcitation is to change radically the original band structure of the material. Because of this change, the photon field causes to increase the band gap of semiconductors. We propose the following two experiments for the measurement of band gap of semiconductors under photoexcitation.

(A) A white light with color filter is allowed to fall on a semiconductor and the optical absorption coefficient ($\overline{\alpha}_0$) is being measured experimentally. For different colors of light, $\overline{\alpha}_0$ is measured and $\overline{\alpha}_0$ versus $\hbar\omega$ (the incident photon energy) is plotted and we extrapolate the curve such that $\overline{\alpha}_0 \to 0$ at a particular value $\hbar\omega_1$. This $\hbar\omega_1$ is the unperturbed band gap of the semiconductor. During

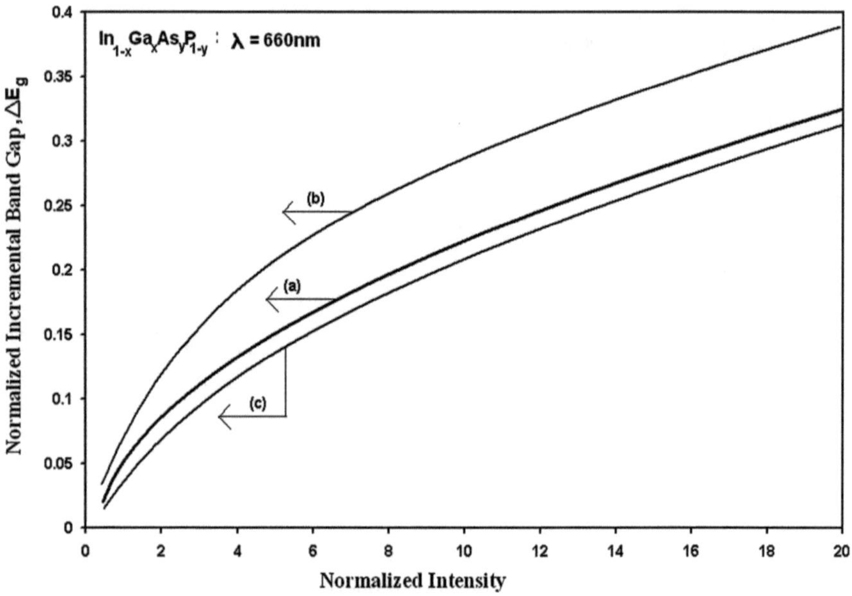

Fig. 7.2 Plots of the normalized incremental band gap (ΔE_g) for $In_{1-x}Ga_xAs_yP_{1-y}$ lattice matched to InP as a function of normalized light intensity for all cases of Fig. 7.1

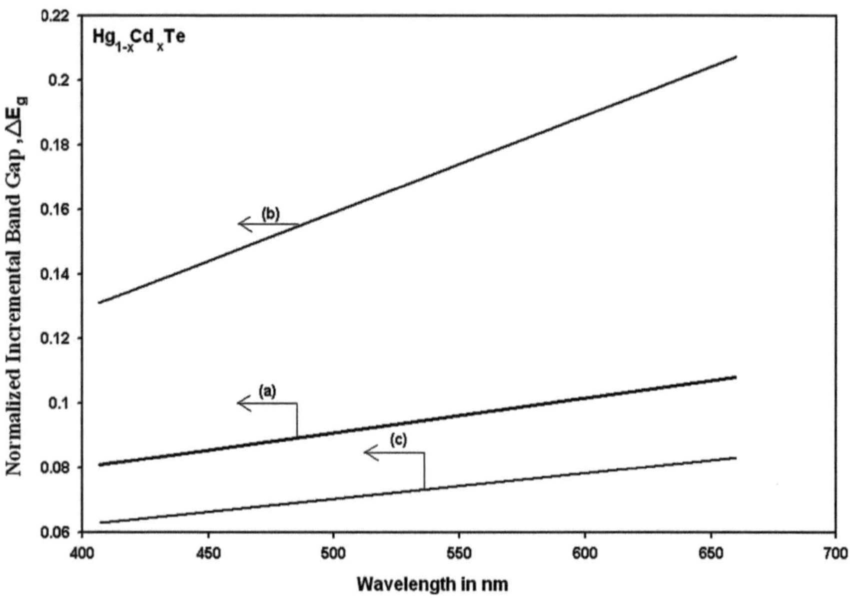

Fig. 7.3 Plots of the normalized incremental band gap (ΔE_g) for $Hg_{1-x}Cd_xTe$ as a function of wavelength for all cases of Fig. 7.1

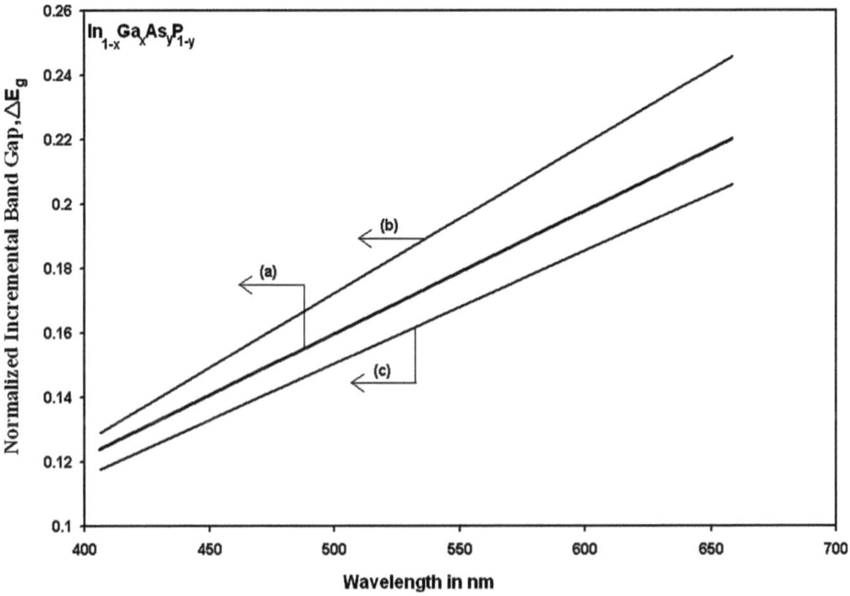

Fig. 7.4 Plots of the normalized incremental band gap (ΔE_g) for $\text{In}_{1-x}\text{Gs}_x$ $\text{As}_y\text{P}1-y$ lattice matched to InP as a function of wavelength for all cases of Fig. 7.2

this process, we vary the wavelength with fixed I_0. From our present study, we have observed that the band gap of the semiconductor increases for various values of λ when I_0 is fixed (from Figs. 5.3 and 5.4). This implies that the band gap of the semiconductor measured (i.e., $\hbar\omega_1 = E_g$) is not the unperturbed band gap E_{g_0} but the perturbed band gap E_g; where $E_g = E_{g_0} + \Delta E_g$, ΔE_g is the increased band gap at $\hbar\omega_1$. Conventionally, we consider this E_g as the unperturbed band gap of the semiconductor and this particular concept needs modification. Furthermore, if we vary I_0 for a monochromatic light (when λ is fixed) the band gap of the semiconductor will also change consequently (Figs. 5.1 and 5.2). Consequently, the absorption coefficient will change with the intensity of light [95]. For the overall understanding, the detailed theoretical and experimental investigations are needed in this context for various materials having different band structures.

(B) The conventional idea for the measurement of the band gap of the semiconductors is the fact that the minimum photon energy $h\nu$ (ν is the frequency of the monochromatic light) should be equal to the band gap E_{g_0} (unperturbed) of the semiconductor, i.e.,

$$h\nu = E_{g_0} \tag{7.105}$$

In this case, λ is fixed for a given monochromatic light and the semiconductor is exposed to a light of wavelength λ. Also the intensity of the light is fixed. From Figs. 7.3 and 7.4, we observe that the band gap of the semiconductor is not E_{g_0} (for a minimum value of $h\nu$) but E_g, the perturbed band gap. Thus, we can rewrite the above equality as

$$h\nu = E_g \tag{7.106}$$

Furthermore, if we vary the intensity of light (Figs. 7.1 and 7.2) for the study of photoemission, the minimum photon energy should be

$$h\nu_1 = E_{g_1} \tag{7.107}$$

where E_{g_1} is the perturbed band gap of the semiconductor due to various intensities of light when ν and ν_1 are different.

Thus, we arrive at the following conclusions:

(a) Under different intensities of light, keeping λ fixed, the condition of band gap measurement is given by

$$h\nu_1 = E_{g_1} = E_{g_0} + \Delta E_{g_1} \tag{7.108}$$

(b) Under different colors of light, keeping the intensity fixed, the condition of band gap measurement assumes the form

$$h\nu = E_g = E_{g_0} + \Delta E_g \tag{7.109}$$

and not the conventional result as given by (7.105).

7.2.6 Diffusion Coefficient of the Minority Carriers

This particular coefficient in quantum-confined lasers can be expressed [82] as

$$D_i / D_0 = dE_{Fi} / dE_F \tag{7.110}$$

where D_i and D_0 are the diffusion coefficients of the minority carriers both in the presence and in the absence of quantum confinements and E_{Fi} and E_F are the Fermi energies in the respective cases. It appears then that the formulation of the above ratio requires a relation between E_{Fi} and E_F, which, in turn, is determined by the appropriate carrier statistics. Thus, our present study plays an important role in determining the diffusion coefficients of the minority carriers of quantum-confined lasers with materials having arbitrary band structures. Therefore, in the investigation of the optical excitation of the optoelectronic materials, which leads to the study of the ambipolar diffusion coefficients, the present results contribute significantly.

7.2.7 Nonlinear Optical Response

The nonlinear response from the optical excitation of the free carriers is given by [96]

$$Z_0 = \frac{-e^2}{\omega^2 \hbar^2} \int_0^{\infty} \left(k_x \frac{\partial k_x}{\partial E} \right)^{-1} f(E) N(E) \mathrm{d}E \tag{7.111}$$

where ω is the optical angular frequency, $N(E)$ is the density of states function. From the various E–k relations of different materials under different physical conditions, we can formulate the expression of $N(E)$ and from band structure we can derive the term $(k_x \frac{\partial k_x}{\partial E})$ and thus by using the density of states function as formulated, we can study the Z_0 for all types of materials as considered in this monograph.

7.2.8 Third-Order Nonlinear Optical Susceptibility

This particular susceptibility can be written as [97]

$$\chi_{NP}(\omega_1, \omega_2, \omega_3) = \frac{n_0 e^4 \langle \varepsilon^4 \rangle}{24 \omega_1 \omega_2 \omega_3 (\omega_1 + \omega_2 + \omega_3) \hbar^4} \tag{7.112}$$

wnere $n_0 \langle \varepsilon^4 \rangle = \int_0^{\infty} \frac{\partial^4 E}{\partial k_z^4} N(E) f(E) \mathrm{d}E$ and the other notations are defined in [97]. The term $(\frac{\partial^4 E}{\partial k_z^4})$ can be formulated by using the dispersion relations of different materials as given in appropriate sections of this monograph. Thus, one can investigate the $\chi_{NP}(\omega_1, \omega_2, \omega_3)$ for all materials as considered in this monograph.

7.2.9 Generalized Raman Gain

The generalized Raman gain in optoelectronic materials can be expressed as [98]

$$R_G = \overline{I} \left(\frac{16\pi^2 c^2}{\hbar \omega \rho g \omega_s^2 n_s n_p} \right) \left(\frac{\Gamma_\rho}{\Gamma} \right) \left(\left(\frac{e^2}{mc^2} \right)^2 m^2 R^2 \right) \tag{7.113}$$

where $\overline{I} = \sum\limits_{n.t_z} [f_0(n, k_z \uparrow) - f_0(n, k_z \downarrow)]$, $f_0(n, k_z \uparrow)$ is the Fermi factor for spin-up Landau levels, $f_0(n, k_z \downarrow)$ is the Fermi factor for spin down Landau levels, n is the Landau quantum number and the other notations are defined in [98]. It appears then the formulation of R_G is determined by the appropriate derivation of

I which in turn requires the magneto-dispersion relations. By using the formulas of the Chaps. 5,6,7, 11, 12, and 13, the band structure as derived in the said chapters R_G can, in general, be investigated.

7.3 Brief Review of Experimental Works

The experimental aspect of FNFE is very wide and it is virtually impossible to even highlight the major developments in a single chapter. It may be noted that FNFE from carbon nanotubes has been studied in the literature without considering the fact that in the presence of intense electric field, the band structure of the carbon nanotubes changes profoundly. For the purpose of condensed presentation, the Sect. 7.3.1 contains the investigation of FNFE from carbon nanotubes in the presence of intense electric field, the optimization of Fowler–Nordheim (FN) field emission current from nanostructured materials is given in Sect. 7.3.2, and the very brief summary of the experimental studies of FNFE from nanostructured materials have been discussed in Sect. 7.3.3.

7.3.1 Field Emission from Carbon Nanotubes in the Presence of Strong Electric Field

The $E - k_y$ relation for the arm chair and zigzag carbon nanotubes throughout the entire Brillouin zone can be expressed as

$$E = t_c[1 + 4\cos(\upsilon\pi/n)\cos(k_y a_c\sqrt{3}/2) + 4\cos^2(k_y a_c\sqrt{3}/2)]^{\frac{1}{2}} \tag{7.114a}$$

and

$$E = t_c[1 + 4\cos(\upsilon\pi/n)\cos(k_y a_c 3/2) + 4\cos^2(\upsilon\pi/n)]^{\frac{1}{2}} \tag{7.115a}$$

where t_c is the C–C bonding energy, $\upsilon = 1, 2, 3, \ldots, 2n$ and a_c is the nearest neighbor C–C bonding distance.

For armchair and zigzag carbon nanotubes, the energy dispersion relations, in the presence of electric field, assume the forms

$$k_x = \frac{2}{\sqrt{3}a_c} A_1(E, E_i, F) \tag{7.114b}$$

and

$$k_x = \frac{2}{3a_c} A_2(E, E_i, F) \tag{7.115b}$$

where $A_1(E, E_i, F) = \cos^{-1}\left[-\frac{1}{2}\left[\frac{\rho_0^2(E_i, F)}{t_c^2} - 5\right] + \frac{1}{2}\left[\left(\frac{\rho_0^2(E_i, F)}{t_c^2} - 5\right)^2 - \left(1 - \frac{\rho_0^2(E, F)}{t_c^2}\right)\right]^{1/2}\right]$,

$$E_i = \frac{|3i - m + n|}{2}(t_c a_c / r_0), \rho_0(E, F)$$
$$= \left[E - (\hbar^2 F^2/12 m_r E_g^2)\left(1 + \frac{2m_c E}{m_r E_g}\right)^{-3/2}\right], \quad \rho_0(E_i, F)$$
$$= \left[E_i - (\hbar^2 F^2/12 m_r E_g^2)\left(1 + \frac{2m_c E_i}{m_r E_g}\right)^{-3/2}\right]$$

and

$$A_2(E, E_i, F) = \cos^{-1}\left[\frac{\rho_0^2(E, F) - 1 - 4\varphi^2(E_i, F)}{4\varphi(E_i, F)}\right],$$
$$\varphi(E_i, F) = \frac{1}{2}\left[\frac{\rho_0(E_i, F)}{t_c} - 1\right]$$

The electron statistics can, respectively, be expressed as

$$n_0 = \frac{8}{\pi a_c \sqrt{3}} \sum_{i=0}^{i_{\max}} [A_1(E_{F1}, E_i, F) + A_3(E_{F1}, E_i, F)] \tag{7.116}$$

and

$$n_0 = \frac{8}{3\pi a_c} \sum_{i=0}^{i_{\max}} [A_2(E_{F2}, E_i, F) + A_4(E_{F2}, E_i, F)] \tag{7.117}$$

where

$$A_1(E_{F1}, E_i, F) = \cos^{-1}\left[-\frac{1}{2}\left[\frac{\rho_0^2(E_i, F)}{t_c^2} - 5\right] + \frac{1}{2}\left[\left(\frac{\rho_0^2(E_i, F)}{t_c^2} - 5\right)^2 - \left(1 - \frac{\rho_0^2(E_{F1}, F)}{t_c^2}\right)\right]^{1/2}\right]$$

E_{F1} is the Fermi energy in the present case,

$$A_3(E_{F1}, E_i, F) = \sum_{R=1}^{R_0}\left[2(k_B T)^{2R}(1 - 2^{1-2R})\xi(2R)\frac{\partial^{2R}}{\partial E_{F1}^{2R}}[A_1(E_{F1}, E_i, F)]\right],$$
$$A_2(E_{F2}, E_i, F) = \cos^{-1}\left[\frac{\rho_0^2(E_{F2}, F) - 1 - 4\varphi^2(E_i, F)}{4\varphi(E_i, F)}\right]$$

E_{F2} is the Fermi energy in the present case, $\rho_0(E_{F1},F) = [E_{F1} - (\hbar^2 F^2/12m_r E_g^2)\left(1+\frac{2m_c E_{F1}}{m_r E_g}\right)^{-3/2}]$,

$$\rho_0(E_{F2},F) = \left[E_{F2} - (\hbar^2 F^2/12m_r E_g^2)\left(1+\frac{2m_c E_{F2}}{m_r E_g}\right)^{-3/2}\right],$$

$$A_4(E_{F2},E_i,F) = \sum_{R=1}^{R_0}\left[2(k_B T)^{2R}(1-2^{1-2R})\xi(2R)\frac{\partial^{2R}}{\partial E_{F2}^{2R}}[A_3(E_{F2},E_i,F)]\right]$$

The field emitted current for armchair and zigzag nanotubes can, respectively, be expressed as

$$I = \frac{4ek_B T}{h}\sum_{i=0}^{i_{\max}} F_0(\eta_{24})\exp(-\theta_{24}) \tag{7.118}$$

and

$$I = \frac{4ek_B T}{h}\sum_{i=0}^{i_{\max}} F_0(\eta_{25})\exp(-\theta_{25}) \tag{7.119}$$

where,

$\eta_{24} = \frac{E_{F1}-\varepsilon_{i1}}{k_B T}$, ε_{i1} is the root of the equation

$$0 = \cos^{-1}\left[-\frac{1}{2}\left[\frac{\rho_0^2(E_i,F)}{t_c^2}-5\right]+\frac{1}{2}\left[\left(\frac{\rho_0^2(E_i,F)}{t_c^2}-5\right)^2 - \left(1-\frac{\rho_0^2(\varepsilon_{i1},F)}{t_c^2}\right)\right]^{1/2}\right]$$

$$\rho_0(\varepsilon_{i1},F) = \left[\varepsilon_{i1} - (\hbar^2 F^2/12m_r E_g^2)\left(1+\frac{2m_c\varepsilon_{i1}}{m_r E_g}\right)^{-3/2}\right]$$

$$\theta_{24} = \frac{2}{\sqrt{3}a_c}\left[\frac{A_1^2(V_0,E_i,F)}{FA_1'(V_0,E_i,F)}\right], A_1'(V_0,E_i,F) = \frac{L_9'(V_0,E_i,F)}{\sqrt{1-L_9^2(V_0,E_i,F)}}$$

$$A_1(V_0,E_i,F) = \cos^{-1}\left[-\frac{1}{2}\left[\frac{\rho_0^2(E_i,F)}{t_c^2}-5\right]+\frac{1}{2}\left[\left(\frac{\rho_0^2(E_i,F)}{t_c^2}-5\right)^2 - \left(1-\frac{\rho_0^2(V_0,F)}{t_c^2}\right)\right]^{1/2}\right]$$

$$L_9(V_0, E_i, F) = \left[-\frac{1}{2}\left[\frac{\rho_0^2(E_i, F)}{t_c^2} - 5\right] + \frac{1}{2}\left[\left(\frac{\rho_0^2(E_i, F)}{t_c^2} - 5\right)^2 - \left(1 - \frac{\rho_0^2(V_0, F)}{t_c^2}\right)\right]^{1/2}\right]$$

$$L_9'(V_0, E_i, F) = \frac{1}{2}\left[\left[\frac{\rho_0(V_0, F)\rho_0'(V_0, F)}{t_c^2}\right]\left[\left(\frac{\rho_0^2(E_i, F)}{t_c^2} - 5\right)^2 - \left(1 - \frac{\rho_0^2(V_0, F)}{t_c^2}\right)\right]^{-1/2}\right.$$

$$\rho_0'(V_0, F) = \left[1 + (\hbar^2 F^2 m_c / 4m_r^2 E_g^2)\left(1 + \frac{2m_c V_0}{m_r E_g}\right)^{-5/2}\right] \quad (7.120)$$

$\eta_{25} = \frac{E_{F2} - \varepsilon_{i2}}{k_B T}$, ε_{i2} is the root of the equation

$$0 = \cos^{-1}\left[\frac{\rho_0^2(\varepsilon_{i2}, F) - 1 - 4\varphi^2(E_i, F)}{4\varphi(E_i, F)}\right]$$

$$\theta_{24} = \frac{2}{3a_c}\left[\frac{A_2^2(V_0, E_i, F)}{FA_2'(V_0, E_i, F)}\right],$$

$$A_2'(V_0, E_i, F) = \left[\frac{L_{10}'(V_0, E_i, F)}{\sqrt{1 - L_{10}^2(V_0, E_i, F)}}\right],$$

$$L_{10}(V_0, E_i, F) = \left[\frac{\rho_0^2(V_0, F) - 1 - 4\varphi^2(E_i, F)}{4\varphi(E_i, F)}\right],$$

$$L_{10}'(V_0, E_i, F) = \frac{\rho_0(V_0, F)\rho_0'(V_0, F)}{2\varphi(E_i, F)} \quad (7.121)$$

and

$$\rho_0(\varepsilon_{i2}, F) = \left[\varepsilon_{i2} - (\hbar^2 F^2 / 12 m_r E_g^2)\left(1 + \frac{2m_c \varepsilon_{i2}}{m_r E_g}\right)^{-3/2}\right].$$

Using the appropriate equations together with the energy band constants as given in Table 1.1, we have plotted the field emitted current as function of electric field for different electron degeneracies in Figs. 7.5 and 7.6 respectively. For the purpose of simplicity, we have assumed a chiral independent work function of 4.2 eV for single-wall carbon nanotubes (SWCNs). It appears that the threshold current starts for field strength of about 10^4V m^{-1} which is abruptly low as compared with the threshold field for metals, which usually occurs beyond 10^9 V m^{-1}. We note that as

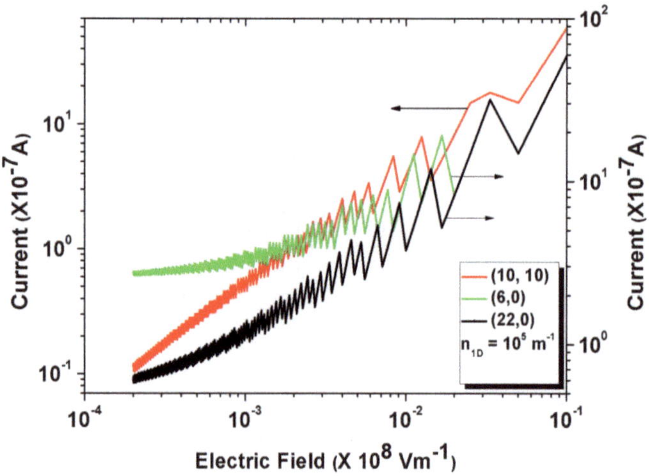

Fig. 7.5 Plot of the field emitted current as function of electric field for different chiral single-walled carbon nanotubes for $n_0 = 10^5 \mathrm{m}^{-1}$

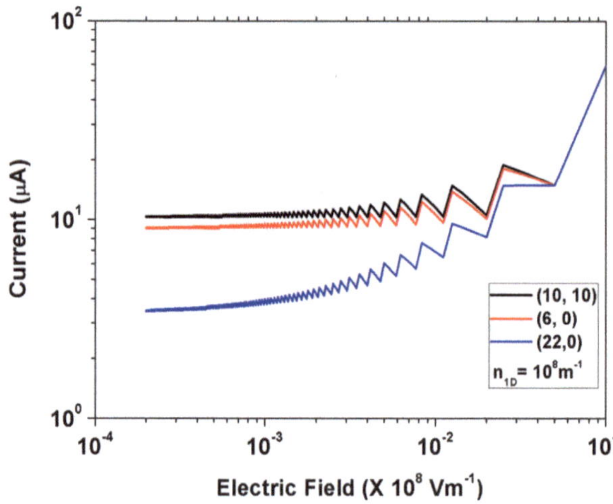

Fig. 7.6 Plot of the field emitted current as function of electric field for different chiral single-walled carbon nanotubes for $n_0 = 10^8 \mathrm{m}^{-1}$

the carrier degeneracy increases, the magnitude of the field current also increases to as large as a few tens of microamperes as also reported elsewhere [99]. With the increase of carrier degeneracy, the magnitude of the field emitted current increases. Moreover, an oscillating current is found for both the carrier degeneracies with the increase in the electric field. The reason for these oscillations is due to the appearance of the van Hove singularities. The influence of quantum confinement is immediately being apparent from all the figures, since the field emission from nanotubes depends strongly on the subband energies owing to Born von Karman boundary conditions in contrast with bulk specimens. It appears from the Figs. 7.5 and 7.6 that the emission current exhibit spikes for particular values of electric field where the singularity occurs. This will be different for other SWCNs due to different spectra constants. The field emission from nanotubes can be more than that of the quantum wires of different compounds and can become several orders of magnitude larger than of bulk specimens of the same materials, which is also a direct signature of quantum confinement. The appearance of the discrete jumps in the figures is due to the redistribution of the electrons among the quantized energy levels when the size quantum number corresponding to the highest occupied level changes from one fixed value to the others.

With varying electric field, a change is reflected in the field emission through the redistribution of the electrons among the quantized levels as noted already. It may be remarked that at the transition zone from one subband to another, the height of the peaks between any two subbands decreases with the increase in the degree of quantum confinement and is clearly shown in Figs. 7.5 and 7.6. Besides, the rate of change of emission current of different SWCNs are totally band-structure dependent. The numerical values of the field emitted current in all (m, n) cases vary widely, and are determined thoroughly by the chiral indices and diameter of the SWCNs. From the said figs, we can assess the influence of chiral index numbers on the emission current from SWCNs, and it further appears that the numerical values of the current from SWCNs are the greatest together with the fact that the oscillatory dependence is due to the crossing over of the Fermi level by the quantized level due to van Hove singularities.

This oscillatory dependence will be less and less prominent with increasing nanotube radius and carrier degeneracy respectively. Ultimately, for larger diameters, the current will be found to be less prominent resulting in monotonic increasing variation. In this context, it may be noted that there has been enormous amount of experimental works on the field emission from different types of nanotubes and we hope that investigations on the different physical properties of graphene will be accelerated by the fact that the Nobel prize in Physics 2010 was awarded jointly to Andre Geim and Konstantin Novoselov *"For Groundbreaking Experiments Regarding the Two-Dimensional Material Graphene."*

7.3.2 Optimization of Fowler–Nordheim (FN) Field Emission Current from Nanostructured Materials

It is important to find out the possible values of the energy band constants of the different materials having different band structures for the maximization of the FN field emission current from the one-dimensional wires since an experimentalist desires to optimize the same from nanowire systems. The accurate answer of this practical aspect is both difficult and deep since the FN field emission current is a

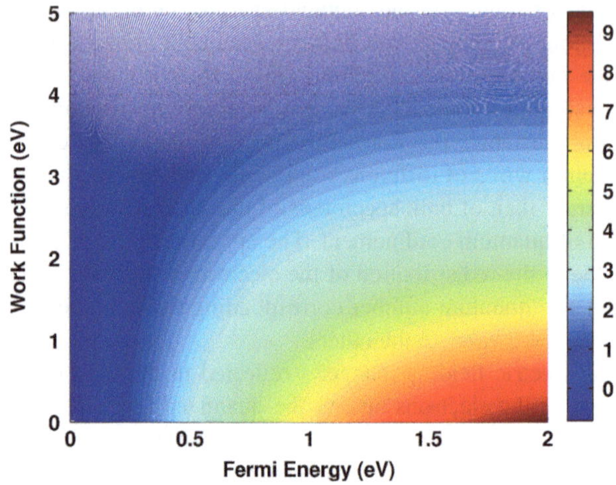

Fig. 7.7 Figure exhibits the surface contour plot of the field emission current (in 10^{-5} A) as function of Fermi energy and work function for n-InSb quantum wire following a basic parabolic energy dispersion relation at the lowest subband level. The cross-sectional dimensions are 25 nm × 20 nm at a field strength of 5×10^9 V m^{-1}

function of the Fermi energy, the electric field (which is the essential singularity of the tunneling probability at the zero value of it) and other spectrum constants like effective mass, band gap, the work function of the material (as can be seen throughout the book), and also since many physical variables depend on the film thickness. Mathematically the problem lies in the domain of nonlinear programming and the computer simulation of the Lagrange method of undetermined multipliers will not throw light into it. All the figures as presented in this book have been dealt with the bulk value of the corresponding work function for the simplification of numerical computation. Our numerical simulation is a simplified one and thus to obtain a rather good match with the experimental data, one has to consider the size dependent material parameters. Since the experimental data of the material constants for the quantum confined counterparts of semiconductors are not available for all the materials as considered in this book, we in turn present in Figs. 7.7–7.9, a "thumb rule" for guessing the FN field emission current at low temperatures. We have taken only one subband since, at low temperatures, almost all the carries occupies the lowest possible subband where the quantum effects become prominent.

From Fig. 7.7, we observe that in order to get a field emitted current of few tenths of microampers, one has to dope the system such that a Fermi energy of about 1.5–2 eV is reached for a dramatic less value in the n-InSb work function. Similar case has been shown in Fig. 7.8 for n-GaAs. We observe from Fig. 7.8 that by assuming that the effective mass to be constant, comparatively a low doping corresponding to a Fermi energy of 0.7 eV is sufficient to obtain a few tenths of microampere FN field emission current, although the magnitude of current in n-InSb is much higher.

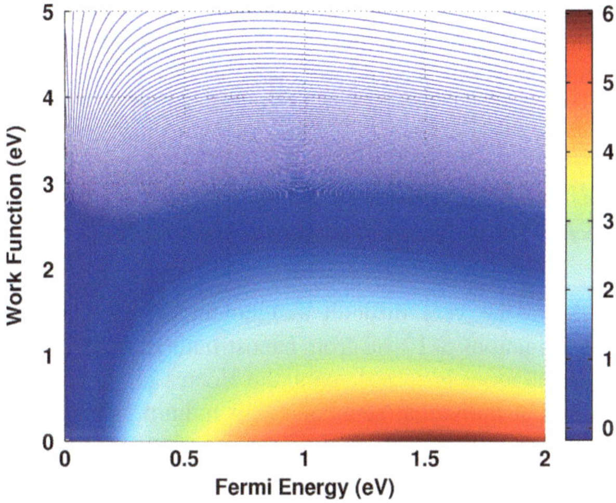

Fig. 7.8 Figure exhibits the surface contour plot of the field emission current (in 10^{-5} A) as function of Fermi energy and work function for n-GaAs quantum wire following a basic parabolic energy dispersion relation at the lowest subband level. The cross-sectional dimensions are 25 nm × 20 nm at a field strength of 5×10^9 Vm^{-1}

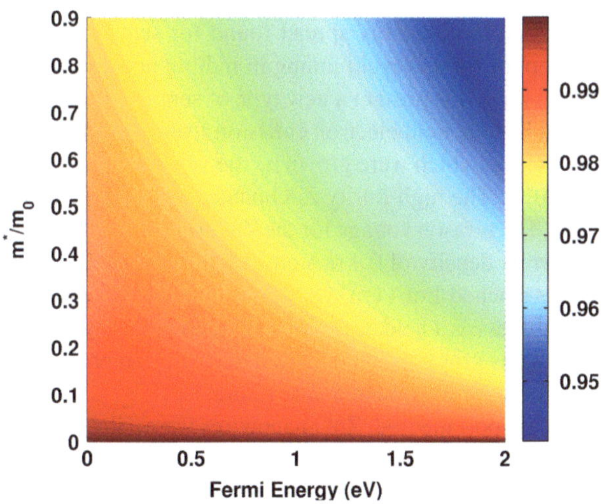

Fig. 7.9 Figure exhibits the surface contour plot of the field emission current (in 10^{-5} A) as function of Fermi energy and effective mass for n-InSb quantum wire following a basic parabolic energy dispersion relation at the lowest subband level. The cross-sectional dimensions are 25 nm × 20 nm at a field strength of 5×10^9 V m^{-1}

Figure 7.9 presents the surface contour plot of the field current as functions of Fermi energy and electron effective mass for quantum wires of n-InSb. Thus, we observe that to obtain a higher value of the field emission current, one has to reduce the value of the effective mass and consider the materials with low values of effective electron mass are necessary together with the fact that within the desired operating zone of any field emitting device, an extremely low value of the electron effective mass apparently dominates over the Fermi level.

7.3.3 Very Brief Description of Experimental Results of FNFE from Nanostructured Materials

We now embark on the very difficult assignment of brief highlighting of the experimental investigations of FNFE from nanostructured materials for the purpose of condensed presentation. In recent years, the Heeres group [100] has developed a procedure to mount individual semiconductor indium arsenide nanowires onto tungsten support tips to serve as electron field emission sources. The electron emission properties of the single nanowires were precisely determined by measuring the emission pattern, current–voltage curve, and the energy spectrum of the emitted electron beam. The two investigated nanowires showed stable, Fowler–Nordheim-like emission behavior and a small energy spread. Their morphology was characterized afterward using transmission electron microscopy. The experimentally derived field enhancement factor corresponded to the one calculated using the basic structural information. The observed emission behavior contrasts the often unstable emission and large energy spread found for semiconductor emitters and supports the concept of Fermi-level pinning in indium arsenide nanowires. Indium arsenide nanowires may thus present a new type of semiconductor electron sources. Lee et al. [101] studied the field electron emission from vertically well-aligned zinc oxide ~ZnO nanowires, which were grown by the vapor deposition method at a low temperature of 550°C. The high-purity ZnO nanowires showed a single crystalline wurtzite structure. The turn-on voltage for the ZnO nanowires was found to be about $6.0\,\mathrm{V\,mm^{-1}}$ at current density of $0.1\,\mathrm{mA\,cm^{-2}}$. The emission current density from the ZnO nanowires reached $1\,\mathrm{mA\,cm^{-2}}$ at a bias field of $11.0\,\mathrm{V\,mm^{-1}}$, which could give sufficient brightness as a field emitter in a flat panel display. Therefore, the well-aligned ZnO nanowires grown at such low temperature can promise the application of a glass-sealed flat panel display in a near future.

Using a vapor transport method, ZnO nanowires were selectively synthesized by Dong et al. [102] both on tungsten tips as electron field emitters and on tungsten plates with designed patterns. Control of the growth locations of the nanowires was accomplished by selectively positioning a thin film of Au catalyst. The angular intensity and fluctuation of the field emission current from the ZnO nanowires synthesized on tungsten tips have been demonstrated to be similar to those of carbon nanotubes. A self-destruction limit of $0.1\,\mathrm{mA\,sr^{-1}}$ for angular intensity was observed, and the power spectra showed a $1/f3/2$ characteristic from 1 Hz to 6 kHz. The ZnO nanowires synthesized by vapor–liquid–solid growth mechanism with Cu and Au as the catalyst were investigated by Li et al. [103]. The principal differences in morphology between Cu and Au catalyzed ZnO nanowires are observed and lead to significant differences in their field emission and photofluorescent characteristics. The Cu catalyzed ZnO nanowires with a high-quality wurtzite structure were grown vertically on p-type Si~100 substrate along [0002] direction. A strong ultraviolet emission at 381 nm is observed. These ZnO nanowires show excellent field emission properties with turn-on field of 0.83 V mm^{-1} and corresponding current density of 25 mA cm^{-2}. The emitted current density of the ZnO nanowires is 1.52 mA cm^{-2} at a bias field of 8.5 V mm^{-1}. The large field emission area factor, b, arising from the morphology of the nanowire field emitter, is partly responsible for the good emission characteristics. The ZnO nanowires with high emission current density and low turn-on field are expected to be used in field emission flat panel display.

The group of Khademi [104] reported the field emission properties of molybdenum oxide nanowires grown on a silicon substrate and its emission performance in various vacuum gaps. A new kind of molybdenum oxides named nanowires with nanoscale protrusions on their surfaces were grown by thermal vapor deposition with a length of ~1 μm and an average diameter of ~50 nm. The morphology, structure, composition, and chemical states of the prepared nanostructures were characterized by scanning electron microscopy (SEM), high-resolution transmission electron microscopy (HRTEM), X-ray diffraction (XRD), and X-ray photoelectron spectroscopy (XPS). According to XRD, XPS, and TEM analyses, the synthesized samples were composed of MoO_2 nanowires formed over a thin layer of crystalline Mo_4O_{11}. TEM observation revealed that these nanowires have some nanoscale protrusion on their surface. These nanoprotrusions resulted in enhancement of field emission properties of nanowires comprising nanoprotrusions. The turn-on emission field and the enhancement factor of this type of nanostructures were measured as 0.2 V μm^{-1} and 42991 at the vacuum gap of 300 μm, respectively. These excellent emission properties are attributed to the special structure of the nanowires that have potential to be utilized in vacuum nanoelectronic and microelectronic applications. It may be noted that Wang et al. [105] investigated the field emission properties of SnO_2 nanowires fabricated by chemical vapor deposition with metallic catalyst assistance. For the as-fabricated SnO_2 nanowires, the turn-on and threshold field were 4.03 and 5.4 V μm^{-1}, respectively. Considerable enhancement of field emission of SnO_2 nanowires was obtained by a postannealing process in oxygen at high temperature. When the SnO_2 nanowires were postannealed at 1,000°C in oxygen, the turn-on and threshold field were decreased to 3.77 and 4.4 V μm^{-1}, respectively, and the current density was increased to 6.58 from 0.3 mA cm^{-2} at the same applied electric field of 5.0 V μm^{-1}. The group of Lai [106] prepared wormhole-like mesoporous tungsten oxide nanowires on a Cu-tape/Si substrate, and explored the field emission performances. The wormhole-like mesoporous tungsten oxide nanowires of 20 nm diameter exhibited excellent field emission properties with extremely low turn-on and threshold fields (emission current density of 10 μA cm^{-2} and 10 mA cm^{-2}) of 0.083 and 1.75 Vμm^{-1}, respectively, as well as current stability of about 1,400 μA cm^{-2} at a fixed field of 0.67 Vμm^{-1}. This approach provides an efficient methodology for fabricating a field emitter that is expected to work at low voltage and can be used in field emission displays.

The explanations of several nanoscale phenomena such as the field enhancement factor in field emission, the large decay length of the adhesion force between a metallic tip and a surface, and the contact resistance in a nanowire break junction have been provided by Zhang and Pantelides [107]. They developed an analytical theory of Thomas–Fermi screening in nanoscale structures and demonstrated that nanoscale dimensions give rise to an effective screening length that depends on the geometry and physical boundary conditions. The electromagnetic response of linear carbon chains to external fields have been studied by Lorenzoni et al. [108], making use of ab initio methods. It is found that the associated emission currents, plotted as a function of bias potential, follow Fowler–Nordheim intensity–voltage curves typical of field emission of metallic tips. Under standard bias conditions, linear carbon chains of 1 nm of length are expected to deliver currents of the order of 1 μA. These systems behave, furthermore, as conducting spheroidal particles in photoabsorption processes. Linear carbon chains are thus likely to constitute the ultimate atomic-scale realization of metallic wires. Silicon carbide (SiC) nanowires on a silicon substrate were prepared by Wong et al [109] using hot-filament-assisted chemical-vapor deposition with a solid silicon and carbon source. The SiC nanowires show good field-emitting properties as revealed by the current–voltage characteristics. Together with its ease of preparation, these SiC nanowires are shown to have great potential in the area of electron field-emitting devices.

Field emission studies on Si nanowires (Si NWs) grown by the vapor–liquid–solid (VLS) technique are presented by Kulkarni et al. [110]. The field emission properties of the Si NWs were characterized in ultrahigh vacuum following several postgrowth processes such as catalyst etching, in situ annealing, and cesiation. The average threshold field of cesiated Si NWs was found to be $\sim 7.76 \pm 0.55\,\mathrm{V\mu m^{-1}}$ and showed a significant improvement over that of as-grown NWs (average threshold field $\sim 11.58\,\mathrm{V\mu m^{-1}}$). The superior field emission characteristics are attributed to the combination of cesiation and quality of the NWs' surface grown via hydrogen reduction of silicon tetrachloride. Silicon carbide (SiC) nanowires were grown directly by Senthil and Yong [111] on Si substrates by thermal evaporation of WO_3 and graphite powders at high temperature using NiO catalyst. The densities of the nanowires were controlled by varying the NiO catalyst concentration. The morphology, structure and composition of the nanowires were characterized by scanning electron microscopy (SEM), X-ray diffraction (XRD), Raman, FTIR, transmission electron microscopy (TEM) and energy-dispersive X-ray spectroscopy (EDX) measurements. The synthesized nanowires were single crystalline β-SiC oriented along the [111] direction. Based on the experimental results, a possible growth mechanism was explained on the basis of solid–liquid–solid (SLS) growth model. Field emission measurements showed that the emission efficiency was strongly dependent on the density of SiC nanowires. Lowest turn-on field of $1.8\,\mathrm{V\mu m^{-1}}$ and highest field enhancement factor of 5.9×10^3 was observed for the medium density SiC nanowire sample. Positive ac dielectrophoresis (DEP) has been used by Zhou et al. [112] to rapidly align ensembles of CdSe semiconductor nanowires (NWs) near patterned microelectrodes. Due to their large geometric aspect ratio, the induced dipole of the wires is proportional to their conductivity, which can be drastically enhanced under super-band-gap illumination by several orders of magnitude, with a corresponding increase in the wire DEP mobility. This optical enhancement of conductivity occurs because of the generation of mobile electrons and holes and is verified by a photocurrent measurement. The linear nanowire alignment exhibits a high degree of fluorescent polarization anisotropy in both absorption and emission. An unexpected observation is a reversible, factor of $\sim$4, electric-field-induced, and frequency-dependent enhancement of the nanowire emission near 10 Hz. Such illumination-sensitive, field-enhanced, and frequency-dependent alignment and emission phenomena of NWs suggest an electrical-optical platform for fabricating CdSe nanowire devices for polarization-sensitive photodetection and biosensing applications.

Yeong and Thong [113] reported the field emission properties of ultrathin tungsten nanowires of 5 nm diameter and several hundred nanometer length. Fowler–Nordheim plots of field emission current–voltage measurements of such nanowires show marked deviation from linearity. After flashing, cold field emission current stability with standard deviation of better than 1% has been observed for periods of at least 30 min at a vacuum level of 10^{-9} mbar. Beyond this, field emission current noise was found to mainly comprise current step jumps and current spikes. At high emission current densities in the order of 10^6 A cm^{-2}, the noise changes into flicker noise. Field emission at high current density induced surface diffusion and crystallization of the disordered nanowire tip due to temperature rise at the field emitting tip. Further increase in the emission current density initiated local arc destruction which caused shortening of the nanowire length. The growth and characteristics of metallic nanowires formed by field emission in the presence of organometallic precursors have been studied by Oon et al. [114]. At low growth currents, single nanowires can be formed, which allows a systematic study of the growth characteristics, and wire morphology, structure, and composition. The major role of metal ion deposition in forming the metallic core is demonstrated experimentally, while the formation of the carbonaceous overcoat results from the deposition of neutral atoms from the precursor dissociation process. Transmission electron microscope analysis of tungsten nanowires shows that the core is polycrystalline, with columnar grains dominating the microstructure for thin wires, while larger diameter nanowires are straddled by multiple grains with a wider range of sizes. The axial and radial growth rates of tungsten nanowires as a function of growth current were studied and can be accounted for by assuming a situation in which the rate of ion formation just ahead of the growing tip is supply rate limited. At higher growth currents, forking and branching phenomena were found to be increasingly probable, and hence a key to the growth of single, well-defined nanowires is to keep the growth current low. Thermal decomposition of the precursor can also contribute to nanowire growth, and evidence for this mechanism was found in the cases of precursors where autocatalytic decomposition is known to result in metal deposition at relatively low temperatures.

Conducting poly (3,4-ethylenedioxythiophene) nanowires were synthesized by Kim et al. [115] by using an electrochemical polymerization method with a nanoporous template. Scanning and transmission electron microscopy confirmed the formation of conducting polymer nanowires (CPNWs) with an open end. The formation and the electrical properties of the CPNWs formed were dependent on synthetic conditions, such as the doping level, the polymerization time, and the applied current. The measured electrical conductivity of a single strand of CPNW was $\sim 3.4 \times 10^{23}$ S cm^{-1} at room temperature. From the ultraviolet and visible absorbance spectra, we observed a p–p* transition at ~2.1 eV for the de-doped systems. A field emission cell of CPNW nanotips was fabricated. The turn-on field of the CPNWs was $3.5 \sim 4 V \mu m^{-1}$ at $10 \mu A$ cm2 and the current density increased up to $100 \mu A\ cm^{-2}$ at $\sim 4.5 V \mu m^{-1}$. The field enhancement factor of CPNW nanotips was ~1,200, which is comparable with those of carbon nanotubes. Bunches of needle-shaped silicon carbide (SiC) nanowires were grown by Wu et al. [116] from commercially available SiC powders in thermal evaporation process and using iron as catalyst. Their structure and chemical composition were studied by Raman spectroscopy and high-resolution transmission electron microscopy. The powder of these nanowires may be easily dispersed and was used to form samples of field electron emitters. The needle shape of individual nanowires is well suited to field electron emission. Stable emission with current density of 30.8 mA cm^{-2} was observed at fields as low as 9.6 $V \mu m^{-1}$, and current density of up to 83 mA cm^{-2} was recorded.

Using a simple method of direct heating of bulk copper plates in air, oriented CuO nanowire films were synthesized by Zhu et al [117] on a large scale. The length and density of nanowires could be controlled by growth temperature and growth time. Field emission (FE) measurements of CuO nanowire films show that they have a low turn-on field of 3.5–4.5 V μm^{-1} and a large current density of 0.45 mA cm^{-2} under an applied field of about $7 V \mu m^{-1}$. By comparing the FE properties of two types of samples with different average lengths and densities ($30 \mu m$, $10^8 cm^{-2}$, and $4 \mu m$, $4 \times 10^7 cm^{-2}$, respectively), we found that the large length–radius ratio of CuO nanowires effectively improved the local field, which was beneficial to field emission. Verified with finite element calculation, the work function of oriented CuO nanowire films was estimated to be 2.5–2.8 eV.

Semet et al. [118] studied the field emission properties of LaS nanoprotrusions called nanodomes, formed by pulsed laser deposition on porous anodic alumina films, have been analyzed with scanning anode field emission microscopy. The voltage necessary to produce a given field emission current is ~3.5 times less for nanodomes for thin films. Assuming the same workfunction for LaS thin films and nanoprotusions, that is, ~1 eV a field enhancement factor of ~5.8 is extracted for the nanodome emitters from Fowler–Nordheim plots of the field emission data. This correlates well with the aspect ratio of the tallest nanodomes observed in atomic force micrograph measurements. Well-aligned arrays of ZnO nanoneedles were fabricated by Zhu et al. [119] using a simple vapor phase growth. The diameters of the nanoneedle tips are as small as several nanometers, which is highly in favor of the field emission. Field emission measurements using the nanoneedle arrays as cathode showed emission current density as high as 2.4 mA cm^{-2} under the field of 7 V mm^{-1}, and a very low turn-on field of 2.4 V mm^{-1}. Such a high emission current density is attributed to the high aspect ratio of the nanoneedles. The high emission current density, high stability, and low turn-on field make the ZnO nanoneedle arrays one of the promising candidates for field emission displays.

Bhattacharjee and Chowdhury [120] performed an experimental investigation of the transition from Fowler–Nordheim (FN) field emission to space-charge-limited (SCL)flows in a nanogap is presented. Electrodes with gap size D(30–70 nm) corresponding to D_o up to a maximum of (2×10^3), where D_o is the de Broglie wavelength of the space-charge-electrons, are experimented. The transition from the FN field emission to the classical SCL flow is a function of the applied bias and lies in the range 5–15 V. The equilibrium transmitted current density for the 50 nm sample indicates a transition from the FN to the quantum SCL flow at 0.4 V with D_o of 35 and then gradually to the classical SCL behavior as the voltage is increased beyond 9 V. The experiments indicate no sharp demarcation between the different regimes. Kher et al. [121] reports the experimental verification of the recently predicted phenomenon that the electric field emission current from a negatively charged surface gets enhanced by incidence of light (even of frequency below the photoelectric threshold) on the cathode.

Horváth et al. [122] present a method for the evaluation of the Schottky barrier height φ_{b0}, Richardson constant A^*, characteristic energy E_{00}, and bias dependence of the barrier height β from the temperature-dependent current–voltage characteristics of Schottky junctions using the thermionic field emission (TFE) theory. The application of this method to experimental current–voltage characteristics of epitaxial $Al/n - Al_{0.25}Ga_{0.75}As$ ($N = 1.4 \times 10^{17}\,cm^{-3}$) barriers shows that the current flow through these junctions is dominated by TFE with anomalously high E_{00}. We conclude that this anomaly may be partly connected with the electric field enhancement at the periphery of the diodes, and with the multistep tunneling through deep levels. Gallium-doped nanostructural zinc oxide fibers have been fabricated by Xu et al. [123] using vapor-phase transport method of heating the mixture of zinc oxide, gallium oxide, and graphite powders in air. The zinc oxide fibers grew along [002] direction, forming a vertically aligned array that is predominantly perpendicular to the substrate surface. With a gallium doping concentration of 0.73%, the corresponding carrier concentration and resistivity were $3.77 \times 10^{20}\,cm^{-3}$ and $8.9 \times 10^{-4}\,\Omega$ cm, respectively. The field of these vertically aligned ZnO fiber arrays showed a low field emission threshold ($2.4 V\mu m^{-1}$ at a current density of $0.1\,\mu A\,cm^{-2}$), high current density, and high field enhancement factor (2,317). The dependence of emission current density on the electric field followed Fowler–Nordheim relationship. The enhanced field emission is attributed to the aligned structure, good crystal quality, and especially, the improved electrical properties (increased conductivity and reduced work function) of the nanofibers due to gallium doping.

Ahmed et al. [124] have observed low-macroscopic field electron emission from wide bandgap nanocrystalline Al doped SnO_2 thin films deposited on glass substrates. The emission properties have been studied for different anode sample spacings and for different Al concentrations in the films. The turn-on field and approximate work function were calculated, and we have tried to explain the emission mechanism from this. The turn-on field was found to vary in the range

5.6–7.5 Vμm^{-1} for a variation of anode sample spacing from 80 to 120 μm. The turn-on field was also found to vary from 4.6 to 5.68 Vμm^{-1} for a fixed anode sample separation of 80 μm with a variation of Al concentration in the films 8.16–2.31%. The Al concentrations in the films have been measured by energy dispersive X-ray analysis. Optical transmittance measurement of the films showed a high transparency with a direct bandgap ~3.98 eV. Due to the wide band-gap, electron affinity of the film decreased. This, along with the nanocrystalline nature of the films, enhanced the field emission properties. Staryga et al. [125] studied field emission from diamond and diamond-like carbon thin films deposited on silicon substrates. The diamond films were synthesized using hot filament chemical vapor deposition technique. The diamond-like carbon films were deposited using the radio frequency chemical vapor deposition method. Field emission studies were carried out using a sphere-to-plane electrode configuration. The results of field emission were analyzed using the Fowler–Nordheim model. It was found that the diamond nucleation density affected the field emission properties. The films were characterized using standard scanning electron microscopy, Raman spectroscopy, and electron spin resonance techniques. Raman spectra of both diamond and diamond-like films exhibit spectral features characteristic of these structures. Raman spectrums for diamond films exhibit a well-defined peak at 1,333 cm^{-1}. Asymmetric broad peak formed in diamond-like carbon films consists of D-band and G-band around 1,550 cm^{-1} showing the existence of both diamond (sp^3 phase) and graphite (sp^2 phase) in diamond-like carbon films.

Amorphous diamond nanorod arrays with excellent field emitting have been fabricated first by Yan et al. [126] on the AAO template by the filtered cathodic arc plasma technique. Microscopic analysis has displayed that the nanorods are very uniformly distributed, and the density is very high up to ~10^{9}cm^{-2}.The nanorod arrays are found to have an extremely low turn-on field 0.16 Vμm^{-1}, which is lower than other reported materials, and a high-emission current density of 180 mA cm^{-2} under an applied field of 2 Vμm^{-1} can also be obtained. Scanning electron microscopy (SEM), Fourier transform infrared spectroscopy (FTIR), and filed emitting tester are employed to characterize the nanorod arrays. A new type of linear field emission cathode with ZnO nanostructure grown on nickel wires was prepared by hydrothermal approach by Lin et al. [127]. The obtained ZnO nanotapers were characterized by scanning electron microscopy (SEM) and X-ray diffraction (XRD). The results indicated that the ZnO nanotapers with sharp tips were high-quality single crystals, and grow along (002) direction. The field emission properties were investigated by ZnO nanotapers on nickel wire as the cathode in the centre of a cylindrical ITO anode. The field enhancement factor β was about 2.23×10^4 cm^{-1}, which improved greatly for the cylindrical configuration and sharp geometry of the ZnO nanotapers tip. Propeller-like ZnO nanostructures are fabricated by a physical vapor deposition method by Yan et al. [128]. This structure exhibits a good field emission characteristic with a turn-on field of 4.36 Vμm^{-1}. Cathodoluminescence studies suggest that the high current density in the surface is attributed to the main reason causing good field emission characteristic.

Silver-tetracyanoquinodimethane (Ag-TCNQ) nanostructured arrays with different morphologies were grown by Ye et al. [129] by an organic vapor-transport reaction under different conditions. The field emission properties of nanostructured arrays were studied systematically. Their morphology and crystal structure were characterized by SEM and XRD, respectively. It was found that the field emission properties were strongly dependent on the reaction temperature and the initial Ag film thickness. The lowest turn-on field with 10-nm-thick silver film is about $2.0\,\mathrm{V\mu m^{-1}}$, comparable with that of carbon nanotubes. The film crystal structure and the morphology are contributed to the final emission performance. Large-scale tetrapod-like ZnO nanostructures have been synthesized by Chen et al. [130] using a thermal chemical vapor deposition method on a silicon substrate. The high-purity nanotetrapods show sharp tips geometry with a wurtzite structure. The field emission properties of the uniform ZnO nanostructural material are investigated at different anode–cathode distances. The turn-on field for the ZnO nanotetrapods is found to be about $3.7\,\mathrm{V\mu m^{-1}}$ at a current density of $1\ \mathrm{\mu A\ cm^{-2}}$. The field emission behavior obeys Fowler–Nordheim relationship. More importantly, the field emission properties are improved after annealing in hydrogen, and therefore high emission current and low turn-on field are obtained. These results indicate that tetrapod-like ZnO nanostructures are a promising candidate for cold cathode emitters. ZnO nanorod arrays are prepared by Qian et al. [131] on a silicon wafer through a multistep hydrothermal process. The aspect ratios and densities of the ZnO nanorod arrays are controlled by adjusting the reaction times and concentrations of solution. The investigation of field emission properties of ZnO nanorod arrays revealed a strong dependency on the aspect ratio and their density. The aspect ratio and spacing of ZnO nanorod arrays are 39 and 167 nm (sample C), respectively, to exhibit the best field emission properties. The turn-on field and threshold field of the nanorod arrays are 3.83 and $5.65\,\mathrm{V\mu m^{-1}}$, respectively. Importantly, the sample C shows a highest enhancement of factor β, which is 2,612. The result shows that an optimum density and aspect ratio of ZnO nanorod arrays have high efficiency of field emission.

Wurtzite stalactite-like quasi-one-dimensional ZnS nanoarrays with ZnO protuberances were synthesized by Li et al. [132] through a thermal evaporation route. The structure and morphology of the samples are studied and the growth mechanism is discussed. X-ray diffraction (XRD) results show both the ZnS stem and the ZnO protuberances have wurtzite structure and show preferred [001]-oriented growth. The photoluminescence and field emission properties have also been investigated. Room temperature photoluminescence result shows that it has a strong green light emission, which has potential application for green light emitter. Experimental results also show that the stalactite arrays have a good field emission property, with turn-on field of $11.4\,\mathrm{V\mu m^{-1}}$, and threshold field of $16\,\mathrm{V\mu m^{-1}}$. The ZnO protuberances on the ZnS stem might enhance the field emission notably. Lee et al. [133] describes an experimental study on field emission characteristics of individual graphene layers for vacuum nanoelectronics. Graphene layers were prepared by mechanical exfoliation from a highly oriented pyrolyzed graphite block and placed on an insulating substrate, with the resulting field emission behavior investigated using a nanomanipulator operating inside a scanning electron microscope. A pair of tungsten tips controlled by the nanomanipulator enabled electric connection with the graphene layers without postfabrication. The maximum emitted current from the graphene layers was 170 nA and the turn-on voltage was 12.1 V. Shigeo et al. [134] report experiments conducted to review several factors closely related to emission quality required for flat panel displays. Using the measurement of the emission of fabricated Spindt-type emitters, the dependence of current density on the distance from the tip of emitter cone to the upper surface of the gate was investigated. It was also confirmed that the shape of the emitter cone was largely affected by the gate hole diameter and material of one. The maximum half angle of emission on anode from a tip was compared between the simulated electron beam spread and actual measurements made for emitter arrays.

Ruskell et al. [135] report an improved method for characterizing thin oxide films using Fowler–Nordheim field emission. The method uses a conducting tip atomic force microscope with dual feedback systems, one for the topography and a second for the field emission bias voltage. Images of the voltage required to maintain a 10 pA emission current through a 3 nm oxide film thermally grown on p-type Si (100) demonstrate a spatial resolution of 8 nm. Well-ordered titanium nitride nanorods were fabricated by Chen et al. [136] reactive ion etch using titanium oxide nanodots as the mask, which were prepared using the anodic aluminum oxide templation method. The TiN nanorods exhibited a concave top surface with a protruding edge. Due to the protruding top edge and a high aspect ratio, the TiN nanorods showed a low turn-on voltage of $1.6\,\mathrm{V\mu m^{-1}}$. The ellipsoidal cylinder model was used to evaluate the field-enhancement effect of the protruding edge, and an underestimation by $\sim$26% was found as compared with the enhancement factor derived from the Fowler–Nordheim plot. Nonlinearity has been observed by Xu et al. [137] in Fowler–Nordheim (FN) plots of field emission from nondoped and nitrogen-doped amorphous diamond films. Based on a unified electron emission equation a detailed analysis is carried out. The results from numerical calculation of the unified equation are consistent with the experimental data. It is shown that the nonlinearity in the FN plot originates from a transition from thermionic emission to field emission as the applied field increases. The electrical field ranges are derived in which the field emission and thermionic emission approximation applies. Temperature dependence of the field emission characteristics has been investigated by Sugino et al. [138] for the phosphorus(P)-doped polycrystalline diamond film in comparison with that of the boron(B)-doped one. The threshold voltage decreases with increasing temperature for the P-doped diamond film, while no variation in the threshold voltage occurs for the B-doped diamond film. It is considered that an increase in the ionized donor concentration with increasing temperature leads to a reduction in the tunnel barrier width at the interface between the diamond and the cathode, resulting in an enhancement of the emission current. Field emission characteristics in the higher voltage region are featured by the space charge limited current. The activation energy estimated from the Arrhenius plot of the emission current suggests the upward band bending at the diamond surface.

Theoretical analyses have been performed by Litovchenko et al [139] of the quantum-size (QS) resonance tunneling in the field emission (FE) phenomenon for different models of the emitting structures. Such experimentally observed peculiarities have been considered as the enhancement of the FE current, the deviation from the Fowler–Nordheim law, the appearance of sharp current peaks, and a negative resistance. Different types of FE cathodes with QS structures (quantized layers, wires, or dots) have been studied experimentally. Resonance current peaks have been observed, from which the values of the energy-level splitting can be estimated. Theoretical and experimental investigations of electron field emission from silicon-based resonance-tunneling layered structures have been performed by Litovchenko et al. [140]. Numerical simulation of resonant and nonresonant field emission in Si–SiO_2–Si*–SiO_2 multilayer cathodes (MLCs) with quantum well (QWs) which takes into account the tunneling process of electrons from the three-dimensional electron density state of the emitter conductive band has been carried out. The influence of the external electric field, temperature, MLC parameters, and emitter doping on the resonant characteristics of the current was analyzed. Computer simulation has shown that the peak current density of MLCs with optimal thin barriers and sufficiently wide QW layers at a resonant value of the electric field can sometimes exceed the current density of conventional cathodes. If the width of the QW is increased, the number of current resonant maxima (CRM) is multiplied. The CRM is shifted towards the lower electric field values and become more narrow if both the QW and the potential barrier widths are increased. With temperature reduction the CRM becomes contrasted due to an increase in the electron impulse relaxation time and redistribution of the electron state density in the emitter conduction band. Experimental multilayer structures with $Si^*\delta$-doped layer Si–SiO_2–Si*–SiO_2 have been formed on silicon using low-pressure chemical vapor deposition of ultrathin SiO_2 and Si* films. In some cases, the first ultrathin SiO_2 layer was grown on silicon with thermal oxidation. The multilayer structures were formed both on flat silicon wafers and on silicon tip arrays. Measurements of electron field emission into vacuum were performed in a diode (cathode–anode) system. The resonant peaks of current density from MLCs have been observed experimentally for the first time. The value of these peaks is more than two times of that of the background curves. A comparison of experimental and theoretical results has been performed to evaluate the fundamental parameters of the field emission resonance process.

Johnson et al. [141] have performed theoretical and experimental studies of field emission from nanostructured semiconductor cathodes. Resonant tunneling through electric-field-induced interface bound states is found to strongly affect the field emission characteristics. Our analytical theory predicts power law and Lorentzian-shaped current–voltage curves for resonant-tunneling field emission from three-dimensional substrates and two-dimensional accumulation layers, respectively. These predicted line shapes are observed in field emission characteristics from self-assembled silicon nanostructures. A simple model describes formation of an accumulation layer and of the resonant level in these systems. Important characteristics of quantum well infrared photodetectors are determined by Vinter et al. [142] almost entirely by the photoionization rate of electrons out of the quantum well (QW) and the recapture into the QWs. To elucidate these processes microscopically, we have made structures in which the QWs are isolated from one contact by a completely blocking barrier, so that the steady state current vanishes. The transient current induced by photoionization out of the QWs gives a direct measurement of the photoionization cross section and the escape probability of a photoexcited electron. We have found that the variation of the latter with the electric field may be described by a simple barrier lowering model combined with statistical fluctuation of the QW width. The capture process has been studied by impedance spectroscopy in samples containing only one well. The capture velocity thus measured is found to decrease with increasing applied electric field but within experimental uncertainties it does not depend on the width of the well for well widths between 3 and 7.5 nm. Theoretical results on optical phonon mediated transitions in the applied field from barrier to well states show a generally good agreement with experiment at low fields but less dependence on the field.

Field emission cathodes fabricated by Spindt et al. [143] using thinfilm techniques and electron beam microlithography are described, together with effects obtained by varying the fabrication parameters. The emission originates from the tip of molybdenum cones that are about 1.5 μm tall with a tip radius around 500 Å. Such cathodes have been produced in closely packed arrays containing 100 and 5,000 cones as well as in single. Maximum currents in the range 50–150 μA per cone can be drawn with applied voltages in the range 100–300 V when operated in conventional ionpumped vacuum at pressures of 10^{-9} torr or less. In the arrays, current densities (averaged over the array) of above $10\,\text{A cm}^{-2}$ have been demonstrated. Life tests with the 100cone arrays drawing 2 mA total emission (or $3\,\text{A cm}^{-2}$) have proceeded in excess of 7,000 h with about a 10% drop in emission current. Studies are presented of the emission characteristics and current fluctuation phenomena. It is tentatively concluded that the emission arises from only one or a few atomic sites on the cone tips. Han et al. [144] investigated field emission characteristics of nitrogen-doped diamond films, which were grown using microwave plasma-enhanced chemical vapor deposition. Nitrogen-doped films showed low turn-on voltages below $2\,\text{V}\mu\text{m}^{-1}$. Secondary ion mass spectroscopy was used to compare nitrogen concentrations in the films. Morphologies, Raman spectra, resistivities, and surface roughness of the films were changed as the nitrogen concentrations varied. The field emission properties of heavily nitrogen-doped diamond films were related to the film resistivity, surface morphologies, and Raman characteristics. AlGaN/GaN quantum well (QW) structures are grown by Grandjean et al. [145] on c-plane sapphire substrates by molecular beam epitaxy. Control at the monolayer scale of the well thickness is achieved, and sharp QW interfaces are demonstrated by the low photoluminescence line width. The QW transition energy as a function of the well width evidences a quantum-confined Stark effect due to the presence of a strong built-in electric field. Its origin is discussed in terms of piezoelectricity and spontaneous polarization. Its magnitude versus the Al mole fraction is determined. The role of the sample structure geometry on the electric field is exemplified by changing the thickness of the AlGaN barriers in multiple QW structures. Straightforward electrostatic arguments well account for the overall trends of the electric field variations.

Fu et al. [146] observed that the application of moderate electric fields to ZnSe/(Zn,Mn)Se quantum wells yields distinct spectral shifts of the recombining exciton luminescence. This shows that confinement effects in this heterostructure are sufficient to increase the exciton ionization threshold. At high applied fields and low temperatures, injection of hot electrons from the n^+GaAs/ZnSe heterojunction at our substrate/buffer layer excites yellow luminescence from internal transitions of the Mn ion in (Zn,Mn)Se layers. Field-emission scanning electron microscopy (FE-SEM) has been used by Perovic et al. [147] to study several semiconductor multilayer heterostructures. Compositional superlattices based on Ge_xSi_{1-x}/Si and $Al_xGa_{1-x}As$/GaAs have been studied in both cross-sectional and oblique plan views after indentation. Secondary and back scattered electron images reveal strong atomic number contrast which is primarily structural in origin. Secondly, for the first time, heterostructures containing n- and p-doping have been directly imaged at low voltages (0.5–1 kV) including: (1) Si- and Be-doped GaAs layers and (2) B- and As- doped Si layers. Secondary electron images reveal strong contrast at doping concentrations as low as $10^{17}\,cm^{-3}$. The results have been interpreted in terms of energy band-bending effects between n- and p-doped layers.

In this monograph, we have studied the FNFE from quantum-confined nonlinear optical, III–V, II–VI, GaP, Ge, $PtSb_2$, stressed materials, bismuth, GaSb, IV–VI, tellurium, II–V, Bi_2Te_3, III–V, II–VI, IV–VI, and HgTe/CdTe quantum wire superlattices with graded interfaces, III–V, II–VI, IV–VI, and HgTe/CdTe effective mass superlattices under magnetic quantization, quantum-confined effective mass superlattices and superlattices of optoelectronic materials under intense electric field and light waves with graded interfaces on the basis of appropriate carrier energy spectra. Finally, it may be noted that although we have considered the FNFE from a plethora of quantized materials having different band structures theoretically, the detailed experimental works are still needed for an in-depth study of the FNFE from such low-dimensional systems as functions of externally controllable quantities which, in turn, will add new physical phenomenon in the regime of the electron emission from nanostructured materials and related topics.

7.4 Open Research Problem

(R7.1) Investigate experimentally the FNFE for all the systems as discussed in this monograph.

References

1. R.B. Dingle, Phil. Mag. **46**, 813 (1955)
2. D. Redfield, M.A. Afromowitz, Phil. Mag. **18**, 831 (1969)
3. H.C. Cassey, F. Stern, J. Appl. Phys. **47**, 631 (1976)

4. M. Mondal, K.P. Ghatak, Phys. Lett. **102A**, 54 (1984)
5. P.K. Chakraborty, G.C. Datta, K.P. Ghatak, Phys. Script. **68**, 368 (2003)
6. B. Mitra, D.K. Basu, B. Nag, K.P. Ghatak, Nonlinear Opt. **17**, 171 (1997)
7. K.P. Ghatak, S. Bhattacharya, J. Appl. Phys. **102**, 073704 (2007)
8. K.P. Ghatak, S. Bhattacharya, H. Saikia, D. Baruah, A. Saikia, K.M. Singh, A. Ali, S.N. Mitra, P.K. Bose, A. Sinha, J. Comput. Theor. Nanosci. **3**, 727 (2006)
9. E.O. Kane, Solid State Electron. **8**, 3 (1985)
10. T. Ando, A.H. Fowler, F. Stern, Rev. Mod. Phys. **54**, 437 (1982)
11. P.K. Basu, *Optical Processes in Semiconductors*(Oxford University Press, Oxford, 2001)
12. A.N. Chakravarti, D. Mukherjee, Phys. Lett. **53A**, 403 (1975)
13. A.N. Chakravarti, S. Swaminathan, Phys. Stat. Sol. (a) **23**, K191 (1974)
14. A.N. Chakravarti, Phys. Stat. Sol (a) **25**, K105 (1974)
15. K.P. Ghatak, S. Bhattyacharya, *Thermoelectric Power in Nanostructured Materials: Strong Magnetic Fields*, *Springer Series in Materials Science*, vol 137 (Springer, Berlin, 2010)
16. A.K. Sreedhar, S.C. Gupta, Phys. Rev. B **5**, 3160 (1972)
17. R.W. Keyes, IBM. J. Res. Dev. **5**, 266 (1961)
18. R.W. Keyes, Solid State Phys. **20**, 37 (1967)
19. S. Bhattacharya, S. Chowdhury, K.P. Ghatak, J. Comput. Theor. Nanosci. **3**, 423 (2006)
20. S. Choudhury, L.J. Singh, K.P. Ghatak, Phys. B **365**, 5 (2005)
21. L.J. Singh, S. Choudhary, A. Mallik, K.P. Ghatak, J. Comput. Theor. Nanosci. **2**, 287 (2005)
22. K.P. Ghatak, J.Y. Siddiqui, B. Nag, Phys. Lett. A **282**, 428 (2001)
23. K.P. Ghatak, J.P. Banerjee, B. Nag, J. Appl. Phys. **83**, 1420 (1998)
24. B. Nag, K.P. Ghatak, Nonlinear Opt. **19**, 1 (1998)
25. K.P. Ghatak, B. Nag, Nanostruct. Mater. **10**, 923 (1998)
26. B. Nag, K.P. Ghatak, J. Phys. Chem. Sol. **58**, 427 (1997)
27. K.P. Ghatak, D.K. Basu, B. Nag, J. Phys. Chem. Solids **58**, 133 (1997)
28. K.P. Ghatak, J.P. Banerjee, B. Goswami, B. Nag, Nonlinear Opt. Quant. Opt. **16**, 241 (1996)
29. K.P. Ghatak, J.P. Banerjee, D. Bhattacharyya, B. Nag, Nanotechnology **7**, 110 (1996)
30. K.P. Ghatak, J.P. Banerjee, M. Mitra, B. Nag, Nonlinear Opt. 17, 193 (1996)
31. B. Nag, K.P. Ghatak, Phys. Scr. **54**, 657 (1996)
32. K.P. Ghatak, B. Mitra, Phys. Scr. **46**, 182 (1992)
33. K.P. Ghatak, Int. J. Electron. **71**, 239 (1991)
34. K.P. Ghatak, B. De, S.N. Biswas, M. Mondal, Mechanical behavior of materials and structures in microelectronics, MRS Symposium Proceedings, Spring Meeting, **216**, 191 (1991)
35. K.P. Ghatak, B. De, MRS Symp. Proc. **226**, 191 (1991)
36. K.P. Ghatak, B. Nag, G. Majumdar, MRS Symp. Proc. **379**, 109 (1995)
37. D. Baruah, S. Choudhury, K.M. Singh, K.P. Ghatak, J. Phys. Conf. Series **61**, 80 (2007)
38. S. J. Adachi, J. Appl. Phys. **58**, R11 (1985)
39. R. Dornhaus, G. Nimtz, *Springer Tracts in Modern Physics*, vol. 78 (Springer, Berlin, 1976)
40. W. Zawadzki, *Handbook of Semiconductor Physics*, ed. by W. Paul, vol. 1 (CRC, Amsterdam, 1982), p. 719
41. I.M. Tsidilkovski, Cand. Thesis Leningrad University SSR (1955)
42. F.G. Bass, I.M. Tsidilkovski, Ivz. Acad. Nauk Azerb SSR **10**, 3 (1966)
43. I.M. Tsidilkovski, *Band Structures of Semiconductors* (Pergamon, London 1982)
44. B Mitra, K.P. Ghatak, Phys. Script. **40**, 776 (1989)
45. S.K. Biswas, A.R. Ghatak, A. Neogi, A. Sharma, S. Bhattacharya, K.P. Ghatak, Phys. E **36**, 163 (2007)
46. K.P. Ghatak, S.N. Biswas, Nonlinear Opt. Quant. Opts. **4**, 347 (1993)
47. A.N. Chakravarti, A.K. Choudhury, K.P. Ghatak, S. Ghosh, A. Dhar, Appl. Phys. **25**, 105 (1981)
48. K.P. Ghatak, M. Mondal, Z. F. Physik B **B69**, 471 (1988)
49. M. Mondal, K.P. Ghatak, Phys. Lett. **131A**, 529 (1988)
50. K.P. Ghatak, A. Ghoshal, B. Mitra, Nouvo Cimento **14D**, 903 (1992)
51. B. Mitra, A. Ghoshal, K.P. Ghatak, Nouvo Cimento D **12D**, 891 (1990)

52. K.P. Ghatak, S.N. Biswas, Nonlinear Opt. Quant. Opts. **12**, 83 (1995)
53. B. Mitra, K.P. Ghatak, Solid State Electron. **32**, 177 (1989)
54. K.P. Ghatak, S.N. Biswas, Proc. SPIE **1484**, 149 (1991)
55. M. Mondal, K.P. Ghatak, *Graphite Intercalation Compounds: Science and Applications, MRS Proceedings*, ed. by M. Endo, M.S. Dresselhaus, G. Dresselhaus, MRS Fall Meeting, **EA**16, 173 (1988).
56. M. Mondal, N. Chattapadhyay, K.P. Ghatak, J. Low Temp. Phys. **66**, 131 (1987)
57. A.N. Chakravarti, K.P. Ghatak, K.K. Ghosh, S. Ghosh, A. Dhar, Z. Physik B. **47**, 149 (1982)
58. V.K. Arora, H. Jeafarian, Phys. Rev. B. **13**, 4457 (1976)
59. M. Singh, P.R. Wallace, S.D. Jog, J.J. Erushanov, J. Phys. Chem. Solids **45**, 409 (1984)
60. W. Zawadski, Adv. Phys. **23**, 435 (1974)
61. K.P. Ghatak, M. Mondal, Z. Nature A **41A**, 881 (1986)
62. T. Ando, A.H. Fowler, F. Stern, Rev. Modern Phys. **54**, 437 (1982)
63. H. Kroemer, IEEE Trans. Electron. Devices **25**, 850 (1978)
64. R.W. Lade, Proc. IEEE **51**, 743 (1964)
65. S.N. Mohammed, J. Phys. C Solid State Phys. **13**, 2685 (1980)
66. P.T. Landsberg, Eur. J. Phys. **2**, 213 (1981)
67. P.T. Landsberg, Proc. R. Soc. A **213**, 226 (1952).
68. S.A. Hope, G. Feat, P.T. Landsberg, J. Phys. A Math. Gen. **14**, 2377 (1981)
69. C.H. Wang, A. Neugroschel, IEEE Electron. Dev. Lett. **ED-11**, 576 (1990)
70. I.-Y. Leu, A. Neugroschel, C.H. Wang, A. Neugroschel, IEEE Trans. Electron. Dev. **ED-40**, 1872 (1993)
71. F. Stengel, S.N. Mohammad, H. Morkoc, J. Appl. Phys. **80**, 3031 (1996)
72. H.J. Pan, W.C. Wang, K.B. Thai, C.C. Cheng, K.H. Yu, K.W. Lin, C.Z. Wu, W.C. Liu, Semicond. Sci. Technol. **15**, 1101 (2000)
73. S.N. Mohammad, J. Appl. Phys. **95**, 4856 (2004)
74. V.K. Arora, Appl. Phys. Lett. **80**, 3763 (2002)
75. S.N. Mohammad, J. Appl. Phys. **95**, 7940 (2004)
76. S.N. Mohammad, Phil. Mag. **84**, 2559 (2004)
77. S.N. Mohammad, J. Appl. Phys. **97**, 063703 (2005)
78. S.G. Dmitriev, Yu.V. Markin, Semiconductors **34**, 931 (2000)
79. M. Tao, D. Park, S.N. Mohammad, D. Li, A.E. Botchkerav, H. Morkoc, Phil. Mag. B **73**, 723 (1996)
80. D.G. Park, M. Tao, D. Li, A.E. Botchkarev, Z. Fan, S.N. Mohammad, H. Morkoc, J. Vac. Sci. Technol. B **14**, 2674 (1996)
81. Z. Chen, D.G. Park, S.N. Mohammad, H. Morkoc, Appl. Phys. Lett. **69**, 230 (1996)
82. K.P. Ghatak, S. Bhattacharya, D. De, *Einstein Relation in Compound Semiconductors and Their Nanostructures, Springer Series in Materials Science*, vol. 116 (Springer-Verlag, Berlin, 2009)
83. K.P. Ghatak, S.N. Biswas, Nanostruct. Mater. **2**, 91 (1993)
84. K.P. Ghatak, S.N. Biswas, J. Appl. Phys. **70**, 4309 (1991)
85. K.P. Ghatak, B. Mitra, M. Mondal, Ann. Physik. **48**, 283 (1991)
86. B. Mitra, K.P. Ghatak, Phys. Scrip. **42**, 103 (1990)
87. B. Mitra, K.P. Ghatak, Phys. Letts. **135A**, 397 (1989)
88. M. Mondal, K.P. Ghatak, Ann. Physik. **46**, 502 (1989)
89. K.P. Ghatak, D. Bhattacharyya, Phys. Lett. A **184**, 366 (1994)
90. A.N. Charkravarti, D.P. Parui Phys. Lett. **40A**, 113 (1972)
91. A.N. Charkravarti, D.P. Parui Phys. Lett. **43A**, 60 (1973)
92. A.N. Charkravarti, D.P. Parui Phys. Stat. Sol (a) **14**, K23 (1972)
93. A.N. Charkravarti, D.P. Parui Canad. J. Phys. **51**, 451 (1973)
94. S.N. Banik, K.P. Ghatak Canad. J. Phys. **67**, 72 (1989)
95. P.K. Chakraborty, S. Bhattacharya, K.P. Ghatak, J. Appl. Phys. **98**, 053517 (2005)
96. A.S. Filipchenko, I.G. Lang, D.N. Nasledov, S.T. Pavlov, L.N. Radaikine, Phys. Stat. Sol. (b) **66**, 417 (1974)

97. M. Wegener, *Extreme Nonlinear Optics*(Springer-Verlag, Berlin, 2005)
98. B.S. Wherreff, W. Wolland, C.R. Pidgeon, R.B. Dennis, S.D. Smith, *Proceedings of the 12th International Conference of the Physics of the Semiconductors*, ed. by M.H. Pilkahn, R.G. Tenbner (Staffgard, 1978), p. 793
99. S. Zaric, G.N. Ostojic, J. Kono, J. Shaver, V.C. Moore, M.S. Strano, R.H. Hauge, R.E. Smalley, X. Wei, Science **304**, 1129 (2004)
100. E.C. Heeres, P.A.M. Bakkers, A.L. Roest, M. Kaiser, T.H. Oosterkamp, N. Jonge, Nano Lett. **7**, 536 (2007)
101. C.J. Lee, T.J. Lee, S.C. Lyu, Y. Zhang, H. Ruh, H.J. Lee, Appl. Phys. Lett. **81**, 3648 (2002)
102. L. Dong, J. Jiao, D.W. Tuggle, J.M. Petty, S.A. Elliff, M. Coulter, Appl. Phys. Lett. **82**, 1096 (2003)
103. S.Y. Li, P. Lin, C.Y. Lee, T.Y. Tseng, J. Appl. Phys. **95**, 3711 (2004)
104. A. Khademi, R. Azimirad, Y.T. Nien, A.Z. Moshfegh, J. Nanoparticle Res., doi:10.1007/s11051-010-0009-0.
105. J.B. Wang, K. Li, X.L. Zhong, Y.C. Zhou, X.S. Fang, C.C. Tang, Y. Bando, Nanoscale Res. Lett. **4**, 1135, doi:10.1007/s11671-009-9367-x
106. W.H. Lai, M.H. Hon, L.G. Teoh, Y.H. Su, J. Shieh, C.K. Chen, J. Electron. Mat. **37**, 1082, doi:10.1007/s11664-008-0474-8
107. X.G. Zhang, S.T. Pantelides, Nano Lett. **9**, 4306 (2009)
108. A. Lorenzoni, H.E. Roman, F. Alasia, R.A. Broglia, Chem. Phys. Lett. **276**, 237 (1997)
109. K.W. Wong, X.T. Zhou, F.C.K. Au, H.L. Lai, C.S. Lee, S.T. Lee, Appl. Phys. Lett. **75**, 2918 (1999)
110. N.N. Kulkarni, J. Bae, C.K. Shih, S.K. Stanley, S.S. Coffee, J.G. Ekerdt, Appl. Phys. Lett. **87**, 213115 (2005)
111. K. Senthil K. Yong, Mat Chem Phys **112**, 88 (2008)
112. R. Zhou, H.C. Chang, V. Protasenko, M. Kuno, A.K. Singh, D. Jena, H. Xing, J. Appl. Phys. **101**, 073704 (2007)
113. K.S. Yeong, J.T.L. Thong, J. Appl. Phys. **100**, 114325 (2006)
114. C.H. Oon, S.H. Khong, C.B. Boothroyd, J.T.L. Thong, J. Appl. Phys. **99**, 064309 (2006)
115. B.H. Kim, M.S. Kim, K.T. Park, J.K. Lee, D.H. Park, J. Joo, S.G. Yu, S.H. Lee, Appl. Phys. Lett. **83**, 539 (2003)
116. Z.S. Wu, S.Z. Deng, N.S. Xu, J. Chen, J. Zhou, J. Chen, Appl. Phys. Lett. **80**, 3829 (2002)
117. Y.W. Zhu, T. Yu, F.C. Cheong, X.J. Xu, C.T. Lim, V.B.C. Tan, J.T.L. Thong, C.H. Sow, Nanotechnology **16**, 88 (2005)
118. V. Semet, Vu.T. Binh, M. Cahay, K. Garre, S. Fairchild, L. Grazulis, J.W. Fraser, D.J. Lockwood, S. Pramanik, B. Kanchibotla, S. Bandyopadhyay, J. Nano Mater. **2008**, Article ID 682920, p. 4 (2008), doi:10.1155/2008/682920
119. Y.W. Zhu, H.Z. Zhang, X.C. Sun, S.Q. Feng, J. Xu, Q. Zhao, B. Xiang, R.M. Wang, D.P. Yu, Appl. Phys. Lett. **83**, 144 (2003)
120. S. Bhattacharjee, T. Chowdhury, Appl. Phys. Lett. **95**, 061501 (2009)
121. S. Kher, A. Dixit, D.N. Rawat, M.S. Sodha, Appl. Phys. Lett. **96**, 044101 (2010)
122. Z.J. Horváth, A. Bosacchi, S. Franchi, E. Gombia, R. Mosca, A. Motta, Mat. Sci. Eng. B **28**, 429 (1994)
123. C.X. Xu, X.W. Sun, B.J. Chen, Appl. Phys. Lett. **84**, 1540 (2004)
124. Sk.F. Ahmed, P.K. Ghosh, S. Khan, M.K. Mitra, K.K. Chattopadhyay, Appl. Phys. A **86**, 139, doi:10.1007/s00339-006-3734-6
125. E. Staryga, D. Jarzynska, K. Fabisiak, A. Banaszak, J. Superhard Mat. **29**, 189, doi:10.3103/S1063457607030161
126. P. Yan, X. Li, J. Xu, X. Li, C. Li, Y. Liu, Sci. China E Tech. Sci. **49**, 156, doi:10.1007/s11431-006-0156-9
127. Z. Lin, Y. Ye, Y. Zhang, T. Guo, J. Mat. Sci. Mat. Electron., doi:0.1007/s10854-010-0063-5
128. H.L. Yan, J.B. Wang, X.L. Zhong, J. Mat. Sci.: Mat. Electron., doi:10.1007/s10854-010-0200-1

129. C. Ye, K. Zheng, W. You, G. Chen, Nanoscale Res. Lett. **5**, 1307, doi:10.1007/s11671-010-9643-9
130. H.S. Chen, J.J. Qi, Y. Zhang, Q.L. Liao, X.M. Zhang, Y.H. Huang, Chinese Sci. Bull. **52**, 1287, doi:10.1007/s11434-007-0190-4
131. X. Qian, H. Liu, Y. Guo, Y. Song, Y. Li, Nanoscale Res. Lett. **3**, 303, doi:10.1007/s11671-008-9154-0
132. J. Li, G.J. Fang, C. Li, L.Y. Yuan, L. Ai, N.S. Liu, D.S. Zhao, K. Ding, G.H. Li, X.Z. Zhao, Appl. Phys. A Mat. Sci. Proc. **90**, 759, doi:10.1007/s00339-007-4352-7
133. S.W. Lee, S.S. Lee, E.H. Yang, Nanoscale Res. Lett. **4**, 1218, doi:10.1007/s11671-009-9384-9
134. I. Shigeo, W. Teruo, O. Kazuyoshi, T. Masateru, U. Satoshi, N. Norio, J. Vac. Sci. Technol. B **13**, 487 (2009)
135. T.G. Ruskell, R.K. Workman, D. Chen, D. Sarid, S. Dahl, S. Gilbert, Appl. Phys. Lett. **68**, 93 (1996)
136. T.M. Chen, J.Y. Hung, F.M. Pan, L. Chang, J.T. Sheu, S.C. Wu, Electrochem. Solid State Lett. **11**, K40 (2008)
137. N.S. Xu, J. Chen, S.Z. Deng, Appl. Phys. Lett. **76**, 2463 (2000)
138. T. Sugino, K. Kuriyama, C. Kimura, S. Kawasaki, Appl. Phys. Lett. **73**, 268 (1998)
139. V. Litovchenko, A. Evtukh, Yu. Kryuchenko, N. Goncharuk, O. Yilmazoglu, K. Mutamba, H.L. Hartnagel, D. Pavlidis, J. Appl. Phys. **96**, 867 (2004)
140. V.G. Litovchenko, A.A. Evtukh, Yu.M. Litvin, N.M. Goncharuk, V.E. Chayka, J. Vac. Sci. Technol. B **17**, 655 (1999)
141. S. Johnson, U. Zulicke, A. Markwitz, J. Appl. Phys. **101**, 123712 (2007)
142. B. Vinter, F. Luc, P. Bois, L. Thibaudeau, E. Rosencher, Solid State Electron. **37**, 773 (1994)
143. C.A. Spindt, I. Brodie, L. Humphrey, E.R. Westerberg, J. Appl. Phys. **47**, 5248 (2009)
144. I.T. Han, N. Lee, S.W. Lee, S.H. Kim, D. Jeon, J. Vac. Sci. Technol. B **16**(4), 2052 (1998)
145. N. Grandjean, B. Damilano, S. Dalmasso, M. Leroux, M. Laugt, J. Massies, J. Appl. Phys. **86**, 3714 (1999)
146. Q. Fu, A.V. Nurmikko, L.A. Kolodziejski, R.L. Gunshor, J.W. Wu, Appl. Phys. Lett. **51**, 578 (2009)
147. D.D. Perovic, M.R. Castell, A. Howie, C. Lavoie, T. Tiedje, J.S.W. Cole, Ultramicroscopy **58**, 104 (1995)

Chapter 8
Conclusion and Future Research

This monograph deals with the FNFE from various types of quantum wires, effective mass superlattices, and superlattices with graded interfaces under different physical conditions, in the presence of quantizing magnetic field and external photoexcitation and also under strong electric field altering profoundly the basic band structures which, in turn, generate pin-pointed knowledge regarding FNFE from various semiconductors and their nanostructures having different carrier energy spectra. The in-depth experimental investigations covering the whole spectrum of solid state and allied science in general are extremely important to uncover the underlying physics and the related mathematics. The FNFE is basically electric field-dominated electron emission phenomena, and we have formulated the simplified expressions of FNFE for few quantized structures together with the fact that our investigations are based on the simplified $\boldsymbol{k}\cdot\boldsymbol{p}$ formalism of solid state science without incorporating the advanced field theoretic techniques. In spite of such constraints, the role of band structure behind the curtain, which generates, in turn, new concepts, is discussed throughout the text.

Finally, we present the last set of open research problems in this particular area of electron emission from solids.

(R8.1) Investigate the FNFE in the presence of a quantizing magnetic field under exponential, Kane, Halperin, Lax, and Bonch–Bruevich band tails [1] for all the problems of this monograph of all the materials whose unperturbed carrier energy spectra are defined in Chap. 1 by including spin and broadening effects.

(R8.2) Investigate all the appropriate problems after proper modifications introducing new theoretical formalisms for the problems as defined in (R8.1) for negative refractive index, macromolecular, nitride, and organic materials.

(R8.3) Investigate all the appropriate problems of this monograph for all types of quantum-confined p-InSb, p-CuCl, and semiconductors having diamond structure valence bands whose dispersion relations of the carriers in bulk materials are given by Cunningham [2], Yekimov et al. [3], and Roman et al. [4], respectively.

S. Bhattacharya and K.P. Ghatak, *Fowler–Nordheim Field Emission*, Springer Series in Solid-State Sciences 170, DOI 10.1007/978-3-642-20493-7_8,

(R8.4) Investigate the influence of defect traps and surface states separately on the FNFE for all the appropriate problems of all the chapters after proper modifications.
(R8.5) Investigate the FNFE under the condition of nonequilibrium of the carrier states for all the appropriate problems of this monograph.
(R8.6) Investigate the FNFE for all the appropriate problems of this monograph for the corresponding p-type semiconductors and their nanostructures.
(R8.7) Investigate the FNFE for all the appropriate problems of this monograph for all types of semiconductors and their nanostructures under mixed conduction in the presence of strain.
(R8.8) Investigate the FNFE for all the appropriate problems of this monograph for all types of semiconductors and their nanostructures in the presence of hot electron effects.
(R8.9) Investigate the FNFE for all the appropriate problems of this monograph for all types of semiconductors and their nanostructures for nonlinear charge transport.
(R8.10) Investigate the FNFE for all the appropriate problems of this monograph for all types of semiconductors and their nanostructures in the presence of strain in an arbitrary direction.
(R8.11) Investigate all the appropriate problems of this monograph for semiconductor clathrates in the presence of strain.
(R8.12) Investigate all the appropriate problems of this monograph for quasicrystalline materials in the presence of strain.
(R8.13) Investigate all the appropriate problems of this monograph for strongly correlated electron systems in the presence of strain.
(R8.14) Investigate the FNFE for all the appropriate problems of this monograph for all types of transition metal silicides in the presence of strain.
(R8.15) Investigate the FNFE for all the appropriate problems of this monograph for all types of electrically conducting organic materials in the presence of strain.
(R8.16) Investigate the FNFE for all the appropriate problems of this monograph for all types of functionally graded materials in the presence of strain.
(R8.17) Investigate the FNFE from all types of available super conductors in the presence of strain.
(R8.18) Investigate all the appropriate problems of this chapter in the presence of arbitrarily oriented photon field and strain.
(R8.19) Investigate all the appropriate problems of this monograph for paramagnetic semiconductors in the presence of strain.
(R8.20) Investigate all the appropriate problems of this monograph for boron carbides in the presence of strain.
(R8.21) Investigate all the appropriate problems of this monograph for all types of argyrodites in the presence of strain.
(R8.22) Investigate all the appropriate problems of this monograph for layered cobalt oxides and complex chalcogenide compounds in the presence of strain.

(R8.23) Investigate all the appropriate problems of this monograph for all types of nanotubes in the presence of strain.
(R8.24) Investigate all the appropriate problems of this monograph for various types of half-Heusler compounds in the presence of strain.
(R8.25) Investigate all the appropriate problems of this monograph for various types of pentatellurides in the presence of strain.
(R8.26) Investigate all the appropriate problems of this monograph for Bi_2Te_3–Sb_2Te_3 superlattices in the presence of strain.
(R8.27) Investigate the influence of temperature-dependent energy band constants for all the appropriate problems of this monograph.
(R8.28) Investigate the FNFE for $Ag_{(1-x)}Cu_{(x)}TITe$ for different appropriate physical conditions as discussed in this monograph in the presence of strain.
(R8.29) Investigate the FNFE for p-type SiGe under different appropriate physical conditions as discussed in this monograph in the presence of strain.
(R8.30) Investigate the FNFE for different metallic alloys under different appropriate physical conditions as discussed in this monograph in the presence of strain.
(R8.31) Investigate the FNFE for different intermetallic compounds under different appropriate physical conditions as discussed in this monograph in the presence of strain.
(R8.32) Investigate the FNFE for GaN under different appropriate physical conditions as discussed in this monograph in the presence of strain.
(R8.33) Investigate the FNFE for different disordered conductors under different appropriate physical conditions as discussed in this monograph in the presence of strain.
(R8.34) Investigate the FNFE for various semimetals under different appropriate physical conditions as discussed in this monograph in the presence of strain.
(R8.35) Investigate all the appropriate problems of this monograph for $Bi_2Te_{3-x}Se_x$ and $Bi_{2-x}Sb_xTe_3$, respectively, in the presence of strain.
(R8.36) Investigate all the appropriate problems of this monograph for all types of skutterudites in the presence of strain.
(R8.37) Investigate all the appropriate problems of this monograph in the presence of crossed electric and quantizing magnetic fields.
(R8.38) Investigate all the appropriate problems of this monograph in the presence of crossed alternating electric and quantizing magnetic fields.
(R8.39) Investigate all the appropriate problems of this monograph in the presence of crossed electric and alternating quantizing magnctic fields.
(R8.40) Investigate all the appropriate problems of this monograph in the presence of alternating crossed electric and alternating quantizing magnetic fields.
(R8.41) Investigate all the appropriate problems of this monograph in the presence of arbitrarily oriented pulsed electric and quantizing magnetic fields.
(R8.42) Investigate all the appropriate problems of this monograph in the presence of arbitrarily oriented alternating electric and quantizing magnetic fields.

(R8.43) Investigate all the appropriate problems of this monograph in the presence of crossed inhomogeneous electric and alternating quantizing magnetic fields.

(R8.44) Investigate all the appropriate problems of this monograph in the presence of arbitrarily oriented electric and alternating quantizing magnetic fields under strain.

(R8.45) Investigate all the appropriate problems of this monograph in the presence of arbitrarily oriented electric and alternating quantizing magnetic fields under light waves.

(R8.46) Investigate all the appropriate problems of this monograph in the presence of arbitrarily oriented pulsed electric and alternating quantizing magnetic fields under light waves.

(R8.47) Investigate all the appropriate problems of this monograph in the presence of arbitrarily oriented inhomogeneous electric and pulsed quantizing magnetic fields in the presence of strain and light waves.

(R8.48) (a) Investigate the FNFE for all the problems of this monograph in the presence of many body effects, strain, and arbitrarily oriented light waves, respectively.
(b) Investigate the influence of the localization of carriers for all the appropriate problems of this monograph.
(c) Formulate the minimum tunneling, Dwell, and phase tunneling, Buttiker and Landauer and intrinsic times for all types of systems as discussed in this chapter.
(d) Investigate all the appropriate problems of this chapter for the Dirac electron.
(e) Investigate all the problems of this monograph by removing all the physical and mathematical approximations and establishing the respective appropriate uniqueness conditions.

The FNFE is the consequence of electric field-induced electron emission phenomena of solid state science and all the assumptions behind the said phenomena are also applicable to FNFE. The formulation of FNFE for all types of semiconductors and their quantum confined counterparts after removing all the assumptions is, in general, a challenging problem. Such investigations covering the total spectrum of materials of modern solid state science require insight. In total, 200 open research problems have been presented in this monograph and we hope that the readers not only will solve them but also will generate new concepts, both theoretical and experimental. In the mean time, our research interest has been shifted and we are leaving this particular topic with the hope that (R8.48) alone is sufficient to draw the attention of the researchers from diverse fields.

References

1. B.R. Nag, *Electron Transport in Compound Semiconductors*, *Springer Series in Solid State Sciences*, vol. 11 (Springer-Verlag, Berlin, 1980)
2. R.W. Cunningham, Phys. Rev. **167**, 761 (1968)
3. A.I. Yekimov, A.A. Onushchenko, A.G. Plyukhin, Al.L. Efros, J. Exp. Theor. Phys. **88**, 1490 (1985)
4. B.J. Roman, A.W. Ewald, Phys. Rev. B **5**, 3914 (1972)

Material Index

Antimony, 43, 61

Bismuth, 41
Bi_2Te_3, 42

Cadmium Arsenide, 40
Cadmium diphosphide, 42, 62
Carbon nanotubes, 42
$CdGeAs_2$, 40
CdS, 41
CdSb2, 42
CuCl, 61

GaAs, 40
$GaAs/Ga_{1-x}Al_xAs$, 41
GaP, 42, 53
GaSb, 41
Germanium, 42
Graphite, 42, 60
Gray tin, 62

$Hg_{1-x}Cd_xTe$, 41
HgTe, 41, 58

InAs, 40
$InxGa_{1-x}As/InP$, 41
$In_{1-x}Ga_xAs_yP_{1-y}$, 41
InSb, 41, 61

$Pb_{1-x}Ga_xTe$, 60
$Pb_{1-x}Ge_xTe$, 42
$Pb_{1-x}Sn_xSe$, 49
PbSe, 43
PbTe, 46
PtSb2, 42

Stressed n-InSb, 41

Tellurium, 42, 60

Zinc diphosphide, 42, 62
ZnSe, 43

S. Bhattacharya and K.P. Ghatak, *Fowler–Nordheim Field Emission*, Springer Series in Solid-State Sciences 170, DOI 10.1007/978-3-642-20493-7,

Subject Index

Antimony, 61
Area quantization, 109

Band, 58–62
Band gap measurement, 299
Band structure, 13
 II-V compounds, 38, 139
 II-VI compound, Hopfield model, 18
 IV-VI compounds, 24, 124
 bismuth, 19, 23, 119
 bismuth telluride, 32, 135
 carbon nanotube, 304
 gallium antimonide, 37, 138
 gallium phosphide, 29, 131
 germanium, 34, 136
 Newson and Kurobe Model, 15
 nonlinear optical, 7
 Palik model, 16, 117
 parabolic band, 14, 115, 187, 189
 platinum antimonide, 31, 133
 Stillman model, 14, 116
 stressed materials, 27, 129
 tellurium, 28, 130
 three band Kane, 11, 112, 216, 221
 two band Kane, 13, 114, 187, 189, 244
Born-Von Karman condition, 309

Carbon nanotubes (CNTs), 304
CuCl, 61
Cyclotron resonance, 109

Debye screening length (DSL), 281–284
de Haas-Van Alphen oscillations, 109
Density-of-states (DOS), 9, 189, 245, 273
 quantum wire, 11
Diamagnetic resonance, 109
Diffusion coefficient, 302
Diffusivity-mobility ratio, 114, 295, 296, 298, 299
Dispersion, 53, 58–61

Effective electron masses, 242, 291–295, 310
Elastic constants, 284–290
Electric field, 233

Fermi energy, 9
Fermi-Dirac integral, 11
Fermi-Dirac probability factor, 9
Fowler Nordheim field emission, 3, 9
 optimization, 309

Graphite, 60

Heavy hole, 188, 234
HgTe, 58

Interband transition matrix element, 235, 241

Kane, 273

Landau subbands/levels, 109, 268
Light waves, 187

S. Bhattacharya and K.P. Ghatak, *Fowler–Nordheim Field Emission*, Springer Series in Solid-State Sciences 170, DOI 10.1007/978-3-642-20493-7,

Magnetic field, 205
Magnetic field/quantization, 109
Magnetic quantum limit, 141, 142

n-GaSb, 58
Non linear optical response, 303

$Pb_{1-x}Ga_xTe$, 60
Potential well, 226

Quantization, 222
Quantum size effect, 4
Quantum wire, 3
Quantum wire effective mass superlattices, 72
Quantum wire superlattices, 72

Raman gain, 303
Reduced effective mass, 234

Spin, 271
Stress, 62
Superlattice (SL), 72, 157, 194, 249, 253, 256, 257, 262
  II-VI, 77, 92, 161, 173
  III-V, 71, 92, 157, 172, 194, 200, 204, 210
  IV-VI, 82, 96, 165, 175
  HgTe/CdTe, 87, 99, 168, 177

Thermoelectric power, 282
Third order nonlinear optical susceptibility, 303
Thomas-Fermi screening, 314
Two-band model, 274

Ultrathin films, 3

van Hove singularity, 308